Cybernetics and Systems

Society is now facing challenges for which the traditional management toolbox is increasingly inadequate. Well-grounded theoretical frameworks, such as systems thinking and cybernetics, offer general level interpretation schemes and models that are capable of supporting understanding of complex phenomena and are not impacted by the passage of time.

This book serves the knowledge society to address the complexity of decision making and problem solving in the 21st century with contributions from systems and cybernetics.

A multi-disciplinary approach has been adopted to support diversity and to develop inter- and trans-disciplinary knowledge within the shared thematic of problem solving and decision making in the 21st century. Its conceptual thread is cyber/systemic thinking, and its realisation is supported by a wide network of scientists on the basis of a highly participative agenda.

The book provides a platform of knowledge sharing and conceptual frameworks developed with multi-disciplinary perspectives, which are useful to better understand the fast-changing scenario and the complexity of problem solving in the present time.

Sergio Barile is Full Professor of Business Management at Sapienza, University of Rome.

Raul Espejo is President of the World Organization of Systems and Cybernetics.

Igor Perko is Director-General of the World Organization of Systems and Cybernetics.

Marialuisa Saviano, PhD, is Full Professor of Business Management at the University of Salerno, Italy.

Routledge-Giappichelli Systems Management
Series Editors: Sergio Barile
Full Professor of Business Management at Sapienza, University of Rome
and
Marialuisa Saviano
Full Professor of Business Management at the University of Salerno, Italy

The *Systems Management* series contributes to the advancement of knowledge in a context of growing complexity in which managers and policy makers are expected to be able to face both the challenges of profitable organizations and a more sustainable, equitable and inclusive world. The *systems approach* can offer an incomparable contribution to this aim providing a unitary and coherent body of knowledge rooted in *systems thinking*. Accordingly, the *Systems Management* series collects new research-level academic books such as monographs, textbooks and contributed volumes that represent the outcomes of original studies and impactful research conducted by scholars and researchers, as well as experts in various managerial areas by adopting a *systems approach*.

The *Systems Management* series welcomes contributions that are scientifically well-argued and rigorous, as well as equipped with practical examples and cases studies, in order to make knowledge accessible and useful for research, teaching and management purposes. A coherent combination of theoretical and conceptual perspectives, on the one hand, and practical examples and case studies as well as empirical research evidences, which cover a variety of topics that deal with complex management issues hard to face with the traditional toolbox, represent ideal contributions to the series.

Cybernetics and Systems
Social and Business Decisions
Edited by Sergio Barile, Raul Espejo, Igor Perko and Marialuisa Saviano with Francesco Caputo

For more information about this series, please visit: www.routledge.com/ Routledge-Giappichelli-Systems-Management/book-series/RGSYSTEMSMAN

Cybernetics and Systems
Social and Business Decisions

Edited by Sergio Barile, Raul Espejo,
Igor Perko and Marialuisa Saviano

Assistant Editor: Francesco Caputo

Routledge
Taylor & Francis Group

LONDON AND NEW YORK

IUSTITIAM COLIMUS

G. Giappichelli Editore

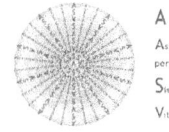

ASvSA

Associazione
per la ricerca sul
Sistemi
Vitali

WOSC2017
CONGRESS

First published 2019
by Routledge

2 Park Square, Milton Park, Abingdon, Oxfordshire OX14 4RN
52 Vanderbilt Avenue, New York, NY 10017

Routledge is an imprint of the Taylor & Francis Group, an informa business

First issued in paperback 2020

British Library Cataloguing-in-Publication Data
A catalogue record for this book is available from the British Library

Library of Congress Cataloging-in-Publication Data
A catalog record for this book has been requested

ISBN: 978-1-138-59728-0 (hbk)
ISBN: 978-0-367-66324-7 (pbk)

Typeset in Galliard
by Apex CoVantage, LLC

Contents

vi *Contents*

Figures

Tables

Foreword

Raul Espejo[1]

WOSC 2017, the 17th Congress of the World Organisation of Systems and Cybernetics, was planned as a conversational space about social dynamics underpinned by systems and cybernetics. Its conceptual thread has been made alive by a network of scientists, practitioners, and researchers from varied countries around the world (see Authors). Multiple conversations helped to integrate a variety of themes that are elaborated in this foreword and developed in over a hundred extended abstracts. The construction of its agenda was highly participative; efforts were made to support diversity within the shared thematic of Science with and for Society-Contributions of Cybernetics and Systems. This document is offered as a platform for the conversations we expect to hold in Rome from the 25th to the 27th of January 2017. We all agree that we are in an increasingly interconnected world. As expected our world is becoming more complex and therefore increasingly beyond our traditional response capabilities. The challenge is how we account for this complexity in the quest for a sustainable world underpinned by inclusion and fairness. More effective problem solving requires constructing a better world which enables individuals to offer their potentials and develop their capabilities. Dealing with an overwhelming complexity requires imaginative propositions and innovative behaviours that we believe can benefit from cybersystemic problem solving.

Systemic thinking helps us visualise wholeness and avoid dysfunctional fragmentation; cybernetics helps us understand how to maintain stability in the interactions between people, institutions, and organisations. Systemic thinking gives us methodological tools; cybernetics gives us tools to manage the complexity of situations from the local to the global. Interactions among us and with the environment may reach regulatory capacity limits in which the situational complexity scales down or up as required by the need to maintain sustainability and viability. WOSC 2017 provided a platform to discuss the nature of our social and individual interactions and the constraints impinging in their success. Contributions to the Congress aim at exploring, debating, and clarifying what kinds of complexity management strategies are more likely to focus our societal and individual actions on desirable directions of sustainability, inclusion, mutual respect, democracy, effective organisation, and quality of life. The challenge is making these strategies functional to societal interests and not to the interests of the few.

We started proposing three major themes for the Congress:

A   People, technology, and governance for sustainability
B   Democracy, interactions, and organisation
C   Cyber-systemic thinking, modelling, and epistemology

Theme A included tracks about human aspect of management, technology and innovation, and governance for sustainability; Theme B explored democracy and transparency, interactions, law, and organisation; and the last one offered the space for cyber-systemic thinking and modelling.

Proposed tracks for the first theme were human aspects of managing systems, smartness and sustainability, smart technologies and Big Data, the brain of the future, and governance in the Anthropocene. The human aspects of managing systems opened the space for psychological and social reflexions about interactions among us and with others in the environment. Interactions within organisations and between them and their environments give us the platform to discuss today's lack of fitness of governance in the Anthropocene; our society and its institutions need effective learning mechanisms, so far lacking, to cope with a fragile environment. Current information technology offers tools to foster smartness in the quest for sustainability. Effective communications and smart technologies make organisational structures sensitive to changing environments and able to distribute scalable responses from the local to the global. The application of digital technologies in organisations is disrupting their structures and increasing their capabilities to deal with complexity and interactions. These technologies are offering the potential for better organisational structures and more effective processes of governance. Cognitive machines, algorithms, deep learning, Big Data are all contributing to the brain of the future; while they amplify human capabilities, they may help introduce constraints in the environment to avoid overwhelming us with irrelevant complexity.

Beyond managerial and institutional matters, the second theme – democracy, interactions, and organisation – was proposed as a bottom-up theme that related the local decision making at multiple structural levels; this was a theme about democratic processes to align local needs and ecologies with organisations and political processes. This was a space to discuss self-organisation, transparency, and local influence in the global. Focus on the local opens reflections about interactions, complexity management, and constraints. The interactions revolution was thought of as emerging from, among others, social networks and the internet; it is about the increasing social and human fluidity allowed by digital technologies and also by the conceptual frameworks of systemic and cybernetic thinking. Ethical issues derive from the development of technologies in the world of e-government and law. Bottom-up processes provide a new perspective of organisational systems, including non-directive approaches to manage the complexity of social systems and the emergence of constraint in complex adaptive systems. These issues are proposed as book tracks in the list below. Additionally, in this second theme we included the role of the educational system to respond to the challenges of sustainable development.

The third theme of WOSC 2017 was methodological and epistemological. Systems thinking in the 21st century reflects on applying systems thinking to current problems, such as social inequality, migration, global warming, and many more. Studying and dealing with these problems requires methodologies and tools; implying contribution to the practice of information and communication systems as well as to the systemic conceptualisation of 21st-century problems. Modelling problematic situations with the support of System Dynamics has proved to be a useful tool in a variety of situations. Formalisms of quantum mechanics have offered a novel platform to discuss modelling complex situations, such as the economy, organisations, psychology, artificial intelligence, human language, cognition, information retrieval, biology, and political science. These formalisms offer the opportunity for transdisciplinary insights and knowledge. Finally, at an epistemological level, this theme included contributions that challenged the scientific basis of traditional science; it referred to conversations about reflexivity, second order science, and conceptualisation of context.

These original three themes evolved into nine, which provided the scaffolding for WOSC 2017 and for this book of abstracts:

- Theme I: human aspects of managing systems
- Theme II: sustainability and the Anthropocene
- Theme III: smartness and Big Data
- Theme IV: democracy, transparency, and social dynamics
- Theme V: interactions revolution
- Theme VI: knowledge and organisation
- Theme VII: systems thinking and system dynamics
- Theme VIII: quantum modelling
- Theme IX: reflexivity, second order science, and context

The challenge for WOSC 2017 was their debate and discovering relations across these themes. Authors may read contributions to all nine themes, and hopefully reflect in their final papers the cross fertilisation of ideas. Abstracts related to all these themes are available on the Congress Website and authors can access them with the support of the Padlets collaborative tool (Padlet.com), which allows leaving comments and developing conversations. Overall our purpose is to provide scientific contributions to the governance, organisation, and administration of social, environmental, economic business issues, with the support of systems and cybernetic ideas. This is referred in the book as the cyber-systemic approach, which aims at a holistic view of all these issues with a focus on the regulation, control, and communications among and between the multiple agents constituting their complexity. Regulation, control, and communications are the outcome of interactions and it is through these interactions that stable forms of relations emerge that constitute organisations.

The themes are connected and one way or the other all of them contribute to the overarching concern for a cyber-systemic appreciation of societal issues. Are together the abstracts submitted to these themes contributing to a richer and more powerful understanding of societal and organisational issues? This is a matter of individual and group discoveries. We would like to think that the nine themes have the potential to enable discoveries in the direction of making sense of the huge complexity that is overwhelming humankind currently. Hopefully WOSC 2017 turns out to be a small contribution to science with and for society.

Authors' contributions to the Congress were made in the context of nine keynote talks, two roundtables, and one experimental session to connect theatre and cybernetics.

A connecting line for the keynotes is the Congress aim to improve our under-standing of the relationships between people, organisations, and their environments. It is widely acknowledged that we are affecting our environment in ways that anticipate significant problems for future generations. Dealing with these problems requires clarifying what are acceptable levels of balance between us, our organisations, and their environments; working out essential variables and related performance criteria at multiple levels of aggregation, from the local to the global, is necessary to support action towards a sustainable world. Since it is actions that are affecting the stability of our world, we need to produce repairing efforts, which imply distributed governance and action.

The Congress's first two keynotes in this book offer insights about these relation-ships. The first of these, "Sustainability Science: Linking Science, Policy and Society for a Sustainable Future" (by Kazuhiko Takeuchi), offers a framework for a sus-tainable future. The second, "Governance in the Anthropocene: Cybersystemic Possibilities?" (by Ray Ison), questions whether in the epoch of the Anthropocene, we are making sufficient use of governance options, technology, and education in the construction of a more sustainable world. What are possible cybersystemic contributions in this construction? The discussion on the governance of the Anthropocene, supported by sustainability principles and structural requirements for effective distributed governance is debated in Theme II "Sustainability and Anthropocene". This theme addresses specific aspects of smart use of technolo-gies, distributed governance towards achieving global and local sustainability targets, and the role of education for sustainable development.

Human aspects of managing systems are many-sided. Among others, they relate to policy processes, people's behaviour, social dynamics, distribution of power, and ethics and norms. These human aspects shape our social relations. We had two keynotes, focused on democracy, participation, and current technolo-gies. First, the keynote "Digital Bill of Rights" (by Richard Barbrook) opens the debate about regulating individual use of online activities not only in their inter-est but also the common interest, introducing the age of cybernetic citizens with rights and responsibilities in the world of algorithms and autonomous machines, and the second "Artificial Intelligence and Law: What Perspective?" (by Daniele Bourcier). It addressed ethical considerations and the potential of soft law to take into account the sensitive and evolutionary aspects of artificial intelligence applica-tions in some areas of human activity (medicine, security, e-government, military robots, and law). These aspects affect the balance between individual rights and the commons. Cybernetic ideas can help us to discuss freedom of expression, autonomy, and respect for each other. Related to these keynotes, three themes provide discussion in more depth issues of participation and social and individual behaviour in the cybersystemic age: first "Human Aspects of Managing Systems", second "Democracy, Transparency, and Social Dynamics", and third "Interactions Revolution".

Present technological developments are enablers as well as threats to democracy. They are amplifying the scope of interactions in orders of magnitude, something that hugely increases social complexity and therefore the potentials for more participation and guided self-organisation, but also for exploitation, abuses of power, and fragmentation. These technologies have the potentials to connect social, economic, and sustainability issues; they pose challenges and open opportunities, but they also increase regulatory difficulties. These issues are explored by three keynotes. The "The Brain of the Future" (by Alexandre Perez Casares) explores how cognitive machines are altering social relationships and increasing the relevance of artificial intelligence in economic activities, to the point that human jobs may be replaced by inference machines, dominated by performative algorithms rather than by simplifying mappings. The second keynote "Knowledge for development vision 2030" (by Elias G. Carayannis) offers a reflection about the societal developments such as smart cities which connect technology, innovation, organisation, economic and social trends; and the third, "Design of a Regional System" (by Alfonso Reyes), provides, at a micro level, considerations for designing enablers for self-organisation and regional development. Three Congress themes enhance the debates offered by these plenaries: "Smartness and Big Data", "Knowledge and Organisation", and "Systems Thinking and System Dynamics".

A most important concern of this publication is exploring the contributions that systems and cybernetics can make towards improving policy processes and policy making. A particular policy issue of great social significance was discussed in the keynote "From Precision Medicine to Systems Medicine: Clinical and Social Implication" (by Christian Pristipino). This particular policy arena, in the age of Big Data and real-time streaming, can offer a holistic approach to medicine of great significance for the future of health services. The transformation from reductionist medicine to holistic medicine is precisely an arena to discuss ethically appropriate individual-patient treatments with the support of systems thinking and cybernetic modelling. The keynote "Recognizing the Dangers of Simplicity Addiction" (by Michael Lissack), introduced the issue of cybernetics and policy issues. This keynote made apparent that though simplification is necessity to cope with complexity, too much simplification is dangerous and argues for requisite variety to achieve effective performance. It opens the space for innovative contributions to modelling and epistemological issues. Two related themes, "Quantum Modelling" and "Reflexivity, Second Order Science, and Context", open discussion on related topics.[1]

Note

Below, in the Overview of this book readers can have a more detailed presentation of the nine themes of WOSC 2017.

1 President of World Organisation of Systems and Cybernetics.

Preface

Nowadays, society is facing challenges for which the traditional management toolbox is increasingly inadequate. Well-grounded theoretical frameworks, such as systems thinking and cybernetics, offer general level interpretation schemes and models that are capable to support understanding of complex phenomena and suffer less from the passing of time.

This book is the outcome of the WOSC 2017, the 17th edition of the Congress of the World Organisation of Systems and Cybernetics, a world organisation that invites scientists, policy-makers, professionals, and students across the globe to contribute to the debate of the dynamics underpinning contemporary societal problems from cybersystemic perspectives. WOSC offers a space for conversations about social dynamics from multiple points of view. A multi-disciplinary approach has been adopted to develop inter- and trans-disciplinary knowledge within the shared thematic of problem solving and decision making in the 21st century. Its realisation is supported by a wide network of scientists in a highly participative agenda.

The book provides a platform of knowledge and conceptual frameworks developed with multi-disciplinary perspectives, which are useful to better understand our fast-changing scenarios and the complexity of problem solving in the 21st century.

The book promotes a boundary-crossing knowledge creation process highlighting the necessity of adopting a multi-disciplinary approach to address the challenges of decision making and problem solving in the 21st century. Under the common view of the general frameworks offered by systems and cybernetics, the book addresses three macro-themes: first, people, technology, and governance for sustainability; second, democracy, interactions, and organisation; and third, cybersystemic thinking, modelling, and epistemology. The contributions collected in the book are the outcomes of knowledge sharing processes and articulated methodologies and perspectives.

Sergio Barile
Raul Espejo
Igor Perko
Marialuisa Saviano

An overview

WOSC 2017 invited contributions to 16 tracks, which were finally structured into nine themes. Some of the tracks have maintained their identity while others have evolved constituting larger themes, incorporating two or three tracks. Some tracks, even if the number of received submissions were small, have remained. Here we introduce the reader to the nine Congress themes. This introduction relies on the abstracts proposed by the tracks and themes coordinators and abstract submissions.

Theme I: human aspects of managing systems

This theme initiated by Sergio Barile, Marialuisa Saviano and Francesco Caputo invited theoretical and empirical contributions, centred on the debate of expected and potential disruptions and modifications in enterprises resulting from current technological and human changes. Direct and indirect modifications in social dynamics and the relevance of people (driving new consumer behaviour and manufacturers' attitudes) are central to evolving enterprises. Contributions were expected to define and interpret the impacts of the changing role and behaviour of people in complex organisations (enterprises, etc.) and also impacts of changing business and management systems on the behaviour of individuals. The proposed focus was mostly on human aspects and human processes that can be analysed at behavioural, cognitive, neurocognitive and affective levels (i.e. values, planning, decision making, memory, problem solving).

Received submissions emphasised aspects of intrinsic motivation and self-organisation in firms. They emphasised the importance of training to bridge cognitive gaps between employers and employees as well as of gender equality to have smart economies and to trigger ethical differences. They highlighted change management in social and labour relations and the emergence of new relations in the network economy, such as equity crowdfounding, dynamic capabilities and successful projects.

This theme is a pivot for many of the Congress contributions. Structural aspects such as centralisation are seen as inadequate for good decision making, highlighting the advantages of heterarchies and autonomy to support flexible adaptable learning organisations in which everyone learns from others. Digital data and networking complexity are growing faster than people's data absorption capacity, triggering data handling weaknesses. This idea relates to Theme VI, "Knowledge and Organisation". Resilience and antifragility are necessary to enable self-organisation and self-regulation. Organisations need anticipatory smart systems as well as dynamic capabilities to reinforce interactions with their environments and structural adaptation to changes. These are all issues also related to Theme III, "Smartness and Big Data".

Acceleration of communications in organisations is associated to digital technology. This quicker pace of communications is increasing interactive imbalances, triggering sustainability issues in organisations (Theme II). The fastest, most agile, survives. Communication professionals' burn out in the digital era is increasing. Organisations need stronger response capacity (Theme V).

Human aspects in project management were highlighted emphasising the softer people aspects, raising questions about balancing cognition and interactions, for instance how to make use of managers' unavoidable bounded rationality to work out constraints and strategies to manage complexity. Management at the edge of chaos requires considering constraint as well as augmented reality.

Related to Theme VII, "Systems Thinking and System Dynamics," the need for soft systems methodologies is proposed. Reality is complex; it is socially constructed and a product of people's ongoing interactions (Theme V, "Interactions Revolution"). This interpretivist ontology triggers the ideas of grounding epistemology in ontology.

New institutional perspectives are emerging from changes in technology as well as better appreciation of human aspects of managing systems. There are trade-offs between internal efficiency and external legitimacy. Strategies to remain competitive could imply mimicking others' strategies to gain legitimacy but this mimicking may affect negatively intra-organisational relationships that are necessary to strengthen viability. The Viable Systems Model and Approach are used as two conceptual frameworks for management. Requisite holism is argued.

Theme II: sustainability and the Anthropocene

Sustainability is required as a must for human activities from the most local to the global. It is at this global level that the Anthropocene is understood as a new geological epoch in which humanity is affecting the earth's dynamics. This second theme, as proposed by Marialuisa Saviano, Hernan Lopez-Garay, Sandro Schlindwein, Markus Schwaninger and Ray Ison, acknowledges that local and global are deeply intertwined. It covers all aspects of sustainability.

Smartness, as discussed in Themes III and I, is grounded in technology and recognises digitalised processes and telematic interactions in human activities. Sustainability is focused on successful human activities as well as in processes and interactions encompassing social, environmental and economic perspectives. How can we make smartness unavoidably linked to sustainability?

Unfortunately, we live in an ever-increasing unsustainable world. Systems thinking helps disclosing sustainability problems because of its focus on interactions and relationships. These are problems deeply rooted in our social structures and culture. Unsustainability relates to human, organisational and environmental imbalances in the management of critical variables. Instabilities are disclosing critical situations. The challenge of sustainability requires transforming research and education from a descriptive-analytic mode towards a transformative one. Distributed educational programs are necessary to build capacity for understanding the systemic nature of our current unsustainable world and the real-world sustainability problems. These programs are necessary to educate future generations of decision makers, problem solvers, change agents and transition managers. We need to reflect upon and learn the competences, pedagogies and assessment tools needed for these purposes. In WOSC 2017 this theme wants to shed light into programs that use cybersystemic tools to educate a new breed of human being aware of the historical conditions we are in. The aim is people caring for the development of a harmonious sustainable world.

As proposed by Ray Ison, earth scientists are claiming that we have entered a new geological epoch: human influences have become so great that they are affecting "whole Earth dynamics" through a range of biophysical and social processes leading to complex global changes. Unfortunately, these changes are the outcome of happenstance more than of reflective governance systems. This calls for a critical reflexion of our past thinking, practices, institutions, patterns of investment and governance.

The Paris Agreement in December 2015 was a call for national, organisational and institutional commitments to take action towards a satisfactory balance between human activity and earth sustainability. This theme wants to reflect upon cybersystemic governance of social-ecological systems.

Contributions to "smartness and sustainability" are closely related to those to Theme III, "Smartness and Big Data". Aspects of value co-creation between providers and customers, willingness to adopt data management technology to support sustainable behaviours are being proposed in both themes. In this context a major topic for debate is "smart cities". Actors, smart technologies, designers and governance come together in sustainable cities; What is a sustainable city? How can systems thinking inform the design and management of smart cities? Can non-conventional conversational places, such as the Cybersyn operations rooms help in the ongoing design of smart cities? What can smart grids, as enabling technologies for interactions, do? These are questions that are also relevant to the themes "Knowledge and Organisation" and "Systems Thinking and System Dynamics". Smartness and sustainability relate environmental, economic and social dynamics. A challenge for the Congress is debating an integrated framework for sustainability: algorithms, hard and soft approaches that weave sustainability and technology.

The Viable System Model offers a conceptual framework for the governance of local and global systems. Its reliance on the Law of Requisite Variety makes it relevant to the debates of the "Systems Thinking and System Dynamics" theme (Theme VIII).

Governing socio-ecological systems in the Anthropocene reveals the high dependency between nature and society. "Socio-ecological systems" are complex and need to be adaptive. They need resilience to absorb change. The structural coupling between ecology and society is a wicked problem. We need models and tools to appreciate better the structural coupling of social systems and their ecosystems. They should help disclose critical variables and the relationships that need to be conserved over time. Contributions to this theme may enable debates about the governance in the Anthropocene using the metaphor of a triple helix between academy, government and industry. Universities should play a significant role not only in research but also in education for sustainability.

It can be appreciated that this theme also weaves several others.

Theme III: smartness and Big Data

This theme, as proposed by Fabio Orecchini, Francesco Polese and Igor Perko, offers the opportunity for a more focused debate about Big Data. This is a disruptive technology in nearly all the domains of our interest. Big Data support interactions, decision processes, production and in general all kinds of services delivery activities fostering the co-evolution of agents of all kinds. Theme III aims at analysing the state-of-the-art and future of Big Data and smart technology's definition, design, exploitation, potentials and limits.

This theme is focused not only on smart cities but also on other kinds of social, economic and production services. Reflections on the smart city ecosystem, the disruptive nature of its relevant technologies, the disruption of traditional market balances to support the emergence of more efficient, effective and sustainable cities are ideas discussed in this theme. Smart cities are the arena for stakeholders' interactions (Theme V, "Interactions Revolution"). Contributors to this theme see cities as complex adaptive systems.

It is based on data collection, analytics and measurements for social, economic and technical systems. Big Data is disrupting the interactions within and between enterprises of all kinds, and over time is supporting the emergence of new relationships between firms and public organisations, between R&D and universities, and indeed the emergence of virtual networks and organisations. Discussions about the nature and characteristics of these relationships should help to visualise new social forms. Contributions to this theme have emphasised triple helix articulations between universities, the economy and society. Smartness helps to focus on stakeholders of all kinds, like youth, elderly, professionals, interest groups, unemployed and so forth. Big Data, analytics and measurement are relevant to each, and many more, of these groups to disclose communication channels relevant to them.

It is necessary to understand the capabilities of internet and data management companies. Enterprises like Amazon, Google, Facebook are shaping our interactions and relationships. Applications like the internet of things and the automation of transport activities will be shaping the way we live. Ubiquitous smart devices are increasing data collection at staggering speed. This growth in data collection capabilities will trigger further technological changes, like for instance quantum computers, linking this theme to the "Quantum Modelling" theme (Theme VIII).

To define a technology as smart it must be aware of its environment and must have capacity to co-evolve with environmental agents and self-configure its capabilities. This theme discusses the viability of smart service systems.

This theme, as well as the "Knowledge and Organisation" theme (Theme VI), is responding to the challenge of knowledge management in the 21st century. Big is not necessarily smart. Data management has the potential of being the pathway towards a smarter planet with better capabilities for solving social, economic or/and other human issues. This is a connection with the "Sustainability and Anthropocene" theme (Theme II). But also, this theme wants to discuss matching managerial capabilities to data, a topic connected to the "Interactions Revolution" (Theme V).

Theme IV: democracy, transparency and social dynamics

Amanda Gregory, Daniele Bourcier, Raul Espejo and Zoraida Mendiwelso-Bendek are coordinating this theme.

The theory of self-organisation, based on the premise that some form of overall order or coordination can arise out of the local interactions between the component parts of a system, has a long history and a wide field of application. Self-organisation is relevant to social systems constituted through local or virtual interactions. It reflects the assumption that self-organisation will emerge through the realisation of the benefits of autonomy, a positive orientation towards self-empowerment and purposeful action. Self-organisation happens in interactions of unequal complexity such as instances of societies facing extreme environmental problems, or dire economic pressures, or challenges of accommodating different values and norms of behaviour. While some self-organisation efforts flourish, others fail as economic, environmental, political and social factors can pose challenges that undermine them. Self-organisation may be dysfunctional to the point of threatening the fairness, justice and even the viability of the interacting agents. These are undesirable instance of social dynamics. Contributions to Theme V – the "Interactions Revolution" – and Theme VI – "Knowledge and Organisation" – also discuss aspects of self-organisation.

One of the contributions addresses the issue of the peace process in Colombia, which suggests the need for national cohesion and the enabling of self-organisation to overcome the entrenched corruption in some of the regions of the country. Corruption makes transparency and democracy unachievable. Citizens science is proposed as a tool to improve communications between citizens, experts and politicians. Democratic deficits, even in the most advanced democracies, pose the need for reflection and this is one of the aims of this theme.

Barbrook is proposing in this Congress the creation of a Digital Bill of Rights to avoid the threat of authoritarian states and monopolistic businesses in the age of the internet. This proposition requires attention to interaction spaces enabling self-organisation. Cybernetic principles of complexity management can enable the democratisation of internet services. As proposed by Espejo, Beer's Viable System Model is one such framework for complexity management and the Law of Requisite Variety offers a heuristic for the democratisation and transparency of these processes. The Liberty Machine, which underpinned Beer's design and development of the projects Cybersyn and Cyberfolk in the Chile of the early 1970s, offered an extraordinary vision of a utopian society in which the people could express themselves, individually or collectively, with autonomy. Today these projects can be undertaken with more sophisticated methodologies and epistemologies, which should support better processes of social construction (Theme IX: "Reflexivity, Second Order Science, and Context"). Also, as is discussed by the "Smartness and Big Data" theme (III), we have hugely more developed information and communication technologies, which should allow us to deal with complexity and support interactions at scales that were unthinkable 40 years ago. These conceptual and technological developments give us opportunities to transform our societies into more democratic and transparent ones. This theme has been proposed as a platform for contributions to discuss and illustrate the application of cybernetic principles and digital technologies to the design of participative and inclusive processes in the public, private and third sector spheres.

As proposed by Bourcier in this Congress the movement of Commons is not only protecting and improving the management of (scarce) resources but also offering the opportunity to renew democracy through a redefection of state and market interactions with the support of systemic ideas. This theme wants to discuss this renewal from the "making of law" perspective. How to reach a consensus on regulation? The notion of Common good highlights the need for a different normative; it suggests a re-thinking of the legal framework towards the production of Common good. It is necessary to experiment building a normative system on local, global and interpersonal relations between social actors, and to implement collaborative legal tools, scalable, inclusive, integrative and experimental. This theme focuses on how have the Commons or Common good been integrated into the legislative texts? And, how can we rethink a normative process based on the social dynamics of the Commons? It is proposed to rethink existing normative tools in our democracies; how have new methods for writing norms emerged from the Commons movement?

Finally, this theme proposes to rethink democracy through legal tools designed in a cooperative and decentralised process. Further developments invite case studies in the various types of legal norms (constitutions, statutes, by-laws, customary laws), in soft law (ethics, charts, guides, Creative Commons licences, . . .) and smart contracts.

Scholte discusses in one of the submissions to this theme: "Varieties of Systems Theatre" is offered as a reflection about interactions and reflexivity. It is experimental in nature. He discusses, following Ross Ashby's assertion that "the discovery that two fields are related leads to each branch helping in the development of the other", how theatre experiments offer an opportunity for cross fertilisation between theatre and second order cybernetics. He is looking for cross fertilisation between "varieties systems theatre" and the cybernetics and systems community. First, he argues that theatre has cybernetic underpinnings; particularly it can be an instance of applied epistemology. Second, it links modes of cybernetic explanation to processes of agent-based modelling (patterns of interactions). Third, it offers reflections about the theatre of the oppressed, which involves the audiences.

Theme V: interactions revolution

This theme originally proposed by Espejo and Dominici wants to enable theoretical and empirical contributions about the increased speed of self-organising processes in today's world. Its focus is fluid interactions in the age of digital technologies. Its worldview is interactions and relationships between systems and their environments and also interactions within organisations. Its emphasis is dynamic and structural couplings and co-evolution rather than the study of the parts of a system. Today organisations and social systems need increased fluidity and agility to respond to environmental, social and economic pressures. Particularly these are requirements focused on the future and therefore this theme is connected to the Keynote "The Brain of the Future". The aim is to investigate opportunities and challenges for organisational and business systems in future scenarios that will require new strategies and policies to face increased fluidity in social structures.

It offers a space for digital technologies and transdisciplinary conceptual frameworks to disclose self-organised organisational forms that are more likely to anticipate and respond to environmental, social and economic changes.This is an 'interactions revolution' towards distributed value co-production and guided self-organisation. Problem solving is at the core of these self-organising processes; which interaction strategies increase response capacity and help, making sense of an often overwhelmingly complex surrounding? These are aspects related to requisite variety. Also, which interactive strategies help to maintain dynamic stability with this environment?

Contributions to this theme have not necessarily recognised this conceptual framework, but at its core are forms of interaction diagnosing, designing and discovering problem-solving strategies. Instances illustrate how people can benefit from current technologies in today's digital world to make their interactions – often dominated by information asymmetry, domination, abuses of power, inequality, lack of mercy and so forth – more humane and acceptable. The emergence of mercy in interactions – eleogenetics – as one of the abstracts proposes is not only a spiritual matter but also a matter of communicative capacity, say, between a mother and a child. Equally, lack of trust, corruption, unfairness relate to communicative capacity. There is much scope for an increased transparency, as well as exposure, in interactions. In a technical sense, algorithms and software code can break through murky communications.

A particular aspect of the abstracts to this theme is their foci on instances, such as a country's banking for sustainable development, language development to improve communications, smart cities to strengthen cohesion and improve quality of life, 3D printing technologies for personalised medication and health workers' interactions with patients to improve their services and so forth. One way or the other interactions are the currency of most of the other Congress themes; however in this theme the emphasis is instances of problem solving and relationships. In the cyber-systemic perspective, this is a theme about variety engineering as proposed by the Viable System Model.

Theme VI: knowledge and organisation

One of the topics proposed by Sergio Boria is knowledge to manage the knowledge society. Another, proposed by Andrée Piecq and Claude Lambert, is the role of emergence within organisations.

Together, they suggest that directive and invasive regulation is inappropriate to deal with complex systems, which far from being deterministic, are chaotic and unpredictable. Organisations don't lend themselves to precise analytical modelling but to learning processes through action. Organisations are complex systems which emerge from human interactions and create knowledge; these interactions may be enabled by management and training. To survive these organisations must be self-regulating and self-organising. Emergence, complexity, self-regulation are three inseparable concepts in all organisations.

Topics discussed by the abstracts range from technological unemployment as frictional unemployment to the extent that machines will replace the toil of work. Professional creativity and problem-solving capabilities will become ever more important in the future. Technological innovation will increase opportunities for new forms of work. One of the abstracts contrasts male and female brains and their contributions to organisation. Another contribution discusses building intellectual capital for sustainability in rural Ghana, and proposes sustainability as a source of knowledge. Fragmented communities fail generating knowledge. Yet, another argues that mainstream management is not the only way to manage social systems and suggests discussing new organisational forms. Consistent with Theme V one contribution suggests moving away from objects into relations and proposes quantum governance of human systems to overcome the idea of steering organisations as machines. The spirit of this theme is bottom-up, democratic, non-directive organisational learning.

Theme VII: systems thinking and system dynamics

This theme opens debates about two interrelated topics, the first "how can systems thinking help to bring solutions to humankind problems in the 21st century?" and the second is about system dynamics. Organisations are subject to contradictory forces. Contradictions stimulate organizational change.

The first topic was mostly articulated by Marie Noelle Sarget from France: though at a first glance societal problems may appear mainly economic, political, geopolitical, environmental or demographic, our failure to solve them may be because of systemic obstacles requiring a systemic approach, and an understanding of their complexity with an interdisciplinary vision and action. The early years of the 21st century are proving increasingly complex. We have many challenges including a population explosion, an unprecedented population migration, a failure managing our finite resources, the wholesale abuse of our environment, dysfunctional financial services, poorly regulated economies, inequality and corruption. We have still to grasp the full impact of technology upon human civilisation. If we, the collective human race, are to develop we should reflect upon how we organise ourselves and how to create policies relevant to our survival and how we turn them into feasible practicality. We believe that systems thinking and related information and communication tools have a significant part to play in policy development and to this end this topic is intended to surface ideas of value addressing some of the above-identified global challenges.

Authors were invited to submit papers addressing concrete problems, offering problem-solving practices going beyond general statements about systems concepts and abstract use of methodology. Each contribution should show how a systemic approach could contribute to understand situations better, bringing more effective solutions in the long term.

For issues like combating poverty the proposal is going beyond modelling economies, which are unlikely to solve problems, into systemic understanding of the roots of poverty. What are the limits to parametric economic modelling? Current trends in international trade: how to think systemically about trade? How can systemic obstacles to the inclusion of disabled people be overcome?

For observers to think systemically about a situation is different to accepting that they are part of a situation that is systemic. The latter is more about reflection and participation, as is proposed by Theme IX; the former could be more about using systemic thinking to gain advantage and power over those operating with a fragmented view of the world. Systems archetypes can be used as instruments to gain advantage in interactions; they may be used as competitive rather than collaborative tools.This reflection connects this topic to system dynamics modelling.

Didier Cumenal proposed the topic, "Systems thinking and System dynamics".

Stefano Armenia, from Italy, and Didier Cumenal, from France, contributed to the articulation of this topic. The impacts of policy/strategy implementation have historically been very difficult to anticipate, due to the many complex and interconnected phenomena acting in the related situations. Decision-making accuracy and effectiveness are seriously affected by the use of old tools for new problems. This implies that policy making requires theories, models and tools that support complexity.

The topic wants to address the general issue of model-based governance, aiming at establishing whether systems thinking and system dynamics can constitute the correct mix of tools in order to effectively and accurately deal with complexity in policy making. Policy makers are increasingly challenged by Big Data (Theme III); modelling tools like system dynamics can help make sense of situations of uncertainty and unpredictability.

Are systems thinking and system dynamics effective tools for policy making? One of the contributions explores the meaning and purpose of the concept "reality of a theory". It offers epistemological reflections; is a theory a deep approximation to reality? Contributions include reflections about teams for new product development, application of system dynamics to games, to space exploration, prison systems and malnutrition. However, in the end, a few submissions are more about economics than systems thinking.

Theme VIII: quantum modelling

François Dubois, from France, proposed this theme. It is a challenging and innovative proposition. It relates to other themes of this Congress, in particular to Theme V, "Interactions Revolution". Ashby proposed variety as a measure of complexity, a concept which relates to entropy and in its turn this is a concept that relates to quantum laws of information; how can quantum strategies be, and should be, developed by regulatory processes to reach a better fitness? Or in the language of Theme V, better interactions?

The formalism of quantum mechanics offers a general framework to consider a variety of areas outside of the natural remit of physics, such as economics, organisations and social interaction, psychology, artificial intelligence, human language, cognition, information retrieval, biology, political science. The application areas addressed typically operate at a macroscopic scale. They may share many key properties with quantum systems including non-commutativity of measurement, indeterminacy, non-separability, contextuality, proposals to test temporal nonlocality in perception and cognition, study of non-commutative structures in learning behaviour.

What can we learn from quantum physics regarding the holism that systems thinkers postulate for social economic systems? How can quantum modelling help to produce learning curves for policy making? Does the present global trend towards deregulated markets threaten the long-range effectiveness of the energy system? Quantum models of semantic memories, eigen logic, interpretable quantum observables are instances relevant to this theme. A document can be represented as a semantic memory in the form of a matrix. Linear operators to semantic memories help determining meaning. Another contribution includes action amplification by stimulated emission of social energy . . . this is a contribution aimed at discussing the social consequences of a social laser model.

Theme IX: reflexivity, second order science and context

This theme is supporting the encounter of about a dozen scientists from different parts of the world. Stuart A. Umpleby, from the USA, and Vladimir Lepskyi, from Russia, have spearheaded it. As proposed by Umpleby,

> several ideas are being used to describe a renewed emphasis in cybernetics on reflection and self-awareness. In addition to an early concern with controlling events in the world around us, cyberneticians have considered the role of the observer in creating descriptions and have conceived the scientist as an actor who participates in systems as well as observes them. There has been a strong

concern with ethics. Recently there have been challenges to the assumptions underlying science and efforts to improve, even reconceptualize the application of scientific ideas. Goals and purposes are now increasingly linked to the conceptualization of context.

Keynotes contributions

Knowledge for development vision 2030 (K4Dev_Vision 2030)

A systems perspective challenges and opportunities for theory, policy and practice

Elias G. Carayannis[1]

Keywords: *Quadruple and Quintuple Innovation Helix Models; knowledge for development; Vision 2030; systems perspective*

At the end of WWI (a century ago), President Wilson framed the vision to "make the world safe for democracy". A century later and fifteen years after 091101, the Arab Spring events and side effects (such as the Syrian conflict and even ISIS) demonstrated the need to "make democracy safe for the world". The future and sustained peace, prosperity and security of the WORLD require that we pursue and accomplish a reasonable modicum of BOTH of those visions and Knowledge for Development (K4Dev) and its related proposed roadmap (K4Dev_Vision 2030) based on the concepts of Glocal (Global/Local) Network of Real and Virtual Incubators (G_RVIN) (Carayannis et al., 2005) as well as the concepts of Strategic Knowledge Arbitrage and Serendipity (SKARSE ©) that play a central role in this set of challenges and opportunities.

The emerging *gloCalising* (globalizing-localizing) frontier of converging systems, networks and sectors of innovation that is driven by increasingly complex, non-linear and dynamic processes of knowledge creation, diffusion and use, confronts us with the need to re-conceptualize, if not to re-invent, the ways and means that knowledge production, utilization and renewal takes place in the context of the knowledge economy and society. Perspectives from and about different parts of the world and diverse human, socio-economic, technological and cultural contexts are interwoven to produce an emerging new worldview on how specialized knowledge, that is embedded in a particular socio-technical context, can serve as the unit of reference for stocks and flows of a hybrid, public/private, tacit/codified, tangible/virtual good, that represents the building block of knowledge economy, society and policy. Carayannis (2001) argues that the *"Mode 3" model is the knowledge production system architecture that engages actively higher order learning* (learning, learning to learn, learning to learn how to learn), in a multi-lateral, multi-nodal, multi-modal and multi-layered manner involving thus entities from government, academia, industry and civil society, as well as driving *coopetition (competition-cooperation)*, *co-specialization* and *coevolution* resource generation, allocation and appropriation processes (C3) that result in

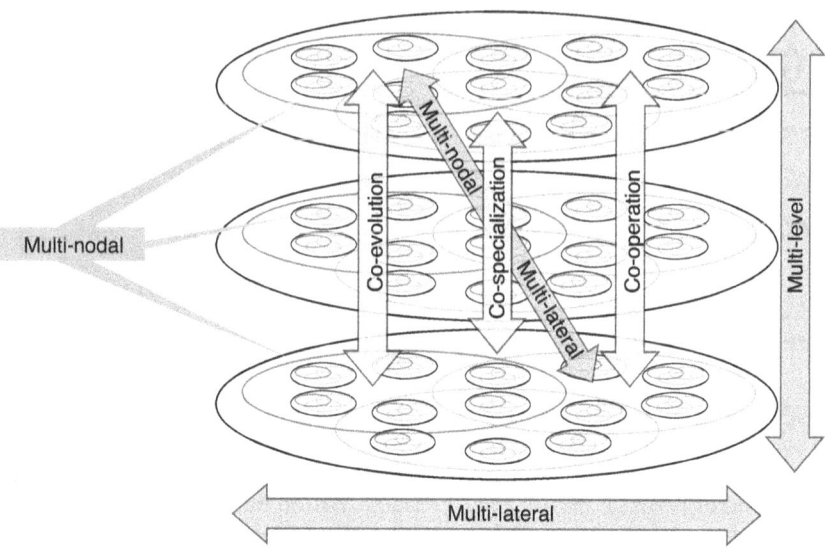

Figure 1.1 Strategic knowledge, serendipity and arbitrage: multi-modal, multi-nodal, multi-lateral, multi-level C3 processes

Sources: Carayannis, E.G., GWU Lectures and Journal Publications, 2001–2016

the formation of modalities such as innovation networks and knowledge clusters (Figure 1.1).

Strategic Knowledge Arbitrage and Serendipity (SKARSE ©) are real option drivers of K4Dev that drive the C3 (see above). Strategic knowledge serendipity refers to the unintended benefits of enabling knowledge to "spill over" between employees, groups and functional domains (*"happy accidents" in learning*). More specifically, it describes the capacity to identify, recognize, access and integrate knowledge assets more effectively and efficiently to derive, develop and capture non-appropriable, defensible, sustainable and scalable pecuniary benefits, while strategic knowledge arbitrage refers to the ability to distribute and use specific knowledge for applications other than the intended topic area. It refers to the capacity to create, identify, reallocate and recombine knowledge assets more effectively and efficiently to derive, develop and capture non-appropriable, defensible, sustainable and scalable pecuniary benefits. To operationalize the K4Dev_VISION 2030, we propose a set of initiatives and policies supporting smart, sustainable and inclusive growth via social innovation and leveraging social media modalities such as *crowd-sourcing*, *crowd-funding* and *crowd-storming* in complementarity to and synergistically with other broader initiatives. These initiatives and policies would be focused primarily but not exclusively at the places where most of the Earth's population is accumulating, namely cities, aiming to transform as many of these urban locales and as profoundly as possible into *"smart cities"* organized around the concepts outlined above and especially based on the *Quadruple and Quintuple*

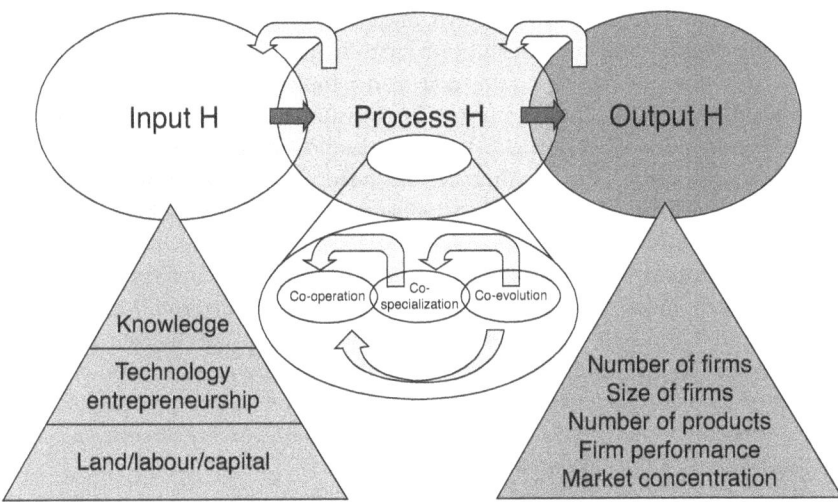

Figure 1.2 Heterogeneity dynamics – input, process, output
Source: Carayannis and Provance, 2008

Innovation Helix models (Carayannis et al., 2003, 2012) which empower civil society on a triple-bottom-line-centric basis (i.e. environmentally, financially and socially sustainable manner).

The *Quadruple Innovation Helix* bridges social ecology with knowledge production (Mode 3) and innovation. The most important constituent element of the quadruple helix – apart from the active "human agents" – is the resource of knowledge, which through a circulation known as circulation of knowledge, between social subsystems, changes to innovation and know-how in a society and for the economy. The quadruple helix, thereby, visualizes the collective interaction and exchange of knowledge in a state by means of the following four subsystems:

- Political System – formulates the direction of where the state/country is heading in the present and future, its laws, etc. (political and legal capital) (GOVERNMENT SECTOR)
- Education System – refers to academia, universities, higher education systems and schools (human capital) (UNIVERSITY SECTOR)
- Economic System – consists of industry/industries, firms, services and banks (economic capital) (INDUSTRY SECTOR)
- Civil Society (Media based/Culture based) – integrates and combines two forms of capital: culture-based public (tradition, values, etc. [social capital]) and media-based public (television, internet, newspapers [capital of information]) (CIVIL SOCIETY SECTOR)

Quadruple Innovation Helix models place a stronger focus on cooperation in innovation, and in particular, the dynamically intertwined processes of coopetition, coevolution and co-specialization, within and across regional and sectoral innovation ecosystems that could serve as the foundation for diverse smart specialization strategies. The European Commission RIS3 guide outlines a set of general principles as to how S3 strategies should be developed at the regional level and recognizes the significance and need for the Quadruple Innovation Helix approach by proposing to add a fourth group to a classical Triple Helix model. *The Quadruple Innovation Helix complements the Quadruple Innovation Helix by adding the (natural) environment as the fifth dimension and ensuring the sustainability of any bottom-up initiatives and top-down policies.* Moreover, the Quadruple/Quintuple Innovation Helix model puts innovation users at its heart and encourages the development of innovations that are pertinent for users (civil society). Users or citizens here own and drive the innovation processes which is the essence of *Social Innovation and in fact Sustainable Social Innovation bottom-up initiatives and top-down policies and practices.*

Related initiatives, policies and practices that could enact and enable the K4Dev_Vision 2030 Roadmap we propose, could encompass networked, global/local, micro-enterprise and SME formation and growth as well diaspora, refugees and inner-city projects via the crowd-storming, crowd-sourcing and crowd-funding modalities as well as the engagement of all sectors of the Quadruple Innovation Helix model within the overarching Quintuple Innovation Helix sustainability framework (see Figure 1.3 as a conceptual reference model as well as the closing questions from the 2005 book that remain as valid today as they were more than ten years ago (Carayannis and Sipp, 2005).

- Should a developing economy adopt a development pathway that aims at fulfilling the macro-level development objectives and goals (following the transitioning economies' example), or would it be *"smarter" for the developing economy in question to focus on fulfilling the meso-level development objectives and goals?*
- In the case where the meso-level focus is chosen, *could this approach lead the developing country to "skip" the transitioning stage, "leap-frogging" towards developed economy status and what, if any, trade-offs might there be* (including the risk of technological, socio-economic and even cultural divides forming within a country as a result of a potentially uneven and unbalanced, accelerated development)? Moreover, *what would the implications be for how the developmental stage of countries is defined and enacted?*
- *Could such accelerated development trajectories necessitate the re-definition of what a transitioning economy is and could such a re-assessment and re-definition of objectives and goals lead to a "smarter" approach to development* (by enabling more functional congruence between development conditions on the ground and development goals and objectives in the numerous reports of the MDAs and the MDGs)?

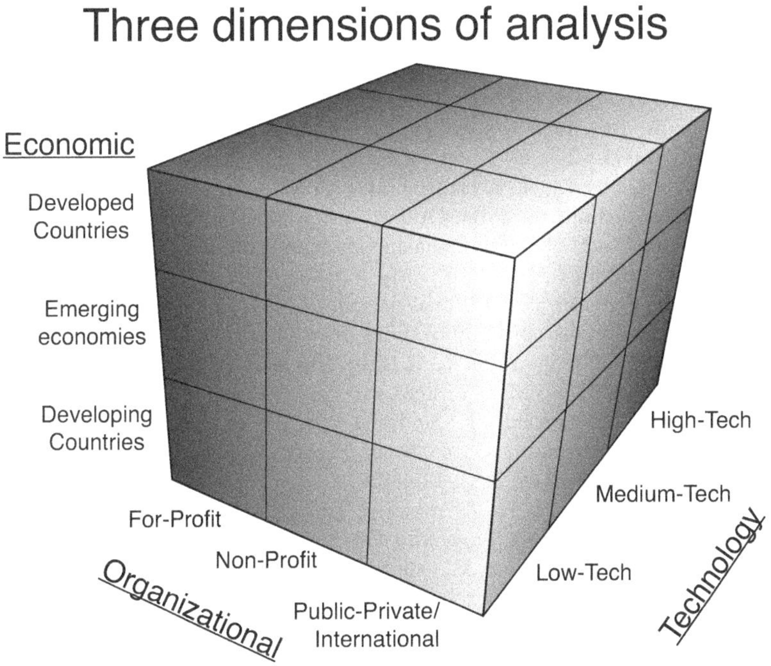

Three dimensions of analysis

Economic
- Developed Countries
- Emerging economies
- Developing Countries

Organizational
- For-Profit
- Non-Profit
- Public-Private/ International

Technology
- High-Tech
- Medium-Tech
- Low-Tech

Figure 1.3 Dynamic cube of e-development intervention strategies. Three dimensions of analysis of e-development in the knowledge economy

Source: Carayannis and Sipp (2005)

Note

1 George Washington University School of Business, Washington, DC, USA.

Bibliography

Carayannis, E. G. (2001). Learning more, better, and faster: A multi-industry, longitudinal, empirical validation of technological learning as the key source of sustainable competitive advantage in high technology firms. *International Journal of Technovation*, 135.

Carayannis, E. G., Barth, T. D., & Campbell, D. F. (2012). The Quintuple Helix innovation model: Global warming as a challenge and driver for innovation. *Journal of Innovation and Entrepreneurship*, 1(1), DOI: 10.1186/2192-5372-1-2.

Carayannis, E. G., & Campbell, D. F. (2006). *Knowledge creation, diffusion, and use in innovation networks and knowledge clusters: A comparative systems approach across the United States, Europe, and Asia*. Greenwood Publishing Group.

Carayannis, E. G., & Campbell, D. F. (2009). 'Mode 3' and 'Quadruple Helix': Toward a 21st century fractal innovation ecosystem. *International Journal of Technology Management*, 46(3–4), 201–234.

Carayannis, E. G., & Campbell, D. F. (2012). Mode 3 knowledge production in quadruple helix innovation systems. In *Mode 3 knowledge production in quadruple Helix innovation systems* (pp. 1–63). Springer, New York.

Carayannis, E. G., & Campbell, D. F. (2014). Developed democracies versus emerging autocracies: Arts, democracy, and innovation in Quadruple Helix innovation systems. *Journal of Innovation and Entrepreneurship*, 3(1), DOI: 10.1186/ s13731-014-0012-2.

Carayannis, E. G., Evans, D., & Hanson, M. (2003). A cross-cultural learning strategy for entrepreneurship education: Outline of key concepts and lessons learned from a comparative study of entrepreneurship students in France and the US. *Technovation*, 23(9), 757–771.

Carayannis, E. G., Gonzalez, E., & Wetter, J. (2003). The nature and dynamics of discontinuous and disruptive innovations from a learning and knowledge management perspective. In Shavinina, L. V. (Ed.), *The international handbook on innovation* (pp. 115–113), Elsevier, Amsterdam.

Carayannis, E. G., Grigoroudis, E., Sindakis, S., & Walter, C. (2014). Business model innovation as antecedent of sustainable enterprise excellence and resilience. *Journal of the Knowledge Economy*, 5(3), 440–463.

Carayannis, E. G., Kwak, Y. H., & Anbari, F. T. (Eds.). (2005). *The story of managing projects: An interdisciplinary approach.* Greenwood Publishing Group.

Carayannis, E. G., & Provance, M. (2008). Measuring firm innovativeness: Towards a composite innovation index built on firm innovative posture, propensity and performance attributes. *International Journal of Innovation and Regional Development*, 1(1), 90–107.

Carayannis, E. G., & Rakhmatullin, R. (2014). The quadruple/quintuple innovation helixes and smart specialisation strategies for sustainable and inclusive growth in Europe and beyond. *Journal of the Knowledge Economy*, 5(2), 212–239.

Carayannis, E. G., & Sipp, C. (2005). *E-development toward the knowledge economy: Leveraging technology, innovation and entrepreneurship for 'smart' development.* Springer, New York.

Carayannis, E. G., & Von Zedtwitz, M. (2005). Architecting gloCal (global-local), real-virtual incubator networks (G-RVINs) as catalysts and accelerators of entrepreneurship in transitioning and developing economies: Lessons learned and best practices from current development and business incubation practices. *Technovation*, 25(2), 95–110.

Governance in the Anthropocene

Cybersystemic possibilities?

Ray Ison[1]

Keywords: *cybersystemic governance; institutions; social-biophysical systems; co-evolution*

The "Anthropocene" is a term formulated by Earth scientists to claim that we have entered a new geological epoch: human influences have become so great that they are affecting "whole Earth dynamics" through a range of biophysical and social processes leading to complex global changes. Congruent with these global changes, there is increasing evidence that current "governance systems" too often are not fit for contemporary circumstances (Ison & Schlindwein 2015). "Living in the Anthropocene", regardless of whether this name is an adequate framing choice, means that we are collectively in a period new to human history. Thus, following Humberto Maturana, when we accept a new explanation our world changes. This novel context calls for an intense period of critical reflexivity in relation to our past thinking, practices, institutions, patterns of investment, and governance: the problems that will arise in this epoch will steadily become more severe, unpredictable, complex, and of a magnitude hitherto unseen. But to date cybersystemic understanding and praxis (theory-informed action) – a way to avoid simplistic, systematic approaches – has been barely present in the unfolding discourse about governance. "Problem-solving" has been reduced to the application of techno-centric knowledge and pseudo-solutions. It seems necessary therefore to take a "design turn" towards more cybersystemic governance of social-biophysical systems (Ison 2010).

Governance itself is a cybersystemic concept from the Greek verb "to steer" which means engaging with, and responding to, feedback from multiple sources in both the social and biophysical worlds so as to chart purposeful courses of action. Successful innovation in the Anthropocene has to be considered in the context of governance but cybersystemic literacy and capability to take the "design turn" is in short supply. The innate cybersystemic sensibility we humans have has also suffered in the face of non-systemic acculturation. These conditions constrain our capacity to understand how underlying systemic processes generate our experienced world, so as to inform the design of transformative practices in public policy and requisite governance arrangements in the Anthropocene (Ison & Shelley 2016; Ison 2016).

In his Encyclical Letter, *Laudato Si' of the Holy Father*, Pope Francis says:

> given the scale of change, it is no longer possible to find a specific, discrete answer for each part of the problem. It is essential to seek comprehensive solutions which consider the interactions within natural systems themselves and with social systems. We are faced not with two separate crises, one environmental and the other social, but rather with one complex crisis which is both social and environmental.
>
> (The Holy See 2015)

The principle emotions that this Encyclical, and this track at WOSC 2017, seeks to engage are those of inquiry (an acceptance of uncertainty) and hope, recognition that enthusiasm and human responsibility are central to human attempts to navigate a co-evolutionary future that is ethically defensible. Research is presented that evidences the need for innovations in: (i) organisational/institutional forms so as to mediate between vertical (i.e., the state) and horizontal governance (i.e., the citizens), especially feedback processes; (ii) practices that pay urgent attention to framing of initial starting conditions for change processes; and (iii) inventing and enacting means to secure the robust institutionalisation of cybersystemic governance.

Note

1 Applied Systems Thinking in Practice (ASTiP) Programme, The Open University UK.

Bibliography

The Holy See (2015), *Encyclical Letter Laudato Si' of the Holy Father Francis on Care for Our Common Home*. Libreria Editrice Vaticana, Rome.

Ison, R.L. (2010), *Systems Practice: How to Act in a Climate-Change World*. Springer, London and The Open University.

Ison, R.L. (2016), Governing in the Anthropocene: What future systems thinking in practice?, *Systems Research & Behavioral Science*, 33(5), 595–613.

Ison, R.L., Alexandra, J., Wallis, P.J. (2018), Governing in the Anthropocene: Cyber-systemic antidotes to the malaise of modern governance. *Sustainability Science*, 13(5), 1209–1223.

Ison, R.L., Schlindwein, S. (2015), Navigating through an 'ecological desert and a sociological hell': A cyber-systemic governance approach for the Anthropocene, *Kybernetes*, 44(6/7), 891–902.

Ison, R.L., Shelley, M. (2016), Editorial: Governing in the Anthropocene: Contributions from systems thinking in practice? ISSS yearbook special issue, *Systems Research & Behavioral Science*, 33(5), 589–594.

The brain of the future

Alexandre Perez Casares[1]

Keywords: *artificial intelligence; technological disruption; global governance; inequality; viable democracy*

The 'Age of the Cognitive Machines' is the most drastic economic transition since the Second Industrial Revolution. This transition is driven by the confluence of multiple technological innovations – such as advanced robotics, machine learning, and the exponential growth of computation capabilities and digital communication bandwidth – which result in the 'Rise of Intelligent Machines', understanding 'Machines' as a concept beyond its physical connotations, and leveraging the change of paradigm in machine intelligence, an evolution from 'Turing Machines' to 'Inference Machines.'

This transition will result in rapid increases of productivity of goods and services, shifts in the structure of our societies and cultures, major disruptions for global commerce and the balance of international power (economic and military), and growing income gaps driven by technological unemployment and the nature of wealth creation. According to the Oxford Martin School, approximately 47% of total current US employment is at high risk of being replaced by computerization over the next two decades, the fastest rate of change in the history of humanity, which will require a significant re-design of the Social Contract together with the transformation of existing education systems, such as the foundations of primary and secondary education and the role of the university.

Beyond the economic opportunities and challenges posed by AI (artificial intelligence), it will transform the role of the human species – and its current organizational structures – and pose significant risks for the systemic viability of western democracies in a world of increasing complexity driven by intelligent machines, requiring a new paradigm of national and global governance.

Note

1 President, The Altius Society at Oxford.

Bibliography

Acemoglu, D., Johnson, S. & Robinson, J. (2001). "The Colonial Origins of Comparative Development: An Empirical Investigation", *American Economic Review* 91, pp. 1369–1401.

Acemoglu, D., Johnson, S. & Robinson, J. (2002). "Reversal of Fortune: Geography and Institutions in the Making of the Modern World Income Distribution", *Quarterly Journal of Economics* 117, pp. 1231–1294.

Barrat, J. (2015). *Our Final Invention: Artificial Intelligence and the End of the Human Era*, 1st edition, Nueva York. Thomas Dunne Books, St. Martin's Griffin.

Beer, S. (1979). *The Heart of Enterprise*, Chichester. Wiley.

Bostrom, N. (2014). *Superintelligence: Paths, Dangers, Strategies*, 1st edition, Oxford. Oxford University Press.

Brynjolfsson, E. & Mcafee, A. (2014). *The Second Machine Age: Work, Progress, and Prosperity in a Time of Brilliant Technologies*, 1st edition, Nueva York. W. W. Norton & Company.

Cowen, T. (2011). *The Great Stagnation: How America Ate All the Low-Hanging Fruit of Modern History, Got Sick, and Will (Eventually) Feel Better*, 1st edition, Nueva York. Penguin Group.

Delong, B. (2015). "Making Do with More", *Project Syndicate, The World's Opinion Page*, February, Prague.

Diamond, J. (1997). *Guns, Germs, and Steel: The Fates of Human Societies*, Nueva York. W. W. Norton & Company.

Engerman, S. & Sokoloff, K. (1997). "Factor Endowments, Institutions, and Differential Paths of Growth among New World Economies: A View from Economic Historians of the United States", in: Harber, Stephen (Ed.), *How Latin America Fell Behind*, Stanford. Stanford University Press, pp. 260–304.

Engerman, S. & Sokoloff, K. (2002). "Factor Endowments, Inequality, and Paths of Development among New World Economies", *National Bureau of Economic Research*, Working Paper, Working Paper Series No. 9259.

Frey, C. B. Y. & Osborne, M. (2013). "The Future of Employment: How Susceptible Are Jobs to Computerization", *Oxford Martin School Publications*, September, Oxford, UK.

Gordon, R. J. (2012). "Is US economy growth over? Faltering innovation confronts the six headwinds", *National Bureau of Economic Research*, Working Paper No 18315, August, Cambridge, MA.

Keynes, J. M. (1930). "Economic Possibilities for Our Grandchildren", in: *Essays in Persuasion*, Nueva York. W. W. Norton & Co., pp. 358–373.

Kurzweil, R. (2012). *How to Create a Mind: The Secret of Human Thought Revealed*, London. Duckworth Overlook.

Maddison, A. (2007). *Contours of the World Economy 1–2030 AD: Essays in Macro-Economic History*, 2007 edition, Oxford. Oxford University Press.

Morris, I. (2010). *Why the West Rules-for Now: The Cycles of History and What They Tell Us about the Future*, 1st edition, Nueva York. Farrar, Straus and Giroux.

Muñiz Villa, M. & Perez Casares, A. (2014). "Progreso tecnológico y orden internacional: hacia una nueva economía y una mejor gobernanza", *Información Comercial Española, Revista de Economía* 880, October, pp. 39–55.

Pérez Ríos, J. (2012). *Design and Diagnosis for Sustainable Organizations: The Viable System Method*, New York, London, Heidelberg. Springer.

Piketty, T. (2014). *Capital in the Twenty-First Century*, 1st edition, Cambridge. Harvard University Press.

Rifkin, J. (2014). *The Zero Marginal Cost Society: The Internet of Things, the Collaborative Commons, and the Eclipse of Capitalism*, 1st edition, Nueva York. Palgrave Macmillan.

Rodrik, D. (2015). "From Welfare State to Innovation State", *Project Syndicate*, January, Princeton.

Soros, G. (1997). "The Capitalist Threat", *The Atlantic*, January.

Thiel, P. (2014). *Zero to One: Notes on Startups or How to Build the Future*, London. Virgin Books.

Design of regional system

Alfonso Reyes[1]

Keywords: *social cybernetics; regional system; local development processes*

Colombia is a rich country with a poor population. It has two oceans (the Caribbean and the Pacific), immense natural resources (the Amazon jungle is part of the border with Brazil as well as the Amazon, the Orinoco and the Magdalena rivers, three of the longest in South America). It has all climates during the whole year because it is located close to the equatorial line. Its topography is very diverse, from a large desert in the north to some of the highest mountains at top of the Andes (5.000 meters high). The biggest open coal mine in the world is there as well as some of the best softer coffees and it is also one of the largest flower producers in the region. Colombia has a strong economy being second in the mean growth of its GDP (per capita) after Chile with inflation below 3% since 2008.

However, Colombia is a country full of contradictions. Although its economy is growing, half of the population lives under the poverty line (52%) and the Gini coefficient is the second worst in South America after Bolivia (0,56). It has a huge extension of land that can be productive (more than 2,5 million hectares) but it has very few industries. It occupies first place in the ease of doing business ranking in South America but most of the productive population have informal jobs. Perhaps corruption, drug trafficking and an internal conflict that has lasted for 50 years are at the bottom of this paradox: a rich country with a poor population.

In November 2016, the FARC (Fuerzas Armadas Revolucionarias de Colombia) and the government finally agreed on a peace process after four years of negotiation. What follows is a long and difficult implementation phase that includes dismantling the armed organization, telling the truth about what happened with people who were kidnapped and never returned to their homes, being judged by a special criminal jurisdiction, forming a new political party and reincorporating to social life. Government estimates that about 8.000 former guerrilla members are going into this process. Some institutional capacity has been created (a new ministry and two new agencies) to handle this complex task. Funds have been gathered and international help canalized to have financial support to implement the agreement.

But to avoid the beginning of a new conflict, many of the structural causes of poverty must be dissolved. One of them is the considerable gap between urban and rural development. About 70% of the population is in the cities whereas people in small municipalities are suffering from unemployment and low quality public services. However, the industrial potential of the country, specially in food production, is there. Who can lead this local development? We think young people have the potential to do it. About 8 million Colombians between 17 and 31 years old live there, but after graduating from high school they do not have further opportunities to pursue their education. Before some of them used to join guerrilla or paramilitary groups; today they do not have that either.

The aim of this key note is to describe, from the perspective of social cybernetics, the design of a regional system to catalyze local development processes by articulating the delivery of technological programs to undergraduates with innovative transformations of local and traditional food production farms. A pilot project is being designed in Chaparral (in a small rural area in the south of Tolima, Colombia) using a German-made technology for aquaponics.

Note

1 Los An des University in Colombia.

Bibliography

Espejo, R. & Reyes, A. (2012). *Organizational Systems: Managing Complexity with the VSM*, London: Springer-Verlag. Co-autor con el professor Raúl Espejo. ISBN: 978-3-642-19108-4; e ISBN: 978-3-642-19109-1.

Espejo, R. & Reyes, A. (2011). "The State of the State", en "Systems Practice and Action Research", *Special Edition, Coauthor with Raúl Espejo*, Vol. 14, No. 2, London, UK, 2001.

Sustainability science

Linking science, policy and society for a sustainable future

Kazuhiko Takeuchi[1]

Key words: *Sustainability science; social and ecological systems; co-evolution; transdisciplinary; SDGs.*

The academic landscape of sustainability science has changed drastically over the years. At the beginning of the twenty-first century, it involved discussions in each specialized field, including engineering, agriculture, health sciences and economics. But now, over a decade into the new century, it has become more interdisciplinary – integrating more disciplines to pursue a comprehensive understanding of social and ecological systems (Takeuchi et al., 2016; Komiyama and Takeuchi, 2006).

Sustainability science also promotes the co-evolution of academia and society. Future Earth, a new international research platform focusing on the global environment, also emphasizes the importance of promoting transdisciplinary approaches that enable the co-design, co-production and co-delivery of knowledge (Future Earth, 2013; Mauser et al., 2013). In order to involve society in conducting research even from the designing stage, measures should be taken to strengthen the interface between science, policy and society.

The international community is now faced with the challenge of achieving the Sustainable Development Goals (SDGs) by 2030. Sustainability science is essential for scientifically explaining the importance of achieving SDGs (Saito et al., 2017). It is expected to play a significant role in systematically revealing relationships with the SDGs and structuring them to meet each objective. Sustainability science needs to present a new vision for a sustainable future that reintegrates social and ecological systems.

Note

1 President, Institute for Global Environmental Strategies Director and Project Professor, Integrated Research System for Sustainability Science (IR3S), The University of Tokyo Institutes for Advanced Study (UTIAS).

Bibliography

Future Earth, 2013. *Future Earth Initial Design: Report of the Transition Team*. Paris: International Council for Science (ICSU).

Komiyama, H., Takeuchi, K., 2006. Sustainability Science: Building a New Discipline. *Sustainability Science*. Volume 1, pp. 1–6.

Mauser, W., Klepper, G., Rice, M., Schmalzbauer, B. S., Hachmann, H. M., Leemans, R., Moore, H., 2013. Transdisciplinary Global Change Research: The Co-Creation of Knowledge for Sustainability. *Current Opinion in Environmental Sustainability*. Volume 5, pp. 420–431.

Saito, O., Managi, S., Kanie, N., Kauffman, J., Takeuchi, K., 2017. Sustainability Science and Implementing the Sustainable Development Goals. *Sustainability Science*. Volume 12, pp. 907–910.

Takeuchi, K., Ichikawa, K., Elmqvist, T., 2016. Satoyama Landscape as Social-Ecological System: Historical Changes and Future Perspective. *Current Opinion in Environmental Sustainability*. Volume 19, pp. 30–39.

Creating a science of purposeful systems

Stuart A. Umpleby[1]

Keywords: *second order science; cybernetics; facilitation methods*

1 Introduction

Creating a science of purposeful systems was an intention in the early years of cybernetics (Foerster et al. 1968; Ackoff & Emery 1972). A key distinction was the difference between goal seeking and goal formulation. A goal-seeking machine, such as a thermostat or a more complicated automatic machine, would have its goal set by the designer or user and the machine would then perform its function repeatedly. But how would one design a machine that would formulate its own goals? How are we to understand purposeful systems such as human beings and organizations? The question has been addressed in several ways in the field of cybernetics.

2 The Viable System Model

Stafford Beer's consulting work with corporations led him to develop the Viable System Model (Beer 1972, 1979, 1985), a model of organization based on the structure of the human nervous system and Ross Ashby's theory of adaptive behavior (Ashby 1952). Ashby noted that in order to be adaptive, an organism, organization or machine needed to have two nested feedback loops. The first operated frequently, made small corrections and enabled learning. Behavior that yielded a successful outcome was repeated. Behavior that yielded an unsuccessful outcome was not repeated. The result was that the system acquired or "learned" a pattern of behavior that was usually successful.

The second feedback loop operated less frequently and enabled adaptation. When the behavior that the organism or machine had learned no longer kept the machine in equilibrium with its environment, that is, when some essential variables went outside their limits, the second feedback loop would erase what had been learned and the machine would learn a new pattern of behavior. For example, when a corporation that had been profitable became unprofitable, the Board of Directors would look for a new Chief Executive Officer. Hence, goal formulation could be defined as a process of maintaining an equilibrial relation between an organism and its environment. Adaptation was defined as changing behavior when necessary so that the organism was always acting in a manner appropriate to the environment.

Cyberneticians working with managers and organizations recognized that a complex system such as an organization has a hierarchy of levels and purposes. For example, part of a manufacturing firm has as its purpose the design and production of quality products; another part of the firm has as its purpose marketing the products to customers. Beer's Viable System Model described a recursive structure for the divisions of an organization. The model has three parts (systems 1, 2 and 3) that produce the current product, a fourth part that scans the environment and designs the next product and a fifth part that regulates the basic values of the organization and decides when to change to the new product.

3   Autopoiesis

Another approach to describing purposeful behavior was contained in the work of Maturana, Varela and Uribe on autopoiesis (Maturana & Varela 1980). Using the Greek words for other, self and production they defined allopoiesis as other production and autopoiesis as self production. To illustrate, when the Ford Motor Company produces a car, it is producing something other than itself. However, the managers of the company not only have the responsibility to produce cars, they also must maintain the company as a viable entity. The processes of self production include hiring and training capable workers and managers.

The term autopoiesis was originally invented to describe a living organism. An organism consists of a set of processes that create molecules that interact so that the molecules and their relations create additional molecules and relations. A living organism is autopoietic both biologically and psychology. That is, biological processes create and maintain the body as a functioning entity. But a human being must also survive within a social system. Hence, a human being must create an image of itself and others so that he or she can function effectively in a social system. As a purposeful system, a human being must create an image of itself, and a role for itself within society, so that it can sustain itself both biologically and psychologically.

4   Reflexivity theory

Vladimir Lefebvre (1982) created a theory of two systems of ethical cognition. Some people grow up thinking that the end justifies the means. Others grow up thinking that the end does not justify the means, that only appropriate means should be used. Throughout one's life the first reaction will be to use the ethical system learned in childhood. However, it is possible for people to learn the other ethical system. Then, if the familiar ethical system is not working, an actor can decide to use the other ethical system. Lefebvre called this process reflexive control. In the 1990s the theory was widely used in Russia for strategic analysis and in education and psychological counseling.

George Soros created a different theory of reflexivity to describe the behavior of purposeful entities (individuals and organizations and some machines) within an economic system. A key concern has been the ability to predict future states, such as stock market values, in a social system. Soros noted that actors in social systems not only observe, they also participate. As they participate, they learn and change their behavior. The future states of such purposeful systems cannot be reliably predicted, hence knowledge of the behavior of purposeful systems within a society is fallible (Soros 2013).

Eric Beinhocker described the nature of a science of social systems when they are viewed as collections of economic agents (Beinhocker 2013). He noted that in Soros's theory there is internal model updating, meaning that purposeful systems can learn or change their minds. Also, social systems are complex in two ways. Social systems have interactive complexity due to multiple interactions among heterogeneous agents, and the system has dynamic complexity due to nonlinearity in feedbacks in the system.

5   Second order science

Another approach to the subject of purposeful systems is to ask how research on social systems should be done given that the elements of social systems (individuals and organizations) have their own purposes and from time to time change their goals and behavior. Since social systems and the elements within them are reflexive and purposeful, how can we create knowledge of social systems? This question has been discussed in several recent publications (Müller 2016; Umpleby 2002, 2014, 2016, 2017).

In the literature on second order science a key question is whether the observer is outside the system or inside the system. Usually in science the observer is thought to be outside the system observed. But if the observer is inside the system, as is usually the case with social systems, additional complexities arise.

In the management literature, mission, vision, goals and objectives are often discussed but less often purposes. Rarely discussed has been whether purposeful systems could be studied with the same methods developed for the physical sciences. Not discussing purpose and not challenging prevailing conceptions of science seemed to go together. But now that science is being reconsidered (Umpleby 2014, 2017), perhaps it is time to revive the idea of developing a science of purposeful systems.

6   Why study purposeful systems?

Both engineering and medicine had to struggle to accept a scientific rather than a craft approach to their field. Perhaps someday management and public administration will have a scientific foundation. The idea dates back at least to Donald T. Campbell (1988), who wrote, "Reforms as Experiments" and "The Experimenting Society." Quality or process improvement methods have made a major contribution to establishing an experimental approach to improving social processes. But a science of social systems seems to require a step further.

One approach is to expand the conception of science to include purposeful systems. The idea that purposeful systems change their goals and their behavior is not a surprise to people in business and government. But it is inconvenient for scientists who assume that if a sample of people behave a particular way one week, a similarly drawn sample of people can be expected to behave similarly the next week. The idea of unchanging elements is carried over from physics, which studies inanimate objects. A different approach to social systems would be to include experimental subjects in the class of experimenters and vice versa. The result would be a conversation. Means exist to hold such conversations using group facilitation methods. See www.gwu.edu/~umpleby/ptp.html.

7   Conclusion

After more than fifty years of work in the field of cybernetics we now have several ways of thinking about purposeful systems. We know that purposeful behavior and adaptation occurs at many levels in organizations (Beer). We know that the primary task of viable systems is to create and maintain their essential components and processes (Maturana). We know that the psychological structure of awareness is different for different persons and that people can change their behavior (Lefebvre). We have learned that earlier models of social systems made unrealistic assumptions about human behavior (Soros). Beinhocker (2013) clarified several assumptions in Soros's theory of reflexivity. Since science is a human activity that rests upon our assumptions about human cognition and human interaction, we have now developed several ideas about how to change our conception of science and our methods for doing science in order to incorporate what we have learned about purposeful systems (Umpleby 2014, 2017).

Note

1 George Washington University, Washington.

Bibliography

Ackoff, R.L. & Emery, F.E. (1972). *On Purposeful Systems.* Aldine-Atherton, Chicago.
Ashby, W.R. (1952). *Design for a Brain: The Origin of Adaptive Behavior.* Chapman & Hall, London.
Beer, S. (1972). *Brain of the Firm: A Development in Management Cybernetics.* Herder and Herder, New York.
Beer, S. (1979). *The Heart of Enterprise.* Wiley, New York.
Beer, S. (1985). *Diagnosing the System for Organizations.* Wiley, New York.
Beinhocker, E. (2013). "Reflexivity, Complexity, and the Nature of Social Science." *Journal of Economic Methodology*, Vol. 20, No. 4, pp. 330–342.
Campbell, D.T. (1988). *Methodology and Epistemology for Social Science: Selected Papers of Donald T. Campbell*, edited by E. S Overman. University of Chicago Press, Chicago.

Foerster, H. von, White, J., Peterson, L. & Russell, J. (eds.) (1968). *Purposive Systems*. Proceedings of the first annual symposium of the American Society for Cybernetics. Spartan Books, New York.

Lefebvre, V. (1982). *Algebra of Conscience: A Comparative Analysis of Western and Soviet Ethical Systems.* D. Reidel Publishing, Dordrecht, Holland.

Maturana, H.R. & Varela, F.J. (1980). *Autopoiesis and Cognition: The Realization of the Living.* Kluwer, Dordrecht.

Müller, K.H. (2016). *Second-Order Science: The Revolution of Scientific Structures.* Edition Echoraum, Vienna.

Soros, G. (2013). "Fallibility, Reflexivity, and the Human Uncertainty Principle." *Journal of Economic Methodology*, Vol. 20, No. 4, pp. 309–329.

Umpleby, S.A. (2002). "Should Knowledge of Management be Organized as Theories or as Methods?" *Janus Head, Journal of Interdisciplinary Studies in Literature, Continental Philosophy, Phenomenological Psychology, and the Arts*, Vol. 5, No. 1, Spring, pp. 181–195.

Umpleby, S.A. (2014). "Second-Order Science: Logic, Strategies, Methods." *Constructivist Foundations*, Vol. 10, No. 1, pp. 16–23. www.constructivist.info/10/1/016

Umpleby, S.A. (2016). "Second Order Cybernetics as a Fundamental Revolution in Science." *Constructivist Foundations*, Vol. 11, No. 3, pp. 455–488. www.univie.ac.at/constructivism/journal/11/3/455.umpleby.pdf

Umpleby, S.A. (2017). "How Science Is Changing," in *Cybernetics and Human Knowing*, forthcoming.

Theme I

Human aspects of managing systems

Sergio Barile, Marialuisa Saviano and Francesco Caputo

Six intelligences making the employees' values systems by knowledge-cum-values management and social responsibility values

Živa Veingerl Čič,[1] *Matjaž Mulej*[2] *and Simona Šarotar Žižek*[3]

Keywords: *intelligence – cognitive, emotional, social, physical and spiritual; leaders; social responsibility; subordinates; values*

1 Introduction

Authors aim to present the findings of their experience-based field research (of a long-lasting informal type) and related desk research about the role of the various and mutually interdependent and different intelligences in overcoming the differences in employee values systems as a source of personal and organizational and societal success. Although there are multiple intelligences we are highlighting in some detail, the cognitive, emotional, spiritual, physical and social intelligences, since these types of intelligences are strongly related to each other in human daily life and work. They determine the behavior of managers and their staff and influence their attitude towards individuals and organizations. These intelligences have the impact on the adopted decisions that can have long-term consequences for all stakeholders in the organization. We will put all five intelligences in the light of the principles of the (corporate) social responsibility as the human attribute/value directing the human application of the available knowledge and equipment. We will model mastering them by knowledge-cum-values management.

Based on systematic qualitative research we have used methodology including desk and informal field research, the Dialectical Systems Theory and its Law of requisite holism.

2 Intelligence and types of intelligence

The concept of intelligence is relatively old. It belongs to the most frequently mentioned concepts in psychological, pedagogical and artistic-essayistic literature. Most definitions of intelligence emphasize the aspects of general intellectual working (Fekonja 2001, 15). The "general" definition of intelligence could be divided into five groups Musek (1993, 186): the abilities

- Of thinking, mental judging, understanding and insight in relations;
- To adapt to new situations, creativity and performance adjustment;
- To use knowledge;
- To solve problems and tasks;
- To learn; and
- Of effectiveness in mutual relations.

There are at least seven types of intelligence, including linguistic, musical, logical-mathematical, spatial, bodily-motor, two personal intelligences – knowledge of oneself and the other (Gardner 1995). Each of the listed human potential intelligences is associated with one of the three basic neural systems in the brain; all intelligences described by Gardner are actually versions of the three basic intelligences – the intellectual, emotional and spiritual intelligence and related neuronal systems (Zohar and Marshall 2000, 14).

First, they equated human intelligence with rational intelligence (IQ). Psychologists developed special tests to measure IQ. Then in the 1960s, they found a contradiction between the intelligence tests and test results. With the IQ they measured only rational, logical, linear intelligence with which people solve certain kinds of logical problems; it is used for certain types of strategic thinking. In the 1990s, Goleman (1997) found that emotions are a very important factor of human intelligence and introduced the concept of emotional intelligence (EQ). Zohar (2006, 91–92) introduced the third type of intelligence and named it spiritual intelligence. There are also the physical and social intelligences; we will brief them later.

Let us add Howard Gardner's theory of multiple intelligences; his early work detected the initial six intelligences as a person's unique aptitude set of capabilities and how one might prefer to demonstrate intellectual abilities; today there are nine intelligences (Gardner 2010):

1   Verbal-linguistic intelligence
2   Logical-mathematical intelligence
3   Spatial-visual intelligence
4   Bodily-kinesthetic intelligence
5   Musical intelligences
6   Interpersonal intelligence
7   Intrapersonal
8   Naturalist intelligence
9   Existential intelligence

3   Discussion

All three authors have several decades of practical experience before and during the academic employment. Thus we detected that: (i) there is not only the cognitive intelligence, which is exposed by the theory of knowledge management that leaves all non-rational human attributes aside as the less important or even negligible ones; (ii) humans with the comparable or even equal formal education, specialization, profession and practical experience can display quite different real behavior; (iii) the latter difference is based on their irrational attributes that may be called the emotional, spiritual, physical and social intelligences; (iv) they have a crucial influence on their owners' application of their knowledge, skills and available equipment; (v) therefore it is not a systemic behavior, if one limits one's effort to master one's organizational working to knowledge management, which is the prevailing attitude that is reported about, modeled and suggested in the prevailing literature on human resources management and organizational development; (vi) thus, we have recently started to publish suggestions that one should become more or even requisitely holistic by switching from the knowledge management concept to the knowledge-cum-values management; (vii) at the same time we worked on social responsibility as human values that are leading to improving the personal and organizational and societal efficiency and effectiveness by practicing responsibility for one's influences on humans and nature, i.e. society in synergy with the practice of interdependence of different persons, organizations, communities and societies, which supports attainment of the requisite holism of approach, insight and requisite wholeness of outcomes of the activity; (viii) these three essential attributes of the socially responsible behavior enjoy support of the principles of the social responsibility, i.e. accountability, transparency, ethical behavior, respect for stakeholders, for the rule of law, for the international norms and for the human rights, (ix) thus, social responsibility can and shall become the sixth intelligence, added to the cognitive, emotional, spiritual, physical and social intelligences; (x) all six intelligences can and shall be considered and practiced in synergy that can and shall enjoy support from the knowledge-cum-values management. This makes the basis of one's personal development.

The integration of one's personal development with one's individual intelligence influences human values systems. Knowledge and developing of various types of intelligence matters: it lets individuals develop faster, in the long run. The higher is one's level of intelligence, the easier one faces one's problems or experiences. Thus, one is becoming a mature personality, who can overcome extreme alternatives to the briefed human values. This process can also receive a meaningful support from the exercise of social responsibility, which is one's responsibility for one's impacts on society, i.e. people and nature.

4 Conclusion

Different intelligences involve using existing learning systems and sensitivity, elaborate and enhance existing knowledge to analyze new situations and develop new solutions to positively impact to the overall performance. The intelligences pointed out can be considered as moderate predictors of job performance and success; employees who lack social intelligences and emotional intelligences perform more poorly than those high in conscientiousness and EQ. Poor social intelligences and emotional intelligences can be strong predictors of employees' and management's failure in their careers. Success of the process depends on "personal requisite holism." The top managers need significantly more emotional and social competences than the others.

The process of developing all six intelligences and the convergence of the leaders and subordinates can receive a meaningful support from the methods of creative collaboration. The same process would also receive a meaningful support from the exercise of social responsibility, which is everyone's responsibility for one's impacts on society, i.e. people and nature.

Work distribution makes the leaders and subordinates differ in prevailing values, too. Mastering of these differences would support business success, survival of jobs included and well-being of coworkers from both groups. Application of the cognitive, emotional and spiritual intelligences might help the organization meet this need. The fourth – physical intelligence – supports ensuring the psychological well-being at work; from it the other mentioned intelligences developed. Mastering of these differences can also receive support from methods of creative cooperation, social responsibility and personal requisite holism; we report about them elsewhere, and point to them only, here. The more holistic intelligences system generates a more socially responsible society.

Notes

1 Faculty of Economics and Business, Maribor, Slovenia.
2 Faculty of Economics and Business, Maribor, Slovenia.
3 Faculty of Economics and Business, Maribor, Slovenia.

Bibliography

Amram, Y. (2009). *The Contribution of Emotional and Spiritual Intelligences to Effective Business Leadership.* CA Palo Alto: Institute of Transpersonal Psychology.

Azizollah, A., Maede-Sadat, R., Narges, M. and Shekoofeh-Sadat, R. (2013). Relationship between Different Types of Intelligence and Student Achievement. *Life Science Journal,* Vol. 10, No. 7s, pp. 128–133.

Borysenko, J. (1997). *Čudežna preobrazba (Miracle Makeover).* Ljubljana: Ganeš.

Campuzano, G. L. and Seteroff, S. S. (2010). A New Approach to a Spiritual Business Organization and Employee Satisfaction. *Eastern Academy of Management: A New Approach,* pp. 1–15.

Covey, R. S. (2000). *Načela uspešnega vodenja (Principles of Successful Management).* Ljubljana: Mladinska knjiga.

Druskat, V. U., Sala, F. and Mount, G. (2006). *Linking Emotional Intelligence and Performance at Work: Current Research Evidence with Individuals and Groups.* Mahwah: Lawrence Erlbaum Associates Publishers.

Ealias, A. and George, J. (2012). Emotional Intelligence and Job Satisfaction: A Correlational Study Research. *Journal of Commerce and Behavioral Science,* Vol. 1, No. 4.

Emmons, R. A. (2000). Is Spirituality an Intelligence? Motivation, Cognition and the Psychology of the Ultimate Concern. *International Journal for the Psychology of Religion,* Vol. 10, No. 1, pp. 3–26.

Fekonja, R. (2001). *Čustvena inteligenca – način za spoznavanje in obvladovanje čustev (Emotional Intelligence: A Way of Understanding and Managing Emotions)*. Cerkvenjak: Diplomsko delo.

Frankovský, M. and Birknerová, Z. (2014). Measuring Social Intelligence: The MESI Methodology. *Asian Social Science*, Vol. 10, No. 6, pp. 90–97.

Fry, W. L. and Slocum,W. J., Jr. (2008). Maximizing the Triple Bottom Line through Spiritual Leadership. *Organizational Dynamics*, Vol. 37, No. 1, pp. 86–96.

Gardner, H. (1995). *Razsežnosti uma, Teorija o več inteligencah (Dimensions of Mind, Theory of Multiple Intelligences)*. Ljubljana: Tangram.

Gardner, H. (2010). Multiple Intelligences. Accessed on www.howardgardner.com/MI/mi.html (20. December 2016).

George, M. (2007). Leadership ter čustvena in duhovna inteligenca (Leadership, Emotional and Spiritual Intelligence). Accesed on www.revija.mojedelo.com/hr/cba-inside-leadership-ter-custvena-in-duhovna-inteligenca-434.aspx (7. September 2016).

Goleman, D. (1997). *Čustvena inteligenca: zakaj je lahko pomembnejša od IQ (Emotional Intelligence: Why It Can Be More Important Than IQ)*. Ljubljana: Mladinska knjiga.

Goleman, D. (2001). *Čustvena inteligenca na delovnem mestu (Emotional Intelligence on the Workplace)*. Ljubljana: Mladinska knjiga.

Goleman, D. (2015). What It Takes to Become a Socially Intelligent Leader. Accesed on www.danielgoleman.info/daniel-goleman-what-it-takes-to-become-a-socially-intelligent-leader/ (7. July 2016).

Goleman, D. and Boyatzis, R. (2008). Social Intelligence and the Biology of Leadership. *Harvard Business Review*, Vol. 86, No. (9), p. 74.

Goleman, D., Boyatzis, R. and McKee, A. (2002). *Prvinsko vodenje (The Primal Management)*. Ljubljana: Gospodarski vestnik.

Hrast, A., Mulej, M. and Lorbek, D. editors (2015). Proceedings of 10 IRDO International Conferences 'Social Responsibility and Current Challenges'. IRDO Institute for development of social responsibility, Maribor, Slovenia. Accesed on www.irdo.si (27. November 2016).

Klemenčič, D. (2002). *Poslovanje s čustveno inteligenco (Dealing with Emotional Intelligence)*. Diplomsko delo. Ljubljana: Ekonomska fakulteta.

Kravitz, S. M. and Schubert, D. S. (2000). *Emotional Intelligence Works*. Thomson Crisp Learning.

Lebe, S. S. and Mulej, M., guest-editors (2014). Social Responsibility and Holism in Tourism. *Kybernetes*, Vol. 43, No. 3–4, pp. 346–666.

Lundin, C. S., Paul, H. and Christensen, J. (2002). *Filozofija po ribje (The Fish Philosophy)*. Ljubljana: Tuma.

Marques, J. F. (2006). *The Spiritual Worker: An Examination of the Ripple Effect That Enhances Quality of Life in-and Outside the Work Environment*. Emerald Group Publishing Limited, pp. 884–895.

Možina, S. (2002). Vodja in vodenje (Leader and Leadership). In Možina, S *Management: nova znanja za uspeh (Management: New Skills for Success)*. Radovljica: Didakta.

Mulej, M. (2016). Knowledge Management or Knowledge-Cum-Values-Management? ENTRENOVA 2016 Proceedings.

Mulej, M., and co-authors (by ABC order). Božičnik, S., Čančer, V., Hrast, A., Jere Lazanski, T., Jurše, K., Kajzer, Š., Knez-Riedl, J., Mlakar, T., Mulej, N., Potočan,

V., Risopoulos, F., Rosi, B., Steiner, G., Štrukelj, T., Uršič, D., Ženko, Z. (2013). *Dialectical Systems Thinking and the Law of Requisite Holism*. Litchfield Park, Arizona: Emergent Publications.

Mulej, M. and Dyck, R., editors and coauthors, with coauthors (2014). *Social Responsibility Beyond Neoliberalism and Charity*. 4 volumes. Sharjah, UAE: Betham Science.

Mulej, M., Hrast, A. and Dyck, R., guest-editors (2013). Social Responsibility: Measures and Measurement. *Systems Practice and Action Research*, Vol. 26, No. 6, pp. 471–588.

Mulej, M., Hrast, A. and Dyck, R., guest-editors (2014). Social Responsibility: A New Socio-Economic Order. *Systems Research and Behavioral Science*, Vol. 32, No. 2, pp. 147–264.

Mulej, M., Merhar, V., Žakelj, V., Hrast, A. and Čagran, B., editors (2016). *Nehajte sovražiti svoje otroke in vnuke, 3 knjige. (Stop Hating Your Children and Grandchildren: Trilogy)*. Slovene: IRDO – Inštitut za razvoj družbene odgovornosti, Maribor in Kulturni center Maribor, Zbirka Frontier Books.

Mulej, M., Šarotar Žižek, S. and Treven, S. (2011). Povezava med psihičnim dobrim počutjem zaposlenih in duhovno inteligentnostjo kot dejavniki primerjave (Relationship between Psychological Well-Being of Employees and Spiritual Intelligence as Factors Comparisons). *Organizacija*, Vol. 44, No. 1.

Musek, J. (1993). *Osebnost pod drobnogledom (Personality Under Scrutiny)*. Maribor: Založba Obzorja.

Novak, B. (2004). Smisel in meje osebnega razvoja odraslih (The Meaning and Limits the Personal Development of Adults). Accesed on www.entra.si (15. December 2016).

Orosová, O. and Gajdošová, B. (2009). The Association of Social Intelligence Factors, Normative Expectations, and Perceived Accessibility with Legal Drug. *Adiktologie*, Vol. 4, pp. 204–211.

Reave, L. (2005). Spiritual Values and Practices Related to Leadership Effectiveness. *The Leadership Quarterly*, Vol. 16, pp. 655–687.

Šarotar Žižek, S. (2012). Vpliv psihičnega dobrega počutja na temelju zadostne in potrebne osebne celovitosti zaposlenega na uspešnost organizacije: doktorska disertacija. (Influence of Psychic Well-Being of Employees, Based on their Requisite Personal Holism, on the Success of Their Organization/PhD Thesis). Maribor: UM, EPF.

Šarotar Žižek, S. and Milfelner, B. (2014). *Vpliv menedžmenta človeških virov na uspešnost organizacij*. Maribor: IRDO.

Šarotar Žižek, S. and Mulej, M. (2013). Social Responsibility: A Way of Requisite Holism of Humans and Their Well-Being. *Kybernetes*, Vol. 42, No. 2, pp. 318–335.

Šarotar Žižek, S., Mulej, M. and Treven, S. (2014a). *Zagotavljanje zadostne in potrebne osebne celovitosti človeka*. Maribor: IRDO.

Šarotar Žižek, S., Mulej, M. and Treven, S. (2014b). *Osebna celovitost človeka (Personal Requisite Holism)*. Maribor: IRDO.

Šarotar Žižek, S. and Treven, S. (2014). *Psihično dobro počutje zaposlenih*. Maribor: IRDO.

Silvera, D. H., Martinussen, M. and Dahl, T. I. (2001). The Tromso Social Intelligence Scale, a Self-Report Measure of Social Intelligence. *Scandinavian Journal of Psychology*, Vol. 42, pp. 313–319.

Simmons, S. and Simmons, J. C. (2000). *Merjenje čustvene intelligence (Measuring of Emotional Intelligence)*. Ljubljana: Mladinska knjiga.

Škoberne, P. (2008). *Telesna inteligenca (Body Intelligence)*. Ljubljana: Center za duhovno kulturo.

Trojnar, F. (2002). *Moč osebne rasti: sezite po uspehu, v osebnem, družinskem in poslovnem življenju (The Power of Personal Growth: Reach for Success in Personal, Family and Business Life)*. Maribor: Trojnar Consulting.

Vaughan, F. (2002). What Is Spiritual Intelligence? *Journal of Humanistic Psychology*, Vol. 42, No. 2, pp. 16–33.

Warner, J. (2001). *Emotional Intelligence, Style Profile*. Amherst Massachusetts: HRD Press inc.

Wigglesworth, C. (2006). Why Spiritual Intelligence Is Essential to Mature Leadership. *Integral Leadership Review* Integral publishers. Accessed on http://integralleadershipreview.com/5502-feature-article-why-spiritual-intelligence-is-essential-to-mature-leadership/

Wilks, F. (2001). *Inteligentna čustva (Intelligent Emotions)*. Kranj: Ganeš.

Wolf, E. (2004). Spiritual Leadership: A New Model. *Healthcare Executive*, Vol. 19, No. 2, pp. 22–25.

Wolman, R. W. (2001). *Thinking with Your Soul: Spiritual Intelligence and Why It Matters*. New York: Harmony.

Zlatanović, D. and Mulej, M. (2015). Soft-Systems Approaches to Knowledge-Cum-Values Management as Innovation Drivers. *Baltic Journal of Management*, Vol. 10, No. 4, pp. 497–518.

Zohar, D. (2006). *Duhovni capital (Spiritual Capital)*. Ljubljana: TOZD.

Zohar, D. and Marshall, I. (2000). *Duhovna inteligenca (Spiritual Intelligence)*. Tržič: Učila.

Prospective memory

Neurocognitive correlates of memory for delayed intentions

Alberto Costa[1,2]

Keywords: *prospective memory; cognitive functions; brain functioning; neurodegenerative diseases*

1 Background

In the field of cognitive and applied neurosciences, cognitive functions that sustain the ability to code, keep in mind and realize intentions in the future are captured by the term *prospective memory* (PM). PM is a crucial ability for autonomous management and functional adaptation of an individual. Attending to appointments, following medical prescriptions, complying with planned commitments are examples of daily living activities that may be severely hampered by a deficit of this complex function. Clinical observations and empirical data also highlight that PM failures are frequent in individuals who complain of memory disorders. Indeed, PM impairments are early detected in persons with mild cognitive impairment (i.e., people who are supposed to be at higher risk to develop dementia) and in individuals with overt neurodegenerative disease such as, for instance, Alzheimer's dementia and Parkinson's disease (Costa et al., 2011a, 2015a). In these individuals PM disorders may significantly affect autonomous living and quality of life.

In line with above observations, the study of cognitive mechanisms and neurobiological processes involved in PM have deserved the growing attention in recent years. Indeed, the understanding of the functional and structural architecture of PM is a relevant issue from both theoretic and applicative perspectives. In fact, both diagnosis and treatment protocols could be improved accordingly. In this vein, in the present report we will review extant data with the aim to discuss two main issues that are currently debated in PM literature. First, should we conceptualize PM as a distinctive cognitive function with dedicated brain networks or, rather, as a component of other well-known cognitive domains (i.e., episodic memory and executive functions)? A second important issue refers to the potential clinical applications of PM assessment and training.

2 Results

Results from studies with brain-damaged individuals show that, within PM, two main components could be dissociated: a properly *prospective component* and a *retrospective component*. The former would refer to the ability to activate the intention to act – at the right moment. The latter component would represent the content of the intention, i.e. the action to be performed and the associative link between intention and the prospective context. The same findings also indicate a significant relationship, on one side, between the prospective component and executive functions, and, on the other side, between the retrospective component and episodic memory functioning (Costa et al., 2011a).

The dissociation between the two PM components is also supported by results from both neuroimaging studies (Burgess et al., 2011) and transcranial magnetic stimulation (TMS) investigations (Costa et al., 2011b, 2013). These findings support the hypothesis that PM is sustained by a distributed brain network, within which the frontal pole (i.e., Brodmann area 10) would play a relevant role in the mediation of the prospective component. At this regard, the hypothesis has been advanced that PM processes are underpinned by a dynamic interplay between mesial ad lateral portions of the frontal pole (Burgess et al., 2011; Barban et al., 2014).

The individuation of the two PM components appears to be quite useful for the purpose of the clinical examination of people with memory complaints. Indeed, recent data document that the assessment of the prospective component may improve, in the elderly, differential diagnosis of early cognitive impairment (Costa et al., 2015a). Results from other studies indicate that PM perfomance might be an interesting outcome index to evaluate the effect of rehabilitative training in people with Parkinson's disease (Costa et al., 2014). The latter finding appears to be particularly relevant in view of data showing, in individuals with Parkinson's disease, that PM impairment is independently associated with decreased autonomous management of daily living activities (Costa et al., 2015b).

3 Conclusions

The present short report focused on neurocognitive mechanisms implicated in PM with the aim to highlight some important theoretic and applicative issues. Available evidence documents that PM is a modular cognitive ability, with functional and structural neural correlates, partially associated with executive and episodic memory functions. Moreover, data showing the potential clinical relevance of assessing and training PM in people with cognitive impairments encourage further research, particularly in the field of dementias.

Notes

1 Niccolò Cusano University, Rome, Italy.
2 IRCCS Fondazione Santa Lucia, Rome, Italy.

Bibliography

Barban, F., Carlesimo, G.A., Macaluso, E., Caltagirone, C., Costa, A. (2014). Functional interplay between stimulus-oriented and stimulus-independent attending during a prospective memory task. *Neuropsychologia*, 53, 203–212.

Burgess, P.W., Gonen-Yaacovi, G., Volle, E. (2011). Functional neuroimaging studies of prospective memory: What have we learnt so far? *Neuropsychologia*, 49, 2246–2257.

Costa, A., Caltagirone, C., Carlesimo, G.A. (2011a). Prospective memory impairment in mild cognitive impairment: An analytical review. *Neuropsychology Review*, 21(4), 390–404.

Costa, A., Oliveri, M., Barban, F., Torriero, S., Salerno, S., Lo Gerfo, E., Koch, G., Caltagirone, C., Carlesimo, G.A.(2011b). Keeping memory for intentions: A cTBS investigation of the frontopolar cortex. *Cerebral Cortex*, 21(12), 2696–2703.

Costa, A., Oliveri, M., Barban, F., Bonnì, S., Koch, G., Caltagirone, C., Carlesimo, G.A. (2013). The right frontopolar cortex is involved in visual-spatial prospective memory. *PLoS One*, 8, e56039.

Costa, A., Peppe, A., Serafini, F., Zabberoni, S., Barban, F., Caltagirone, C., Carlesimo, G.A. (2014). Prospective memory performance of patients with Parkinson's disease depends on shifting aptitude: Evidence from cognitive rehabilitation. *Journal of the International Neuropsychological Society*, 20(7), 717–726.

Costa, A., Fadda, L., Perri, R., Brisindi, M., Lombardi, M.G., Caltagirone, C., Carlesimo, G.A. (2015a). Sensitivity of a time-based prospective memory procedure in the assessment of amnestic mild cognitive impairment. *Journal of Alzheimer's Disease*, 44(1), 63–67.

Costa, A., Peppe, A., Zabberoni, S., Serafini, F., Barban, F., Scalici, F., Caltagirone, C., Carlesimo, G.A. (2015b). Prospective memory performance in individuals with Parkinson's disease who have mild cognitive impairment. *Neuropsychology*, 29(5), 782–791.

Eleogenetics and its applications

Massimo Schinco[1]

Keywords: *mercy; epiphany; emergence; relationships; emotions*

1   What is Eleogenetics?

In the ancient Greek, "eleos" means "mercy". "Eleogenetics" consists in the study of processes creating and facilitating the "epiphany" of mercy. The word "epiphany" also derives from the ancient Greek, and means "manifestation". In the modern language of Systems Theory, we say something similar using the term "emergence", but when we focus on human beings, the word "epiphany" is preferable, so to emphasize the aesthetical, transcendent and ethical aspects of this peculiar kind of emergence.

Why talk about mercy? Usually, people are boggled at hearing or reading this word outside of a religious or a solidarity context, especially when applied to fields like management and organization studies in a systemic and cybernetics key. Although, from an anthropological point of view, "mercy" is a widespread reality, on which especially religions have focused with their own different cut; from a cultural point of view it is a neglected concept. Mercy can be considered as a disposition, a peculiar way to generate the relationship between the subject and the world. Its peculiarity resides in the attention to states of need, vulnerability, wrongness that, more or less, belong to all human beings. Thus, every day we experience mercy and, conversely, its lack, at least to some extent. Eleogenetics approaches mercy with a socio-psychological and a systemic cut, arguing that mercy is crucial, so that relational processes do not smother the autonomy and the reality of communities and individuals intended as subjects.

An eleogenetic-oriented approach to reality takes in the utmost consideration the connection with emotions and feelings, both positive and negative, as well as the full acknowledgement of other conditions and human needs that ground our life experience in actual circumstances and relationships, providing them with meaning and a direction for action, when possible.

The relationship between the epiphanies of mercy and the processes giving rise to them is recursive and features "charmed loops". No act of mercy carries with it a tendency to isolation or fragmentation. Although eleogenetic practices start in a definite situation, they take place simultaneously in different dimensions of reality, including those where our ordinary coordinates of space and time lose their meaning. There is no doubt that the effects of mercy can be macroscopic from a personal and social point of view. Nonetheless, it is peculiar of mercy and of eleogenetic practices to take place in events and interactions whose order of magnitude is small; for example the interaction between mother and child, or the interaction between a therapist and a patient. Eleogenetic practices are future-oriented and trigger epiphanies of Goodness, Beauty and Truth in original and unpredictable forms. The effects of entropic processes, when combined with eleogenetic practices, become similar to old bricks used to realize brand new buildings.

2   Applications of Eleogenetics

The official debut of Eleogenetics was in an extra-clinical context and emphasized the radically different effects of being "person oriented" instead of "problem oriented".[2] However, up to now Eleogenetics has found applications mainly in fields connected to the professional activity of its originator. The establishment of a Scientific Committee and the interest coming from different fields of study makes likely an expansion in different directions in the short term.

• Psychotherapy

Following the Eleogenetic Approach, the duty of the therapist is like "opening breaches", where life can sprout from again. When a breach is open, grass and flowers begin to diffuse themselves everywhere, following their own logic in the interaction with the ground they are growing on. An experience like this can feature a major change in the life of a patient, because it implies to some extent a divestiture, a transfer of control.

• Applications to treatment and care in hospital and other institutions

It is peculiar of mercy and of eleogenetic practices to take place in events and interactions whose order of magnitude is small; usually, events of this kind tend to be underrated and neglected. Meanwhile they strongly affect the quality of the everyday life in health organizations and the destiny of a help relationship. It is crucial to take in consideration these exchanges in accord with the context in which they take place.

• Supervision for psychotherapists and clinical supervision for professionals in the social field (educators, social workers)

Eleogenetics claims that the more people are honest in approaching reality, the more the processes they are part of will have eleogenetic outcomes. In consequence of this, Eleogenetics keeps in the highest consideration the effort of realizing good practices of supervision in the clinical field. The eleogenetic approach to supervision emphasizes a complex style of conduction and work. In particular, much more than focusing on "case reports", sessions probe a detailed examination of a specific, contextualized interactive situation, which by definition is unique.

• Dream study and dream working

Although things might be different in arts and religion, in general human narratives show a tension toward continuity, where discontinuity is experienced rather as a "break" in form of a positive surprise or of an inconvenient. The peculiar narrative of dreams features itself as an intermediate state in this tension, with continuity and discontinuity standing side by side and overlapping. Traditionally, both in theory and clinical practice, nocturnal discontinuities have been approached as if they would reflect not well-elaborated diurnal discontinuities, especially traumas and developmental breakdowns. Eleogenetics proposes to approach discontinuities as information coming straight from the being attracted by the future part of the dreamer. In this frame, the diurnal "discontinuous" events and the nocturnal "discontinuous" contents of a dream appear to be products of the interaction between the peculiar dreamer's entanglement in the past and the peculiar way the dreamer is attracted by the future.

• Vocational guidance in life choices

Defining itself as a theoretical and practical approach to development and change, Eleogenetics bestows the future with great respect. In particular, the Eleogenetic Approach focuses on a series of questions, all relaying on key aspects of the Eleogenetic Approach, such as: encounters, discontinuities, the future, the relational world, original talents, neglected parts of personality.

3   Conclusions

Behind the assumptions of Eleogenetics lies a philosophical background, ranging from Bergson to Marcel, Lévinas, Teilhard de Chardin and others. Nevertheless, as it is written in the webpage introducing its Scientific Committee,

> To be part of it does not imply any sort of dogmatic acceptance of theories and doctrines concerning Eleogenetics. [Rather is requested] fondness for being active parts of wide processes, where all individuals are considered as priceless and original parts of a priceless and unceasingly original wholeness oriented to future.
>
> (www.eleogenetics.cloud, 2016)

From a practical point of view, in the everyday social intercourses, Eleogenetics encourages counterintuitive attitudes, commending "weak" and "failing" aspects of human existence that are usually avoided, rejected and fought with more or less violence. Eleogenetics aspires to demonstrate instead that, in a valid philosophical frame, as well as in the light of the most advanced contributions in the fields of the study of systems, consciousness and clinical psychology, those attitudes are meaningful and effective in promoting a better and sustainable quality of life.

Notes

1 University of Pavia.
2 www.slideshare.net/beinetter/krakow-elcogenetic-2015-online

Bibliography

Bateson, G., Jackson, D. D., Haley, J., Weakland, J. (1956), Toward a theory of schizophrenia, *Behavioral Science*, 1, 251–264.
Bellet, M. (2014, January 1), *La communauté invisible*, retrieved from: http://belletmaurice.blogspot.it/2014/01/la-communaute-invisible.html
Bergson, H. (1934), *La Pensée et le mouvant. Essais et conférences*, Félix Alcan, Paris.
Bohm, D. (1980), *Wholeness and the Implicate Order*, Routledge & Kegan Paul, London.
Bohm, D., Peat, F. D. (1987), *Science, Order and Creativity*, Bantam Books.
Bonhoeffer, D. (1988), *Act und Sein. Transzendentalphilosophie und Onthologie in der systematischen Theologie*, Chr. Kaiser Verlag, München.
Capello, C. (in preparation), *L'operosità misericordiosa nel lavoro di cura – aspetti relazionali*.
Capello, C., Fenoglio, M. T. (2006), *Perché mi curo di te. Il lavoro di cura tra affetti e valori*, Rosenberg & Sellier, Torino.
Cecchin, G. F., Apolloni, T. (2003), *Idee Perfette: Hybris delle prigioni della mente*, Franco Angeli Milano.
Damasio, A. R. (1994), *Descartes' Error: Emotion, Reason and the Human Brain*, Putnam, New York, NY.
Di Corpo, U., Vannini, A. (2015), *Syntropy: The Spirit of Love*, ICRL Press, Princeton, NJ.
Frankl, V. (1982), *Ärztliche Seelsorge*, Franz Deuticke, Wien.
Gardiner, J. (2015), The illusion of fragmentation: The architecture of wholeness and the collective unconscious, in: Owczarski, W., Ziemann, Z. (editors), *Dreams, Phantasms and Memories*, Wydawnictwo Uniwersytetu Gdańskiego, Gdańsk, pp. 81–95.
Giuliani, M. (2009), Contributi della teoria dell'ipertesto a una terapia sistemico-narrativa, in Giuliani, M., Nascimbene, F., *La terapia come ipertesto*, Antigone Edizioni, Torino, pp. 68–110.
Gruen, A. (1988), *The Betrayal of the Self*, Grove Press, New York, NY.
Gruen, A. (2007), *The Insanity of Normality*, Human Development Books, Berkeley, CA.
Hartmann, E. (2011a), *Boundaries, a New Way to Look at the World*, CIRCC Ever-Press, Summerland, CA.
Hartmann, E. (2011b), *The Nature and Functions of Dreaming*, Oxford University Press, New York, NY.
Jaspers, K. (1959), *Allgemeine Psychopathologie*, Springer-Verlag, Heidelberg.
Kasper, W. (2012), *Barmherzigkeit. Grundberiff des Evangeliums – Schlüssel christlichen Lebens*, Verlag Herder GmbH, Freiburg.

Lévinas, E. (1961), *Totalité et Infini*, Martinus Nijhoff, La Haye.

Lévinas, E. (1978), *Autrement qu'être ou au-delà de l'essence*, Martinus Nijhoff Publishers.

Maldiney, H. (1991), *Penser l'homme et la folie*, Million, Grenoble.

Manousakis, E. (2006), Founding quantum theory on the basis of consciousness, *Foundations of Physics*, 36(6), 795.

Marcel, G. (1945), *Homo Viator*, Aubier, Paris.

Meltzer, D., Harris, M. (1994), A psychoanalytic model of the child-in-the-family-in-the-community, in: Hahn, A. (editor), *Sincerity and Other Works: Collected Papers*, Karnac, London.

Mina, F. (2007), *La madre nei sogni del bambino*, Lulu.com, Raleigh, NC.

Peat, F. D. (1987), *Synchronicity, the Bridge between Matter and Mind*, Bantam, New York, NY.

Peat, F. D. (2008), *Gentle Action: Bringing Creative Change in a Turbulent World*, Pari Publishing, Pari.

Schinco, M. (2002), *O Divina Bellezza, O Meraviglia – Uno psicoterapeuta ascolta Turandot*, Carabà Srl edizioni, Milano (reprinted as e-book 2012).

Schinco, M. (2007), La natura dialogica della psicoterapia, in: Capello, C., Gianone, E. (editors), *i non colloqui di Alice*, ISU Università Cattolica, Milano, pp. 115–159.

Schinco, M. (2011), *The Composer's Dream: Essays on Dreams, Creativity and Change*, Pari Publishing, Pari.

Schinco, M. (2014, December 1), *Quattro passi, anzi cinque nella stanza della terapia*, retrieved from: www.academia.edu/9448559/Quattro_passi_anzi_cinque_nella_stanza_della_terapia

Schinco, M. (2015), *Siamo sognati a nostra volta*, Durango Edizioni, Manerbio.

Schinco, M. (2015, March 21), *The establishing manifesto of eleogenetics*, retrieved from: www.eleogenetics.cloud

Schinco, M. (2015, March 28), *Approccio sistemico e identità individuale: conversazioni in sospeso*, retrieved from: www.massimoSchinco.it/wp-content/uploads/2014/04/approccio-sistemico-e-identit%C3%A0-individuale.pdf

Schinco, M. (2015, April 24), *Epiphanies of right, epiphanies of evil: Eleogenetic practices as an alternative to a "problem based" reality*, retrieved from: www.slideshare.net/beinetter/krakow-eleogenetic-2015-online.

Schinco, M. (in press), *Come Home, Doc! Practical Aspects of the Eleogenetic Approach to Therapy*.

Schinco, M., Schinco, S. (2015), Spazi travagliati: tempo, affetti e memoria nei processi di identità territoriale, in: Piselli, A. (editor), *Alteridentità. Luoghi narrati, luoghi taciuti, luoghi comuni*, Durango Edizioni, Manerbio, pp. 15–43.

Schinco, S., Schinco, M. (2016), Attracted by the future, conditioned by the past, shaped by our decisions: That's where a place is, in Lytovka, O. (editor), *The Place of Memory and Memory of Place*, Interdisciplinary Research Foundation Press, Warsaw, pp. 32–41.

Stein, E., Walter, G. (2014), (editors Ales Bello, A., Pellegrino, M. P.), *Incontri Possibili. Empatia, telepatia, comunità, mistica*, Castelvecchi, Roma.

Stern, D. (2010), *Forms of Vitality: Exploring Dynamic Experience in Psychology and the Arts*, Oxford University Press, New York, NY.

Teilhard de Chardin, P. (1963), *L'activation de l'énergie*, Editions du Seuil, Paris.

Thomas Aquinas (2014), *Summa Theologiae I*, Edizioni Studio Domenicano, Bologna.

Ullman, M. (1987), Wholeness and dreaming, in Hiley, B. J., Peat, D. F. (editors), *Quantum Implications*, Routledge, London, pp. 386–395.
Ullman, M. (1996), *Appreciating Dreams: A Group Approach*, Sage Publications, Thousand Oaks, CA; republished 2006 by Cosimo Books, New York, NY.

Websites

www.eleogenetics.cloud

The human side of open innovation

The case of Officina delle Reti

Giuseppe Russo[1] *and Federica Evangelista*[2]

Keywords: *open innovation; human resource management; business network; inter-firm relationship*

1 Introduction

The level of complexity of the competitive environment (Zahra and Bogner 2000; Bierly and Daly 2007; Espejo and Reyes 2011) in contemporary daily acts is greatly influenced by the market globalization (Levitt 1983; Ohmae 1990), the challenging evolution of economy (Dosi 2000; Nelson and Winter 2009; Barile et al. 2015; Di Nauta et al. 2015), technology (Christensen and Rosenbloom 1995; Schilling 2009), society (Del Giudice et al. 2016), and institutions (Osborn and Hagedoorn 1997; Caputo et al. 2016; Evangelista et al. 2016) has highlighted the need for enterprises to enhance their existing knowledge and skills and to create new ones (Nonaka 2008).

In this context, the increasing need to better manage the process of creation and diffusion of innovations has led to the emergence of the Open Innovation paradigm (Chesbrough 2006), as a "philosophy" directed to support enterprises in adopting "external technologies and innovations" in order to encourage the "internal" growth and create value (Polese et al. 2016; Dominici et al. 2017).

This process, however, requires that companies adopt an open business model and that their employees are likely to exchange knowledge (Argote and Ingram 2000; Kale and Singh 2007) and skills with other companies in order to support possible innovative performances.

Over the time, the growing interaction among companies has influenced the development of formalized long-time relationships of trust across the business networks (Khanna et al. 1998). More specifically, the networks can be considered a sort of firms' agglomeration that could have relevant benefits by adopting the Open Innovation model (Mortara et al. 2010).

The relationships (Ring and Van de Ven 1994) among the companies involved in a network allow them to better manage efficient, effective, and sustainable information flows (Caputo 2016) and technology exchanges (Powell et al. 1996; Mowery et al. 1998; Tsai 2001; Lichtenthaler and Lichtenthaler 2010; Caputo and Walletzky 2017).

Building on these reflections, it is possible to state that human resources (McLean 2005; Collins and Smith 2006) have a prominent role for the adoption of collaborative approaches (Lichtenthaler and Ernest 2009) and for the information and technological sharing (Inkpen 1996; Nooteboom 2004; Popadiuk and Choo 2006).

In such a line, the paper investigates the role of human resources in the application of the Open Innovation model and, consequently, on companies' performance (Mavondo et al. 2005).

Stating from the defined conceptual framework, the paper investigates the following hypothesis:

H_1: There is a positive correlation between the organization's knowledge about the Open Innovation and its revenue.

H_2: There is a positive correlation between investment in open innovation and companies' revenue.

H_3: There is a positive correlation between the relevance attributed by the organization to the tools and actors involved in Open Innovation projects and its revenue.

H_4: There is a positive correlation between the relevance attributed by the organization to investments in Open Innovation and its revenue.

2   Research pathway and methodology

The study applies a quantitative methodology that, inspired by the contributions offered by the existing management literature, aims to investigate effects, risks, and opportunities of Open Innovation in the context of business networks (Lee et al. 2010).

The research analyzes a sample of 60 companies involved in a network supported by the technological platform "Officina delle Reti". A question survey was submitted to investigate the human resources' tendency towards the innovation (Beugelsdijk 2008). The responses were then analyzed through a Structural Equation Model (SEM) (Bandalos 2002; Ullman and Bentler 2003; Byrne and Stewart 2006) to test possible relationships among human resources' perception, companies' adoption of the Open Innovation model, and companies' performances (see Figure 10.1).

3   Theoretical and practical implications

The increasing cooperation and the development of strong relationships among several companies have supported the application of open business models (Chesbrough 2007). These models have led to the creation of business networks as a useful and versatile tool able to support the achievement of joint economic and/ or competitive goals (Saviano and Caputo 2012; Simone et al. 2014).

The paper aims to enrich the existing knowledge about the topic of Open Innovation in business networks (Vanhaverbeke and Cloodt 2006) by proposing

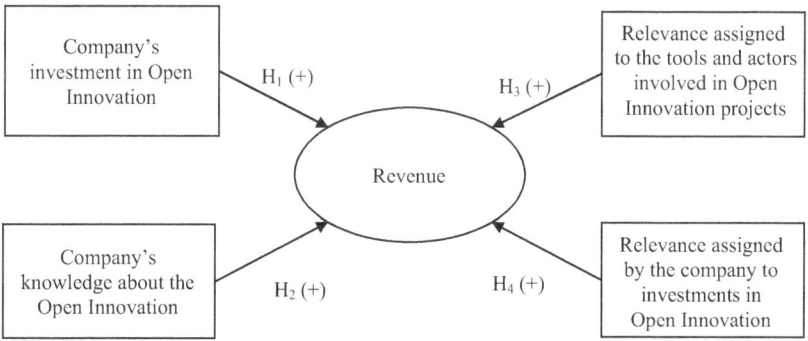

Figure 10.1 The conceptual model

an empirical study about the role of human resources in the application of open business models (Chesbrough and Rosenbloom 2002; Chesbrough 2010).

Reflections and empirical findings herein are relevant in the light of systems perspective (Barile et al. 2014). More specifically, according to the recursion theory, the human resources' propensity to collaborate, exchange information (Espejo 2003; Caputo et al. 2017), and develop innovative ideas affect the companies' ability to innovate (Szeto 2000).

4   Conclusions and future directions for research

The conceptual reflections and empirical results herein enrich previous literature on the topics of corporate networks and Open Innovation.

More specifically, they show the impact of human resource predisposition to Open Innovation (Saviano et al. 2016) on the companies' and networks' innovativeness. In this sense, this proposes possible stimulus of reflection to decision makers interested in the management of business networks by providing evidence about the opportunities arising from an evolved approach to collaboration and innovation.

Notes

1 Department of Economics and Law, University of Cassino and Southern Lazio, Italy.
2 Department of Economics and Law, University of Cassino and Southern Lazio, Italy, f.evangelista@unicas.it (Corresponding Author).

Bibliography

Argote, L., Ingram, P. (2000). Knowledge transfer: A basis for competitive advantage in firms. *Organizational Behavior and Human Decision Processes*, 82(1), 150–169.
Bandalos, D.L. (2002). The effects of item parceling on goodness-of-fit and parameter estimate bias in structural equation modelling. *Structural Equation Modeling*, 9(1), 78–102.

Barile, S., Saviano, M., Caputo, F. (2014). A systems view of customer satisfaction. In National Conference "Excellence in quality, statistical quality control and customer satisfaction", University Campus "Luigi Einaudi", University of Turin, September 18–19.

Barile, S., Saviano, M., Caputo, F. (2015). How are markets changing? The emergence of consumers market systems. In Dominici, G. (ed.) *The 3rd International Symposium Advances in Business Management: "Towards Systemic Approach"* (pp. 203–207). Busyness Systems, E-book Series, Avellino.

Basile, G., Caputo, F. (2017). Theories and challenges for systems thinking in practice. *Journal of Organisational Transformation & Social Change*, 14(1), 1–3.

Beugelsdijk, S. (2008). Strategic human resource practices and product innovation. *Organization Studies*, 29(6), 821–847.

Bierly, P.E., Daly, P.S. (2007). Alternative knowledge strategies, competitive environment, and organizational performance in small manufacturing firms. *Entrepreneurship Theory and Practice*, 31(4), 493–516.

Byrne, B.M., Stewart, S.M. (2006). Teacher's corner: The MACS approach to testing for multigroup invariance of a second-order structure: A walk through the process. *Structural Equation Modeling*, 13(2), 287–321.

Calabrese, M., Iandolo, F., Caputo, F., Sarno, D. (2017). From mechanical to cognitive view: The changes of decision making in business environment. In Barile, S., Pellicano, M., Polese, F. (eds.) *Social Dynamics in a System Perspective* (pp. 223–240). Springer, New York.

Caputo, F. (2016). A focus on company-stakeholder relationships in the light of the Stakeholder Engagement framework. In Vrontis, D., Weber, Y., Tsoukatos, E. (eds.) *Innovation, Entrepreneurship and Digital Ecosystems* (pp. 455–470). EuroMed Press, Cipro.

Caputo, F. (ed.) (2016). *Governing Business Systems: Theories and Challenges for Systems Thinking in Practice 4th Business Systems Laboratory International Symposium* (pp. 124–129). Business Systems Laboratory, Avellino.

Caputo, F. (2017). Reflecting upon knowledge management studies: Insights from systems thinking. *International Journal of Knowledge Management Studies*, 8(3/4), 177–190.

Caputo, F., Del Giudice, M., Evangelista, F., Russo, G. (2016). Corporate disclosure and intellectual capital: The light side of information asymmetry. *International Journal of Managerial and Financial Accounting*, 8(1), 75–96.

Caputo, F., Evangelista, F. (2017). Information sharing and cognitive involvement for sustainable workplaces. In Leon, R.D. (ed.) *Managerial Strategies for Business Sustainability during Turbulent Times* (pp. 122–139). IGI Global, New York.

Caputo, F., Evangelista, F., Perko, I., Russo, G. (2017). The role of big data in value co-creation for the knowledge economy. In Vrontis, S., Weber, T., Tsoukatos, E. (Eds.), *Global and National Business Theories and Practice: Bridging the Past with the Future* (pp. 269–280). EuroMed Press, Cyprus.

Caputo, F., Evangelista, F., Russo, G. (2016). Information sharing and communication strategies: A stakeholder engagement view. In Vrontis, D., Weber, Y., Tsoukatos, E. (eds.) *Innovation, Entrepreneurship and Digital Ecosystems* (pp. 436–442). EuroMed Press, Cyprus.

Caputo, F., Evangelista, F., Russo, G., Buhnova, B. (2017). A systems view of companies' communication in online social environment. *Journal of Organizational*

Transformation & Social Change. http://dx.doi.org/10.1080/14779633.2017. 1291144.

Caputo, F., Perano, M., Mamuti, A. (2017). A macro-level view of tourism sector: Between smartness and sustainability. *Enlightening Tourism: A Pathmaking Journal*, 7(1), 36–61.

Caputo, F., Walletzky, L. (2017). Investigating the users' approach to ICT platforms in the city management. *Systems*, 5(1). doi:10.3390/systems5010001.

Carayannis, E.G., Caputo, F., Del Giudice, M. (2017). Technology transfer as driver of smart growth: A Quadruple/Quintuple innovation framework approach. In Vrontis, S., Weber, T., Tsoukatos, E. (eds.) *Global and National Business Theories and Practice: Bridging the Past with the Future* (pp. 295–315). EuroMed Press, Cyprus.

Chesbrough, H.W. (2006). *Open Innovation: The New Imperative for Creating and Profiting from Technology.* Harvard Business School Press, Boston, MA.

Chesbrough, H.W. (2007). Why companies should have open business models. *MIT Sloan Management Review*, 48(2), 1–22.

Chesbrough, H.W. (2010). Business model innovation: Opportunities and barriers. *Long Range Planning*, 43(2), 354–363.

Chesbrough, H.W., Rosenbloom, R.S. (2002). The role of the business model in capturing value from innovation: Evidence from xerox corporation's technology spin-off companies. *Industrial and Corporate Change*, 11(3), 529–255.

Christensen, C.M., Rosenbloom, R.S. (1995). Explaining the attacker's advantage: Technological paradigms, organizational dynamics, and the value network. *Research Policy*, 24(2), 233–257.

Collins, C.J., Smith, K.G. (2006). Knowledge exchange and combination: The role of human resource practices in the performance of high-technology firms. *Academy of Management Journal*, 49(3), 544–560.

Del Giudice, M., Ahmad, A., Scuotto, V., Caputo, F. (2017). Influences of cognitive dimensions on the collaborative entry mode choice of Small and Medium-Sized Enterprises. *International Marketing Review*, 34(5), 652–673.

Del Giudice, M., Caputo, F., Evangelista, F., (2016). How are decision systems changing? The contribution of social media to the management of decisional liquefaction. *Journal of Decision Systems*, 25(3), 214–226.

Del Giudice, M., Scuotto, V., Khan, Z., Caputo, F., Carayannis, E. (2017). Micro level actions by owners-managers in improving sustainability practices of culture and creative Small and Medium Enterprises: UK-Italy Comparison. *Journal of Organizational Behaviour*, 1–19. doi:10.1002/job.2237.

Di Fatta, D., Caputo, F., Evangelista, F., Dominici, G. (2016). Small world theory and the World Wide Web: Linking small world properties and website centrality. *International Journal of Markets and Business Systems*, 2(2). doi:10.1504/IJMABS.2016.080237.

Di Nauta, P., Merola, B., Caputo, F., Evangelista, F. (2015). Reflections on the role of university to face the challenges of knowledge society for the local economic development. *Journal of Knowledge Economy*, 1–19.

Dominici, G., Yolles, M., Caputo, F. (2017). Decoding the dynamics of value cocreation in consumer tribes: An agency theory approach. *Cybernetics and Systems: An International Journal*, 48(2), 84–101.

Dosi, G. (2000). *Innovation, Organization and Economic Dynamics: Selected Essays.* Edward Elgar Publishing, Cambridge, MA.

Espejo, R. (2003). Social systems and the embodiment of organisational learning. In Mitleton, K.E. (ed.) *Complex Systems and Evolutionary Perspectives on Organisations: The Application of Complexity Theory to Organizations* (pp. 53–69). Oxford: Pergamon.

Espejo, R., Reyes, A. (2011). *Organizational Systems: Managing Complexity with the Viable System Model.* Springer Science & Business Media, Verlag, Berlin, Heidelberg.

Evangelista, F., Caputo, F., Russo, G., Buhnova, B. (2016). Voluntary corporate disclosure in the Era of Social Media. In Caputo, F. (ed.) *The 4rd International Symposium Advances in Business Management: Towards Systemic Approach* (pp. 124–128). Business Systems, Avellino.

Formisano, V., Caputo, F., D'amore, R. (2015). Tratti evolutivi della società della conoscenza: il contributo degli studi sulle reti nella prospettiva sistemica. *Esperienze d'impresa*, 2, 73–94.

Inkpen, A.C. (1996). Creating knowledge through collaboration. *California Management Review*, 39(1), 123–140.

Kale, P., Singh, H. (2007). Building firm capabilities through learning: The role of the alliance learning process in alliance capability and firm-level alliance success. *Strategic Management Journal*, 28(10), 981–1000.

Khanna, T., Gulati, R., Nohria, N. (1998). The dynamics of learning alliances: Competition, cooperation, and relative scope. *Strategic Management Journal*, 19(3), 193–210.

Lee, S., Park, G., Yoon, B., Park, J. (2010). Open innovation in SMEs: An intermediated network model. *Research Policy*, 39(2), 290–300.

Levitt, T. (1983). The globalization of markets. *Harvard Business Review*, 61(May–June), 92–102.

Lichtenthaler, U., Ernest, H. (2009). Opening up the innovation process: The role of technology aggressiveness. *ReD Management LDT*, 39(1), 38–54.

Lichtenthaler, U., Lichtenthaler, E. (2010). Technology transfer across organizational boundaries: Absorptive and desorptive capacity. *California Management Review*, 53(1), 154–170.

Martins, E., Terblanche, F. (2003). Building organizational culture that stimulates creativity and innovation. *European Journal of Innovation Management*, 6(1), 64–74.

Mavondo, F.T., Chimhanzi, J., Stewart, J. (2005). Learning orientation and market orientation: Relationship with innovation, human resource practices and performance. *European Journal of Marketing*, 39(11/12), 1235–1263.

McLean, L.D. (2005). Organizational culture's influence on creativity and innovation: A review of the literature and implications for human resource development. *Advances in Developing Human Resources*, 7(2), 226–246.

Mortara, L, Slacik, I, Napp, J., Minshall, T. (2010) Implementing open innovation: cultural issues *International Journal of Entrepreneurship and Innovation Management* 11(4), 369–397.

Mowery, D.C., Oxley, J.E., Silverman, B.S. (1998). Technolgical overlap and inter-firm cooperation: Implication for the resource-base view in the firm. *Research Policy*, 27(5), 507–523.

Nelson, R.R., Winter, S.G. (2009). *An Evolutionary Theory of Economic Change.* Harvard University Press, Boston, MA.

Nonaka, I. (2008). *The Knowledge-Creating Company.* Harvard Business Review Press, Boston, MA.

Nooteboom, B. (2004). *Inter-Firm Collaboration, Learning & Networks: An Integrated Approach*. Routledge, London.

Ohmae, K. (1990). *The Borderless Word*. Harper Business, New York, NY.

Osborn, R.N., Hagedoorn, J. (1997). The institutionalization and evolutionary dynamics of interorganizational alliances and networks. *Academy of Management Journal*, 40(2), 261–278.

Polese, F., Caputo, F., Carrubbo, L., Sarno, D. (2016). The value (co)creation as peak of social pyramid. In Russo-Spena, T., Mele, C. (eds.) *Proceedings 26th Annual RESER Conference, "What's Ahead in Service Research: New Perspectives for Business and Society"* (pp. 1232–1248). RESER, University of Naples "Federico II".

Popadiuk, S., Choo, C.W. (2006). Innovation and knowledge creation: How are these concepts related? *International Journal of Information Management*, 26(4), 302–312.

Powell, W.W., Koput, K.W., Smith-Doerr, L. (1996). Interorganizational collaboration and the locus of innovation: Networks of learning in biotechnology. *Administrative Science Quarterly*, 41(1), 116–145.

Ring, P.S., Van de Ven, A.H. (1994). Developmental process of co-operative interorganizational relationships. *Academy of Management Review*, 19(1), 90–118.

Saviano, M., Barile, S., Caputo, F. (2017). Re-affirming the need for systems thinking in social sciences: A viable systems view of smart city. In Vrontis, S., Weber, T., Tsoukatos, E. (eds.) *Global and National Business Theories and Practice: Bridging the Past with the Future* (pp. 1552–1567). EuroMed Pres, Cyprus.

Saviano, M., Barile, S., Spohrer, J., Caputo, F. (2017). A service research contribution to the global challenge of sustainability. *Journal of Service Theory and Practice*, 27(5), 951–976.

Saviano, M., Caputo, F. (2012). Le scelte manageriali tra sistemi, conoscenza e vitalità. Management senza confini. Gli studi di management: tradizione e paradigmi emergenti. In XXXV Convegno annuale AIDEA, University of Salerno, October 4–5 (pp. 1–21).

Saviano, M., Nenci, L., Caputo, F. (2017). The financial gap for women in the MENA region: A systemic perspective. *Gender in Management: An International Journal*, 32(4), 203–217.

Saviano, M., Polese, F., Caputo, F., Walletzký, L. (2016). A T-shaped model for rethinking higher education program. In 19th Toulon-Verona International Conference Excellence in Services-Conference Proceedings (pp. 425–440).

Schilling, M. (2009). *Strategic Management of Technological Innovation*. McGraw Hill Higher Edition, Grandview Hts, OH.

Simone, C., Polese, F., Iandolo, F., Caputo, F. (2014). Alla ricerca di un possibile principio evolutivo della teoria e della pratica d'impresa. Il percorso degli studi dell'economia d'impresa. In Atti del XXVI Convegno annuale di Sinergie "Manifattura: quale futuro?", University of Cassino and Southern Lazio, 13–14 November.

Slacik, I., Napp, J.J., Minshall, T. (2010). Implementing Open Innovation (OI): Cultural issues. *International Journal of Entrepreneurship and Innovation Management*, 11(4), 369–397.

Szeto, E. (2000). Innovation capacity: Working towards a mechanism for improving innovation within an inter-organizational network. *The TQM Magazine*, 12(2), 149–158.

Tronvoll, B., Barile, S., Caputo, F. (2017). A systems approach to understanding the philosophical foundation of marketing studies. In Barile, S., Pellicano, M., Polese, F. (eds.) *Social Dynamics in a System Perspective* (pp. 1–18). Springer.

Tsai, W. (2001). Knowledge transfer in intraorganizational networks: Effects of network position and absorptive capacity on business unit innovation and performance. *Academy of Management Journal*, 44(5), 996–1004.

Ullman, J.B., Bentler, P.M. (2003). *Structural Equation Modelling.* John Wiley & Sons, Inc, New York, NY.

Vanhaverbeke, W., Cloodt, M. (2006). Open innovation in value networks. In Chesbrough, H.W., Vanhaverbeke, W., West, J. (eds.) *Open Innovation: Reasearching a New Paradigm* (pp. 258–281). Oxford University Press, Oxford.

Zahra, S.A., Bogner, W.C. (2000). Technology strategy and software new ventures' performance: Exploring the moderating effect of the competitive environment. *Journal of Business Venturing*, 15(2), 135–173.

The human dimension in project management

Véronique Gignoux-Ezratty[1]

Keywords: *project management; informal system; cognitive mechanisms; sponsorship*

1 Purpose

A project is a system that evolves during its lifecycle. The main bodies of knowledge on project management are based on the same approach, "planning, execution, control," and focus on the formal aspects of the system. But, there is a consensus that the soft and informal aspects are as important as the formal ones.

This paper proposes a new approach to deal with soft aspects. The approach consists in focusing on the cognitive mechanisms of the project manager. We theorize that the most important tool of the project is the project manager's mind. With the support of the findings in cognitive psychology, this paper looks at how to make a better use of this mind. The project manager's mind is a very efficient tool but driven by human motivation and, as Herbert Simon stated, rationally bounded. This paper concludes that people in charge of governance and sponsorship of projects should adapt the constraints they give and the communication they make by taking into account the cognitive mechanisms of the project manager.

2 Context

In the first bodies of knowledge on project management, the purpose of the project was to realize deliverables in compliance with requirements. In a new approach, it is often considered that the purpose of the project is to create value for the sponsor organization and, sometimes, for other key stakeholders.

The literature provides various approaches to take into account these uncertainties within the project and the impact of human factors. They are based on sense-making (Thiry 2001; Alderman et al. 2005), soft systems (Yeo 1993; Neal 1995; Winter & Cleckland 2003), competencies of management teams (Crawford 2005; Müller & Turner 2010) and constructivist approaches (Bellini & Canonico 2008; Jackson & Klobas 2008; Pellegrinelli 2011).

The project manager acts in accordance with the "framework" given by the organization and explained by the sponsor. This framework contains the mission of the project manager and the constraints of this mission. Within this framework, the project manager acts as a strategist that has bounded rationality, and is driven by human motivation. The project manager acts with the support of partial information on the projects. She or he constructs a cognitive representation of the project with the information from the formal monitoring and with the information retrieved in the discussion with project team members and other stakeholders.

Because she or he is human, she (or he) acts in her (or his) own interests. It is in the project manager's interest to realize projects with efficiency and to ensure that her or his sponsor is satisfied. But she or he has much freedom of action within the project. Therefore, the information represents power. An individual will not give valuable information if this means losing freedom of action (Crozier & Friedberg 1977–1980).

3   Human behavior in projects

For the project manager, the project is the work he or she has to organize, direct and oversee with the support of the sponsor in order to achieve the project objectives and to satisfy the sponsor. The satisfaction of the sponsor depends on many criteria. There are the measurable criteria such as the achievement of objectives concerning cost, time and quality. There are the competencies of the project manager. In real life, the sponsors, as all individuals, look out for their own interests. In some bureaucratic organizations, these last criteria mean that you have to address the topic of subtle human behaviors. The criteria concerning satisfaction of the sponsor should be clarified during the interaction between the project managers and the sponsors.

The project manager acts with the support of partial information on the projects. She or he constructs a cognitive representation of the project with the information she or he has obtained in the formal monitoring and with the information retrieved from the discussion with project team members and other stakeholders.

The formal monitoring is only a small part of the information used by the project manager to apply his strategy and to act within the project. Much information used by the project manager to direct the project is not in the formal monitoring. There are many facts that the project manager knows but that are not in writing. For example, this delivery may be late; there is a requirement in the specification that is not compliant with the need of the customer; a task is a little overestimated. . . . In the assumption of this paper, the appropriate tool for decision-making is not the formal monitoring but the cognitive representation the project manager has built with all the elements at her or his disposal.

4   Conclusion

This paper proposes an original angle to take into account the soft aspects of project management. From the traditional point of view, the soft aspects are in addition to the good use of the processes, techniques and tools of project management. In the angle investigated in this paper, the main aspects for project management are the strategy and tactics deployed by the project manager to achieve her or his personal purposes. The processes, techniques and tools of project management should be aligned with the cognitive mechanisms of the project manager. The framework of a project with mission and constraints should be decided by taking in account these cognitive mechanisms.

Note

1 CeFASP – Cercle Francophone pour l'Application des Sciences des Systèmes aux Projets-France.

Bibliography

Alderman, N., Ivory, C., McLoughlin, I., Vaughan, R. (2005) Sense-Making as a process within complex service-led projects, *International Journal of Project Management*, Vol. 23, 380–385.

Bellini, E., Canonico, P. (2008) Knowing communities in project driven organizations: Analysing the strategic impact of socially constructed HRM practices, *International Journal of Project Management*, Vol. 26, 44–50.

Crawford, L. (2005) Senior management perceptions of project management competence, *International Journal of Project Management*, Vol. 23, 7–16.

Crozier, M., Friedberg, E. (1977–1980) *Actors and Systems: The Politics of Collective Action*, Publisher: University of Chicago Press (Tx) (November 1980), Original Edition: L'acteur et le système, Les contraintes de l'action collective, Editions du seuil, 1977.

Jackson, P., Klobas, J. (2008) Building knowledge in projects: A practical application of social constructivism to information systems development, *International Journal of Project Management*, Vol. 26, 329–337.

Müller, R., Turner, R. (2010) Leadership competency profiles of successful project managers, *International Journal of Project Management*, Vol. 28, 437–448.

Neal, R. A. (1995) Project definition: The soft-systems approach, *International Journal of Project Management*, Vol. 13, No. 1, 5–9.

Pellegrinelli, S. (2011) What's in a name: Project or programme?, *International Journal of Project Management*, Vol. 29, 232–240.

Thiry, M. (2001) Sensemaking in value management practice, *International Journal of Project Management*, Vol. 19, 71–77.

Winter, M., Cleckland, P. (2003) Soft Systems: A fresh perspective for project management, *Civil Engineering*, Vol. 156, 187–192.

Yeo, K. T. (1993) Systems thinking and projects management-time to reunite, *International Journal of Project Management*, Vol. 11, No. 2, May, 111–117.

Websites

http://wosc2017rome.asvsa.org/index.php

A five-step inclusive model to help corporate leaders not to lose their soul in uncertain times

Marie-Laure Blanc[1]

Keywords: *corruption; ethics; model; corporate leaders*

1 Introduction

The corporate world is regularly shaken by corruption scandals. Every industry and country in turn is affected: the gas industry with Enron (US) in 2001, the dairy produce industry with Parmalat (Italy) in 2003 and more recently the banking industry with Société Générale (France) in 2008.

Once the corruption is discovered, the scenario follows a fixed pattern: internal or external regulation entities issue new controls, standards, codes of conduct or ethics charters. The Basel II framework issued in 2004 and implemented in 2006 is one example. It aimed at integrating operational risks (fraud and IT failures) in banking activities. Two years later, a trading hyper fraud happened at Société Générale in France.

Considering the regularity at which these scandals happen, which elements are – with the same regularity – set aside by these entities and standards, therefore paving the way to future corruption scandals?

2 The system manufactures its own catastrophe because of elements set aside

Literature is huge on the subject of systems manufacturing – from inside – the catastrophe they were supposed to protect from: John Gall explained in *Systemantics* how a Cape Canaveral building designed to protect space rockets from adverse weather conditions ended up generating these adverse weather conditions. Ivan Illitch dedicated his life to showing how systems became counterproductive. Some more recent business-oriented publications include *Better Ethics Now: How to Avoid the Ethics Disaster You Never Saw Coming* by Christopher Bauer.

We identified three set-aside and counter-intuitive elements:

1   The ring of Gyges's syndrome

Told by Plato in Book II of the *Republic*, the story of Gyges is the story of an honest man who became dishonest after finding a ring that made him invisible. While the ring (the material cause) can only be regarded as instrumental (a cause that is not sufficient by itself to trigger an action), it does play a role in Gyges's entry into corruption.

Today, do ethical standards act as a ring of Gyges? The fact of having an instrumental cause in the process was addressed in 2012 by economist Patrick Slovik in his OECD working paper on regulation challenges. Patrick Slovik stated: "Bank regulation might have contributed to or even reinforced adverse systemic shocks that materialised during the financial crisis. Capital regulation based on risk-weighted assets encourages innovation designed to circumvent regulatory requirements and shifts banks' focus away from their core economic functions."

2   The current rationale of corporate ethical behaviour

Corporate ethical standards are prescriptive. They take the form of documents corporate leaders have to read, sign and abide by. Two questions can be asked:

- If corporate leaders behave ethically because that is a company policy, how much of their ethics is corrupted because of their behaviour not being freely chosen but prescribed?
- Does such a prescriptive approach entail that at some point, corporate leaders will stop acting ethically because they will no longer want to abide by a behaviour they have not freely chosen?

3   The role of self-confidence in entering into Akrasia

First described by Plato in the *Protagoras,* Akrasia (*a*- for without and *kratos* for power, strength) is defined as acting intentionally against one's own will. A human being therefore enters into Akrasia when he chooses an option – freely and fully aware of what his choice entails – while knowing there is a better option he could choose and implement.

Considering self-confidence is regarded by companies as a positive and sought-after trait in corporate leaders, to what extent does self-confidence trigger an akratic process? Nick Leeson – Barings Bank, 1997, Jérôme Kerviel – Société Générale, 2008[2] seemed to share this common trait: because of a high level of self-confidence, they were convinced their next "gamble" would cover their last loss.

3   A five-step model to shift from prescriptive/ externally imposed ethics to generative/freely chosen ethics

We designed a dynamic, scalable, trans-disciplinary and generative model using the attributes and the physical reality of a Newton Pendulum. The objective is to provide corporate leaders with an inclusive tool aimed at raising awareness on essential questions they are faced with when deciding, such as: how much freedom am I going to take from other people to ensure mine?

Since the tool has a physical reality, corporate leaders can have it before their eyes and use it as an "awareness anchor" while seated at their desk. Our underlying intention of using a Newton pendulum is to "get the whole system in the room" as stated by Marvin Weisbord. While an ethics charter carries the implicit message "we know you will do wrong so we prefer to tell you before not to," the pendulum does not carry this type of message since, historically, it is the typical desktop executive toy.

Each of the five marbles embodies an essential question that cannot be locked up in a discipline. Each marble is then divided into as many sub-pendulums as necessary. The pendulum provides insights in order to try and prevent the five following traits we identified in the corruption scandals we studied.

1   One of the first question arising after the discovery of the corruption: how could such a scandal happen in our company?
2   The speaking collapse before the collapse of the institution: how people language themselves and how they speak to each other?
3   Inconsistencies between the CV, the career pathway of corporate leaders and how they react to situations.
4   The Bernarda Alba syndrome plaguing the corporate communication department. "What I want is a beautiful façade" was the motto of Bernarda Alba, a character of Spanish writer Garcia Lorca. For the sake of maintaining the façade, Bernarda Alba destroys the life of her family.

    To what extent do internal or external communication departments contribute to generating the destruction of the corporation by trying to maintain the "beautiful façade"?
5   The after-the-collapse repentance – be it through books, interviews or conferences – of the people accused of corruption.

In uncertain times, our aim is to enable corporate leaders to shift from a corporate culture where ethics is a widely talked-about subject to a corporate culture where ethics is generated from within their soul and embodied in every word, action and step.

Notes

1 AFSCET.
2 Nick Leeson and Jérôme Kerviel are top on the list of bad traders. They have lost extreme amounts in the name of their banks.

Bibliography

Anquetil, A. (2004, Décembre). Agir intentionnellement à l'encontre de ses valeurs. *Annales des Mines-Gérer et Comprendre*, (78), 4–16.
Bargh, J. A., Chen, M., & Burrows, L. (1996). Automaticity of social behavior: Direct effects of trait construct and stereotype activation on action. *Journal of Personality and Social Psychology*, 71(2), 230.

Crozier, M. (1995). *La crise de l'intelligence, essai sur l'impuissance des élites à se reformer.* InterEditions.

Dupuy, J. P. (2009). *Pour un catastrophisme éclairé: quand l'impossible est certain.* Seuil.

Ferrari, G. R. F. (Ed.). (2000). *Plato: 'The Republic'.* Cambridge University Press.

Gall, J. (1977). *Systemantics: How Systems Work and Especially How They Fail.* Times Books.

Hirschman, A. O. (1970). *Exit, Voice, and Loyalty: Responses to Decline in Firms, Organizations, and States* (Vol. 25). Harvard University Press.

Kerviel, J. (2010). *L'Engrenage: Mémoires d'un trader.* Flammarion.

Le Bret, H. (2014). *La semaine où Jérôme Kerviel a failli faire sauter le système financier mondial.* Les Arènes.

Le Moigne, J. L. (1987). *Qu'est-ce qu'un modèle?* Université d'Aix-Marseille III, Faculté d'économie appliquée.

Mandis, S. (2013). *What Happened to Goldman Sachs: An Insider's Story of Organizational Drift and Its Unintended Consequences.* Harvard Business Review Press.

McLean, B., & Elkind, P. (2013). *The Smartest Guys in the Room: The Amazing Rise and Scandalous Fall of Enron.* Penguin UK.

Pouponneau, F. (2014). *Sociologie des élites délinquantes. De la criminalité en col blanc à la corruption politique.* Armand Colin.

Slovik, P. (2012). Systemically Important Banks and Capital Regulation Challenges, OECD Economics Department Working Papers, No. 916, OECD Publishing.

Sterner, J. (2000). *Other People's Money: The Ultimate Seduction.* Applause Theatre and Cinema Books.

Stroh, D. P. (2015). *Systems Thinking for Social Change: A Practical Guide to Solving Complex Problems, Avoiding Unintended Consequences, and Achieving Lasting Results.* Chelsea Green Publishing.

Verneuil, L. (1932). *La banque Nemo: pièce en trois actes et neuf tableaux* (Vol. 300). L'illustration théâtre.

Weisbord, M. R. (1991). *Productive Workplace: Dignity, Meaning, and Community in the 21st Century.* Jossey Bass.

Human organizational theoretical aspects of the cybernetic management conditional systems of international businesses' competitive strategies, viable with equifinality

Giovanni Tamponi[1]

Keywords: *theoretical viable international businesses with equifinality; theoretical international businesses's strategies; theoretical human international businesses's behavioral organizations; international business management's knowledge; Knowledge's international businesses management; theoretical cybernetic management systems; theoretical contingent general conditional relationships*

The problem defined: "Human organizational theoretical aspects, of the cybernetic management conditional systems of international businesses' competitive strategies, viable with equifinality", examines the theme of the:

> Knowledge's management of theoretical cybernetic conditional systems of theoretical human behavioural organizations, such as collective, social, economic, open, conditional finalized, rational, directional, viable with equifinality, of the international business's competitive strategies, making use of cognitive contributions of the following theoretical perspectives.
> (Beer, 1959; Simon, 1958; Abbagnano, 1971; Cyert and March, 1970; Gregor Mac, 1975; Freeman, 1984; Maturana and Varela, 1984; Usai, 1990; Hanica, 1967; Blau and Scott, 1972; Golinelli G, 2000; Barile, 2008; Pellicelli, 2007; Vallini, 1990; Bertalanffy, 1971; Mackie, 1991; Rullani, 1989; Gambardella, 2014; Ling Hsing Chang and Ching Lin, 2015)

In order to process the theoretical formulations of the particular competitive strategies of peculiarly programmed international businesses, firstly, it may be helpful to formulate a general theoretical model of reference of a management's knowledge of a general international business, as for instance, a general theoretical human behavioral organization finalized of reference, a theoretical management system of a general international business, from which we can conjecture, through a problem-solving cognitive process of strategic planning, theoretical particular cybernetic management's systems of the peculiar human behavioral organizations that may describe each peculiar competitive strategy planned by each specific international business observed (Porter, 1971; Stanton and Varaldo, 1984; Rullani, 1989; Levitt, 1990; Tagliagambe and Usai, 1994; Caselli, 1966; Barile, 2009; Tamponi, 2012).

The theoretical approach to human behavioral in the light of organization and cybernetic studies defines the organization as a set of elements interrelated in a common stucture with few basic elements such as (i) mission and finalities and (ii) a structure of networked (material, immaterial, financial, etc.) elements (Freeman, 1984; Drucker, 1063; Giudici, 2002; Usai, 1990).

The general contingent human behavioral theoretical organization – collective; social; economic; flexible; complex; viable with equifinality; finalized, with bounded rationality, of a general international business, implementing a production of goods and services for the international target market; designed, directed and controlled by human subjects – describes a general theoretical human cybernetic management system – finalized, viable with equifinality, contingent, directional, with bounded rationality, of a general international business.

A general theoretical human behavioral organization, finalized of a general international business in order to describe the theoretical functional finalities and resources which contribute to the achievement of the mission and the structure of its general finalities, could point out the following behavioral network of general management's functions.

1   Structure of general operative management's functions: international marketing, production, supplying, research and development, accounting and finance.
2   Structure of general directional management's functions: information, planning, organization, coordination, business's implementation, communication, control.

    The peculiar theoretical cybernetic management systems of the peculiar human behavioral organization of operating cybernetic international businesses' competitive strategies are designed, implemented and controlled by human subjects, which also implement the production of goods and services for contributing both to satisfy the international target market, and to make logically and concretely as possible the execution of the structure of general fundamental finalities and mission.
3   To define the general fundamental contingent, finalities and mission of the general management's knowledge system of elements of a general international business are elaborated. Firstly, it may be useful to apply the method of formulating the theoretical reference structures of the fundamental general problems and general solutions of a general international business, for instance determined by the structures of the fundamental general problems and general solutions of: customer satisfaction, shareholder satisfaction, corporate social responsibility, and in space survival, survival over time (Stanton and Varaldo, 1984; Levitt, 1990; Pellicelli, 2007; Golinelli, 2009;).

The statement of the contingent knowledge of structures of fundamental general problems and general solutions of: customer satisfaction, shareholder satisfaction, social responsibility, in space survival, survival over time makes it possible to also enunciate the contingent knowledge of:

1   The structure of the fundamental terminal general finalities of: customer satisfaction, shareholder satisfaction, social responsibility, in space survival and survival over time.
2   The structure of the mission of useful values achievement of the fundamental terminal general finalities of: customer satisfaction, shareholder satisfaction, social responsibility, in space survival and survival over time.
3   The structures of the flexible general finalities of: customer satisfaction, shareholder satisfaction, social responsibility, in space survival and survival over time, both of a general theoretical international business and of general structures of operative and directional management's functions of a general theoretical international business observed.

The cybernetic management conditional systems of the theoretical contingent human peculiar behavioral organizations of international businesses's competitive strategies could be qualified cybernetic, even though in condition of bounded rationality, if they are considered potentially effective to realize the mission's structure of the useful values of fundamental terminal general finalities of: customer satisfaction, shareholder satisfaction, corporate social responsibility, in spatial survival and over time survival.

Theoretical cybernetic management system of a general international business, interpreted by means of the contingent knowledge of the theoretical human conditional behavioral organization of a general international business, finalized to the achievement of the structures of general fundamental solutions and of the fundamental terminal general finalities of: customer satisfaction, shareholder satisfaction, corporate social responsibility, in space survival and over time survival, could be directed by the mission structure of useful values fulfillment of the fundamental terminal general finalities: customer satisfaction, shareholder satisfaction, corporate social responsibility, in space survival and over time survival (Stanton and Varaldo, 1984; Golinelli C., 2009; Levitt, 1990; Drucker, 1964; Simon, 1958; Stalk, 1991).

In order to reach the general or peculiar conditional formulations of the theoretical cybernetic human behavioral organizations, of rational finalities and resources relationships, of a general or particular international businesses, we may interpret the human behavioral theoretical organizations, of finalities and resources relationships, of international businesses with the help of the conceptual theoretical cybernetic contingent knowledge system of conditional elements, final and initial, described by a theoretical conditional contingent relationship with equifinality, joined and disjoined, which can be organized by a theoretical contingent system of the general conditional relationships: $q = f(x, y, z, t)$, formed by the final condition contingent q, and conjointly from the initial basic conditions contingent: x, y, z, t (Bertalanffy, 1971; Mackie, 1991; Golinelli, 2011; Tamponi, 2012; Simon, 1958; Beer, 1959; Wiener, 1953; Ashby, 1971; Shannon and Weaver, 1971; Vallini, 1990).

The final condition q observed may be considered a theoretical contingent event if:

1   It is considered a possible event that could arise only under certain coordinates of space and time,
2   It could be produced disjointly by one or more potentially effective complex initial conditions:

(i) f(x, y, z, t); (ii) f(y, z, t); (iii) f(x, z, t); (iv) f(x, y, t); (v) f(x, y, z,), each of them unnecessary, but sufficient in one or more specific circumstances to yield the final condition of the contingent event q observed.
(Abbagnano, 1971; Bertalanffy, 1971; Mackie, 1991; Ashby, 1971; Boscolo, 1969; De Latil, 1962; Tamponi, 2012).

In the abstract is proposed the primary problem: "if theoretical structure of the fundamental terminal general final conditions, of theoretical cybernetic management system -finalized of a general human behavioral conditional theoretical organization, viable with equifinality, finalized, rational of international businesses", is formulated by the fundamental terminal contingent final general conditions of: customer satisfaction corporate social responsibility, survival in space and survival over time, consequently, the mission of international businesses should be formulated as the structure's mission of useful values realization of the general structure of theoretical fundamental terminal general, contingent final conditions of: customer satisfaction, shareholder satisfaction, corporate social responsibility, in space survival and survival over time (Drucker, 1964; Porter, 1971; Stanton and Varaldo, 1984; Levitt, 1990; Stalk, 1991; Caselli, 1966; Pellicelli, 2007; Golinelli, 2009; De Lattre, 1964; Wiener, 1953; Whipp, 2001; Tamponi, 2012; Freeman, 1984; Sciarelli, 2007; Bucley, 1967).

Note
1 University of Cagliari.

Bibliography
Abbagnano, N. (1971), *Dictionary of philosophy, voice cybernetics*, UTET.
Ashby, W.R. (1971), *Introduction to cybernetics*, Einaudi, Turin.
Barile, S. (2008), *L'impresa come sistema*, Giappichelli, Turin.
Barile, S. (2009), *Management sistemico vitale*, Giappichelli, Turin.
Beer, S. (1959), *Cybernetics and management*, Bompiani, Milan.
Bertalanffy, L.V. (1971), *General system theory*, ILI, Milan.
Blau, P.M., Scott, W.R. (1972), *The formal organizations*, Angeli, Milan.
Boscolo, P. (1969), *Cybernetics and education*, Nuova Italia, Florence.

Bucley, W. (1967), *Sociology and modern systems theory*, Prentice Hall.
Caselli, L. (1966), *Organization and enterprise decision-making theory*, Giappichelli, Turin.
Cyert, R., March, J.C. (1970), *Theory of firm behavior*, Angeli, Milan.
De Latil, P. (1962), *Artificial thinking, introduction to cybernetics*, Feltrinelli, Milan.
De Lattre, P. (1964), *Systems theory and epistemology*, Einaudi, Turin.
Drucker, P. (1964), *Managing for results*, Heinemann, London.
Freeman, R. (1984), *Strategic management*, Pitman, Boston.
Gambardella, A. (2014), The innovative entrepreneur as an agent of technical, economic and social development. *Sinergie Italian Journal of Management*, vol. 93, pp. 3–18.
Giudici, E. (2002), *New perspectives for the efficiency and effectiveness of the companies*, Giappichelli, Turin.
Golinelli, C. (2009), *The territory viable system*, Giappichelli, Torino.
Golinelli, G.M. (2000), *L'approccio sistemico vitale al governo dell'impresa*, Cedam, Padua.
Golinelli, G.M. (2011), *The Viable Systemic Approach (VSA): Towards the scientification of government action*, Cedam, Padua.
Gregor Mac, D. (1975), *Enterprise human aspect*, Angeli, Milan.
Hanica, F. (1967), *Towards a science of business management*, ETAS Kompass, Milan.
Levitt, T. (1990), *Marketing imagination*, Sperling & Kupfer, Milan.
Ling Hsing Chang, C., Ching Lin, T. (2015), The role of organizational culture in the knowledge management process. *Journal of Knowledge Management*, vol. 19, Issue, 3, pp. 433–455.
Mackie, J. (1991), Causes and conditions, in Sosa, F., Tooely, M., *Causation*, Oxford University Press, Oxford.
Maturana, H., Varela, F. (1984), *The tree of knowledge*, Garzanti, Milan.
Pellicelli, G. (2007), *International marketing*, V edition, ETAS, Milan.
Porter, M. (1971), *Competitive strategy*, Il Mulino, Bologna.
Rullani, E. (1989), *The theory of the firm, in Rispoli: The industrial company*, Il Mulino, Bologna.
Sciarelli, S. (2007), *Ethics and social responsibility in the enterprise*, Giuffrè, Milan.
Shannon, C.E., Weaver, W. (1971), *The mathematical theory of communication*, ETAS Kompass, Milan.
Simon, H. (1958), *Administrative behavior*, Il Mulino, Bologna.
Stalk, G. (1991), *Competing against time*, Sperling & Kupfer, Milan.
Stanton, W., Varaldo, R. (1984), *Marketing*, Il Mulino, Bologna.
Tagliagambe, S., Usai, G. (1994), *The enterprise between assumptions, myths and reality*, Isedi, Turin.
Tamponi, G. (2012), *The theoretical conditional system, viable and equifinal, international business*, Angeli, Milan.
Usai, G. (1990), *Efficiency in organizations*, UTET, Turin.
Vallini, C. (1990), *Foundations of government and corporate direction: The real business and its teleology*, Giappichelli, Turin.
Whipp, R. (2001), *Making time: Time and management in modern organizations*, Oxford University Press, Oxford.
Wiener, N. (1953), *Cybernetics*, Bompiani, Milan.

Gender equality in economic development – key factor for innovative growth in global economy

Eka Sepashvili[1]

Keywords: *global economy; innovations; economic growth; gender equality*

1 Introduction

The national economic policy of any country aims at economic growth; meanwhile fair distribution of gains stays high on the agenda. Contemporary globalization heavily relies on innovations and knowledge, which is generated by human resources. The development of new technologies and innovations diminished the importance of low-skilled labor and unprecedentedly increased the value of workers able to perform cognitive tasks.

The primary and secondary educational institutions are just fundaments for qualified human resources development; meanwhile, the higher education universities with study programs that adequately respond to the demands of the labor markets, which can deal not only with current challenges, but also predict future prospects, are even more important and represent the sphere for private business in many countries (Porter, 1992).

Human resources are characterized not only by education, but also by gender dimensions. In this regard, women's economic contributions are vital and the ability to realize the full labor potential of women in the national economy is decisive to meet the challenge of attracting highly qualified employees to compete in international markets. Unfortunately, few nations and companies recognize the significance and potential that women hold.

2 Evidences on gender equality

Generally speaking, over the past several decades woman were promoted significantly across the number of countries in terms of access to education, healthcare, involvement in work force, even in decision-making process. Despite this significant progress, that is evidenced by the World Bank Report on Development (World Bank, 2011), women's representation in the top managerial positions is still lagging behind that of men's (Bureau of Labor Statistics, 2007).

The World Bank Report on Development (World Bank, 2011) clearly reports the positive correlation between economic development and gender balance, but cultural and historical factors are also strongly influencing women's promotion. If we compare several macroeconomic data, the emerging picture will give us a different idea of business performance correlation with gender balance.

Despite the difference in measurement methodology and gained results, the trend of gender equality in the top countries studied, distinguished by the high representation of women in top business management, significantly differs from countries that showed the worst results. Access to education is approximately equal in the majority of selected countries. The main factors that impede women's promotion to higher career paths are as follows: maternity leave; parenthood; child and elderly care activities in the family; the double burden on women performing productive and reproductive work, meaning they perform the lion's share of domestic work and hence have less time for education and/or qualification improvement; as well as gender-biased treatment in the society creating glass ceilings and structural discrimination.

3   Arguments for women leadership to achieve better economic growth

Studies conclude that women's involvement in managerial activities of firms/corporations impose direct and positive impacts on a company's risk management results. The first attempt to observe women leadership's impact on business activities was in the late '80s of the last century (Morrison et al., 1987). Almost a decade later the first empirical study (Adler, 1998) showed a strong and positive correlation between company results and women's involvement in managerial positions, which was generating higher profit. Later, another study (Catalyst, 2004) re-confirmed these findings.

More recent studies also (Dezso et al., 2011; Welbourne, 2012) showed that a higher percentage of women's involvement in senior management positions up to a CEO level was absolutely clearly related to better company performance.

Policy measures are essential to promote gender balance in decision-making. EU Council Recommendation 96/694/EC directly calls the Member States to adopt a wide and comprehensive, integrated strategy to encourage the balance between women and men in decision-making, as well as to boost the business/private sector to increase the number of women at all levels of decision-making, in particular through the adoption of special programs and strategies, like equality plans and/or positive measure programs.

Note

1 Iv. Javakhishvili Tbilisi State University.

Bibliography

Adler, R.D. (1998). Women in the Executive Suite Correlate to High Profits. *European Project on Equal Pay.*

Borkowski, S.C., and Ugras, Y.J. (2003). Business Students and Ethics: A Meta-Analysis. *Journal of Business Ethics*, 17, 1117–1127.

Bureau of Labor and Statistics (2007). *Women in the Labor Force: A Databook.*

Carli, L.L. (2006). Gender Issues in Workplace Groups: Effects of Gender and Communication on Social Influence. In Barrett, M. and Davidson, M.J. (Editors.), *Gender and Communication at Work.* Burlington, VT: Ashgate.

Catalyst (2004). *The Bottom Line: Connecting Corporate Performance and Gender Diversity.* New York, San Jose and Toronto: Catalyst.

Dezso, C., Ross, D., and Gaddis, D. (2012). Does Female Representation in Top Management Improve Firm Performance? A Panel Data Investigation. *Strategic Management Journal*, 33(9), 1072–1089.

Fagenson, E.A. (1999). Personal Value Systems of Men and Women Entrepreneurs Versus Managers. *Journal of Business Venturing*, 8, 409–430.

Glover, S.H., Burmis, M.A., Logan, J.E., and Ciesla, J.R. (2002). Re-Examining the Influence of Individual Values on Ethical Decision-Making. *Journal of Business Ethics*, 16, 1319–1329.

Hofstede, G. (2001). *Culture's Consequences II: Comparing Values, Behaviors, Institutions and Organizations across Nations.* Beverly Hills, CA: Sage.

International Monetary Fund (2016). World Economic Outlook Database, October 2016, Database updated on 4 October 2016. https://www.imf.org/external/pubs/ft/weo/2016/02/weodata/index.aspx Accessed on 6 October 2016.

Ljunggren, E., and Kolvereid, L. (1996). New Business Formation: Does Gender Make a Difference? *Women in Management Review*, 11, 3–12.

Loko, B., and Diouf, M.A. (2009). Revisiting the Determinants of Productivity Growth: What's New?" *IMF Working Paper 09/225* (Washington).

McDonald, G.M. (1997). Ethical Perceptions, of Expatriate and Local Managers in Hong Kong. *Journal of Business Ethics*, 16(15).

McKinsey and Company (2008). A Business Case for Women. *The McKinsey Quarterly*, September.

Morrison, A.M., White, R.P., and Van Velsor, E. (1987). *Breaking the Glass Ceiling: Can Women Reach the Top of America's Corporations?* New York: Addison-Wesley.

Mueller, C.B., and Van Deusen, C.A. (2002). Gender and Age Differences in Business Goals across 16 Countries. *Proceedings of the Second Annual International Research Seminar, Co-Hosted by the US and Poland*, February 15, pp. 55–69.

Porter, M. (1992). *The Competitive Advantage of Nations.* New York: The Free Press and A Division of Macmillian, Inc.

Sepashvili, E. (2008). The Structure of Time Budget and Welfare of Population in Georgia. *Norwegian Institute of International Affairs (NUPI), Regional Competence-Building for Think-Tanks in the South Caucasus and Central Asia*, Tbilisi, p. 30.

Sepashvili, E. (2011). Success of Business on Global Markets and Gender Aspects of Resources Productivity. *International Conference Paper "Social-Economic Environment of Business"*, February 28, pp. 229–240.

Sepashvili, E. (2012). Economic Aspects of Gender Discrimination in the Georgian Labour Market: Myths and Realities. *Norwegian Institute of International Affairs (NUPI), Regional Competence-Building for Think-Tanks in the South Caucasus and Central Asia.* Accessed on georgica.tsu.edu.ge/files/03-Society/.../Sepashvili-2012. pdf (21.9.2018).

World Bank (2011). *2012 Development Report.* Accessed on https://openknowledge. worldbank.org/handle/10986/4391

World Economic Forum (2016). *Global Gender Gap 2016.* Accessed on http:// reports.weforum.org/global-gender-gap-report-2016/results-and-analysis

About the importance of brand as a system

Maia Seturi[1] and Ekaterine Urotadze[2]

Keywords: *brand; branding; brand image; marketing research*

1 Introduction

Success of produced products on the market can be assessed with their sales figures (indicatora). Sales growth is impossible if users do not have information about the company's products (brands) on their characteristics and advantages. The proper management of brand is very important for modern companies. Brand is presented as a complex system of interconnected elements.

It is very important for our country's Georgian products to be presented not only in the domestic market, but also to increase its exports. A significant place in the country's total exports is held by Georgian wine, mineral water, alcoholic beverages, carbonated soft (nonalcoholic) drinks, etc. (National Statistics Office of Georgia, 2016).

It should be noted that branding was applied in Georgia in certain forms in the past as well. There existed lots of Georgian-made consumer products that had the benefit of their high reputations between consumers not only in Georgia, but also abroad.

The tasks of this research were: to determine consumer evaluations at Tbilisi (the capital of Georgia) consumers' market about Georgian brands; to find out the attitude of consumers to the advantages of Georgian brands. The results of the research determined weaknesses which prevent Georgian brands from success. In the conclusive part of the work are given research results, conclusions, and some recommendations.

2 Literature review

Modern marketing systems attach great importance to human aspects. This is revealed in the fact that some companies place customers first. Successful marketing companies invert the chart, placing customers at the top; next in importance are frontline people who meet, serve, and satisfy customers; under them are the middle managers, whose job is to support the frontline people so they can serve customers well; and at the base is top management, whose job is to hire and support good middle managers. Managers at every level must be personally involved in knowing, meeting, and serving customers. With the rise of digital technologies such as the Internet, increasingly informed consumers today expect companies to do more than connect with them, more than satisfy them, and even more than delight them. They expect companies to listen and respond to them (Kotler & Keller, 2012, p. 124).

Branding plays an important role in creating loyal customers. Satisfied buyers can purchase products they like again without excessive risk. Loyalty demonstrated towards brand could cause the readiness for even the payment of higher prices (Kotler & Armstrong, 2015, p. 258).

Customer assessments of product performance depend on many factors, especially the type of loyalty relationship the customer has with the brand. Consumers often form more favorable perceptions of a product with a brand they already feel positive about (Kotler & Keller, 2012, p. 128).

Building brand equity depends on three main factors: (i) The initial choices for the brand elements or identities making up the brand; (ii) the way the brand is integrated into the supporting marketing program; and (iii) the associations indirectly transferred to the brand by links to some other entity (the company, country of origin, channel of distribution, or another brand) (Kotler & Keller, 2012, p. 268).

To maintain the success for a long time, the company's management should realize well the importance of this issue. They should consider in brand development the following aspects: consistency, clarity, constancy, awareness, originality (Kotler & Pfertch, 2007, pp. 220–353).

Brand strength depends on what customers think about it (brand). The brand equity is a very important strategic bridge between the past and future. This is the accumulation of values associated with the product (service) by consumers. These associations can further increase brand value, more than in the past. They can be found with the help of Keller CBBE (Customer-Based Brand Equity model) (Kotler & Pfertch, 2007, p. 222).

A successful brand is a system that consists of certain components: high-quality goods, differential characteristics, and additional values (Doyle & Stern, 2007, pp. 216–223). Thus, a successful brand contains broad and profound meaning.

3   Methodology

In the work there are used the concepts of marketing theories, data analysis statistical methods, results of the carried out research, information existing on web pages of certain Georgian organizations, etc. Most marketing research projects do include some primary-data collection (Todua & Urotadze, 2013, pp. 8–12). As an object of the research was selected the consumer market of Tbilisi, the attitude of consumers to Georgian brands and marketing aspects related thereto. In February–March 2016 we carried out the research using a quantitative method for the marketing research, a questionnaire survey.

4   Findings

As a result of the research, the most recognizable Georgian brand at Tbilisi consumer market is Barambo (sweets: chocolate, candy, ice cream) – 25% of respondents; 9% of respondents named Borjomi (mineral water); 9% of respondents named Nikora (meat foods); 8% Natakhtari (lemonade, beer); 7% Kula (fruit juice); 10% Georgian wine brands, etc. Although "Barambo" is not a brand with a long-time history, it was able to gain high brand recognition (awareness) on the Georgian market rather quickly.

As our study (research) showed, one of the high-profile Georgian brands is Borjomi (mineral water). Unlike Barambo, Borjomi has a long history. It was very popular in the past and now it is known in Georgia as well as abroad, especially in the post-Soviet countries. "Borjomi" is unique as long as nature itself provides it for people and its composition makes it special. Special features of "Borjomi" are specified at the web page of Sakpatenti. Among them we can read:

> Unlike many sodium bicarbonate waters, Borjomi spring water does not have time to cool before reaching the surface at a temperature of 38–41°C. On its journey upwards, the rocks of the Caucasian mountains enrich the water with over 60 different mineral compounds.

The mineral spring of Borjomi water is located in Georgia, in a narrow gorge along the river Mtkvari in Borjomi town. The natural pressure of carbon dioxide pushes Borjomi water to the surface from a depth of 8–10km (Sakpatenti, 2013).

Ten percent of respondents named various Georgian wine products as the most recognizable Georgian brands. It should be noted that Georgia is considered as one of the oldest wine-producing countries (The National Wine Agency, 2014). The most well-known Georgian wine brands are: Teliani Valley and Badagoni.

The brand's awareness is stimulated by the correct and effective selection of the brand's elements (the color, the symbol, etc.). So, for example, 77% of Barambo's consumers remembered the brand-related colors correctly, while they were asked to remember the colors related to the "Georgian Brand". To the following question: Do you remember a Georgian brand's TV commercial or any other type of an advertisement, including its slogan? – 36% of the Barambo's consumers were able to remember the slogan correctly, while the others (64%) were not (Seturi, 2016b, p. 132).

The image of the brand emerges by the psychological satisfaction of the buyer. As a result of the research it was found out that the majority of respondents who had named some concrete brands as the most recognizable for them, buy these brands in most cases and are their consumers. To the question: "If you do not buy the famous Georgian brand you have named, what is the reason?" – respondents named:

* High price as compared with foreign analogues;
* Low quality as compared with foreign analogues.

(Seturi, 2016a, p. 216)

The brands can be considered successful if consumers think they have different and distinctive features or even unique characteristics from the other similar products. As the research has revealed, despite "Barambo" having a high awareness, with the above-mentioned indicators it lags behind other Georgian brands. Thirty-seven percent of respondents (the respondents for whom the brand "Barambo" is the most well-known Georgian brand) think that "Barambo" has no different or distinctive characteristics from other similar brands. As for any unique features or positions, 89% of respondents think they are not characterized by this brand.

For uniqueness and exclusiveness of the brand, users are given the highest assessments of Georgian wine brands: Teliani Valley and Badagoni. Respondents believe that these Georgian wine brands have not only different, but also unique features as well.

5   Conclusions

Thus, the marketing approaches of considered Georgian brands are characterized by a number of shortcomings. To overcome them we have the following recommendations:

- Consumers can't often see the difference between a particular Georgian brand and its similar products. Therefore, it is necessary companies add to their brands some specific features and advantages characteristic only for this particular brand. And then demonstrate its advantages with advertising and other marketing activities.
- Georgian companies operating on the consumers market should pay much more attention not only to brand management, but also give a greater role and significance to marketing in their organizations, which has no place nowadays. The importance of consumers is first in modern organizations. It's very important to gain and then maintain consumers' trust and confidence, which positively retains customers on the company's side. Organizations can't implement it unless they stop false promises and do not justify expectations of consumers.
- From our point of view, the systematical methods of approach and the complex use of recognized principles deserve a very high attention in branding, and all of it has to foresee and be in accordance with the changes that take place in the constantly mobile market environment.

Notes

1 Iv. Javakhishvili Tbilisi State University.
2 Iv. Javakhishvili Tbilisi State University.

Bibliography

Business Press News, 14 November. (2014). Ten Oldest Georgian Brands. Retrieved from ww.bpn.ge/biznesi/7514-athi-udzvelesi-qarthuli-brendi.html?lang=ka-GE

Doyle, P., & Stern, P. (2007). *Marketing Management and Strategy* (4th ed.). Sankt Peterburg, Piter (translate in Russion).

Kotler, P., & Armstrong, G. (2015). *Principles of Marketing* (14th ed.). Translate in Georgian. Tbilisi, Sulakauri Publishing.

Kotler, P., & Keller, K.L. (2012). *Marketing Management* (14th ed.). Boston, Prentice Hall, Pearson.

Kotler, P., & Pfertch, W. (2007). B2B Brand Management. In *Russion*. Sankt Peterburg, Vershina.

National Statistics Office of Georgia, Agriculture, Environment and Food Safety. (2016). Retrieved from www.geostat.ge/?action=page&p_id=427&lang=geo

The National Wine Agency. (2014). History. Retrieved from http://georgianwine.gov.ge/geo/text/121/

The Newspaper "Kviris Palitra", 25 October. (2010). "Barambo" Sweet as Honey and Useful Products, High Quality: Affordable Price. Retrieved from www.kvirispalitra.ge/economic/5408-qbaramboq.html

Official web page of company "Barambo". About Barambo. Retrieved from www.barambo.ge/en/about

Official web page of company Teliani Valley. History. Retrieved from www.telianivalley.com/

Sakpatenti (The National Intellectual Property Center of Georgia). (2013). Mineral Water "Borjomi". Retrieved from www.sakpatenti.org.ge/index.php?lang_id=GEO&sec_id=332

Seturi, M. (2016a). Some Aspects of Successful Brand (at Tbilisi Consumer's Market). Book of Abstracts of the 4rd Business Systems Laboratory International Symposium: "Governing Business Systems". Vilnius, Lithuania, August 24–26, 2016. Retrieved from http://bslab-symposium.net/, 214–218.

Seturi, M. (2016b). Brands Awareness and Image (at Tbilisi Consumer's Market). Book of Abstracts of the International Scientific Symposium: "Economics, Business & Finance". (Iris-Alkona). Jurmala, Latvia. July 5–9, 2016. Retrieved from http://irissymposium.wixsite.com/alkona/proceedings, 130–134.

Todua, N., & Urotadze, E. (2013). *Principles of Marketing Research, Textbook, Iv.* Tbilisi, Javakhishvili Tbilisi State University.

The role of the influencer in innovation adoption

Beatrice Orlando,[1] *Antonio Renzi,*[2]
Giuseppe Sancetta[3] *and Maria Antonella Ferri*[4]

Keywords: *innovation adoption; bandwagon behavior; herd behavior; influencer; waterfall effect*

The current paper seeks to investigate the role of the influencer in innovation adoption. Specifically, this study explores the social dynamics of innovation adoption, considering how the influencer contributes to foster a cascade of adoptions over time. We argue that the influencer triggers a herd behavior within the social community. We consider herd behaviors in the evolutionary perspective, as the emulative response which occurs semi-unconsciously as a means to increase survival chances. Simplifying, the influencer can be a trigger for herd behaviors. We originally introduce a specific type of herd behavior, named the waterfall effect. A waterfall effect occurs when an individual is called to action (and he/she actually acts) after observing others behaviors. It impacts on intuitive thinking of individuals; and, thus, they need less information and less time than usual to make a decision. In sum, we argue the influencer triggers a cascade of innovation adoptions. Individual adopters are biased by the waterfall effect when making the decision. As a result, the influencer fosters herd behaviors at the collective level. The model was tested on a sample of 60 individuals, through using the decision gambles method. Results of our explorative analysis confirm the likelihood of the model: individuals do use the influencer behavior as a shortcut for making the decision on innovation adoption. Thus, they are subjected to the waterfall effect, because they consider the influencer behavior as a heuristic, without searching for further information. At the firm level, this model sheds light on the decision-making process in innovation adoption: managers are more likely to adopt innovations when other critical players, for instance the incumbent, have already adopted it. At the consumer level, the model explains the critical role of influencers to foster the diffusion of new products.

1 Introduction

"The boundary between perception and judgment is fuzzy and permeable" (Kahneman and Frederick, 2002). The influencer impacts the very folds of perception, ultimately affecting judgments. The current study aims to investigate what is the role played by the influencer in innovation adoption. We argue that the influencer allows to create a heuristic for decision making in innovation adoption: people use the influencer behavior as a shortcut in their judgments, so that they adopt the innovation, without wasting time in the search for further information. As for that reason, we conclude that the influencer triggers, at social level, a cascade of adoptions. The motivation of collective adoptions is explained in terms of herd behaviors. However, we further specify the bias which affects individual judgments, originally introducing the waterfall effect. Especially at the digital platform level, individuals make decisions very quickly, without searching and processing a ton of news; as far afield as they don't look carefully to the entire information that was offered, but they just retrieve some parts of it, and they frame their own knowledge consequently.

People can make decisions very fast, especially when they are exposed to news deriving from someone who is largely considered trustworthy and influential. This dynamic is particularly emblematic in the case of innovation adoption. Innovations are surrounded by a halo of uncertainty and ambiguity, as instance, with concern to their effectiveness, usefulness, easiness of use, etc. In such case, people are more exposed to external judgments they consider superior and relevant.

On the other hand, the continuing growth of the digital environment has changed the dynamics of innovation adoption and diffusion. Thus, we must re-think in a more creative manner both processes.

"The diffusion process explains and predicts the time path of adoption of new products and technologies in a market, and is based on concepts and theories of communication and interaction among users" (Mahajan, Muller, and Wind, 2000, p. 76). The digital environment can influence both adoption behavior and the diffusion process in significant ways, e.g. word-of-mouth and marketer-controlled communications (Rangaswamy and Gupta, 2000).

In general, the internet can be considered as a vast repository of information that can be dynamically organized and retrieved in a multiplicity of ways according to the needs of individual users. Online communities and platforms take part in this process: they further speed adoptions, mostly thanks to the imitation mechanism. For such reasons, the presence of the influencer can markedly alter and affect, in the very first place individual adoption by users. Hence, the influencer behavior could decide the fate of one innovative product diffusion, indeed. It is commonly accepted that new product diffusion is often driven by a social contagion (Van den Bulte and Stremersch, 2004). So, although diffusion models often describe new product diffusion patterns over time quite well, it is unclear what kind of contagion process happens, especially on digital platforms. Adopting the ecosystem and evolutionary perspectives, we can reinvoke to the concept of the waterfall effect.

Technically, the waterfall effect is a visual illusion created by watching a moving object such as flowing water, then looking at a stationary object (Barlow, 1972): neurons coding a particular movement reduce their responses with time of exposure to a constantly moving stimulus; this is neural adaptation. Neural adaptation also reduces the spontaneous, baseline activity of these same neurons when responding to a stationary stimulus. We argue that digital users mostly rely on their intuitive thinking (Kahneman, 2011) and process less information to make a decision. So, influencers can steer the social narrative and lead to fast innovation adoptions by users, in reason of an imitation mechanism. We refer to imitation as that semi-unconscious mechanism, implemented to increase survival chances. In particular, we consider the influencer as the innovator, which initial behavior is able to trigger further and later adoptions by an indefinite number of users. From the initial community, or the target reached by the influencer, the effect spreads in other communities, thanks to the action of followers.

2   Literature framework: a brief overview

People use shortcuts when making decisions: information deducted by cognitive narratives shared by community that are influential both for individual convictions and emotions (Kahneman and Tversky, 1973; Zaloom, 2009; Mar and Oatley, 2008; Chong and Tuckett, 2015).

The imitation effect and its positive role in innovation adoption (or rejection) is the central concern in different studies and models: the Contagion Model (Mansfield, 1961), Rogers' Diffusion Model (Rogers, 1962); institutional isomorphism (March, 1981); the Threshold Model (Granovetter, 1978); fad's role and institutional bandwagons (Abrahamson, 1991; Abrahamson, 1996; Abrahamson and Rosenkopf, 1993; Abrahamson and Rosenkopf, 1997; Abrahamson and Fairchild, 1999); adaptive emulation (Strang and Macy, 2001; Standing, Sims, and Love, 2009); the Coping Model of User Adaptation, in relation to fashion waves (Beaudry and Pinsonneault, 2005).

3   Experiment design and empirical analysis

As stimulus material, we have used a questionnaire, designed using the technique of decision gambles: each subject has to evaluate a case scenario and opt for one of the alternatives. Answers were equated for desirability. The material was used on a sample of 60 individuals, almost in their twenties.

Instructions were given before the experiment, and written also on the top of the questionnaire. Respondents were asked to answer quickly, after the reading of each case.

Answers were collected and tabulated, then analyzed with the use of descriptive statistics and multivariate analysis. Hypotheses were also tested for non parametric independence.

4   Conclusions

Early research findings confirm our insights. Future studies should prove the validity of the model with more extensive analyses, on larger samples, and longitudinal observations. The model allows to view more clear and in a more systematic manner the characteristics of behavioral cascades and of individual mimetic response to external triggers. Our study has relevant managerial and practical applications. At managerial level, it explains decision-making dynamics in innovation adoption, when in presence of the influencer. We originally introduce the waterfall effect, as a bias deriving from intuitive thinking. One relevant practical application concerns new products and technologies adoption: in fact, it clearly explains how to speed up their diffusion. Future investigations should also extend the field of practical applications.

Notes

1 Sapienza University of Rome.
2 Sapienza University of Rome.
3 Sapienza University of Rome.
4 Mercatorum University.

Bibliography

Abrahamson, E. (1991). Managerial fads and fashions: The diffusion and rejection of innovations. *Academy of Management Review*, 16(3), 586–612.

Abrahamson, E. (1996). Management fashion. *Academy of Management Review*, 21(1), 254–285.

Abrahamson, E., & Fairchild, G. (1999). Management fashion: Lifecycles, triggers, and collective learning processes. *Administrative Science Quarterly*, 44(4), 708–740.

Abrahamson, E., & Rosenkopf, L. (1993). Institutional and competitive bandwagons: Using mathematical modeling as a tool to explore innovation diffusion. *Academy of Management Review*, 18(3), 487–517.

Abrahamson, E., & Rosenkopf, L. (1997). Social network effects on the extent of innovation diffusion: A computer simulation. *Organization Science*, 8(3), 289–309.

Barlow, H. B. (1972). Single units and sensation: A neuron doctrine for perceptual psychology? *Perception*, 1(4), 371–394.

Beaudry, A., & Pinsonneault, A. (2005). Understanding user responses to information technology: A coping model of user adaptation. *MIS Quarterly*, 29(3), 493–524.

Chong, K., & Tuckett, D. (2015). Constructing conviction through action and narrative: How money managers manage uncertainty and the consequence for financial market functioning. *Socio-Economic Review*, 13(2), 309–330.

Granovetter, M. (1978). Threshold models of collective behavior. *American Journal of Sociology*, 83(6), 1420–1443.

Kahneman, D. (2011). *Thinking, fast and slow*. New York: Farrar, Straus & Giroux.

Kahneman, D., & Frederick, S. (2002). Representativeness revisited: Attribute substitution in intuitive judgment. *Heuristics and biases: Extensions and Applications*. New York: Cambridge University Press.

Kahneman, D., & Tversky, A. (1973). On the psychology of prediction. *Psychological Review*, 80(4), 237.

Mahajan, V., Muller, E., & Wind, Y. (Eds.). (2000). *New product diffusion models* (Vol. 11). Springer Science & Business Media.

Mansfield, E. (1961). Technical change and the rate of imitation. *Econometrica: Journal of the Econometric Society*, 9(4), 741–766.

Mar, R. A., & Oatley, K. (2008). The function of fiction is the abstraction and simulation of social experience. *Perspectives on Psychological Science*, 3(3), 173–192.

March, J. G. (1981). Footnotes to organizational change. *Administrative Science Quarterly*, 26(4), 563–577.

Rangaswamy, A., & Gupta, S. (2000). Innovation adoption and diffusion in the digital environment: Some research opportunities. *New product diffusion models*, 75–96. Springer Science & Business Media.

Rogers, E. M. (1962). *Diffusion of innovations*. New York: Free Press.

Standing, C., Sims, I., & Love, P. (2009). IT non-conformity in institutional environments: E-marketplace adoption in the government sector. *Information & Management*, 46(2), 138–149.

Strang, D., & Macy, M. W. (2001). In search of excellence: Fads, success stories, and adaptive emulation 1. *American Journal of Sociology*, 107(1), 147–182.

Van den Bulte, C., & Stremersch, S. (2004). Social contagion and income heterogeneity in new product diffusion: A meta-analytic test. *Marketing Science*, 23(4), 530–544.

Zaloom, C. (2009). How to read the future: The yield curve, affect, and financial prediction. *Public Culture*, 21(2), 245–268.

A conceptual framework to combine Maturana's theory of autopoiesis and Checkland's soft systems methodology to explore human activity systems

Alberto Paucar-Caceres[1] and Bruno Jerardino-Wiesenborn[2]

Keywords: *autopoiesis;. soft systems methodology; checkland; maturana; biology of cognition; accommodation; decision process.*

Amongst the systemic methodologies available to systems practitioners, Soft Systems Methodology (SSM) (Checkland, 1981, 1999; Checkland and Scholes, 1990; Munro and Mingers, 2002) is one of the most used problem structuring methods. However, some critics have argued that it has some shortcomings particularly in the initial phases when SSM practitioners structure the situation and when deciding which areas of a problematic situation are deemed to be selected as relevant. SSM tools, e.g. rich pictures and the three SSM analyses, are only sketched guidelines/models and in some cases not useful and arguably difficult to operationalise. Also, during the process debating the set of changes, when SSM advises to implement 'culturally desirable' and 'systemically feasible' changes and offers the concept of 'accommodation' a key and subtle feature of SSM, the researcher is left with a vague idea as to how to use it, leaving a frustrating gap in the methodology.

As it has been widely reported in the management science and system literature, Soft Systems Methodology operates under what is called the interpretivism paradigm (Jackson, 1991, 2003; Mingers, 1984). The main tenets of this paradigm are that reality is complex, it is socially constructed, and a product of continued people's interactions (interpretivist ontology). Under this paradigm, the aim of any intervention is therefore to understand reality through an interpretative process in which meaning is attributed (anti-positivist epistemology). No perspective exhausts the richness of reality or distorts the nature of things; each view is unitary not global.

The work of Maturana and Varelaon on the nature of living, the biological nature of cognition and knowledge have been having a far-reaching influence on the systems and various others fields (Maturana and Varela, 1980, 1988; Maturana, 1988a, 1988b; and Maturana and Bunnell, 1998). While the Checkland approach lies certainly in the interpretivist camp, the philosophical implications of Maturana's work are more difficult to frame. Maturana's theories of cognition (ToC) imply certainly an antirealist ontological position. Epistemologically, he claims that the world as we experience (or constitute) it depends on a subject and that that objective knowledge (or transcendental knowledge as he labels it) is impossible. For some commentators, his position is inconsistent and rather than confining him into constructivism, he can be better understood as critical realist. For others his radical claims denying the existence of any independent reality make him a candidate of radical constructivism. In this paper, and for the purposes of contrasting the two approaches and seeking synergies between them, we will adopt the most wide argument of placing him in the constructivism camp.

The SSM general mode starts when a problem situation is perceived and somehow structured. From this perception, the stakeholders will select relevant systems and express them in basic root definitions. A model-building construction follows as a means for predication of conceptual models. All activities of a purposeful action are carried out by individuals. These then will be compared with the perceived situation before taking action.

This paper attempts to address SSM limitations and proposes a theoretical framework informed by ToC concepts/ideas, by exploring how key concepts from Maturana's ToA and BoC, i.e., Structured-Determined Systems and Organizational Closure might help to expand and complement Checkland's SSM process. We propose to use the ToC concepts as metaphors but endeavour to apply as much rigour as possible to avoid a 'naïve' application of ToC to SSM process. We also draw from Brocklesby (2007) to explore further SSM theoretical underpinnings (Vickers's epistemology enquiring systems) and Maturana's ideas. We develop a framework in which we deploy the possible consequences of applying ToA and BoC; these ideas are discussed and posed against the stages of SSM mode 1 & 2 application types. We summarise this in a set of questions, which form points for reflexion at various SSM stages. It is hoped that reflection on these questions will yield lessons to overcome SSM limitations and enhance and refine its application. This is a work in progress and in this paper, we present the framework; we hope to use the model in a real-world situation later on.

Notes

1 Manchester Metropolitan University.
2 Universidad de Santiago de Chile.

Bibliography

Brocklesby, J. (2007) The Theoretical Underpinnings of Soft Systems Methodology: Comparing the Work of Geoffrey Vickers and Humberto Maturana, *Systems Research and Behavioral Science* 24, 157–168.
Checkland, P.B. (1981, 1999) *Systems Thinking, Systems Practice*, Wiley.
Checkland, P.B., and Holwell, S. (1998) *Information, Systems and Information Systems*, Wiley, Chichester.

Checkland, P.B., and Poulter, J. (2006) *Learning for Action: A Short Definitive Account of Soft Systems Methodology, and Its Use for Practitioners, Teachers and Students*, Wiley.

Checkland, P.B. and Scholes, J. (1990) *Soft Systems Methodology in Action*, Wiley.

Jackson, M.C. (1991) *Systems Methodology for the Management Sciences*, Plenum Press, New York.

Jackson, M.C. (2003) *Systems Thinking: Holism for Managers*, Wiley, Chichester.

Maturana, H. (1988a) Reality: The Search for Objectivity or the Quest for a Compelling Argument, *Irish Journal of Psychology* 9, 25–82.

Maturana, H. (1988b) Ontology of Observing: The Biological Foundations of Self Consciousness and the Physical Domain of Existence. www.inteco.cl/biology/ontology/index.htm.

Maturana, H., and Varela, F. (1980). *Autopoiesis and Cognition: The Realization of the Living*, Reidel, Dordrecht.

Maturana, H., and Varela, F. (1988) *The Tree of Knowledge: The Biological Roots of Human Understanding*, Shambala, Boston.

Maturana, H., and Bunnell, P. (1998). Biosphere, Homosphere, and Robosphere, *Society for Operational Learning*. Retrieved 21.9.2018 from https://issuu.com/gfbertini/docs/biosphere__homosphere__and_robosphere_-_living_sys

Mingers, J. (1984) Subjectivism and Soft Systems Methodology: A Critique, *Journal of Applied Systems Analysis* 11, 85.

Munro, I., and Mingers, J. (2002) The Use of Multimethodology in Practice-Results of a Survey of Practitioners, *Journal of Operational Research Society* 59 (4), 369–378.

Implementation of augmented reality in Real Palace Museum of Naples

An organisational perspective

Filomena Izzo[1]

Keywords: *augmented reality; organisational perspective; museum; technology adoption*

1 Introduction

In recent years, there has been a rising demands for cultural institutions to find new ways to deliver enhanced and unique tourist experiences (Izzo et al. 2015, 2016; Solima et al. 2015; Yovcheva et al. 2013). Every day technology pervades our life; this also stresses cultural institutions to provide access to information anytime and anywhere. As a result, it has become more and more necessary for them to adopt new technologies in order to be competitive and attractive to tourists. For this reason Augmented Reality (AR), has emerged as a popular tool to enhance the visitor experience. Despite the novelty of the field, many studies (Taqvi 2013; Di Serio et al. 2013; Roesner et al. 2014; Wu et al. 2013; Leue et al. 2014; Wang et al. 2013; Wasko 2013; Palumbo et al. 2013; Chung et al. 2015; Martínez-Graña et al. 2013; von der Pütten et al. 2012; Kounavis et al. 2012; Han et al. 2014; Leue et al. 2015; Hume and Mills 2011) identify the chance and potential of using AR in tourism and a lot of AR applications were applied to cultural heritage attractions, for example museums and art galleries.

Implementing a new technology innovation needs to assess and understand the organisational perspective and adoption among users. More studies evaluate the determinants of IT adoption to understand the factors that influence acceptance and use (Bharadwaj and Soni 2007; Fuller 1996; Irvine and Anderson 2008; Thong 1999); however, research into the implementation of AR from an organisational perspective is scarce. This study aims to understand the acceptance of AR at an organisational level in the Real Palace Museum of Naples.

2 Literature review

In literature, more studies evaluate the determinants of IT adoption to understand the factors that influence acceptance and use, but there are few studies on support from the organisation that is ready or willing to accept new technologies (Jung et al. 2015; Cranmer et al. 2016). Technology Acceptance Model – TAM – (Davis

1989) is used to identify user acceptance and it is, therefore, not directly relevant at an organisational level.

Diffusion of Innovation theory – DOI – (Rogers 1995) and Technology, Organisation, Environment (TOE) framework (Tomatzky and Fleischer 1990) are generally applied to evaluate at firm-level intention to accept and adopt technologies. Such models are important for managers to understand what factors influence effective and successful technology use. The DOI theory and TOE framework are the only significant models that focus on technology acceptance and adoption at an organisational firm-level and most other relevant models are derived from them (Oliveira and Martins 2011); however there are no studies on AR implementation.

DOI theory suggests the diffusion of an innovation is based upon two principles: (i) characteristics of the technology, (ii) users' perception of the technology. On the other hand, TOE model identifies three elements of an organisation's context that influence the process by which it adopts and implements a technological innovation: (i) technological context (perceived direct/indirect benefits/barriers, technology readiness); (ii) organisational context (perceived financial costs, technical competence, consumer readiness, firm size, financial commitment); (iii) environmental context (market analysis, competitors analysis, internal acceptance).

Despite the fact understanding an organisations' willingness to adopt and accept new technologies like AR is recognised as fundamental for the success and longevity of implementations, this study aims to contribute to knowledge by understanding how the introduction of AR is perceived at organisational level.

The Real Palace Museum of Naples, an international cultural attraction, is used as a case example; in particular the study tries to understand the perception of AR implementation at an organisational level.

The present study starts with the theoretical background, then I define the research methodology, followed by the findings. Finally, I get to the discussion and conclusion showing the practical implications, the research limits and the future research steps.

3 Methods

Existing research attempting to understand the implementation of AR from an organisational perspective is scarce. The case of the Real Palace Museum of Naples is explored to understand the perception towards AR implementation at an organisational level. As a new area of research the study is exploratory; four semi-structured interviews will made during November–December 2016 with a range of internal stakeholders.

The variety of respondents' roles within the Real Palace Museum from management to operations (museum manager, curator, learning development director, restorer) creates a more complete picture, and their variety of perspectives helps to capture and understand the complexity of elements that combine to produce a tourist product.

The small organisation of Real Palace Museum is beneficial; it was possible to gain insight from a variety of perspectives to generate a more complete overview involving the entirety of the organisation (Saunders et al. 2012; Greenfield 2002; Garrod et al. 2006).

Respondents were provided a short video clip example of AR potential application in the Real Palace Museum.

4   Findings

Respondents suggested using AR could create a multi-sensory environment delivering a better experience to all visitors, but especially those visiting without a guide and children. The educational side of Real Palace Museum is a large part of the business and respondents believed AR has the potential to increase Real Palace Museum's appeal to visitors of all ages by enhancing its learning facilities.

Respondents believed that AR could improve visitor experience, attracting more visitors and encouraging repeat visits. They agreed AR would modernise the Real Palace Museum, raising its profile to bring it more into the modern era, providing a positive perception change about the sort of Museum it is. Using AR to engage younger people was identified as particularly important, making educating and guiding more fun and interesting.

5   Discussion and conclusion

Interviews found that each respondent recognises the potential of using AR to enhance and improve both education and the visitor experience. They believed AR implementation would modernise the Real Palace Museum's existing offering.

Naturally, in order to implement successful AR application, it is necessary to obtain organisational support. Hence, it is necessary to further educate internal stakeholders about exactly what AR implementation entails and the expected outcomes. A further study should consider debating the potential limitations and barriers involved with AR implementation, to discern how an organisation would overcome these.

From a managerial and industry perspective, the study helps practitioners to understand an organisational perspective towards AR implementation. This will help practitioners ensure their organisation creates, develops and implements applications that best suit the organisation's strategies as well as providing maximum benefits and added value.

However, this study has more limitations and recommendations for further research. The findings are based on one case example and therefore the findings are not to be generalised to apply to other cultural heritage attractions.

Moreover, the Real Palace Museum of Naples is a small organisation, with a small staff base; the findings cannot be generalised to a larger organisation with more complex stakeholder networks.

This study adopted an exploratory approach; therefore some issues like technical, user and financial implications would require further exploration. Moreover, it would be recommended to conduct interviews or discussions with internal stakeholders throughout the IT implementation to ensure coherence with overall goals and strategies of the organisation.

Another limitation of the present study concerns the focusing on internal stakeholders. Therefore, it is recommended research is extended to understand the perception of external stakeholders (e.g. visitors, business partners, local tour operations and relevant customers), to give a more comprehensive and complete overview about AR implementation.

Note

1 Economics Department, University of Campania "Luigi Vanvitelli", Capua, Italy.

Bibliography

Bharadwaj, P. & Soni, R. (2007). E-commerce usage and perception of e-commerce issues among small firms: Results and implications from an empirical study. *Journal of Small Business Management*, 45(4): 501–521.

Chung, N., Han, H. & Joun, Y. (2015). Tourists' intention to visit destination: Role of augmented reality applications for heritage site. *Computers in Human Behaviour*, 50: 588–599.

Cranmer, E. & Jung, T. (2014, May). *Augmented Reality (AR): Business Models in Urban Cultural Heritage Tourist Destinations: Pacific Council on Hotel, Restaurant and Institutional Education: APacCHRIE, Kuala Lumpur.* Manchester: Research Gate.

Cranmer, E., Jung, T., tom Dieck, M. C. & Miller, A. (2016). Understanding the acceptance of augmented reality at an organisational level: The case of Geevor Tin mine museum. In *Information and Communication Technologies in Tourism.* Springer International Publishing.

Davis, F. (1989). Perceived usefulness, perceived ease of use, and user acceptance of information technology. *MIS Quarterly*, 13(3).

Di Serio, A., Ibanez, M. & Kloos, C. (2013). Impact of an augmented reality system on students' motivation for a visual art course. *Computers & Education*, 68(1): 586–596.

Fuller, T. (1996). Fulfilling IT needs in small businesses: A recursive learning model. *International Small Business Journal*, 14(4): 25–44.

Garrod, B., Wornell, R. & Youell, R. (2006). Re-conceptualising rural resources as countryside capital: The case of rural tourism. *Journal of Rural Studies*, 22(1): 117–128.

Greenfield, T. (2002). *Research Methods for Postgraduates.* London: Arnold.

Han, D. I., Jung, T. & Gibson, A. (2014). Dublin AR: Implementing augmented reality in tourism. In Z. Xiang & I. Tussyadiah (Eds.), *Information and Communication Technologies in Tourism 2014* (pp. 511–523). Vienna: Springer.

Hume, M. & Mills, M. (2011). Building the sustainable Museum: Is the virtual museum leaving our museums virtually empty. *International Journal of Non-Profit & Voluntary Sector Marketing*, 16(3): 275–289.

Irvine, W. & Anderson, A. (2008). ICT peripherally and smaller hospitality businesses in Scotland. *International Journal of Entrepreneurial Behaviour & Research*, 14(4): 200–218.

Izzo, F., Mustilli, M. & Guida, M. (2015). Realtà aumentata e valorizzazione dei beni culturali. Riflessioni sull'offerta culturale casertana. Paper presented at the Sinergie Annual Conference, Termoli, July.

Izzo, F., Mustilli, M., Sasso, P. & Solima, L. (2016). Service orientation and technology innovation in museum: Museo Archeologico Nazionale of Naples case study. 26th Annual RESER Conference, Naples-Italy, September.

Jung, T., Chung, N. & Leue, M. (2015). The determinants of recommendations to use augmented reality technologies: The case of a Korean theme park. *Tourism Management*, 49(1): 75–86.

Kounavis, C., Kasimati, A. & Zamani, E. (2012). Enhancing the tourist experience through mobile augmented reality: Challenges and prospects. *International Journal of Engineering Business Management*, 4(10): 1–6.

Leue, M. C., Jung, T. & tom Dieck, D. (2015). Google glass augmented reality: Generic learning outcomes for art galleries. In I. Tussyadiah & A. Invesini (Eds.), *Information and Communication Technologies in Tourism 2015* (pp. 463–476). Heidelberg: Springer.

Leue, M. C., tom Dieck, D. & Jung, T. (2014). A theoretical model of augmented reality acceptance. *e-Review of Tourism Research*, 5: 1–5.

Martínez-Graña, A., Goy, J. & Cimarra, C. (2013). A virtual tour of geological heritage: Valourising geodiversity using Google Earth and QR code. *Computers & Geosciences*, 61(12): 83–93.

Oliveira, T. & Martins, M. (2011). Literature review of information technology adoption models at firm level. *The Electronic Journal Information Systems Evaluation*, 14(1): 110–121.

Palumbo, F., Dominci, G. & Basile, G. (2013). Designing a mobile app for museums according to the drivers of visitor satisfaction. Recent Advances in Business Management and Marketing: Proceedings of the 1st International Conference on Management, Marketing, Tourism, Retail, Finance and Computer Applications(MATREFC'13), WSEAS Press, Dubrovnik, Croatia.

Roesner, F., Kohno, T. & Molnar, D. (2014). Security and privacy for augmented reality systems. *Communications of the ACM*, 57(4): 88–96.

Rogers, E. (1995). *Diffusion of Innovations.* New York: Everett.

Saunders, M., Lewis, P. & Thornhill, A. (2012). *Research Methods for Business Students.* New York: Pearson.

Solima, L., Della Peruta, M. R. & Del Giudice, M. (2015). Object-generated content and knowledge sharing: The forthcoming impact of the internet of things. *Journal of the Knowledge Economy*, 7(3): 738–752.

Taqvi, Z. (2013). Reality and perception: Utilization of many facets of augmented reality. 23rd International Conference on Artificial Reality and Telexistence (ICAT), IEEE, Houston, TX, USA.

Thong, J. Y. (1999). An integrated model of information systems adoption in small businesses. *Journal of Management Information Systems*, 15(4): 187–214.

Tomatzky, L. & Fleischer, M. (1990). *The Process of Technology Innovation.* Lexington: Lexington Books.

tom Dieck, M. C. & Jung, T. (2015). A theoretical model of mobile augmented reality acceptance in urban heritage tourism. *Current Issues in Tourism*, 18: 1–21.

von der Pütten, A., Klatt, J., Ten Broeke, S., McCall, R., Krämer, N., Wetzel, R., Blum, L., Oppermann, L. & Klatt, J. (2012). Subjective and behavioural presence measurement and interactivity in the collaborative augmented reality game Time-Warp. *Interacting with Computers*, 24(4): 317–325.

Wang, X., Kim, M., Love, P. & Kang, S.-C. (2013). Augmented reality in built environment: Classification and implications for future research. *Automation in Construction*, 32(1): 1–13.

Wasko, C. (2013). What teachers need to know about augmented reality enhanced learning. *Tech Trends*, 57(1): 17–21.

Wu, H.-K., Lee, S. W.-Y., Chang, H.-Y. & Liang, J.-C. (2013). Current status, opportunities and challenges of augmented reality in education. *Computers & Education*, 62: 41–49.

Yovcheva, Z., Buhalis, D. & Gatzidis, C. (2013). Engineering augmented tourism experiences. In L. Cantoni & Z. Xiang (Eds.), *Information and Communication Technologies in Tourism 2013*. Heidelberg: Springer.

Social entrepreneurship and corporate social responsibility in the context of moral economy-dilemma for post-Soviet countries – Georgian case

Ia Natsvlishvili[1]

Keywords: *entrepreneurship; social entrepreneurship; corporate social responsibility; moral economy; post-Soviet countries*

1 Introduction

Social Entrepreneurship (SE) is an important feature of moral economy. SE is seen as one of the best development strategies. SE is a relatively new phenomenon in post Soviet countries where social services were provided only by the state for several decades. After the collapse of the Soviet Union the lack of state support led to the emergence of the third sector – Non Governmental Organizations – in these countries to help to solve many social problems, but due to the limited organizational capacity and low organizational maturity their efforts are not sufficient. The recent situation of civil society development in Georgia is not satisfactory. The state considers the civil society as a competitor and the agreed division of labor between the state and civil society does not exist. Entrepreneurs should find balance between the company's success, the employees' needs, and environmental and social stability. These three priorities form the foundations of Corporate Social Responsibility (CSR). CSR is a certain corporate policy that has to meet two basic requirements: efficient business performance and moral principles such as honesty, fairness and responsibility. Researchers argue that economic history and empirical facts offer evidence that moral standards have macroeconomic and microeconomic positive impacts. As several studies show companies operating in Georgia spend significant funds on social projects and charity although such socially oriented activities are chaotic instead of being integrated in their business plans. Strengthening the CSR could be considered as steps toward development of SE.

2 Companies' corporate social responsibility in Georgia

Recent studies emphasize the economic impact of Christian faith on societies such as: savings from health-related issues (less illness-based absenteeism at work), fiscal benefits (less shadow economy and less tax frauds) and development of education and literacy, for instance the commitment of Christian missions to alphabetization (Haupt, 2015).

Several Georgian researchers aimed to evaluate the role of corporate social responsibility in companies operating in Georgia. According to qualitative studies conducted in 2015, 89% of the responding companies consider themselves as having corporate social responsibility, while 8% do not and 2% are not sure. Types of social responsibility activities in companies were the following: protection of the rights of people and company's employees – 98% of respondents; environmental protection – 56% of respondents; caring about the societal/community needs – 89% of respondents; consumers' rights protection and production responsibility – 87% of respondents; relations with the suppliers and consumers – 67% of respondents; corruption prevention – 80% of respondents; transparency and reporting – 4%; all – 4%. Only 25% of the companies include the CSR in their development strategy. Seventy-five percent of them have spontaneous CSR actions. The factors supporting the CSR development in Georgia are: increasing social responsibility awareness – 75%; introduction of the international standards of social reporting – 67%; introduction of social indices – 35%; stimulation – 95%; including certain allowances by the state in respect of taxes, licenses, export, etc. – 97%. Ninety-five percent of the companies think in terms of various methods, such as awards, societal awareness, advertisement (Chokheli, 2015).

European Commission defines CSR as a "concept where the companies voluntarily consider social and environmental issues in their business operations and in their relations with stakeholders" (Chokheli, Narmania, 2015, p. 3). Corporate social responsibility is the highest among European companies and respectively, three-quarters of companies with CSR are European and one-quarter of companies with CSR are American. Generally in developing countries, including Georgia, social responsibility is still associated (in some companies) with charity (Chokheli, Narmania, 2015, p 3).

Companies in Georgia spend certain funds on social projects but such socially oriented activities are chaotic. They are not systemized and related to the company's priorities and strategies. Some researchers also suggest the use of a social responsibility index in order to describe the quality of social responsibility. This index can be calculated as correlation of enterprise net profit and the volume of spending on social activities (Chiladze, 2015).

Activities of corporate social responsibility take place in the society where the individuals have certain entrepreneurial attitudes. According to "Global Entrepreneurship Monitor – 2014 Georgia Report" Georgians consider successful entrepreneurs to have a high status in society (75.9% of adult population). Motivation to engage in entrepreneurial activities is almost equally distributed between necessity-driven (48.6%) and opportunity-driven (50.6%) entrepreneurship. Compared to efficiency-driven EU and non-EU economies, early stage entrepreneurship activities in Georgia are mainly necessity-driven rather than opportunity-driven (Lezhava et al., 2014).

As surveys show, positive attitudes of the respondents in Tbilisi towards entrepreneurship indicate their self-confidence, feeling of social and political stability, expectations of success of market-oriented economic reforms (Natsvlishvili,

2012). Analysis of results derived from the several research studies shows that in Georgia there is a weak negative attitude towards entrepreneurship. On the background of high unemployment, the desire to be self-employed is quite high. Entrepreneurship is seen as a special form of employability. In Georgia traditionally the share of self-employment among employed people is prevalent (Natsvlishvili, 2016).

3   Social entrepreneurship in global environment

Definitions of Social Responsibility have taken many forms. These definitions

> view social responsibility as a process of creating value by combining resources in new ways. These resource combinations are intended primary to explore and exploit opportunities to create social value by stimulating social change or meeting social need. SE . . . can occur equally well in a new organization or in an established organization, where it may be labeled 'social intrapreneurship.'
>
> (Kuratko, 2014)

Important conclusions can be drawn from the research conducted by The Global Entrepreneurship Monitor (GEM) on social entrepreneurship activity. The average prevalence rate of broad social entrepreneurial activity among nascent entrepreneurs in the start-up phase across all 58 GEM economies is 3.2%. By comparison, the rate of start-up commercial entrepreneurship averages 7.6% in the world. The average prevalence rate of individuals who are currently leading an operating social entrepreneurial activity across all 58 GEM economies is 3.7% (Bosma et al., 2016).

Social enterprise has a dual goal: economic and social. The economic goal serves as a mean for accomplishing the second important goal – the social goal. The scale of social enterprise can vary. Social enterprise can exist in any allowed organizational-legal form. In some countries they have certain legal forms. Today social entrepreneurship is performedwith different practices in different countries and it is differently defined. All these definitions have one common thing – a business approach for social goals. Despite the fact that nonprofit organizations in the USA and Europe have conducted such activities for a long time the concept of social entrepreneurship takes its origins from the '70s, in the past century. Its particular dynamic development started from '90s (The Center for Strategic Research and Development of Georgia, 2013).

The concept of social entrepreneurship first appeared in Europe in the 1990s and widely wqw popularized in many European countries. In some of these countries legal forms of social enterprise were defined by law. In France, Portugal, Spain and Greece, social enterprises have forms of cooperatives. In other countries more open models of social enterprises exist that aren't limited by traditions of cooperatives. In 1996 European Research Network – EMES – was formed. EMES defines a social enterprise as an organization that is initiated by the group of citizens, has a clear purpose of benefiting the community and in which financial interests of capital investors are limited. EMES has drawn certain criteria that social enterprise must satisfy. These criteria are divided into three groups: economic, social and co-participation in management. These are characteristics of "ideal social enterprise" and serve as an instrument for their self-assessment. Economic criteria cover: Continuous manufacturing/supply of goods and services; certain degrees of economic risk; attitude to paid job. Social criteria are: Clear goal that is focused on the wealth of society; initiative comes from certain group of citizens or civil society organization; limited distribution of profit. Criteria of the management are: High degree of independence; decision making is not based on capital ownership; community that promotes high involvement of interested parties (The Center for Strategic Research and Development of Georgia, 2013).

4   Peculiarities of social entrepreneurship and social enterprises in Georgia

Practice of integration of commercial and social goals have been established in organizations for many years. Examples of this are: many charitable and civil society organizations, using commercial and financial indicators for measuring social consequences; a growing number of business companies who care about development of social responsibility in their organizations. Georgia's legislation doesn't define the concept of social entrepreneurship. Entrepreneurial activities including social entrepreneurship can be carried out in different legal forms. Civil society organizations in Georgia are non-entrepreneurial (non-commercial) legal entities (NNLE). Despite the fact that NNLE are created for the purpose of non-commercial activities, Georgia's legislation allows them to conduct entrepreneurial activities of supportive characteristics.

Civil society organizations in Georgia conduct economic activities mainly within their organizations (76%). Twenty-four percent of them have established separate enterprises. Most of the organizations (72.7%) which have established separate enterprises have chosen a form of Limited Liability Company (The Center for Strategic Research and Development of Georgia, 2010b).

5   Conclusion

Generally in developing countries, including Georgia, social responsibility is still associated with charity. Companies in Georgia spend certain funds on social projects but such socially oriented activities are chaotic. They are not systemized and related to the company's priorities and strategies. Companies lack a system of social responsibility projects as part or their business plan. From 1990s certain civil society organizations in Georgia have conducted some revenue-generating activities for diversifying funding sources but the number of new economic activities significantly increased from 2005. Georgia's legislation doesn't define the concept of social entrepreneurship and hence, there are no special norms for Georgia's social enterprises. Entrepreneurial activities including social entrepreneurship can be carried out in a different legal forms. Non Governmental Organizations in these countries help to solve many social problems, but due to the limited organizational capacity and low organizational maturity, their efforts are not sufficient. On one hand the recent situation of civil society development in Georgia is not satisfactory and it is reflected in economic indicators with social and political implications. The state considers the civil society as a competitor and the agreed division of labor between the state and civil society does not exist.

Note

1 Ivane Javakhishvili Tbilisi State University.

Bibliography

Bosma, N. S., Schott, Th., Siri, A. T., Penny, K. (2016). Global Entrepreneurship Monitor2015 to 2016: Special Report on Social Entrepreneurship. Global Entrepreneurship Research Association. Retrieved from http://gemconsortium.org/report/49542

Bygrave, W. D., Zacharakis, A. (2014). *Entrepreneurship*. Third Edition, Wiley.

Chiladze, I. (2015). Business Social Responsibility and Christianity. Christianity and Economics. Proceedings of the 8th International Scientific Conference Papers. Batumi, 169–175 (in Georgian).

Chokheli, E. (2015). The Role of Social Responsibility and the Growth Perspectives in Business (The Case of Georgia). Retrieved from www.iises.net/proceedings/international-academic-conference-rome/front-page

Chokheli, E., Narmania, D.(2015). Impact of Corporate Social Responsibility on the Competitiveness of Companies. Retrieved from www.researchgate.net/publication/309375314

Haupt, R. (2015). Business and Christian Faith: A Contradiction? Christianity and Economics. Proceedings of the 8th International Scientific Conference Papers. Batumi, 187–190.

Kuratko, D. F. (2014). *Entrepreneurship: Theory, Process, Practice*. Ninth Edition. South-Western, CENGAGE Learning.

Lezhava, B., Brekashvili, P., Melua, I. (2014). Global Entrepreneurship Monitor, 2014 Georgia Report. Retrieved from www.cu.edu.ge/images/caucasus_university/docs/csb/GEM%20Georgia/Georgia_GEM_National_Report_2014%20final_CS.pdf

Natsvlishvili, I. (2012). Entrepreneurial Attitudes in Post Communist Countries (Case of Georgia). *International Journal of Business and Management Studies, USA*. 1(3): 451–457.

Natsvlishvili, I. (2016). Social Attitudes toward Starting Business and Challenges of Entrepreneurial Activities in Georgia. Strategic Imperatives of Modern Management (SIMM-2016). Proceedings of the III International Scientific and Practical Conference). Kyiv, 64–71.

Parker, S. C. (2009). *Economics of Entrepreneurship*. Cambridge: Cambridge University Press.

Dynamic capabilities models in systems thinking

Marcello Sansone,[1] *Roberto Bruni*[2]
and Annarita Colamatteo[3]

Keywords: *dynamic capabilities; systems thinking; relationships; context; systems*

A system is an entity made up of parts (elements), not only connected in a mutual relation and interaction, but also able to share a specific aim (Golinelli, 2005; Barile, 2009). Environment, society and any relation or interaction can be investigated as systems; the system emerges from the activity of the structure and by the activity of adaptation to the turbulence and variability of the environment. Could be difficult to survive in the complexity and, in particular, in the unpredictability of the markets for systems (individuals, companies and other organizations).

Particular capabilities – *dynamic capabilities* – could be useful to system survival in the complexity, perceiving the environment, adapting the value proposition and re-organizing the structure with plasticity; dynamic capabilities could increase the possibilities to perceive the environment and the opportunities to activate relations between the actors.

Several research works about the topic of dynamic capabilities are in the research streams of company organization and management (Eisenhardt, Martin, 2000; Teece, 2007; Easterby-Smith, Prieto, 2008; Ali, Peters, Lettice, 2012; Sicotte, Drouin, Delerue, 2014; Helfat, Peteraf, 2014). In this work, the focus is on the models that are able to highlight the strategic perspective of the microfoundations of the dynamic capabilities (Teece, Pisano, 1994; Teece, Pisano, Shuen, 1997; Eisenhardt, Martin, 2000; Teece, 2007, 2014).

Dynamic capabilities models in systems thinking could be useful to study the role of *sensing*, *seizing* and *transforming* of organizations in the service ecosystems. In a recent work (Barile et al., 2016), the relations between systems, networks and ecosystems are well explained and the service ecosystem is proposed as an integration between systems and network systems perspectives, as a relatively self-contained, self-adjusting system of resource-integrating actors or entities, connected by shared institutional logics and mutual value creation through the service exchange (Vargo, Lusch, 2011).

It seems that it is possible to reduce the complexity through the increasing knowledge and informative variety (Barile, 2009); therefore, the relevance of the dynamic capabilities are based on the knowledge and on the sensitivity, to define the opportunity and to know perspectives, environment variability and conditions to change.

In the actual environment, the perception, the adaptation and the reorganization of the value proposition are relevant. Based on these forewords, interpreting dynamic capabilities models on systems thinking perspective seems beneficial for planning future strategies, perceiving the environment from self-perspective and from the specific context – *particular representation of the environment in a constructivist logic* – by the other actors. That is not an adaptation by a third-party strategy but the recognition of the environmental variability, consequently managing complexity by dynamic capabilities of other actors.

Notes

1 University of Cassino and Southern Lazio.
2 University of Cassino and Southern Lazio.
3 University of Cassino and Southern Lazio.

Bibliography

Ali, S., Peters, L. D., & Lettice, F. (2012). An organizational learning perspective on conceptualizing dynamic and substantive capabilities. *Journal of Strategic Marketing*, 20(7), 589–607.

Barile, S. (2009). *Management Sistemico Vitale*, Giappichelli, Torino.

Barile, S., Lusch, R., Reynoso, J., Saviano, M., & Spohrer, J. (2016). Systems, networks, and ecosystems in service research. *Journal of Service Management*, 27(4), 652–674.

Easterby Smith, M., & Prieto, I. M. (2008). Dynamic capabilities and knowledge management: An integrative role for learning?*. *British Journal of Management*, 19(3), 235–249.

Eisenhardt, K. M., & Martin, J. A. (2000). Dynamic capabilities: What are they? *Strategic Management Journal*, 21, 1105–1121.

Golinelli, G. M. (2005). *L'approccio sistemico al governo dell'impresa. L'impresa sistema vitale*, CEDAM, Padova.

Helfat, C. E., & Peteraf, M. A. (2014). Managerial cognitive capabilities and the microfoundations of dynamic capabilities. *Strategic Management Journal*, 36, 831–850.

Sicotte, H., Drouin, N., & Delerue, H. (2014). Innovation portfolio management as a subset of dynamic capabilities: Measurement and impact on innovative performance. *Project Management Journal*, 45(6), 58–72.

Teece, D. J. (2007). Explicating dynamic capabilities: The nature and microfoundations of (sustainable) enterprise performance. *Strategic Management Journal*, 28(13), 1319–1350.

Teece, D. J. (2014). A dynamic capabilities-based entrepreneurial theory of the multinational enterprise. *Journal of International Business Studies*, 45(1), 8–37.

Teece, D. J., & Pisano, G. (1994). The dynamic capabilities of firms: An introduction. *Industrial and Corporate Change*, 3(3), 537–556.

Teece, D. J., Pisano, G., & Shuen, A. (1997). Dynamic capabilities and strategic management. *Strategic Management Journal*, 509–533.

Vargo, S. L., & Lusch, R. F. (2011). Service-dominant logic: Looking ahead. *Presentation at the Naples Forum on Service*, June 14–17, Isle of Capri, Italy.

Navigating managerial decision-making processes between competition, asymmetric information, isomorphism, legitimacy, decoupling, manipulation, power and cynicism

Claudio Nigro[1] *and Enrica Iannuzzi*[2]

Key words: *uncertainty; legitimacy; isomorphism; decoupling*

For more than three centuries, scholars and practitioners have sought the effective key factors useful for any business organizations operating under uncertain conditions – about the trend of served markets – and participating in several competitive processes, both horizontal (at the sectorial level) and vertical (at the chain level). One result of this investigation has been that any organization aims at driving their actions through the main principle (a kind of *gründ norm*) of value creation for the shareholders by means of the fulfilment of most relevant stakeholders.

Nowadays, scholars and practitioners are trying to recognize and realize the role of business organizations in the structuring and changing of both social and economic general systems. Indeed, institutions, organizations, academies, scholars, associations belonging to the economic and social disciplines have been focused for more than two decades about models and strategies for addressing complex problems and issues: environmental degradation and sustainability, climate change, expansion and effects from the digital economy, protection of fundamental human rights, market transparency, security and privacy and so on.

New-institutional theory has been focusing, for several decades, on the dynamics into a so-called 'organizational field', emphasizing diffusion of *practices* and *norms* across organizations. One of the most important results of this diffusion is a kind of 'materialization' of *institutional logics* aimed to define generally accepted 'rules of competition' within very diverse spheres of social and economic systems.

Moving from these considerations, we have made efforts in order to analyze some issues that are consistent with and coherent within the new-institutionalist perspective.

Given these premises, the work aims at coming up with an overall framework in which a business organization acts, introducing at an epistemological level some conceptual categories sometimes classified as *politically incorrect*. This choice is based on the idea that bringing together both several concepts and very diverse logics and objectives, organizations could experience conflict among norms, external and/or internal expectations and their own ends.

In attempting to reconcile concepts, logics, objectives, norms, expectations, ends and, at the same time, withstanding the competitive pressures, any organization may fall prey either to *conflict, instability* or *confusion* – increasing the overall risk of business – or to an *isomorphic pull*, consistent with the norms of a specific organizational field – in order to obtain *legitimacy*. This is because, if unpredictability, instability and a high degree of dependence on few key resources (characterizing the inter-organizational relationships) are elements defining the competitive environment – contributing to the determination of the level of uncertainty perceived by the actors – in contrast, according to the literature, the culture of 'risk aversion', and then, the attempt to contain it, pushes the actor to take isomorphic mimetic behaviours, rather than searching 'rational' solutions. Hence, the level of effectiveness in problem-solving processes could be not related with solutions taken by the actor, but it would be borrowed from routinized and taken-for-granted beliefs coming from the so-called *institutional context*. Furthermore, the cognitive limitation, that characterizes any actor, affects the business confidence and risk appetite, seen as *prerequisites* for an isomorphic process.

Therefore, the uncertainty governed through isomorphic processes has led many scholars to question a possible *trade-off* between the search for internal organizational efficiency and external legitimacy. The uncertainty of context, in fact, would drive the actor's decisions toward a final goal that the new-institutionalist literature defines as the '*research of social legitimacy*'. Where uncertainty represents the limit to be overcome to contain the '*risk exposure*' which qualifies the individual to act in a constant evolving world, the search for social legitimacy could embody one of the objectives of the strategic action. The concept of *social legitimacy* will be analyzed in its own multidimensional characteristics: it is socially constructed and related on the observers' perception (especially if leaders) over the other actors' behaviour; condition for the actors seeking consensus and legitimation from other players, both internal and external to the context of action, is the promotion and strengthening of their image in terms of status, reputation and thus reliability.

In this scenario, according to the new-institutionalism literature, the research of legitimacy and isomorphic conducts would play a decisive role in shaping organizational structures and strategic options. As fallout, it could produce, by contrast, diseconomies relying on the necessity of making compatible the legitimacy (as 'purpose') and an isomorphic behaviour (as 'mean') with the real internal processes of an entrepreneurial organization.

New-institutionalism theorists have made efforts to investigate this trade-off between the research of legitimacy and internal efficiency. Indeed, when an organization faces competing pressures, it can engage in *compromise, avoidance, defiance, decoupling* and/or *manipulation*. On the other hand, when an organization faces internal competing identities/values and some factions (internal groups) adhering to those identities, managers could attempt *deletion, compartmentaliza-tion, aggregation* and *synthesis.*

Through doing so, organizations may *pro-tempore* pursue both legitimacy and internal efficiency without breaking down to the supremacy of just one logic.

The introduction of these conceptual categories reopens the debate on the nature of the competition between organizations and the introduction/enforce-ment of rules aimed at limiting the adoption of *incorrect behaviours* by organiza-tions in competition.

On a conceptual level, the work attempts to provide a key to understanding the relationship between the value that can be generated through the respect of shared rules and the value derived by the respect of shared rules and the value that can be created through the production of 'proprietary pockets of knowledge', or, also, by *information asymmetries.* The modern managerial sciences, however, while trying to drop out those theories and models, accusing them of too much simplicity, at the same time would underestimate the logical connection, far from insignificant and meaningless, between information (a)symmetry and presence/ absence of a surplus value.

In this work, we would try to speculate that the enterprises might seek to gen-erate and maintain an asymmetric knowledge, incorporating it in its own 'supply system' more and more articulated in order to prevent it from imitation processes made by other players.

The effectiveness in keeping in time the behaviours based on information asymmetries may be, however, explained by the introduction and analysis of the *political model*, based on the concept of *power*, that is, in a broad sense, the agent's ability to produce and maintain degrees of freedom of action deemed appropriate and consistent with the objectives pursued, with the characteristics of the context in which the agent operates, and, with the peculiarity of own resources. This conceptual category plays a central role in the managerial practice, involving sev-eral aspects: from intra-organizational relationships (e.g. co-optation, resources allocation and so on) to the inter-organizational relationships (dynamics that occur in the organizational field and aimed at building the *normative* and *symbolic institutional framework*).

In fact, the opportunity of combining different aims is given from the possibil-ity to describe these phenomena in terms of power struggles and negotiations among external and internal constituencies, attributed to very diverse institu-tional logics.

This work, in an attempt to systematize all the previous categories, tries to focus on a last conceptual category: the *cynicism.* Driven by an ethical premise, it could be plausible that the term might prepare the reader to a kind of hostility or dis-like; but this standing is not allowed at a scientific or epistemological level. Over the time the concept has been narrowed, assuming the role of a sort of 'ability/ attitude' of an agent to solve the difficulties that might pose obstacles to his own objectives. Such as, we may return to the starting point, linking this conceptual category to the competition, trying, albeit briefly, to understand its role in the competitive dynamics affecting the organizations.

Notes

1 University of Foggia, Italy.
2 University of Foggia, Italy.

Bibliography

Ashforth, B. E., and Gibbs, B. W. (1990). The double-edge of organizational legitimation. *Organization Science*, 1(2), 177–194.

Barley, S. R., and Tolbert, P. S. (1997). Institutionalization and structuration: Studying the links between action and institution. *Organization Studies*, 18(1), 93–117.

Berger, P. L., and Luckmann, T. (1991). *The social construction of reality: A treatise in the sociology of knowledge* (No. 10), Penguin UK.

Carolillo, G., Mastroberardino, P., and Nigro, C. (2013). The 2007 financial crisis: Strategic actors and processes of construction of a concrete system. *Journal of Management & Governance*, 17(2), 453–489.

Dean, J. W., Brandes, P., and Dharwadkar, R. (1998). Organizational cynicism. *Academy of Management Review*, 23(2), 341–352.

Deephouse, D. L., and Suchman, M. (2008). Legitimacy in organizational institutionalism. *The Sage Handbook of Organizational Institutionalism*, 49, 77.

Dowling, J., and Pfeffer, J. (1975). Organizational legitimacy: Social values and organizational behavior. *Pacific Sociological Review*, 122–136.

Eisenhardt, K. M., and Zbaracki, M. J. (1992). Strategic decision making. *Strategic Management Journal*, 13(S2), 17–37.

Epstein, E. M., and Votaw, D. (1978). Legitimacy. *Rationality, Legitimacy, Responsibility: Search for New Directions in Business and Society*, 69–82.

Fields, C. R. I. O., Dimaggio, P. J., and Powell, W. W. (1983). The iron cage revisited: Institutional isomorphism and collective rationality in organizational fields. *American Sociological Review*, 48(2), 147–160.

Friedberg, E. (1994). Il potere e la regola. *Dinamiche dell'azione organizzata, Etas Libri, Milano*, 65.

Greenwood, R., Raynard, M., Kodeih, F., Micelotta, E. R., and Lounsbury, M. (2011). Institutional complexity and organizational responses. *The Academy of Management Annals*, 5(1), 317–371.

Jay, J. (2013). Navigating paradox as a mechanism of change and innovation in hybrid organizations. *Academy of Management Journal*, 56(1), 137–159.

Kraatz, M. S., and Block, E. S. (2008). Organizational implications of institutional pluralism. *The Sage Handbook of Organizational Institutionalism*, 840, 243–275.

Lawrence, T. B., and Suddaby, R. (2006). 1.6 institutions and institutional work. *The Sage Handbook of Organization Studies*, 215.

Leca, B., and Naccache, P. (2006). A critical realist approach to institutional entrepreneurship. *Organization*, 13(5), 627–651.

Mastroberardino, P. (2011). La governance del sistema impresa tra istituzionalizzazione e azione del soggetto imprenditoriale. *Sinergie Italian Journal of Management*, (81), 135–171.

Meyer, J. W., & Rowan, B. (1977). Institutionalized organizations: Formal structure as myth and ceremony. *American Journal of Sociology*, 340–363.

Neilsen, E. H., and Rao, M. H. (1987). The strategy-legitimacy nexus: A thick description. *Academy of Management Review*, 12(3), 523–533.

Nigro, C., Iannuzzi, E., and Carolillo, G. (2012). Comunicazione e strutturazione di un quadro istituzionale. Riflessioni sulla recente crisi del sistema finanziario. *Sinergie Italian Journal of Management*, 109–130.

Nigro, C., Iannuzzi, E., and Petracca, M. (2016). Isomorphic and decoupling processes: An empirical analysis of governance in Italian state museums. *Sinergie Italian Journal of Management*, 99, 253–274.

Oliver, C. (1991). Strategic responses to institutional processes. *Academy of Management Review*, 16(1), 145–179.

Pache, A. C., and Santos, F. (2010). When worlds collide: The internal dynamics of organizational responses to conflicting institutional demands. *Academy of Management Review*, 35(3), 455–476.

Pfeffer, J. (1981a). *Management as symbolic action: The creation and maintenance of organizational paradigms*. Graduate School of Business, Stanford University.

Pfeffer, J. (1981b). *Power in organizations*, Ballinger Publishing Company, Cambridge, MA.

Pratt, M. G., and Foreman, P. O. (2000). Classifying managerial responses to multiple organizational identities. *Academy of Management Review*, 25(1), 18–42.

Roberto, M. A. (2004). Strategic decision-making processes beyond the efficiency-consensus trade-off. *Group & Organization Management*, 29(6), 625–658.

Suchman, M. C. (1995). Managing legitimacy: Strategic and institutional approaches. *Academy of Management Review*, 20(3), 571–610.

Westphal, J. D., and Zajac, E. J. (2001). Decoupling policy from practice: The case of stock repurchase programs. *Administrative Science Quarterly*, 46(2), 202–228.

Zucker, L. G. (2000). Il ruolo dell'istituzionalizzazione ai fini della persistenza culturale. Dimaggio, P., Powell, W. W. (a cura di), *Il neoistituzionalismo nell'analisi organizzativa*, Edizioni di Comunità, Torino.

How dynamic capabilities matter for the implementation of a successful equity crowdfunding campaign

Nicola Del Sarto[1] *and Domitilla Magni*[2]

Keywords: *equity crowdfunding; dynamic capabilities; Web 2.0; crowdsourcing; crowdfunding*

1 Introduction

The advent of Internet has dramatically changed the landscape in which firms compete, create value and innovate (Wirtz et al., 2010). Web 1.0 technologies allowed companies and expert individuals to create contents and display them through the Web, reaching a swarm of passive internauts and displaying information about products and services. Technological evolution has boosted the transition from Web 1.0 toward Web 2.0; a new environment in which information is "many to many", allowing communication between active users (Berthon et al., 2012; Mangold and Faulds, 2009). Web 2.0 has been identified as a prerequisite for the development of crowdsourcing (Brabham, 2008). The development of Web 2.0 technology, the popularity of crowdsourcing and the global financial crisis, which has increased the cost of traditional financial resources, has been recognized as three main drivers of the current crowdfunding phenomenon (Dushnitsky et al., 2016; Lukkarinen et al., 2016). Crowdfunding is an umbrella term used to describe different forms of fundraising, alternatives to traditional ones (bank, venture capitalist, angel, accelerators). Under this umbrella is included equity crowdfunding that consists of an open call in which entrepreneurs sell a specified amount of equity or bond-like shares in a company on the Internet, hoping to attract a large group of investors. Due to the novelty of this interesting topic, the mechanisms and dynamics of equity crowdfunding are not yet well understood (Griffin, 2012). In particular, the preliminary work of the authors found that emerging literature is focused on the analysis of three dimensions: crowdfunding platforms (Dushnitsky et al., 2016; Giudici et al., 2013), crowdfunding investors (Cholakova and Clarysse, 2015; Hornuf and Neuenkirch, 2016) and crowdfunding campaigns (Ahlers et al., 2015; Lukkarinen et al., 2016). The aim of the work is to propose a new dimension of evaluation of the equity crowdfunding phenomenon, the organization. This paper use a dynamic capabilities view (Teece and Pisano, 1994) in order to understand if the characteristics of a crowdfunding campaign, and hence its probability of failure or success, are related to a firm's capabilities.

2 Design/methodology/approach

In order to reach our goal we chose to adopt a conceptual approach. Given that this paper tries for the first time to establish a link between the topics of dynamic capabilities and equity crowdfunding, this is the most suitable approach. Our methodology consisted then in a review of the two topics separately; after that, thanks to the extensive knowledge acquired we were able to build propositions assessing the importance of dynamic capabilities in enhancing the success rate of an equity crowdfunding campaign.

3 Dynamic capabilities as determinant of equity crowdfunding project success

3.1 Dimensions of dynamic capabilities underlying equity crowdfunding projects

In order to study the impact of dynamic capabilities on equity crowdfunding projects, it is useful to abstract from specific routines and processes and to consider broader composite dimensions. This paper distinguishes three dimensions: coordinating/integrating activities, learning and strategic competitive response processes. It is thought that these dimensions constitute distinct and significant drivers that lead the development of new configurations of functional competences. Coordination/integration capability describes the firm's ability to assess the value of existing resources and integrate them to shape new competences (Iansiti and Clark, 1994; Amit and Schoemaker, 1993). Teece et al. (1997) suggest that the lack of efficient coordination and combination of different resources and tasks may explain why slight technological changes have overwhelming effects on incumbent firms' competitive positions in a market. Learning capability can be conceived as a principal means of attaining strategic renewal. Renewal requires that organizations explore and learn new ways while at the same time exploit what they have already learned (March, 1991). Strategic competitive response capability is based on the extended definition of dynamic capabilities proposed by Eisenhardt and Martin (2000) that include the creation of market change as well as the response to exogenous change (Helfat, 2007). This capability can be conceptualized as the ability of the firm to scan the environment, identify new opportunities, assess its competitive position and respond to competitive strategic moves.

3.2 The role of sensing, seizing and transforming capabilities within an equity crowdfunding process

Strong dynamic capabilities allow firms to challenge competitors and competition in a changing environment. According to Teece, dynamic capabilities can be broken down into three categories: sensing, seizing and transforming and each one plays a different role in the recourse to equity crowdfunding. The sensing capabilities is important because they determine a firm's ability to understand the importance of equity crowdfunding as a new source of capital alternative to more traditional ones. Seizing capabilities enables firms to allocate the correct resource to an equity crowdfunding process; in this way the firm can efficiently manage the entire process without missing other interesting financial opportunities. With the transforming capabilities firms are able to renew their structure, their capabilities and their routines, in order to successfully manage the new equity crowdfunding process. The interaction of those capabilities helps firms to successfully manage the environmental changes, represented in this case by the introduction of a new source of capital.

4   Discussion and conclusion

The advent of the Internet, Web 2.0 technology and the financial crisis has led to a change in the environment in which firms compete. These three factors are recognized as important drivers also for the evolution of the financing cycle of firms, with the emergence of new available sources of capital (Dushnitsky et al., 2016; Lukkarinen et al., 2016). This paper focuses on one of these new sources, crowdfunding. This term is considered as an umbrella, under which is included equity crowdfunding (Belleflamme et al., 2014) that consists of an open call in which entrepreneurs sell a specified amount of equity or bond-like shares in a company on the Internet, hoping to attract a large group of investors. The purpose of this paper is to provide a new dimension of analysis in order to facilitate the understanding of the equity crowdfunding phenomenon for the firm. This dimension is added to the ones existing in the literature: crowdfunding platforms, crowdfunding investors, crowdfunding campaigns and it's analyzed under the theoretical lenses of dynamic capabilities. Dynamic capabilities theory in fact analyzes the ability of firms to maintain and sustain their competitive advantage by actively adapting to rapid environmental changes; when the environment evolves rapidly and unpredictably, firms can reach and maintain their competitive edge through continuous development of resources (Teece et al., 1997) and routines (Eisenhardt and Martin, 2000; Winter, 2003). This paper theoretically links the dynamic capabilities of firms with the implementation of a successful equity crowdfunding campaign and analyzes what is the role of different categories of dynamic capabilities in the implementation of such a campaign. Our findings are twofold:

1   The process of coordination, learning and strategic competitive responses are activities that facilitate change within an organization. These three processes are all important for the implementation of an equity crowdfunding process, since it is considered as an environmental change.

2   Sensing, seizing and transforming capabilities play a different role in the equity crowdfunding process. Sensing helps firms to understand the importance of this new source of capital, seizing enables firm to allocate resources correctly and transforming is crucial to renew the structure and make it fit with equity crowdfunding needs.

To the best of our knowledge this is the first paper that attempts to link equity crowdfunding with the theory of dynamic capabilities, arguing that crowdfunding can be considered as an environmental change and that possessing dynamic capabilities help firms to adapt to this change. This paper has certain limitations, due to the fact that is exploring a new topic and that there are no references on the interception between equity crowdfunding/dynamic capabilities. Future analysis must attempt to find empirical evidence of this link, by analyzing firms that successfully implement an equity crowdfunding campaign.

Notes

1 Scuola Superiore Sant'Anna.
2 Sapienza, University of Rome.

Bibliography

Ahlers, G. K., Cumming, D., Günther, C., & Schweizer, D. (2015). "Signaling in equity crowdfunding". *Entrepreneurship Theory and Practice*, 39(4), 955–980.

Amit, R., & Schoemaker, P. J. (1993). "Strategic assets and organizational rent". *Strategic Management Journal*, 14(1), 33–46.

Belleflamme, P., Lambert, T., & Schwienbacher, A. (2014). "Crowdfunding: Tapping the right crowd". *Journal of Business Venturing*, 29(5), 585–609.

Berthon, P. R., Pitt, L. F., Plangger, K., & Shapiro, D. (2012). "Marketing meets Web 2.0, social media, and creative consumers: Implications for international marketing strategy". *Business Horizons*, 55(3), 261–271.

Brabham, D. (2008). "Crowdsourcing as a model for problem solving an introduction and cases". *Convergence: The International Journal of Research into New Media Technologies*, 14(1), 75–90.

Cholakova, M., & Clarysse, B. (2015). "Does the possibility to make equity investments in crowdfunding projects crowd out reward-based investments?". *Entrepreneurship Theory and Practice*, 39(1), 145–172.

Dushnitsky, G., Guerini, M., Piva, E., & Rossi-Lamastra, C. (2016). "Crowdfunding in Europe". *California Management Review*, 58(2), 44–71.

Eisenhardt, K. M., & Martin, J. A. (2000). "Dynamic capabilities: What are they?". *Strategic Management Journal*, 21(10–11), 1105–1121.

Giudici, G., Guerini, M., & Rossi Lamastra, C. (2013). "Crowdfunding in Italy: State of the art and future prospects". *Economia e Politica Industriale-Journal of Industrial and Business Economics*, 40(4), 173–188.

Griffin, Z. J. (2012). "Crowdfunding: Fleecing the American masses". *Case W. Res. Journal of Law Technology & the Internet*, 4(2), 375.

Helfat, C. E. (2007). "Stylized facts, empirical research and theory development in management". *Strategic Organization*, 5(2), 185–192.

Hornuf, L., & Neuenkirch, M. (2016). "Pricing shares in equity crowdfunding". *Small Business Economics*, 1–17.

Iansiti, M., & Clark, K. B. (1994). "Integration and dynamic capability: Evidence from product development in automobiles and mainframe computers". *Industrial and Corporate Change*, 3(3), 557–605.

Lukkarinen, A., Teich, J. E., Wallenius, H., & Wallenius, J. (2016). "Success drivers of online equity crowdfunding campaigns". *Decision Support Systems*, 87, 26–38.

Mangold, W. G., & Faulds, D. J. (2009). "Social media: The new hybrid element of the promotion mix". *Business Horizons*, 52(4), 357–365.

Teece, D. J., & Pisano, G. (1994). "The dynamic capabilities of firms: An introduction". *Industrial and Corporate Change*, 3(3), 537–556.

Teece, D. J., Pisano, G., & Shuen, A. (1997). "Dynamic capabilities and strategic management". *Strategic Management Journal*, 509–533.

Winter, S. G. (2003). "Understanding dynamic capabilities". *Strategic Management Journal*, 24(10), 991–995.

Wirtz, B. W., Schilke, O., & Ullrich, S. (2010). "Strategic development of business models: Implications of the Web 2.0 for creating value on the internet". *Long Range Planning*, 43(2), 272–290.

The temporal acceleration of communication in organizations

The ICT workers, enablers or victims?

Solveig Beyza Evenstad,[1] *Sylvie Alemanno*[2] *and Nicolas Pélissier*

Keywords: *acceleration; temporality; time; communication; culture; stress; burn-out; ICT workers*

1 Introduction

Researchers in Information and Communication Sciences (ICS), particularly in the field of organizations, consider communication practices as 'temporalization practices' which implement social control through discourses of flexibility, adaptation, responsiveness and sustainability (Carayol, 2004). However, this comes with a cost. Pesqueux (2008) notes that the ability to rapidly change, responsiveness to the passing of time, the permanence of constant adaptability causes pain of all kinds: psychological, emotional, cognitive, structural and organizational.

2 Acceleration of life and businesses

An acceleration of the speed of life has effects on individuals' experience of time: it causes people to consider time as scarce, to feel under time pressure and stress. These feelings seem to have increased over recent decades because of the 'digital revolution' and the processes of globalization. Information technologies make it possible to exceed the boundaries in time and space. The technology also allows people to squeeze more and more different activities into a time unit and 'multi-tasking' is a new skill. Efficiency, profitability and competitiveness are the carriers of positive value in the business world. The speed is a marker of status and those who want to succeed must be fast and flexible. According to some recent publications (Aubert, 2003; Finchelstein, 2011), the workers and the contemporary society will be 'sick' of the constant sense of emergency and this 'infernal spiral' of acceleration.

This acceleration manifests itself in various forms in the business world. The first, short-term orientation or short-termism refers to an excessive focus on short-term results at the expense of long-term interests. An excessive short-term focus by some corporate leaders combined with insufficient regard for long-term strategy may jeopardize the long-term sustainability of the organization. Such short-term strategies are often based on accounting-driven metrics and profit maximization. The most prevalent accounting-driven metric is the earnings per share (EPS). The Value Based Management (VBM) model that has been a dominant philosophy will maximize shareholder value, which is EPS. In an effort to reduce costs and maximize profits, organizations are putting too much pressure on employees through work acceleration and intensification, which jeopardize the long-term organizational sustainability.

Secondly, the work processes have been accelerated during the last forty years, which resulted in work intensification. We have witnessed a fundamental transformation driven by the new management and ICT tools, which enable new business models. Competition, technological advancement and the increasing sophistication of consumers' needs have brought forward new competitive paradigms such as 'time-based competition' in the 1990s (Stalk & Hout, 1990). Time is to the 1990s what quality was to the 1980s. Time-based competition involves recognition and careful management of time as a limited resource while continuously removing waste or non-value-added activities. Managerial discourses that promote speed resulted in development of management *dispositifs* such as just-in-time (JIT), flexible manufacturing systems (FMS), the total quality management philosophy (TQM), enterprise resource planning (ERP), agile manufacturing (AM). These techniques aim at improving the efficiency of the production process by reducing lead time, lowering product cost, improving product quality and enhancing product innovation.

3   Acceleration in the ICT sector

ICT companies have adopted the agility paradigm: it is not the biggest that survives, but the fastest. They developed agile methodologies that borrow values and principles from their business counterpart. Rapid application development (RAD) has been described as a tool, methodology and an attitude (Martin, 1991). RAD espoused getting something – anything – into production rapidly and developers had to cut corners with the design and documentation, which are cornerstones of sustainable software development. Lean software development (LSD) that followed RAD is a translation of lean manufacturing and lean IT principles and practices to the software development domain. By adopting ITIL framework, the ICT service operations are optimized and accelerated. According to McGee, Khirallah, & Lodge (2000, p. 50),

> IT work has always been stressful. Projects have short deadlines and implementation schedules. IT operations and support require 24-by-7 availability. And every new IT paradigm shift brings with it the need for new skills. The volume and acceleration of IT-work in business is pushing more stressed-out IT workers towards burnout.

The acceleration in the ICT sector has profound consequences on the well-being of ICT workers.

4 Consequences

The consequences of the urgency of professional actions are deconstructing the communication and altering the individual. Rosa (2013) refers to the hamster wheel "the wheels of acceleration stay in place, but wheels don't always propel us forward: they can also spin around endlessly along their own axes without getting us anywhere . . . the hamster wheel is about to become the icon of late-modernity". Many researchers are bringing critical dimensions including technical accelerations at the cost of individual health (Chaudet, 2013; Heller, Huët, & Vidaillet, 2013; Sarrouy, Patrascu, & Lonneux, 2014).

Knowledge intensive industries such as Information and Communication Technology (ICT) have been going through profound changes because of globalization and disappearance of space and time constraints thanks to the internet technologies. The whole ICT industry in parallel with the business world they provide technological services to, operates with the 'do more with less' motto. As profit margins are squeezed, managers 'squeeze' the ICT workers to perform more and faster. High workload levels, long working hours and high demands for knowledge and results, and project work with tight deadlines add up to the stress of ICT workers. ICT workers who enable acceleration of the business world become victims of acceleration and suffer from stress and burnout.

A study employing interpretative phenomenological analysis comparing a group of ICT workers in France and Norway confirms this trend. Our participants have experienced acceleration as constant time pressure, intensified work processes, hyper connectivity, frequent organizational changes, short-term oriented managers and rapid pace of technological change. Consequences were stress, burnout, work-family conflicts. They felt alienated and dehumanized at work. There were some significant differences between French and Norwegian groups' experiences, revealing the intimate link between temporality and culture.

Notes

1 University of Nice Sophia Antipolis, Laboratoire i3M.
2 CNAM, Laboratoire DICEN-IDF.

Bibliography

Aubert, N. (2003). *Le Culte de l'urgence : La Société Malade du Temps.* Paris, Flammarion.
Carayol, V. (2004). *Communication organisationnelle: une perspective allagmatique.* Paris, L'Harmattan.
Chaudet, B. (2013). "Modernité et rationalisation: Usage de la théorie critique dans l'analyse d'une plateforme collaborative ", in Heller, T., Huet, R. et Vidaillet, B. (dirs.), *Communication et Organisation: Perspectives Critiques.* Presses du Septentrion.
Finchelstein, G. (2011). *La dictature de l'urgence.* Paris, Fayard.
Heller, T., Huët, R., & Vidaillet, B. (2013). " Introduction. Du sens et du rôle de la critique ", in V. d'Ascq (dir.), *Communication et Organisation: Perspectives Critiques*, Presses du Septentrion, 21–28.

Martin, J. (1991). *Rapid Application Development*. Indianapolis, Macmillan Publishing Co.

McGee, M. K., Khirallah, D. R., & Lodge, M. (2000). "Backlash," *Information Week*, *805*, 50–56.

Pesqueux, Y. (2008). "Un modèle organisationnel du changement?," *Communication et organisation*, *33*, 86–92.

Rosa, H. (2013). *Accélération. Une critique sociale du temps*. Paris, La Découverte.

Sarrouy, O., Patrascu, M., & Lonneux, C. (2014). "Dispositifs numériques mobiles et restructuration des espaces-temps professionnels," in Parrini-Alemanno, D. S. (dir.), *Communication organisationnelle, management et numérique*. Paris, France, L'Harmattan, 399–406.

Stalk, G., Jr., & Hout, T. M. (1990). *Competing against Time: How Time-Based Competition Is Reshaping Global Markets*. New York, Free Press.

Enactive management

A nurturing technology enabling fresh decision-making to cope with conflict situations

*Osvaldo García D,[1] Patrick Humphreys[2]
and Maria Soledad Saavedra U[3]*

Keywords: *enactive management; conflict management; interactions; CLEHES; Enactive Laboratory*

1 Introduction

The focus of this paper is observation, self-observation, and enactive management of organizational conflict situations whereby a community, an organization, or a human being has the possibility of recognizing their resources and generating changes in their practices if they so desire, and making fresh decisions, in the sense that different ontological dimensions are involved. These are Body-Language-Emotions-History-Eros-Silence which configure a nurturing technology called CLEHES©. This tool has been applied for diverse people, groups, communities, and organizations that need and wish to develop their own skills to inquire conflict practice resolutions, in order to learn as a human decision support system. Conflict situations are understood as interactions, a breakdown in-between CLEHES from the individual or social standpoints.

This tool allows observing the boundaries of conflict situations and building an observer system with the ability to manage, solve, or attenuate the situation, enabling fresh decision-making attending to the context in which the organization moves. This learning process happens in a special place called Enactive Laboratory where strategies are developed to cope with the domains and context in the perceived individual and human activities systems. We present a case study focusing on a National Family Mediation System.
[[We have renumbered below headings as per standard treatment]]

2 The approach and the ontological tool

We postulate that the rupture, breakage, or conflict situations within the human activity system that configures an organization are unprecedented spaces that open states of possibilities by questioning and opening the distinction schemes with which one decides and acts, i.e., it questions the human and organizational cognitive blindness and silence spaces. In these terms, the rupture of the interaction choreographies constitutes a unique source to design, create, and re-create *that* reality (Garcia and Saavedra, 2014). This leads to the discovery or emergence of new modes of interaction that can reconfigure the system's identity. Through this voyage between enactor and context, possibility states emerge, (Maturana, 2006; Maturana and Varela, 1984) and, more than being captured intellectually, they configure a creative emotional growth and learning experience of the system. While the language of rationality, of scientific objectivity, has hidden the human being in its existential and ontological condition and in its complex contradiction, our perspective proposes to recover and look through the human beings that we are, recognizing as a principle that *being* a manager *is a mode of being a human being*.

A conflict situation exposes and brings up all the ontic complexity that we observe in human beings: their contradictions and their limits. If we take the place of the manager, the question that arises is how to take actions in rupture or conflict situations that ensure the system's viability and learning from it.

We postulate that decision support systems are effective only if they are configured freshly *within* human activity systems, as knowledge that is embodied in the human beings that constitute the system, that arises from a recursive process of observation and self-observation, and that this self-reference process occurs when the distinctions horizon is expanded within those who are in charge and are ethically responsible for the organizational care.

We conceive CLEHES as a nurturing technology in terms of the recursive and recurrent learning that this tool generates in the diverse experiences and contexts where it has been applied. Upon unveiling human beings ontologically, affective historical structural drift arises as the greatest complexity to be observed and managed: a matter that amazes, moves and challenges our own learning. It is a technology because it treats the body as a device that instigates the carrying out of observations and, with that, it changes the conversations: ergo, it changes the action.

The CLEHES technology exhibits two meanings: (i) as a way of ontologically understanding human beings from the six dimensions that it comprises and unveiling the observer that has been constituted along with the learning, and (ii) as a technology that allows *moving* the observer, generating different tools for self-observation and for the observation of others.

3   National Family Mediation System

The case that we present is the design of the National Family Mediation System, an educational program that incorporated the CLEHES laboratory to promote conflict-resolving practices in agreement with new demands generated by this integrated social device in the country.

The aesthetic conversational movement that arises from the laboratory experience as an expression of the re-designing of conflict-resolving practices and innovation in decision-making is seen as a learning of the system oriented at opening its possibilities and declaring its limits and resistances. This reconfiguration emerges from the perturbation generated by the CLEHES tool in the ways of distinguishing the conflict situations and the associated meanings (individually and collectively), and by daring to enact emergent situations and come to fresh decisions by moving the distinctions schemes.

Notes

1 Universidad de Santiago de Chile.
2 The London School of Economics and Political Sciences.
3 Universidad de Santiago de Chile.

Bibliography

Argyris, C. (2012). *Organizational Traps: Leadership, Culture, and Organizational Design*, Oxford, England: Oxford University Press.

Bateson, G. (1972). *Steps to an Ecology Mind*, Londres, Inglaterra: Ballantine Books.

Bauman, Z. (2010). *El arte de la vida. De la vida como obra de arte*, Buenos Aires, Argentina: Paidós.

Bauman, Z. (2013). *Sobre la educación en un mundo líquido*, Buenos Aires, Argentina: Paidós.

Colombetti, G. (2009). Enaction: Towards a new paradigm for cognitive science, In *Enaction, Sense-Making and Emotion*, Cambridge, MA: MIT Press.

Espejo, R. and Harden, R. (1989). *The Viable System Model: Interpretations and Applications of Stafford Beer's VSM*, Great Britain: Wiley.

Espejo, R. and Reyes, A. (2016). *Sistemas Organizacionales. El Manejo de la Complejidad con el Modelo del Sistema Viable*, Bogotá, Ibagué, Colombia: Uniandes.

Froese, T. and Di Paolo, E. (2012). The enactive approach: Theoretical sketches from cel to society, *Pragmatics and Cognition*, Vol. 19, pp. 1–36.

García, O. (2009). Human re-engineering for action: An enactive educational management program, *Kybernetes*, Vol. 38, No. 7/8, pp. 1329–1340.

Garcia, O. (2017). O Technology. Working Paper. Universidad de Santiago de Chile.

García, O. and Laulié, L. (2010). The CLEHES-MOOD: An enactive technology towardeffective and collaborative action, *System Research and Behavioral Science*, Vol. 27, pp. 319–335.

Garcia, O. and Orellana, R. (1997). Decision support systems: Structural, conversationaland emotional adjustments: Breaking and taking of organizational care. In P. Humphreys and S. Ayestaran (eds.), *Decision Support in Organizational Transformation*, London, England: Chapman & Hall.

Garcia, O. and Mendoza, C. (2011). Enactive management: Application of CLEHES and VIPLAN, *Kybernetes*, Vol. 40, pp. 439–453.

Garcia, O. and Orellana, R. (2008). Metasystemic reengineering: An organizational intervention, In F. Adam and P. C. Humphreys (eds.), *Encyclopedia of Decision Making and Decision Support Technologies*, Vol. 2, Hershey, MA: IGI Global.

Garcia, O. and Saavedra, M. (2006). Self management: An innovative tool for enactive human design. In F. Adam, P. Brezillon, S. Carlsson, and P. Humphreys (eds.), *Creativity and Innovation in Decision Making and Decision Support*, London, England: Decision Support Press (pp. 195–214).

Garcia, O. and Saavedra, M. (2014). Enactive management: Dancing with uncertainty and complexity, *Kybernetes*, Vol. 43, No. 8.

Humphreys, P. and Jones, G. (2006). The evolution of group decision support systems to enable collaborative authoring of outcomes, *World Futures: Journal of General Evolution*, Vol. 62, pp. 193–222.

Jones, A. and Issroff, K. (2005). Learning technologies: Affective and social issues in computer supported collaborative learning, *Computer & Education*, Vol. 44, pp. 395–408.

Le Breton, D. (2009). *El sabor del mundo. Una antropología de los sentidos*, Buenos Aires, Argentina: Nueva Visión.

Maturana, H. (2006). Self-consciousness: How? When? Where?, *Constructivist Foundations*, Vol. 1, No. 3, pp. 91–102.

Maturana, H. and Bernhard, P. (2004). *From Being to Doing: The Origins of Biology of Cognition*, Phoenix, USA: Zeig, Tucker & Theisen, Inc.

Maturana, H. and Varela, F. (1984). *El árbol del conocimiento*, Santiago de Chile: Universitaria.

Noe, A. (2004). *Action in Perception*, Cambridge, MA: MIT Press.

Scharmer, O. (2009). *Theory U: Leading from the Future as It Emerges*, San Francisco, USA: Berrett-Kochler Publisher, Inc.

Schwager, V. (2010). The Role of Play in Enhancing Decision-Making in Innovation Creativiy Environments. PhD Thesis, London, England: London School of Economics and Political Science.

Senge, P., Scharmer, O., Jaworski, I. and Flowers, B. (2005). *Presence: Exploring Profound Change in People, Organizations and Society*, London, England: Nicolas Brealy Publishing.

Stewart, J., Gapenne, O. and Di Paolo, E. (2010). *Enaction: Toward a New Paradigm for Cognitive Science*. 1st Ed, Cambridge, MA: MIT Press, pp. 267–387.

Varela, F. (1988). *Connaître: Les Sciences Cognitives, Tendences et Perspectives*, Paris, France: Editions du Seuil.

Varela, F. (2000). *El Fenómeno de la Vida*, Santiago, Chile: Dolmen.

Varela, F., Thompson, E., and Rosch, E. (1991). *The Embodied Mind: Cognitive Science and Human Experience*, Cambridge, MA: MIT Press.

Group dynamics and systems thinking

Interdisciplinary roots, metaphors and applications

Sergio Barile,[1] Xhimi Hysa,[2] Mario Calabrese[3] and Laura Riolli[4]

Keywords: *group dynamics; systems thinking; Tavistock Institute; Mental Research Institute*

The term "group dynamics" was coined and popularized for the first time by Kurt Lewin in the 1930s with the scope to describe the way groups and individuals act and react to changing circumstances (Lewin, 1936). Fundamentally, the dynamics of a group conceptually derives from the continuous interaction (resonance) between its members. For Lewin, the principle of interactionism in his field theory is expressed by the formula: B = f(P, E) which means that the behavior (B) of an individual (i.e. group member) is a function (f) of the interaction between personal attributes (P) and environmental factors (E) (Lewin, 1951). Said with Lewin's words: "every psychological event depends upon the state of the person and at the same time on the environment, although their relative importance is different in different cases" (Lewin, 1936, p. 12).

Even though Lewin is recognized by the scientific community as the founder of group dynamics both as a subject matter and a scientific discipline of study, other predecessors have written about the topic. In the late 1800s and in the early 1900s various disciplines were concerned about the behavior of individuals within small or large groups.

Therefore, the first part of this work deals with the interdisciplinary roots of group dynamics that can be summarized in Table 25.1.

Next, the second part offers a metaphorical perspective of groups by using some organizational metaphors typical of the Viable Systems Approach. Thus, shifting the focus on the behavior within organizations, other interesting perspectives (with a common denominator on systems theory) are to be taken into consideration. According to Golinelli (2010, pp. 27–35), a firm can be seen as a mechanical, organic, cybernetic, autopoietic, cognitive (including the emotional dimension), and viable system. These metaphors and analogies are important for understanding the dynamics of groups within business firms and other organizations. In addition, only through metaphors and analogies the science makes progress and creates new paradigms, taking always into account

Table 25.1 Group analysis from different viewpoints

Disciplines and branches	Main contributors	Relevant topics
Social Psychology	Kurt Lewin	Group Dynamics
Psychoanalysis	Sigmund Freud	Identification
Social Psychology	Henri Tajfel	Social Identity Theory
Physiological Psychology	Ivan Pavlov	Classical Conditioning
Sociology and Social Psychology	Gustave Le Bon	Psychology of Crowds
Sociology	Emile Durkheim	Collective Consciousness
Political Philosophy	Jean-Jacques Rousseau	Social Contract
Humanistic Psychology	Abraham Maslow	Hierarchy of Human Needs
Organizational Psychology	Elton Mayo	Hawthorne Effect
Organizational Psychology	Frederick Herzberg	Two-factor Theory
Organizational Psychology	Douglas McGregor	Theory X & Y
Organizational Theory	Gareth Morgan	Metaphors of Organization
Organizational Theory	C. Argyris and D. Schon	Organizational Learning
Organizational Theory	Peter Senge	Learning Organization
Management	Peter Drucker	Knowledge Worker
Knowledge Management	Michael Polanyi	Tacit Knowing
Knowledge Management	I. Nonaka and H. Takeuchi	Knowledge Conversion Model
Knowledge Management	Karl Weick	Sense-making
Psychology	Daniel Goleman	Emotional Intelligence
Mathematics	Norbert Wiener	Cybernetics
Cybernetics and Systems Theory	William Ross Ashby	Law of Requisite Variety
Living Systems Theory	H. Maturana and F. Varela	Autopoietic Systems

Source: Authors' elaboration

the limits of an exaggerated vocabulary composed by metaphors and analogies (Golinelli, 2010).

The last part of this paper deals directly with groups in systems thinking by underlying, first, the system's properties of groups and their dynamics, and second by exploring the applicative contexts. In reference to the applications, two main directions are followed: the socio-technical perspective and the socio-psychological perspective. The first one is represented by the research handled near the Tavistock Institute of Human Relations: "Coal Mining Studies" (Trist and Bamforth, 1951); "Indian Textile Mills Studies"; "Socio-Technical Approach" (Emery and Trist, 1960). For the socio-psychological view (Trist and Murray, 1990), it is relevant to mention the studies of Tavistock Clinic with Bowlby's Attachment theory (Bowlby, 1958, 1959, 1960) and Mental Research Institute of Palo Alto with the Bateson Project (Bateson et al., 1963).

From the methodological standpoint, the present research type is a conceptual research based on the interpretivist paradigm. From the ontological viewpoint, this research relies on constructivism and relativism, emphasizing the role of the observer. From the epistemological standpoint, the present research focuses on non-dualism, subjectivism, holism, quest of the possible. From the methodological perspective, the focus is on constructivism and constructed realities. In summary, this study uses the qualitative methodology and the methods of literature review and theory development.

1   According to Barile and Iannuzzi (2006, p. 50), there is a difference between a metaphor and an analogy. A metaphor enables, using the simulation of a concept through a specific word, expressing a defined experience referring to another.An analogy goes further: it aims to extend the knowledge background of a particular phenomenon or entity and the behavioral properties of that phenomenon/entity to another one which seems to be "similar". For example, seeing a group as a brain is a metaphor, instead seeing it as a cognitive system is an analogy because it explains how the brain works. So, a metaphor is a structural concept; it is static (e.g. a photo). At the other hand an analogy is a systemic concept, it is dynamic (e.g. a video); one expresses the anatomy and the other the physiology of the phenomenon.
2   For a deeper analysis on the origin of group dynamics see: Cartwright, D., Zander, A. (Eds.). (1968). *Group dynamics: Research and theory*, 3rd Ed. New York: Harper & Row; Cartwright, D., Zander, A. (2000). "Origins of Group Dynamics". Group Facilitation: A Research and Applications Journal, 2, pp. 40–55.

Notes

1 Sapienza University of Rome.
2 Epoka University.
3 Sapienza University of Rome.
4 California State University, Sacramento.

Bibliography

Argyris, C. (1964). *Integrating the Individual and the Organization.* New York: Wiley.
Argyris, C., & Schön, D.A. (1974). *Theory in Practice: Increasing Professional Effectiveness.* San Francisco: Jossey-Bass.
Barile S., & Iannuzzi E. (2006). L'impresa rappresentata come un sistema. *L'impresa come sistema. Contributi sull'approccio Sistemico Vitale*, pp. 45–61. Torino: Giappichelli.
Barile, S. (2008). *L'impresa come sistema: Contributi sull'Approccio Sistemico Vitale*, 2nd Ed. Torino: Giappichelli.
Barile, S. (2009a). *Management Sistemico Vitale: Decidere in contesti complessi.* Torino: Giappichelli.

Barile, S. (2009b). "Verso la qualificazione del concetto di complessità sistemica". *Sinergie*, 79, pp. 47–76.

Barile, S. (2011). *Management Sistemico Vitale: Decisioni e scelte in ambito complesso.* Avellino: International Printing Editore.

Barile, S., Bassano, C., Calabrese, M., Confetto, M., Di Nauta, P., Piciocchi, P., Polese, F., Saviano, M., Siano, A., Siglioccolo, M., & Vollero, A. (2011). *Contributions to Theoretical and Practical Advances in Management: A Viable Systems Approach (VSA).* Avellino: International Printing Editore.

Barnard, C. (1938). *The Functions of the Executive.* Cambridge, MA: Harvard University Press.

Bateson, G. (2008). *Verso un'Ecologia della Mente*, 25th Ed. Roma: Adelphi.

Bateson, G., Jackson, D., Haley, J., & Weakland, J. (1963). "A note on the double bind: 1962". *Family Process*, 2 (1), pp. 154–161.

Bowen, M. (1985). *Family Therapy in Clinical Practice.* New York: Jason Aronson.

Bowlby, J. (1958). "The nature of the child's tie to his mother". *International Journal of Psycho-Analysis*, 39, pp. 1–23.

Bowlby, J. (1959). "Separation anxiety". *International Journal of Psycho-Analysis*, 41, pp. 1–25.

Bowlby, J. (1960). "Grief and mourning in infancy and early childhood". *The Psychoanalytic Study of the Child*, 15, pp. 9–52.

Bresnahan, C., & Mitroff, I. (2007). "Leadership and attachment theory". *American Psychologist*, 62 (6), pp. 607–608.

Cartwright, D., & Zander, A. (Eds.). (1968). *Group Dynamics: Research and Theory*, 3rd Ed. New York: Harper & Row.

Cartwright, D., & Zander, A. (2000). "Origins of group dynamics". *Group Facilitation: A Research and Applications Journal*, (2), pp. 40–55.

Carver, C., & Scheier, M. (2002). "Optimism". In: C.R. Snyder & S. Lopez (Eds.), *Handbook of Positive Psychology*, pp. 231–243. Oxford, UK: Oxford University Press.

Connors, J., & Caple, R. (2005). "A review of group systems theory". *Journal for Specialists in Group Work*, 30 (2), pp. 93–110.

Davenport, T., & Prusak, L. (1998). *Working Knowledge: How Organizations Manage What They Know.* Boston: Harvard Business School Press.

Drucker, P.F. (2002). *Il management, l'Individuo, la Società.* Milano: FrancoAngeli.

Durkheim, É. (1997). *The Division of Labor in Society.* New York: Free Press.

Emery, F.E., & Trist, E.L. (1960). "Socio-technical systems". In: C.W. Churchman & M. Verhurst (Eds.), *Management Science, Models and Techniques*, Vol. 2, pp. 83–97. London: Pergamon Press.

Emery, F.E., & Trist, E.L. (1965). "The causal texture of organizational environments". *Human Relations*, 18 (1), pp. 21–32.

Freud, Z. (1922). *Group Psychology and the Analysis of the Ego.* New York: Boni and Liveright.

Golinelli, G.M. (2010). *Viable Systems Approach (VSA): Governing Business Dynamics.* Padova: Kluwer (Cedam).

Haley, J. (1973). *Uncommon Therapy: The Psychiatric Techniques of Milton H. Erickson MD.* New York: Norton & Company.

Härtel, Ch.E.J., Zerbe, W.J., & Ashkanasy, N.M. (Eds.). (2005). *Emotions in Organizational Behavior.* Mahwah, NJ: Lawrence Erlbaum Associates.

Le Bon, G. (2001). *The Crowd: A Study of the Popular Mind.* Kitchener: Batoche Books.

Lewin, K. (1936). *A Dynamic Theory of Personality.* New York: McGraw-Hill.

Lewin, K. (1951). *Field Theory in Social Science: Selected Theoretical Papers.* New York: Harper & Row.

Maslow, A. (1954). *Motivation and Personality.* New York: McGraw-Hill.

Minuchin, S. (1974). *Families and Family Therapy.* Cambridge: Harvard University Press.

Polanyi, M. (1966). *The Tacit Dimension.* London: Routledge.

Satir, V. (1983). *Conjoint Family Therapy.* Palo Alto, CA: Science and Behavior Books.

Selvini, M. (Ed.). (1988). *The Work of Mara Selvini Palazzoli.* New York: Jason Aronson.

Tajfel, H. (Ed.). (1978). *Differentiation Between Social Groups: Studies in the Social Psychology of Intergroup Relations.* London: Academic Press.

Tajfel, H., & Turner, J.C. (1979). "An integrative theory of intergroup conflict". In: W.G. Austin & S. Worchel (Eds.), *The Social Psychology of Intergroup Relations,* pp. 33–47. Monterey, CA: Brooks/Cole.

Trist, E.L., & Bamforth, K. (1951). "Some social and psychological consequences of the Longwall method of coal-getting". *Human Relations,* 4, pp. 3–38.

Trist, E.L., & Murray, H. (1990). *The Social Engagement of Social Science: A Tavistock Anthology, Volume 1: The Socio-Psychological Perspective.* Philadelphia: University of Pennsylvania Press.

von Bertalanffy, L. (1950). "The theory of open systems in physics and biology". In: F.E. Emery (Ed.). *Systems Thinking,* pp. 70–85. Harmondsworth: Penguin.

Watzkawick, P., Weakland, J., & Fisch, R. (1974). *Change: Sulla Formazione e la Soluzione dei Problemi.* Roma: Astrolabio.

In search of a possible evolutionary principle of management theory and practice

Sergio Barile,[1] Marialuisa Saviano,[2]
Mario Calabrese[3] and Antonio La Sala[4]

Keywords: *methods; tools; techniques; management evolution; systems thinking*

1   Introduction and aims

Starting from the view of economy as the field in which the sustainable dynamics of a society are expressed (Barile et al., 2013, 2014; Espejo, 2014), the aim of this work is to investigate the hidden and sensitive relation which links together the evolution of the managerial theories and that of production systems, integrating an empirical and a theoretical approach.

The main question of research is summarized as follows:

> *Is there (and if there is, is it possible to discover and isolate) a general low through which explains the conjunct evolution of the managerial theories and the production systems?*

The work applies the conceptual framework of the *Viable Systems Approach* (*VSA*) (Barile, 2000, 2009; Golinelli, 2000, 2010; Barile et al., 2012; Basile and Caputo, 2017; Calabrese et al., 2017; Tronvoll et al., 2017) merging a theoretical and a practical perspective. In particular, two studies are mentioned in order to create the conceptual context from which to start the analysis: Simone et al. (2014) focus on the 'epistemological element' which relies under and connotes the birth and the development of an economic theory (from merchandise and industrial and commercial technique to management); Massaroni et al. (2014) give to their analysis an empirical connotation, focusing on the evolution of the industrial models of production with reference to some key sectors of the economy, underlining the passage from craft production and mass production to modular production. The work aims to make interpretative hypothesis of the evolutionary dynamics theorized and observed by discussing them conceptually and highlighting theoretical and practical implications: from these studies, it seems to emerge a cause-effect relation, chained to the firm-environment dynamic, which connotes the development of productive models and the evolution of managerial theories.

2 Discussion

Our findings lead us to the formulation of a possible evolutionary principle that, taking together the area of business economics and the productive systems, is able to highlight an autopoietic process (Maturana and Varela, 1980). This autopoietic process starts from the creation of a useful tool to solve a specific problem, leads first to the definition of a technical approach to manage categories of similar problems, and then to the generalization of a method, useful in analogous problematic situations (Figure 26.1).

Therefore, the tool-technique-method dynamic is a potential evolutionary path: it has no possibility to start if the focus remains on the traditional and dominant concept of technical and technological knowledge as unidirectional and vertically specialized. This evidence offers solutions to a wide variety of specific problems. Method is the capability to look through efficiency, focusing on the fundamental dynamics of an effective and at the same time sustainable development, realizing a relevant integration among the social, economic and environmental context (Barile et al., 2014; Caputo, 2016a, 2017). The study discovers the dialectical increase in information and in knowledge (cognitive variety), stressing its non-linearity and identifying a way to manage its intrinsic complexity (Di Nauta et al., 2015; Formisano et al., 2015; Del Giudice et al., 2017).

The transition from the pair tool-technique to method can be associated to the transition from explicit to tacit knowledge (Nonaka and Takuchi, 1995): in particular, while techniques and tools represent a typically explicit knowledge,

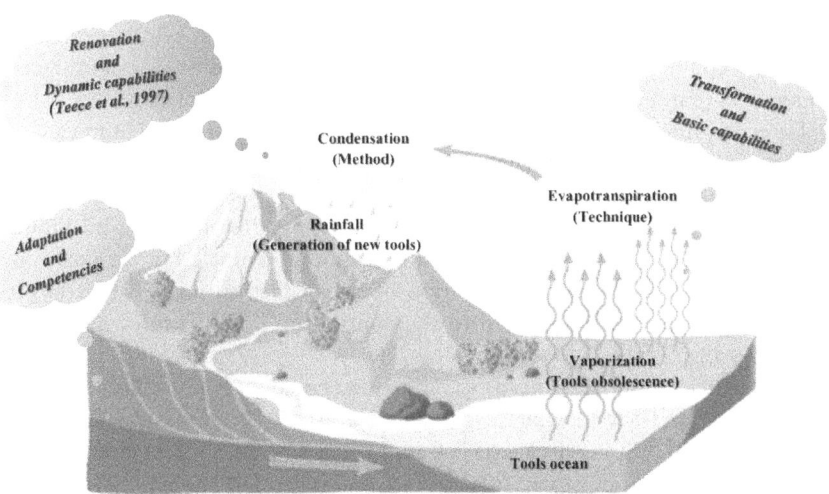

Figure 26.1 The cycle of knowledge tool-technique-method
Source: Our elaboration – www.asvsa.org

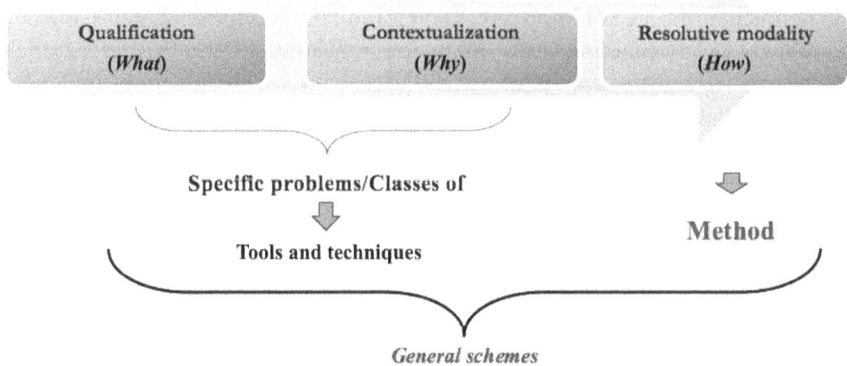

Figure 26.2 The general scheme of decision process
Source: Our elaboration – www.asvsa.org

easily transferred through a process of formation, a method qualifies a typically implicit knowledge, tacit and difficult to encode. In this perspective, it can be said that when it is possible to encode a method, it becomes a technique; when the set of procedures and routines (technique) are standardized, by establishing protocols, the conditions for the creation of a tool. Applications of the conceptual framework developed could be addressed to the deeper understanding of the pair general schema and specific schema, also in relation to other managerial fields (Figure 26.2).

Studies could be addressed to analysis of the behavioural theories in the economic field (e.g. motivation, identity, control, individual performance).

Three studies as well as of the production systems, useful in defining a conscious and responsible management, which acts and takes decisions in order to improve, with a unique action of government, the efficiency, the effectiveness and the sustainability of organizations. The process described can be applied to the analysis of the exchange dynamics among knowledge systems, and can be thought as a new interpretation of the dialectical process that has its antithesis in the problems linked to the continuous emergence of new needs (Table 26.1).

The 'vertical' knowledge needs to be supplemented with a 'horizontal knowledge' (Barile and Saviano, 2013; Macaulay et al., 2010; Prahalad and Hamel, 2006; Caputo and Evangelista, 2017; Caputo et al., 2017). This integration shifts the focus from the endowment of competencies to the development of 'dynamic' capabilities (Teece et al., 1997), required to co-evolve in harmony with a changing environment.

Table 26.1 Resolution approaches: an instrumental, technical and methodological perspective

Resolution approach	Level of interaction	Applicative characteristics	Learning dynamics
Instrumental	Just do	Elasticity	Instruction/Training: Practical abilities
Technical	Understand and do	Flexibility	Education/Formation: Competencies
Methodological	Understand, develop and do	Fluidity	Sharing/Involvement: Capabilities

Source: Our elaboration – www.asvsa.org

3   Conclusions

The study provides an explanation of the evolutionary dynamics under investigation and is directed towards the identification of a general evolutionary principle that is able, according to the *VSA*, to take together concepts like sustainability, value, dynamic capabilities, decision making and others (Polese et al., 2016; Saviano et al., 2017).

It represents one other step ahead in the promotion of a new paradigm that seems now to be at stake.

Notes

1 Sapienza, University of Rome.
2 University of Salerno (*corresponding author*) – msaviano@unisa.it.
3 Sapienza, University of Rome.
4 University of Salerno.

Bibliography

Adler, P.S., Forbes, L.C., Willmott, H. (2007). Critical Management Studies. *The Academy of Management Annals*, 1(1), 119–179.
Argyris, C., Schön, D. (1978). *Organizational learning: A Theory of action perspective*. Addison & Wesley, Reading, MA and Palo Alto, CA.
Barile, S. (2000). *L'impresa come sistema: contributi sull'approccio sistemico vitale (ASV)*. Giappichelli, Torino.
Barile, S. (2009). *Management Sistemico Vitale*. Giappichelli, Torino.
Barile, S., Calabrese, M., Polese, F., Saviano, M. (2013). Il Governo dei Sistemi Complessi Tra Competenze Attuali e Capacità Potenziali. In: Barile, S., Eletti, V., Matteuzzi, M. (a cura di), *Decisioni e scelte in contesti complessi*. Cedam-Kluwer, Lavis.

118 *Sergio Barile et al.*

Barile, S., Franco, G., Nota, G., Saviano, M. (2012). Structure and Dynamics of a 'T-Shaped' Knowledge: From Individuals to Cooperating Communities of Practice. *Service Science, Informs*, 4(2), 161–180.

Barile, S., Saviano, M. (2011). Foundations of Systems Thinking: The Structure-System Paradigm. In: Barile, S., Bassano, C., Calabrese, M., Confetto, M.G., Di Nauta, P., Piciocchi, P., Polese, F., Saviano, M., Siano, A., Siglioccolo, M., Vollero, A. *Contributions to theoretical and practical advances in management: A Viable Systems Approach (VSA)* (pp. 1–25). International Printing Srl Editore, Avellino.

Barile, S., Saviano, M. (2013). Dynamic Capabilities and T-Shaped Knowledge: A Viable Systems Approach. In: Barile, S. (Ed.), *Contributions to theoretical and practical advances in management: A Viable Systems Approach (VSA)* (vol. 2). Aracne, Roma.

Barile, S., Saviano, M. (2017). Complexity and Sustainability in Management: Insights from a Systems Perspective. In Barile, S., Pellicano, M., Polese, F. (Eds.), *Social dynamics in a systems perspective*, New Economic Window Book Series. Springer International.

Barile, S., Saviano, M., Iandolo, F., Calabrese, M. (2014). The Viable Systems Approach and Its Contribution to the Analysis of Sustainable Business Behaviors. *Systems Research and Behavioral Science*, 31(6), 683–695.

Basile, G., Caputo, F. (2017). Theories and Challenges for Systems Thinking in Practice. *Journal of Organisational Transformation & Social Change*, 14(1), 1–3.

Bateson, G. (1972). *Steps to an ecology of mind: Collected essays in anthropology, psychiatry, evolution, and epistemology*. Chicago: University of Chicago Press.

Becker, H.S. (1958). Problems of Inference and Proof in Participant Observation. *American Sociological Review*, 20, 652–660.

Calabrese, M., Iandolo, F., Caputo, F., Sarno, D. (2017). From Mechanical to Cognitive View: The Changes of Decision Making in Business Environment. In: Barile, S., Pellicano, M., Polese, F. (Eds.), *Social dynamics in a system perspective* (pp. 223–240). Springer, New York.

Capra, F. (1997). *La rete della vita. Una nuova visione della natura e della scienza*. Rizzoli, Milano.

Caputo, F. (2016a). A Focus on Company-Stakeholder Relationships in the Light of the Stakeholder Engagement Framework. In: Vrontis, D., Weber, Y., Tsoukatos, E. (Eds.), *Innovation, entrepreneurship and digital ecosystems* (pp. 455–470). EuroMed Press, Cipro.

Caputo, F. (Ed.) (2016b). *Governing business systems: Theories and challenges for systems thinking in practice 4th business systems laboratory international symposium* (pp. 124–129). Business Systems Laboratory, Avellino.

Caputo, F. (2017). Reflecting Upon Knowledge Management Studies: Insights from Systems Thinking. *International Journal of Knowledge Management Studies*, 8(3/4), 177–190.

Caputo, F., Evangelista, F. (2017). Information Sharing and Cognitive Involvement for Sustainable Workplaces. In: Leon, R.D. (Ed.), *Managerial strategies for business sustainability during turbulent times* (pp. 122–139). IGI Global, New York.

Caputo, F., Evangelista, F., Perko, I., Russo, G. (2017). The Role of Big Data in Value Co-Creation for the Knowledge Economy. In: Vrontis, S., Weber, T., Tsoukatos, E. (Eds.), *Global and national business theories and practice: Bridging the past with the future* (pp. 269–280). EuroMed Press, Cyprus.

Caputo, F., Evangelista, F., Russo, G., Buhnova, B. (2017). A Systems View of Companies' Communication in Online Social Environment. *Journal of Organizational*

Transformation & Social Change. http://dx.doi.org/10.1080/14779633.2017. 1291144.

Caputo, F., Perano, M., Mamuti, A. (2017). A Macro-Level View of Tourism Sector: Between Smartness and Sustainability. *Enlightening Tourism: A Pathmaking Journal*, 7(1), 36–61.

Del Giudice, M., Ahmad, A., Scuotto, V., Caputo, F. (2017). Influences of Cognitive Dimensions on the Collaborative Entry Mode Choice of Small and Medium-Sized Enterprises. *International Marketing Review*, 34(5), 652–673.

Del Giudice, M., Scuotto, V., Khan, Z., Caputo, F., Carayannis, E. (2017). Micro Level Actions by Owners-Managers in Improving Sustainability Practices of Culture and Creative Small and Medium Enterprises: UK-Italy Comparison. *Journal of Organizational Behaviour*, 1–19. DOI: 10.1002/job.2237.

Di Nauta, P., Merola, B., Caputo, F., Evangelista, F. (2015). Reflections on the Role of University to Face the Challenges of Knowledge Society for the Local Economic Development. *Journal of Knowledge Economy*, 1–19.

Espejo, R. (2014). Organizational Transparency, Inclusion and Sustainability. *Refereed Proceedings of the Business Systems Laboratory, 2nd International Symposium*, January 23–24, Universitas Mercatorum, Rome (Italy).

Espejo, R., Schuhmann, W., Schwaninger, M., Bilello, U. (1996). *Organizational Transformation and Learning: A Cybernetic Approach to Management.* John Wiley & Sons, Chichester.

Formisano, V., Caputo, F., D'amore, R. (2015). Tratti evolutivi della società della conoscenza: il contributo degli studi sulle reti nella prospettiva sistemica. *Esperienze d'impresa*, 2, 73–94.

Foss, N. (2005). *Strategy, economic organization and the knowledge economy.* Oxford University Press, Oxford.

Fournier, V., Grey, C. (2000). At the Critical Moment: Conditions and Prospects for Critical Management Studies. *Human Relations January*, 53(1), 7–32.

Golinelli, G. M. (2000). *L'approccio sistemico al governo dell'impresa* (Vol. 1). Padova, Italy: Cedam.

Golinelli, G.M. (2010). *Viable systems approach: Governing business dynamics.* Cedam Kluwer, Padova.

Golinelli, G.M., Barile, S., Saviano, M., Farioli, F., Masaru, Y. (2015). Towards a common framework for knowledge co-creation: Opportunities of collaboration between Service Science and Sustainability Science. In: Gummesson, E., Mele, C., Polese, F. (Eds.), *Service dominant logic, network and systems theory and service science: Integrating three perspectives for a new service agenda.* Giannini, Napoli.

Hansen, T., Von oetinger, B. (2001). Introducing "T-shaped" Managers: Knowledge Management's Next Generation. *Harvard Business Review*, 79(3), 106–116.

Ison, R., Schlindwein, S.L. (2015). Navigating through an "Ecological Desert and a Sociological Hell": A Cyber-Systemic Governance Approach for the Anthropocene. *Kybernetes*, 44(6/7), 891–902.

Kriz, J. (1988). *Facts and artefacts in social science: An epistemological and methodological analysis of empirical social science research techniques.* McGraw-Hill Book Company, New York.

Macaulay, L., Moxham, C., Jones, B., Miles, I. (2010). Innovation and Skills: Future Service Science Education. In: Maglio, P., Kieliszeski, C.A., Spohrer, J.C. (Eds.), *Handbook of service science: Research and Innovation in the service economy.* Springer, New York.

Massaroni, E., De Falco, S., Sancetta, G., Cozzolino, A., Bilotta, A., Carrubbo, L. (2014). Alla ricerca di un possibile principio evolutivo della teoria e della pratica d'impresa. Dinamiche di sviluppo delle modalità di produzione industrial. *XXVI Convegno annuale di Sinergie, Manifattura: quale futuro?*, novembre 13–14, Università di Cassino e del Lazio Meridionale.

Maturana, H., Varela, F. (1980). Autopoiesis and Cognition. *Boston Studies in Philosophical Science*, 42.

Mele, C., Polese, F. (2011). Key Dimensions of Service Systems in Value-Creating Networks. In: Hefley, B., Murphy, W. (a cura di), *The science of service systems: Service science: Research and innovations in the service economy* (pp. 37–59). Springer, New York.

Minsky, M. (1975). A Framework for Representing Knowledge. In: Winston, P. (Ed.), *The psychology of computer vision*. McGraw-Hill, New York.

Nonaka, I., Takuchi, H. (1995). *The knowledge-creating company*. Oxford University Press, New York.

Parsons, T. (1960). *Structure and process in modern societies*. The Free Press, New York.

Pels, J., Barile, S., Saviano, M., Polese, F., Carrubbo, L. (2014). The Contribution of VSA and SDL Perspectives to Strategic Thinking in Emerging Economies. *Managing Service Quality: An International Journal*, 24(6), 565–591. http://dx.doi.org/10.1108/MSQ-09-2013-0199.

Polese, F. (2009). Reflections about Value Generation through Networking Culture and Social Relations. *Quaderno di Sinergie*, 16, 193–215.

Polese, F., Caputo, F., Carrubbo, L., Sarno, D. (2016). The Value (Co)creation as Peak of Social Pyramid. *In 26th Annual RESER Conference, "What's ahead in service research: New perspectives for business and society", Università di Napoli "Federico II"*, Settembre, 8–10, Italy.

Prahalad, C.K., Hamel, G. (2006). The Core Competence of the Corporation. In: *Strategische unternehmungsplanung – strategische unternehmungsführung* (pp. 275–292). Springer, Berlin and Heidelberg.

Saviano, M., Barile, S., Caputo, F. (2017). Re-Affirming the Need for Systems Thinking in Social Sciences: A Viable Systems View of Smart City. In: Vrontis, S., Weber, T., Tsoukatos, E. (Eds.), *Global and national business theories and practice: Bridging the past with the future* (pp. 1552–1567). EuroMed Press, Cyprus.

Saviano, M., Barile, S., Spohrer, J., Caputo, F. (2017). A Service Research Contribution to the Global Challenge of Sustainability. *Journal of Service Theory and Practice*, 27(5), 951–976.

Saviano, M., Caputo, F. (2013). Managerial Choices between Systems, Knowledge and Viability. In: Barile, S. (Ed.), *Contributions to theoretical and practical advances in management: A Viable Systems Approach (VSA)* (vol. 2, pp. 219–242). Aracne, Roma.

Saviano, M., Nenci, L., Caputo, F. (2017). The Financial Gap for Women in the MENA Region: A Systemic Perspective. *Gender in Management: An International Journal*, 32(4), 203–217.

Scalia, M., Angelini, A., Farioli, F., Mattioli, G.F., Saviano, M. (2016). The Chariots of Pharaoh at the Red Sea: The Crises of Capitalism and Environment: A Modest Proposal towards Sustainability. *Culture della sostenibilità*, 1, Numero Speciale, 3–63.

Simone, C., Polese, F., Iandolo, F., Caputo, C. (2014). Alla ricerca di un possibile principio evolutivo della teoria e della pratica d'impresa. Il percorso degli studi dell'economia d'impresa. *XXVI Convegno annuale di Sinergie, Manifattura: quale futuro?*, Novembre 13–14, Università di Cassino e del Lazio Meridionale.

Spicer, A., Alvesson, M., Kärreman, D. (2009). Critical Performativity: The Unfinished Business of Critical Management Studies. *Human Relations*, 62(4), 537–560.

Spohrer, J.C., Maglio, P.P. (2010). Toward a Science of Service Systems: Value and Symbols. In: Maglio, P.P., Kieliszewski, C.A., Spohrer, J.C. (Eds.), *Handbook of service science: Research and innovations in the service economy*. Springer, New York.

Teece, D.G., Pisano, P., Shuen, A. (1997). Dynamic Capabilities and Strategic Management. *Strategic Management Journal*, 18(7), 509–533.

Tronvoll, B., Barile, S., Caputo, F. (2017). A Systems Approach to Understanding the Philosophical Foundation of Marketing Studies. In: Barile, S., Pellicano, M., Polese, F. (Eds.), *Social dynamics in a system perspective* (pp. 1–18). Springer.

Von Glasersfeld, E. (1988). *Radical constructivism: A way of knowing and learning*, Studies in Mathematics Education Series: 6. Falmer Press and Taylor & Francis Inc.

Websites

www.asvsa.org

Sustainable supply chain management

A review of tools, enablers and best practices

Ewa Wankowicz,[1] *Alessandra Cozzolino,*[2] *Harwin de Vries*[3] *and Enrico Massaroni*[4]

Keywords: *sustainable supply chain management; green supply chain management; sustainable initiatives; sustainable volunteer practices*

1 Background

Managing globally dispersed supply chains (SCM) in the world of today is extremely complex. Not only should supply chains be optimized with respect to traditional performance measures such as cost, quality, time of delivery. Companies are also increasingly willing or pressured to incorporate sustainability concerns in their management practices. Sustainability is commonly being defined by three core pillars – social, economic, and environmental sustainability – each of which is largely affected by SCM practices. This role is mediated by two elements; the first relates to the current business scenario characterized by the supply chain competition (Gallinaro, 2001; Boyaci and Gallego, 2004; Christopher, 2005; Cozzolino, 2009; Ashby et al., 2012; Rezapour et al., 2014); the second is that the supply chain is present in all stages of the product/service creation and its distribution (Linton et al., 2007). Benefits of sustainable SCM practices have been extensively analyzed. In contrast with common beliefs, they do not only include improved environmental and economic sustainability (Zhu and Sarkis, 2006; Diabat and Govindan, 2011; Abdala et al., 2014; Golicic et al., 2013; Wu et al., 2015), but also increased operational efficiency and effectiveness (Testa and Iraldo, 2010; Golicic et al., 2013). Sustainable companies have been showed to be more profitable, and research has identified several mechanisms that make a causal relationship between the two plausible (Kumar et al., 2012; Kushwaha, 2011; Haanaes et al., 2013). Hence, sustainable practices can assure the competitiveness of the company and support the achievement of sustainability (Hassini et al., 2012; Gandhi et al., 2015), thus positively impact the triple bottom line (Elkington, 1997; Morali and Searcy, 2013).

2   Complexity of sustainability management in supply chains

Sustainability management in cross-boundary supply chains has been shown to be very difficult. For example, Nike was accused of not being able to manage its sub-suppliers and their sustainability when it was involved in the so-called sweatshop phenomenon. The collapse of Rana Plaza in Bangladesh, which killed 1.134 people and injured 2.515, also exemplifies this: globally known clothing companies like Benetton, Mango, Primark, and Walmart outsourced manufacturing of fashion items to unsustainable subcontractors. Undoubtedly, strategic sustainability goals and operational implementation are oftentimes misaligned in the fashion industry (Wankowicz, 2016). Also other sectors have been accused of provoking unsustainable externalities, for example the logistics sector (Rossi et al., 2013). Such incidents teach us at least three lessons. The first is that business models that are based on purely profit-oriented aspirations are not necessarily the most successful ones in the long-term. The second is that implementing sustainability goals requires a company to monitor and affect all tiers of its supply chain. Lastly, it highlights the importance for a company to promote its sustainability-oriented initiatives internally and externally with supply chain partners and stakeholders. Finding the right way to incorporate environmental and social sustainability goals into day to day management is, however, far from trivial.

3   Rationale of the study

Though an abundance of theory on sustainability exists, generally accepted definitions are still lacking. This is at least in part due to difficulties in distinguishing between green SCM and sustainable SCM, which often are treated as synonymous (Ahi and Searcy, 2013; Ashby et al., 2012). Nevertheless, sustainable SCM can be considered an extension of green SCM and defined as

> the management of material, information and capital flows as well as cooperation among companies along the supply chain while taking goals from all three dimensions of sustainable development, i.e., economic, environmental and social, into account which are derived from customer and stakeholder requirements.
>
> (Seuring and Müller, 2008, p. 1700)

Ahi and Searcy (2013, p. 339) go one step further in their definition, by highlighting the role of intrinsic motivation. In their view, the "truly sustainable supply chains" are those created

> "through the voluntary integration of economic, environmental, and social considerations with key inter-organizational business systems designed to

efficiently and effectively manage the material, information, and capital flows associated with the procurement, production, and distribution of products or services in order to meet stakeholder requirements and improve the profitability, competitiveness, and resilience of the organization over the short- and long-term."

In the same vein, Laosirihongthong et al. (2013) classify sustainable SCM practices as reactive or proactive, where the first are guided by legislative and normative pressures and the second are voluntarily promoted by the companies. Research has traditionally mainly focused at the first group. It has even been claimed that sustainability is only achievable only through external incentives (Pagell and Shevchenko, 2014). There are, however, valid reasons to believe that intrinsic motivation (IM) is least as important. Previous studies have shown that acting through IM may have stronger positive impact resulting from the environmental practices (D'Amato et al., 2014). Regulations can improve the sustainability performance in those companies that have little IM (Grant et al., 2015). Literature shows that a wide variety of reactive and proactive sustainable SCM practices exists. To the best of our knowledge, however, no comprehensive, multisectoral review has been performed. This study aims to fill this gap in the literature and thereby to structure this field and to provide companies with an overview of tools, enablers, and (best) practices in sustainable SCM.

4   Methodology

The research methodology used in this paper in based on two-phase approach. First, the systematic literature review (SLR) was conducted. SLR as a methodologically rigorous exercise, permits assessment of the findings of previous related research and then a synthesis of the results (Denyer and Tranfield, 2009). The databases used were Ebsco, Elsevier, and Emerald. We follow the five steps proposed by Denyer and Tranfield (2009): (1) question formulation, (2) locating studies, (3) study selection and evaluation, (4) analysis and synthesis, (5) reporting and using the results.

In the second phase, the exploratory qualitative approach based on the multiple case study (MCS) (Yin, 2003) was performed. MCS "enables comparisons that clarify whether an emergent finding is simply idiosyncratic to a single case or consistently replicated by several cases" (Eisenhardt and Graebner, 2007, p. 27; Eisenhardt, 1991). The empirical investigation was performed through cross-case analysis using the (best) practices found in the literature.

5   Preliminary results

A preliminary overview of practices emerging from our literature review is presented in Table 27.1.

Table 27.1 A preliminary classification of the environmentally sustainable practices

Some of practices identified	References
Supplier environmental audit, supplier certification	Min and Galle (1997); Lee et al. (2012); Laosirihongthong et al. (2013)
Green procurement, green purchasing	Günther and Scheibe (2006)
Environmental purchasing	Carter et al. (2000); Zsidisin and Siferd (2001); Sarkis (2012)
Green logistics	Sarkis (2012); Murphy and Poist (2000)
Packaging and waste reduction, supplier evaluation based on the environmental performance, eco-friendly product development, Emission reduction of CO_2	Walker et al. (2008)
Internal environmental management, green purchasing, cooperation with customers, eco-design, green labeling	Lee et al. (2012)
Waste reduction, saving packaging materials, ISO 14001 certification, lean management, eco-design, production facilities.	Ageron et al. (2012); Gimenez et al. (2012)
Green productivity, reducing energy and consumption, reducing consumption of natural resources, reducing pollution, recycling	Jaggernath and Khan (2015)
Eco-management and audit scheme	Vachon and Klassen (2006, 2008); Sarkis (2012); Zhu et al. (2005); Large and Thomsen (2011); Min and Galle (2001)
Product eco-design, packaging Eco-design	Laosirihongthong et al. (2013)
Environmental collaboration with suppliers, environmentally friendly purchasing practices, waste reduction, decreasing the consumption of hazardous and toxic materials, ISO 14001 certification, Reverse logistics, environmental collaboration with customers, environmentally friendly packaging, working with customers to change product specifications.	Azevedo et al. (2011)
Environmental collaboration, environmental monitoring	Vachon and Klassen (2006)
Environmental certification; pollution prevention, life cycle assessment design for environment	Klassen and Johnson (2004)
Green design, green sourcing/procurement, green operations or green manufacturing, green distribution, logistics/marketing, reverse logistics	Srivastava (2007)

Source: Our elaboration

6   Conclusion

According to ISO (2014) "the sustainability of an individual organization may, or may not, be compatible with the sustainability of society as a whole, which is attained by addressing social, economic and environmental aspects in an inte grated manner". In other words, companies pursuing sustainability need to look beyond their own borders. Our study provides an overview and classification of best practices in sustainable SCM. Such systematization can constitute a valuable platform for management during selection and implementation of those practices, as they can facilitate the achievement of sustainability according to triple bottom line theory (Elkington, 1997).

Notes

1 Sapienza University of Rome.
2 Sapienza University of Rome.
3 Erasmus University Rotterdam.
4 Sapienza University of Rome.

Bibliography

Abdala, E. C., & Barbieri, J. C. (2014). Determinants of sustainable supply chain: An analysis of mensuration models of pressures and socio-environmental practices. *JOSCM: Journal of Operations and Supply Chain Management*, 7(2), 110.

Ageron, B., Gunasekaran, A., & Spalanzani, A. (2012). Sustainable supply management: An empirical study. *International Journal of Production Economics*, 140(1), 168–182.

Ahi, P., & Searcy, C. (2013). A comparative literature analysis of definitions for green and sustainable supply chain management. *Journal of Cleaner Production*, 52, 329–341.

Ashby, A., Leat, M., & Hudson-Smith, M. (2012). Making connections: A review of supply chain management and sustainability literature. *Supply Chain Management: An International Journal*, 17(5), 497–516.

Azevedo, S. G., Carvalho, H., & Machado, V. C. (2011). The influence of green practices on supply chain performance: A case study approach. *Transportation Research Part E: Logistics and Transportation Review*, 47(6), 850–871.

Boyaci, T., & Gallego, G. (2004). Supply chain coordination in a market with customer service competition. *Production and Operations Management*, 13(1), 3–22.

Carter, C. R., Kale, R., & Grimm, C. M. (2000). Environmental purchasing and firm performance: An empirical investigation. *Transportation Research Part E: Logistics and Transportation Review*, 36(3), 219–228.

Christopher, M. (2005). *Logistics and supply chain management: Creating value-added networks*. Pearson Education.

Cozzolino, A. (2009). *Operatori logistici. Contesto evolutivo, assetti competitivi e criticità emergenti nella supplychain*. Padova: Cedam.

Diabat, A., & Govindan, K. (2011). An analysis of the drivers affecting the implementation of green supply chain management. *Resources, Conservation and Recycling*, 55(6), 659–667.

D'Amato, A., Mancinelli, S., & Zoli, M. (2014). Two Shades of (Warm) Glow: Multidimensional intrinsic motivation, waste reduction and recycling (No. 2114). SEEDS, *Sustainability Environmental Economics and Dynamics Studies*. Retrieved 21.9.2018 from https://ideas.repec.org/p/srt/wpaper/2114.html

Denyer, D., & Tranfield, D. (2009). Producing a systematic review. In D. A. Buchanan & A. Bryman (Eds.), *The Sage handbook of organizational research methods* (pp. 671–689). Thousand Oaks, CA: Sage Publications Ltd.

Eisenhardt, K. M. (1991). Better stories and better constructs: The case for rigor and comparative logic. *Academy of Management Review*, 16(3), 620–627.

Eisenhardt, K. M., & Graebner, M. E. (2007). Theory building from cases: Opportunities and challenges. *Academy of Management Journal*, 50(1), 25–32.

Elkington, J. (1997). *Cannibals with forks. The triple bottom line of 21st century*. London: Capstone.

Gallinaro, S. (2001). *Imprese e competizione nell'era della modularità*. Padova: CEDAM.

Gandhi, S., Mangla, S. K., Kumar, P., & Kumar, D. (2015). Evaluating factors in implementation of successful green supply chain management using DEMATEL: A case study. *International Strategic Management Review*, 3(1), 96–109.

Gimenez, C., Sierra, V., & Rodon, J. (2012). Sustainable operations: Their impact on the triple bottom line. *International Journal of Production Economics*, 140(1), 149–159.

Golicic, S. L., & Smith, C. D. (2013). A meta-analysis of environmentally sustainable supply chain management practices and firm performance. *Journal of Supply Chain Management*, 49(2), 78–95.

Grant, D. B., Trautrims, A., & Wong, C. Y. (2015). *Sustainable logistics and supply chain management* (Revised Edition). Kogan Page Publishers.

Günther, E., & Scheibe, L. (2006). The hurdle analysis: A self-evaluation tool for municipalities to identify, analyse and overcome hurdles to green procurement. *Corporate Social Responsibility and Environmental Management*, 13(2), 61–77.

Haanaes, K., Michael, D., Jurgens, J., & Rangan, S. (2013). Making sustainability profitable. *Harvard Business Review*, 91(3), 110–115.

Hassini, E., Surti, C., & Searcy, C. (2012). A literature review and a case study of sustainable supply chains with a focus on metrics. *International Journal of Production Economics*, 140(1), 69–82.

Jaggernath, R., & Khan, Z. (2015). Green supply chain management. *World Journal of Entrepreneurship, Management and Sustainable Development*, 11(1), 37–47.

Klassen, R. D., & Johnson, P. F. (2004). *The green supply chain: Understanding supply chains: Concepts, critiques and futures*, 229–251. New York, NY: Oxford University Press.

Kumar, S., Teichman, S., & Timpernagel, T. (2012). A green supply chain is a requirement for profitability. *International Journal of Production Research*, 50(5), 1278–1296.

Kushwaha, G. S. (2011). Sustainable development through strategic green supply chain management. *International Journal of Engineering and Management Sciences*, 1(1), 7–11.

Laosirihongthong, T., Adebanjo, D., & Choon Tan, K. (2013). Green supply chain management practices and performance. *Industrial Management & Data Systems*, 113(8), 1088–1109.

Large, R. O., & Thomsen, C. G. (2011). Drivers of green supply management performance: Evidence from Germany. *Journal of Purchasing and Supply Management*, 17(3), 176–184.

Lee, S. M., Tae Kim, S., & Choi, D. (2012). Green supply chain management and organizational performance. *Industrial Management & Data Systems*, 112(8), 1148–1180.

Linton, J. D., Klassen, R., & Jayaraman, V. (2007). Sustainable supply chains: An introduction. *Journal of Operations Management*, 25(6), 1075–1082.

Min, H., & Galle, W. P. (1997). Green purchasing strategies: Trends and implications. *Journal of Supply Chain Management*, 33(3), 10.

Min, H., & Galle, W. P. (2001). Green purchasing practices of US firms. *International Journal of Operations & Production Management*, 21(9), 1222–1238.

Morali, O., & Searcy, C. (2013). A review of sustainable supply chain management practices in Canada. *Journal of Business Ethics*, 117(3), 635–658.

Murphy, P. R., & Poist, R. F. (2000). Green logistics strategies: An analysis of usage patterns. *Transportation Journal*, 5–16.

Pagell, M., & Shevchenko, A. (2014). Why research in sustainable supply chain management should have no future. *Journal of supply chain management*, 50(1), 44–55.

Rezapour, S., Farahani, R. Z., Dullaert, W., & De Borger, B. (2014). Designing a new supply chain for competition against an existing supply chain. Transportation Research Part E: *Logistics and Transportation Review*, 67, 124–140.

Rossi, S., Colicchia, C., Cozzolino, A., & Christopher, M. (2013). The logistics service providers in eco-efficiency innovation: An empirical study. *Supply Chain Management: An International Journal*, 18(6), 583–603.

Sarkis, J. (2012). A boundaries and flows perspective of green supply chain management. *Supply Chain Management: An International Journal*, 17(2), 202–216.

Seuring, S., & Müller, M. (2008). From a literature review to a conceptual framework for sustainable supply chain management. *Journal of Cleaner Production*, 16(15), 1699–1710.

Srivastava, S. K. (2007). Green supply-chain management: A state-of-the-art literature review. *International Journal of Management Reviews*, 9(1), 53–80.

Testa, F., & Iraldo, F. (2010). Shadows and lights of GSCM (Green Supply Chain Management): Determinants and effects of these practices based on a multi-national study. *Journal of Cleaner Production*, 18(10), 953–962.

Vachon, S., & Klassen, R. D. (2006). Green project partnership in the supply chain: The case of the package printing industry. *Journal of Cleaner Production*, 14(6), 661–671.

Vachon, S., & Klassen, R. D. (2008). Environmental management and manufacturing performance: The role of collaboration in the supply chain. *International Journal of Production Economics*, 111(2), 299–315.

Walker, H., Di Sisto, L., & McBain, D. (2008). Drivers and barriers to environmental supply chain management practices: Lessons from the public and private sectors. *Journal of Purchasing and Supply Management*, 14(1), 69–85.

Wankowicz, E. (2016). Sustainable fibre for sustainable fashion supply chains: Where the journey to sustainability begins, International Conference on Industrial Logistics, 28 September–1 October, Zakopane (Poland), pp. 342–351.

Wu, K. J., Liao, C. J., Tseng, M. L., & Chiu, A. S. (2015). Exploring decisive factors in green supply chain practices under uncertainty. *International Journal of Production Economics*, 159, 147–157.

Yin, R. K. (2003). *Case study research: Design and methods*. Beverly Hills, CA: Sage Publications.

Zhu, Q., & Sarkis, J. (2006). An inter-sectoral comparison of green supply chain management in China: Drivers and practices. *Journal of Cleaner Production*, 14(5), 472–486.

Zhu, Q., Sarkis, J., & Geng, Y. (2005). Green supply chain management in China: Pressures, practices and performance. *International Journal of Operations & Production Management*, 25(5), 449–468.

Zsidisin, G. A., & Siferd, S. P. (2001). Environmental purchasing: a framework for theory development. *European Journal of Purchasing & Supply Management*, 7(1), 61–73.

Zsidisin, G. A., & Ellram, L. M. (2001). Activities related to purchasing and supply management involvement in supplier alliances. *International Journal of Physical Distribution & Logistics Management*, 31(9), 629–646.

Transnational firms aggregation in agro-food sector

A viable system view

Leonardo Di Gioia,[1] Savino Santovito,[2]
Raffaele Silvestri[3] and Nicola Faccilongo[4]

Keywords: *Association of Producers Organization; VSA; horizontal aggregation*

1 Introduction

The modern food system, due to many factors, such as the consolidation in the industry and in food distribution, the importance of quality and differentiation and the increase of the vertical coordination degree, has gradually abandoned the neoclassical paradigm of perfect competition to establish itself as a comprehensive set of interrelated markets with complex governance structures.

The Producers Organization (PO) is a form of business networking through the horizontal aggregation of small farmers aimed at overcoming the single firm operational inefficiencies and at achieving greater bargaining power toward the trade system.

An additional form of horizontal aggregation at the transnational level (called Association of Producers' Organization – APO) can enable small producers in penetrating foreign markets, thus overcoming barriers to entry, which, otherwise, could not be managed by a single PO.

In order to achieve a better understanding of the organizational models able to improve the agro-food firms market development and competitiveness, our purpose is to analyze the APO moving from the viable system approach.

2 Association of Producers' Organization

We are addressing the question how the horizontal integration among several SMEs in the agro-food industry, such as the Association of Producers' Organization (APO), can affect those firms' competitive strategies for new foreign market penetration, by focusing on the following research questions: which is the main role of the aggregation system within the new foreign potential market? How can the aggregation system support its members in achieving effective alignment with new markets actors?

The research, according to the viable systems approach, focuses on the role of the APO toward the relevant external systems of a new potential market by analyzing both the related resources' criticality and influencing capability. A viable system interpretation of the transnational business horizontal aggregation can enable better formalizing of the industrial relationships within the agro-food supply network.

The APO system aims at managing the transnational commercial dynamics of several single associated agro-food firms, created in order to optimize the transaction costs and the timing of the penetration strategy into new foreign markets; to access the new market relationship assets.

The marketing research, the commercial and financial reliability checks of a potential customer are mutually transferred automatically between the PO of different countries, thus allowing the associated firms to get important information on potential business partners.

In this sense, the APO plays a role of connection between aggregation of business systems which act in different countries, through relationship management with key stakeholders of the distribution system aimed at realizing the international marketing strategies. The APO uses the interactions within the network and the trade agreements or exclusive procurement contracts as structural instruments, to realize the commercial operating synergies between the PO.

Therefore, the APO can be meant as a managing system of two supra-systems: the distribution and legislative ones. Both, highly relevant, in viable system terms, for the implementation of quantitative and qualitative growth strategies of companies in the agro-food sector, indeed, can establish strict access and operative rules into a market, constraints on products traded, and can affect the profitability and financial performance of the incoming players.

The APO system is engaged in erasing the information asymmetry between a wide multitude of small farmers on one hand, and both the distributive and legislative supra-systems of new foreign markets, on the other hand: the information asymmetry concerns the market knowledge (defined as organized and structured information on the market, comprising the competences and know-how centered on customers' characteristics, preferences and needs that firms are requested to satisfy), and the managerial knowledge (defined as the set of competences and know-how necessary to efficiently and effectively coordinate and supervise organizational resources and processes, including operational and applied knowledge as well as more abstract and complex knowledge, such as strategic processes, cross functional competences, etc.). Thus, the APO is involved in enhancing the area of converging interests between the following relevant systems: high-quality agro-food goods production supra-system, distribution supra-system, institutional normative supra-system and consumption supra-system.

The managing action of the APO is aimed at achieving consonance between the expectations of the above-mentioned supra-systems, coherently with the institutional framework in which most of the actors are oriented toward agro-food products with higher quality and profitability: the APO works by leveraging the criticality of the high-quality agro-food resource, through leveling bargaining power between different actors, which allows to maximize the utility for consumers and to get a fair distribution of value.

In the effort to harmonize the interactions between the actors of an extensive and multiform network of actors, which can get and share a much higher value compared to the one achievable singularly, the APO becomes a mediation system in the formulation and implementation of strategic decisions that involve an increase of the degree of systemic openness for the single associated firms, that, in this way, increase the ability to exchange resources with the competitive environment from which arises an inevitable qualitative growth for the operational structure of the managing body.

The APO is committed in searching for consonance, in terms of structural compatibility between the production system (the network of producers in a specific category of food products) with the distribution and normative supra-systems of new foreign markets.

Into a second stage, the role of the APO, aimed at managing and optimizing the physical and information flows between the actors, through harmonizing the interactions between them, may evolve in such a systemic compatibility in the exchange (resonance), or a harmony and coincidence of purposes.

Notes

1 Università di Foggia.
2 Università di Bari.
3 Università di Bari.
4 Università di Foggia.

Bibliography

Ashby, W.R. (1956), *An Introduction to Cybernetics*, Chapman & Hall, London.
Barile, S. (2000), *Contributi sul pensiero sistemico in economia d'impresa*, Arnia, Salerno.
Barile, S. (2009), *Management sistemico vitale*, Giappichelli, Torino.
Barile, S., and Saviano, M. (2011a), "Foundations of systems thinking: The structure-system paradigm", in Various Authors, *Contributions to Theoretical and Practical Advances in Management: A Viable Systems Approach (VSA)*, International Printing, Avellino, pp. 1–25.
Barile, S., and Saviano, M. (2011b), "Qualifying the concept of systems complexity", in Various Authors, *Contributions to Theoretical and Practical Advances in Management: A Viable Systems Approach (VSA)*, International Printing, Avellino, pp. 27–63.
Beer, S. (1972), *Brain of the Firm: The Managerial Cybernetics of Organization*, The Penguin Press, London.
Bogdanov, A. (1922), *Tektologya: Vseobschaya Organizatsionnaya Nauka*, Berlin and Petrograd, Moscow.
Emery, F.E., and Trist, E.L. (1960), "Socio-technical Systems", in Churchman, C.W. and Verhurst, M. (eds.), *Management Science, Models and Techniques*, Vol. 2, Pergamon Press, London, pp. 83–97.
Forrester, J.W. (1994), "System dynamics, systems thinking, and soft OR", *System Dynamics Review*, 10(2–3), 245–256.

Frascarelli, A. (2014), "Il sostegno della Pac tra competitività e beni pubblici", *Relazione convegno Sidea*, Benevento, 18–20 settembre.

Frascarelli, A., and Sotte, F. (2010), "Per una politica dei sistemi agricoli e alimentari dell'UE", *Agriregionieuropa*, 6(21), Giugno, Ancona.

Golinelli, G.M. (2000), *L'approccio sistemico al governo dell'impresa. L'impresa sistema vitale*, 1st Ed., CEDAM, Padova.

Goodhue, R.E. (2011), "Food quality: The design of incentive contracts", *Annual Review of Resource Economics*, 3, 119–140.

Li, T., and Calantone, R.J. (1998), "The impact of market knowledge competence on new product advantage: Conceptualization and empirical examination", *The Journal of Marketing*, 62(4), 13–29.

McCorriston, S. (2002), "Why should imperfect competition matter to agricultural economists?", *European Review of Agricultural Economics*, 29, 349–372.

Rogers, R.T. (2001), "Structural change in U.S. food manufacturing, 1958–1997", *Agribusiness*, 17(1), 3–32.

Russo, C., Goodhue, R.E., and Sexton, R.J. (2011), "Agricultural support policies in imperfectly competitive markets: Why market power matters in policy design", *American Journal of Agricultural Economics*, 93(5), 1328–1340.

Saitone, T.L., and Sexton, R.J. (2010), "Product differentiation and quality in food markets: Industrial organization implications", *Annual Review of Resource Economics*, 2, 341–368.

Sammarra, A., and Biggiero, L. (2008), "Heterogeneity and specificity of Inter-Firm knowledge flows in innovation networks", *Journal of Management Studies*, 45(4), 800–829.

Saviano, M., and Berardi, M. (2009), "Decision making under complexity: The case of SME", in Vrontis, D., Weber, Y., Kaufmann, R., and Tarba, S. (eds.), *Managerial and Entrepreneurial Developments in the Mediterranean Area*, EuroMed Press, Cipro, pp. 1619–1643.

Sexton, R.J. (2013), "Market power, misconceptions, and modern agricultural markets", *American Journal of Agricultural Economics*, 95(2), 209–219.

Von Bertalanffy, L. (1968), *General System Theory: Foundations, Development, Applications*, George Braziller, New York.

Williamson, O.E. (2007), "Pragmatic methodology: A sketch with applications to transaction cost economics", *Journal of Economic Methodology*, 16(2), 145–157.

Interdependences of duality, systems theories and cybernetics in business systems

Vojko Potocan,[1] *Matjaz Mulej*[2] *and Zlatko Nedelko*[3]

Keywords: *duality; systems theories; cybernetics; business systems*

1 Introduction

The modern environment requires organizations to develop new approaches aimed at requisite holism, mastering of entanglement and dynamics, which can enable improvement of their functioning and behaviour (Mintzberg, 1973; Chesbrough, 2003; Potocan and Mulej, 2009; Birkinshaw et al., 2016). Additionally, the partial cognitions from several theories – e.g. from systems theory, cybernetics and management – offer limited insights into reasons for surpassing of one-sided and mono-disciplinary consideration of organizational phenomena (Francois, 2004; Mulej, 2004; Wallis, 2009). In the recent decades researches on organizations have mainly learned from cybernetics and systems theories (Bertalanffy, 1968; Wiener, 1956; Beer, 1985; Flood, 1999); they aim at investigations of holism of thinking, decision-making and actions (Beer, 1985; Francois, 2004; Gibson and Birkinshaw, 2004; Wallis, 2009). However, holism in Systems theories and Cybernetics often remains inside the chosen single discipline, accepting perhaps some impacts from its neighbouring disciplines, but missing the necessary inter – disciplinary understanding of reality (Mintzberg, 1973; Foerster, 1974; Vallee, 2003; Smith and Lewis, 2011). Authors often tend to forget about the crucial basis of systems theory: "Interdisciplinary creative cooperation replacing the current over-specialization" (Bertalanffy, 1968, p. VII in Foreword). Norbert Wiener did exactly that (Wiener, 1956; Francois, 2004). As a mathematician, he worked on a project with natural and engineering scientists, solved the problem and created cybernetics. Problems in business require the same.

Some studies focus more attention on the fundamental concepts, backing the phenomena and providing additional knowledge for the improvement of holism and interdisciplinary working and behaviour of business systems (Mulej and Kajzer, 1998; Sutherland and Smith, 2011). Duality offers possible solutions (Fare and Primont, 1995; Evans and Frankish, 2009).

2 Theoretical starting points for consideration of duality

The idea of duality and using duality for explanation of reality originates from ancient Greece (Black, 1997) – e.g., Aristotle explained: "All, nearly all, thinkers agree that being and substance are composed of contraries; at least all of them name contraries as their first principles" (Aristotle, 1976, p. 29), and in China (Wood, 2000) with the philosophical background of life, yin and yang (Black, 1997; Delgado and Banathy, 1993).

Duality is a universal principle summarizing generally valid cognitions about elementary – e.g., natural, social, technical, etc. – systems; consequently, its understanding is very similar in different sciences (Wood, 2000). From the context viewpoint, duality expresses simultaneous coexistence of paired – i.e. conjugate – components, e.g., the interaction forms the necessary criteria for the existence of an elementary system (Delgado and Banathy, 1993; Wallis, 2009). The general principle of duality is undefeatable; the existence of its counterparts proves this universal principle.

Each scientific discipline utilizes duality to consider purposes and goals inside its selected starting points and environment (Wallis, 2009). Therefore, science uses several different definitions of duality, which emphasize specific purposes and goals of each of them (Afuah, 1998; Dubrovsky, 2001). The most frequently mentioned explanations of duality include physics (Englert et al., 1994), theology (Gordon, 1995; Mann, 1995; Black, 1997), and economics (Diewert, 1974; Mulej, 2004).

Explanation of duality is importantly influenced by cognitions about the existence of different appearance levels of each idea (Black, 1997; Birkinshaw et al., 2016). On the base of results from previous researches, we presume that duality can be defined as a concept, methodology, method, technique and instrument. The modern science defines the terms idea of duality and concept of duality in a relatively unified way. The other levels of duality are differently defined by various sciences, as well as in the frame of any single science (Black, 1997; Wiener, 1956; Wallis, 2009).

In our research, we concentrated considerations on the duality of Cybernetics and Systems theory in organizations as business systems – rather than natural, sociological, engineering, etc. ones. In the last nearly five decades we researched business systems with two basic and interdependent disciplines: Cybernetics (Wiener, 1956), and System(s) theory (Mulej, 2004). Both disciplines are aimed at holism of thinking, decision-making and action (Checkland, 1981; Mulej, 2004; Potocan et al., 2005). Business Cybernetics considers humans and their tools producing benefits for their customers and themselves in interdependent basic, management and information processes (Wiener, 1956; Checkland, 1981; Afuah, 1998).

History has shown that both disciplines have been right and still are so, but the 'necessary and sufficient' = 'requisite' holism can rarely be attained by the General Systems Theory's notion of isomorphism, because it causes crucial oversights, e.g. isomorphism lets specialists leave aside their mutual impacts with specialists of different orientations (Checkland, 1981; Afuah, 1998).

Holism, limited to isomorphism, remains inside a single discipline accepting perhaps some impacts from its neighbouring disciplines, offering an insufficient step forward toward humankind's survival. One must add interdisciplinary behaviour to several types of isomorphism (Mulej, 2004).

Utilization of the concept of duality can improve holism of Business Cybernetics and General Systems Theory toward the requisite holism (Mulej, 2004; Potocan et al., 2005). This matches previous studies, reporting that duality can support achievement of holism in business systems. If holism is limited into a single discipline, it is fictitious due to this limitation; the real, i.e. total holism includes all existing attributes with no limitation to any single discipline; this reaches beyond human capacities. Hence, the requisite holism is the realistic way between the both briefed extremes; it is based on interdisciplinary creative cooperation generating a "dialectical system" (Mulej, 2004; Potocan et al., 2005).

Fewer researches inquire if and how duality can support the interdisciplinary behaviour of business systems. Superficial considerations of duality – as simultaneous coexistence of paired (conjugate) components, such as the interaction, forms the necessary criteria for the existence of elementary systems (Afuah, 1998; Dubrovsky, 2001; Birkinshaw et al., 2016); one can conclude that duality does not enable the interdisciplinary behaviour. Nevertheless, duality tries to define well enough, and to study, events as simultaneous paired (conjugate) components. While doing this, duality does not study the event under consideration from the viewpoint of a single pair, but it rather repeats its consideration until it detects most of the selected attributes under consideration, or all attributes found essential by researchers.

Consequently, this causes a set, but not a system, of dual considerations and their outcomes; we can call them the partial- or sub-systemic considerations and their outcomes. In which way can one attain comparability and/or understanding of them as parts of a synergetic entity?

In terms of contents, we thus face two rather basic issues related to the dilemma of how to detect whether or not the duality-based trial's consideration and outcome are interdisciplinary. Cognitions from the relevant literature (Afuah, 1998; Wallis, 2009; Birkinshaw et al., 2016), and from our previous contributions (Potocan et al., 2005; Potocan and Mulej, 2009), post two research questions for our work.

RQ1: How to find out whether or not the duality is an interdisciplinary one?
RQ2: How to understand duality characteristics in the utilization of Business Cybernetics and Systems Theory in business systems?

3   Duality in business cybernetics and systems theory

Duality and systems theory

Bertalanffy created a new worldview called the General Systems Teaching/Theory (Bertalanffy, 1968; Davidson, 1983; Mulej, 2004); one of Bertalanffy's crucial sentences says that humankind has a poor chance to survive, if we do not think and behave as citizens of the world rather than single countries only, and if we do not consider the entire biosphere as one whole (Bertalanffy, 1968).

Some authors put 'system' equal to the object under consideration. On the other hand, systems are only mental pictures of the considered object; systems are made by the authors of this object from their selected viewpoints in order to expose the attributes of the object that they find the most important (Mulej, 2004; Vallee, 2003; Birkinshaw et al., 2016).

Thus, systems differ from the represented object, and models represent systems (as mental pictures limited to observers' mental capacity and interest rather than reality with all existing attributes). Despite this mental capacity, limited for natural reasons, we humans anyway try to control/manage/create the world, although we have a very limited insight into its reality; to overcome this natural lack of capacity, humans specialize. Therefore, we need interdisciplinary creative cooperation of specialists who are mutually different (Davidson, 1983; Mulej, 2004; Vallee, 2003), and interdependent counterparts. This reminds us of the duality concept.

When one tries to understand the concept of Duality in terms of systems theory, one comes across issues concerning systems theory, e.g. systems theory is a science about: holistic thinking, similarities/isomorphism of features, interdisciplinary behaviour, synthesis and integration of systems. The multiple systems approaches enable different levels of systemically or systemic achievement of their synergy: their scopes can be complementary and are hence interdependent, making a case of duality. This case is perhaps not crucial in all human activities, but it is crucial in dealing with business systems (see e.g., Vallee, 2003; Mulej, 2004). The finding that these approaches are interdependent makes us think of the ancient yin and yang from Ancient China and the less (but still) ancient philosophy of dialectics, being the ancient Greek word for interdependence.

Hence, the duality concept can provide steps from a one-sided and therefore fictitious holism toward a realistic consideration of the law of the requisite holism, or even toward the Bertalanffian concept of the total holism (Mulej, 2004). In other words, a link exists between the concepts of duality and Systems Theory as a science of a (requisitely) holistic thinking and behaviour.

Duality and business cybernetics

The term Cybernetics originated in 1947 when Norbert Wiener used it to name a discipline apart from, but touching upon, such established disciplines as electrical engineering, mathematics, biology, neurophysiology, anthropology and psychology (Wiener, 1956; Beer, 1985; Francois, 2004; Birkinshaw et al., 2016).

Cybernetics is originally created on an interdisciplinary basis: they were trying to make a synergy of knowledge of biology, mathematics and technology to produce novelties, supposedly (and frequently able) to become even innovations as novelties used beneficially by their users.

They used the word system to denote that they are dealing with a complex and complicated feature as a whole rather than one-sidedly (Wiener, 1956; Bertalanffy, 1968; Francois, 2004). They were fearful of doing something similar in relation to social sciences, and they were fearful of letting models become self-sufficient rather than a simplifying tool helping humans to understand and master their selected parts of attributes of the object under consideration (Wiener, 1956; Beer, 1985; Wallis, 2009). Business Cybernetics is specializing in organizations and individuals as the so-called business systems emphasizing the so-called business viewpoints rather than the natural, social, technical or technological viewpoints of consideration of objects, features, events and processes making the real life (Beer, 1985; Francois, 2004; Wallis, 2009).

Calling them 'business systems' rather than 'features, events, and processes considered from the viewpoint/s of business sciences', may mean that the requisite holism of consideration and action is consciously or subconsciously limited to the selected viewpoint/s, and therefore one-sided, although the term 'system' suggests holism. In this case, like in all other cases when one uses the word system, one should describe quite explicitly what viewpoint/s and content/s one has in mind – in order to avoid mutual misunderstanding and the resulting lack of capacity and possibility to cooperate creatively on an interdisciplinary basis.

Five concepts of cybernetics are available, at least: (1) The zero order, (2) The first order, (3) The second order, (4) The third order cybernetics and (5) The cybernetics of the conceptual systems (Wallis, 2009; Potocan and Mulej, 2009).

The concept of business cybernetics is close to uniting the cybernetics' concepts 3, 4 and 5 in order to enable the requisite holism of management and working in the business systems. Thus, this concept adds the duality of the concepts of cybernetics of conceptual systems and of cybernetics of the observation, decision-making and impacting; these are phases of the same process.

4 Conclusion

Duality can be seen as a case of interdependence, concentrating on two interdependent attributes or features, if the word is taken (too) literarily; like in a dialogue, several entities can be considered, too. In reality, interdependence is the background of interaction, so is duality; there is no interaction with no interdependence.

Hence, the Cybernetics' impact reaches far beyond single organizations and humans as business systems/persons; it does so by the working of interdependence as the way out from the traditional relations in organizations. So does duality.

Multiple authors define Systems Theory differently because they use different approaches on the basis of their different selected viewpoints. So do authors on cybernetics. If one takes the traditional definition that a system is an entity made of a set of components and of a set of relations, and if one takes the first order cybernetics, duality is visible, but limited to the relations inside the entity/system under consideration/impact.

If one adds, in order to be less abstract and closer to reality, the consideration of the influential role of the selected viewpoint/s, one comes closer to the array of the different less traditional systems theories, as well as to the traditional concept of cybernetics. In this case, duality is visible again, but it is extended to the relations between the object under consideration and the humans dealing with it.

Notes

1 Faculty of Economics and Business, University of Maribor.
2 Faculty of Economics and Business, University of Maribor.
3 Faculty of Economics and Business, University of Maribor.

Bibliography

Afuah, A. (1998). *Innovation Management: Strategies, Implementations, and Profits.* New York: Harvard Press.

Aristotle (1976). *Metaphysics (Books IV).* Oxford: Clarendon Press.

Beer, S. (1985). *Diagnosing the System for Organizations.* London: John Wiley.

Bertalanffy, L. (1968). *General Systems Theory: Foundations, Development, Applications.* New York: Brazillier.

Birkinshaw, J., Crilly, D., Bouquet, C., & Lee, S. Y. (2016). How Do Firms Manage Strategic Dualities? A Process Perspective. *Academy of Management Discoveries,* 2(1), 51–78.

Black, J. (1997). *A Dictionary of Economics.* Oxford: Oxford University Press.

Checkland, P. (1981). *Systems Thinking, Systems Practice.* Chichester: Wiley.

Chesbrough, H. (2003). *Open Innovation.* Boston: Harvard Business Press.

Davidson, M. (1983). *Uncommon Sense: The Life and Thought of Ludwig von Bertalanffy Father of General Systems Theory.* Los Angeles: Tarcher.

Delgado, R., & Banathy, H. (1993). *International Systems Science Handbook.* Madrid: Systemic Publications.

Diewert, W. (1974). Theory of the Firm: Applications of Duality Theory. In D. Kendrick (ed.), *Frontiers of Quantitative Economics* (Vol. 2, pp. 106–166). Amsterdam: North-Holland Publishing Company.

Dubrovsky, V. (2001). Systems of Abstract System Principles. In J. Willby & J. Allen (eds.), *Proceedings of the 45th Annual Conference of the International Society for the Systems Sciences* (pp. 1–20). Asilomar, CA: ISSS.

Englert, B., Scully, M., & Walther, H. (1994). The Duality in Matter and Light. *Scientific American,* pp. 86–92, December.

Evans, J., & Frankish, K. (eds.) (2009). *In Two Minds: Dual Processes and Beyond.* New York: Oxford University Press.

Fare, R., & Primont, D. (1995). Theories of the Firm through Duality. In D. Fare & D. Primont (eds.), *Multi-Output Production and Duality: Theory and Applications* (pp. 1–6). New York: Springer.

Flood, F. (1999). *Rethinking the Fifth Discipline-Learning with the Unknowable.* London: Routledge.

Foerster, H. (1974). *Cybernetics of Cybernetics.* Urbana: University of Illinois.

Francois, C. (ed.) (2004). *International Encyclopedia of Systems and Cybernetics.* Munich: K. G. Saur.

Gibson, C., & Birkinshaw, J. (2004). The Antecedents, Consequences and Mediating Role of Organizational Ambidexterity. *Academy of Management Journal*, 47, 209–226.

Gordon, A. (1995). *Family Bible*. Glendale: Clark Company.

Mann, W. (1995). *Book of the Torah: The Narrative Integrity of the Pentateuch*. New York: Westminster John Knox Press.

Mintzberg, H. (1973). *The Nature of Managerial Work*. New York: Harper and Row.

Mulej, M. (2004). How to Restore Bertalanffian Systems Thinking. *Kybernetes*, 33(1), 48–61.

Mulej, M., & Kajzer, S. (1998). Ethics of Interdependence and the Law of Requisite Holism. In M. Rebernik & M. Mulej (eds.), *Proceedings of the 4th International Conference on Linking Systems Thinking, Innovation, Quality, Entrepreneurship and Environment* (pp. 19–32). Maribor: Institute for Systems Research Maribor.

Potocan, V., & Mulej, M. (2009). Business Cybernetics-Provocation Number Two. *Kybernetes*, 38(1/2), 93–112.

Potocan, V., Mulej, M., & Kajzer, S. (2005). Business Cybernetics: A Provocative Suggestion. *Kybernetes*, 34(9/10), 1496–1516.

Smith, W., & Lewis, M. (2011). Toward a Theory of Paradox: A Dynamic Equilibrium Model of Organizing. *Academy of Management Review*, 36, 381–403.

Sutherland, F., & Smith, A. (2011). Duality Theory and the Management of the Change-Stability Paradox. *Journal of Management & Organization*, 17(4), 534–547.

Vallee, R. (2003). History of Cybernetics. *EOLSS Encyclopedia of Life Support Systems, EOLSS*. Retrieved: 14. 07. 2016, from http://www.eolss.net.

Wallis, S. E. (ed.) (2009). *Cybernetics and Systems Theory in Management: Tools, Views, and Advancements*. IGI Global.

Wiener, N. (1956). *The Human Use of Human Beings: Cybernetics and Society*. New York: Doubleday Anchor.

Wood, R. (2000). *Managing Complexity: How Business Can Adapt and Prosper in the Connected Economy*. London: The Economics Books.

Theme II

Sustainability and the Anthropocene

Marialuisa Saviano, Hernan Lopez Garay, Sandro Schlindwein, Markus Schwaninger and Ray Ison

Managing open innovation with public partners
The case of smart cities

Alberto Ferraris,[1] *Stefano Bresciani*[2] *and Armando Papa*[3]

Keywords: *smart city; open innovation; public organizations; public partner; smart city management*

1   Introduction and objectives

The main goal of this paper is to study the challenges found and the strategies followed by firms developing open innovation strategies with public partners in smart cities projects. For the last decade, the management of inbound and outbound knowledge flows is increasingly relevant in corporate strategies. Under the shadow of this new knowledge strategy, academia has tried to understand this topic producing a growing stream of literature on open innovation from Chesbrough's (2003) seminal research. However, most of this research has focused on knowledge partnerships between firms. Both mainstream literature on open innovation and the older research on r&d cooperation have traditionally neglected the r&d cooperation between firms and public organizations (Perkmann et al., 2013; Sandulli et al., 2017).

Nowadays, an increasing number of public organizations are participating in open innovation strategies (Lee et al., 2012) in smart city projects (Sandulli et al., 2017; Scuotto et al., 2016; Ferraris et al., 2017). In fact, some of the open innovation adoption in public organizations is explained by the growing interest of policymakers for smart cities programs. However, partnerships with public organizations in open innovations strategies differ from partnerships with private organizations in a number of ways. In first place, public organizations do not need to compete and therefore have a lower pressure to innovate (Perry and Rainey, 1988). In second place, decision making in public organizations are moderated by political action and strict bureaucratic processes, creating significant organizational slack and misalignment of objectives with private partners (Damanpour and Schneider, 2009). In third place, public organizations have weaker absorptive capacity and therefore their ability to profit from new technologies may be limited, reducing the attractiveness of new technologies (Inkpen and Beamish, 1997). For these reasons, firms need to develop specific open innovation strategies when dealing with public partners.

Within the scope of smart cities programs, firms and public organizations tend to cooperate to develop innovations, which usually are more radical than incremental (Sandulli et al., 2017). Nowadays, firms are increasing the number and the relevance of their alliances within smart cities because modern cities are a great locus of innovation (Florida, 2003; Scuotto et al., 2016; Ferraris and Santoro, 2014). So, the concept of "smart city" has become quite popular between scholars and practitioners. One of the well-known definitions of a smart city is "a city that aims at connecting the physical infrastructure, the IT infrastructure, the social infrastructure, and the business infrastructure to leverage the collective intelligence of the city" (Hollands, 2008).

Firms involved in smart cities projects usually follow a business model experimentation approach because of the high technological risk of a large number of these projects. Most of the literature on smart cities has studied the phenomenon either from a technological perspective or from the perspective of the innovation ecosystem which is closer to the lens and challenges of policymakers (Almirall et al., 2014; Neirotti et al., 2014). In this research we make a relevant contribution focusing on the firms' perspective and on how firms deploy strategies depending on the characteristics of the smart city project. More specifically, we address the following main key challenges for open innovation with public partners in smart cities: (i) the governance of public-private alliances; (ii) the role of public managers; (iii) the management of intellectual property rights with public organizations.

2   Methodology and results of the case studies

This study uses a multiple-case research design. We have studied the strategies of seven large corporations with a clear open innovation strategy in different smart cities. We used multiple data sources. First, we collected and analyzed extensive secondary materials. Then, we conducted deep semi-structured interviews on the different aspects of the project and their open innovation (OI) strategy with each project manager for these smart city projects.

Our data shows that firms used smart cities as testbeds of future technologies and products and services. However, current procurement processes limited the scalability of successful projects to a larger number of citizens. Fund sourcing was critical for assessing the risk of the project for the city. When research funds came from the firm or other governmental bodies (mostly European public bodies), public organizations were more flexible in managing the relationship. However, when funds from the public organization were required, innovation projects were treated as normal procurement projects, creating several pitfalls and disfunctionalities. Risks also defined the role of public managers. For low risk projects, public managers played the role of coordinators of the network of actors involved in the project leaving the leadership of the project to private partners, while for higher risk projects they took the leadership and reinforced the formal controls over the project.

Because of the high uncertainty of smart city exploration projects, most managers in our interviews argued that firms and city governments required more informal alliance governance mechanisms in order to forge a more flexible and adaptive relationship. However, despite the reluctance of public partners, some of the firms in our sample supported the adoption of equity governance mechanisms in smart cities initiatives. Equity governance is useful when partners have concerns over the retention of intellectual property (IP) outcomes. Furthermore, the appropriability regime was stronger when shared knowledge relied on the core competencies of the firm and when experimentation strategies of the firms were more focused on business models than on technology. When risk was much lower, the project was more focused on initial stages of a technology, the distance of the technologies to the market was larger and consequently the appropriability regime was weaker also because firms were looking for cross-fertilization of ideas and projects throughout the smart city ecosystem. For high-risk projects, scalability of initial pilots was only possible under innovation-friendly organizational processes.

Despite this, contractual provisions are the most common governance mechanism in smart city alliances, since they are more flexible than joint ventures and at the same time they may help managing exchange uncertainty in a variety of ways. The high levels of uncertainty especially in most exploration smart city projects require resource flexibility and continuous mutual adaptation. However, public-private contracts are based on traditional modes of public procurement where public partners require to know the solution they need in advance and to work with prescreened vendors. These standard procedures are highly bureaucratic and involve detailed request for proposals, a cumbersome and slow selection process, and messy contract negotiations. Therefore, these procedures are a strong barrier to innovation since they often make it challenging for public partners to work with entrepreneurs, start-ups or SME, as well as to attract private partners for small projects where transaction costs and bureaucratic efforts may overwhelm the expected profits. City governments in several cities recognized public procurement procedures as one of the major challenges in smart city projects and consequently promoted more flexible contracts, longer-term relationships, raised the threshold for the instigation of the official procurement process or relaxed some of the rules around private-public partnerships.

Finally, we found that the development of the innovation ecosystem was built through the own firms' network in the case of private partners in the ecosystem, while public organizations were bringing the support of public partners to the project. Within this respect, OI units and managers are playing an active role in organizational learning in smart city projects of the firms we interviewed. All of them had an oi strategy and specific oi units with well-defined processes and tools for both the recognition of external knowledge and the internal absorption and integration of these external inputs. OI units supported smart city units in developing specific oi mechanisms to scan external knowledge related to urban services and solutions.

3   Concluding remarks

In general, firms' alliances with city governments are growing with the business opportunities arising from the application of technological innovation to urban services. In the face of high technological uncertainty and the significant differences on how each city conceives the smart city concept, firms pursue several experiments in urban environments to map out the best alliance management routines, including partner selection and alliance governance. In this abstract we have briefly discussed how firms can address public-private alliances to increase the success of their projects in smart cities.

Overall, this exploratory research puts forward an in-depth understanding on how open innovation projects in smart cities start, reach a pilot stage and in most do not go further because of the risk aversion of public organizations and the legal constraints of public managers in developing large-scale radical innovation projects. The research provides oi managers in innovative firms with valuable knowledge on what will be the main challenges of public-private open partnerships and suggests some good practices to overcome in a fresh way the challenges of partnering with public administration in innovation projects.

Notes

1 University of Turin.
2 University of Turin.
3 University of Rome "Link Campus."

Bibliography

Almirall, E., Lee, M., & Majchrzak, A. (2014). Open innovation requires integrated competition-community ecosystems: Lessons learned from civic open innovation. *Business Horizons*, 57(3), pp. 391–400.

Chesbrough, H. (2003). *Open innovation: The new imperative for creating and profiting from technology.* Boston, MA: Harvard Business School Publishing.

Damanpour, F., & Schneider, M. (2009). Characteristics of innovation and innovation adoption in public organizations: Assessing the role of managers. *Journal of Public Administration Research and Theory*, 19(3), pp. 495–522.

Ferraris, A., Erhardt, N., & Bresciani, S. (2017). Ambidextrous work in smart city project alliances: Unpacking the role of human resource management systems. *International Journal of Human Resource Management*, pp. 1–22. doi: 10.1080/09585192.2017.1291530.

Ferraris, A., & Santoro, G. (2014). Come dovrebbero essere sviluppati i progetti di social innovation nelle smart city? *Un'analisi Comparativa, Impresa Progetto: Electronic Journal of Management*, 4, pp. 1–15.

Florida, R. (2003). Cities and the creative class. *City & Community*, 2(1), pp. 3–19.

Hollands, R. G. (2008). Will the real smart city please stand up? Intelligent, progressive or entrepreneurial? *City*, 12(3), pp. 303–320.

Inkpen, A. C., & Beamish, P. W. (1997). Knowledge, bargaining power, and the instability of international joint ventures. *Academy of Management Review*, 22(1), pp. 177–202.

Lee, S. M., Hwang, T., & Choi, D. (2012). Open innovation in the public sector of leading countries. *Management Decision*, 50(1), pp. 147–162.

Neirotti, P., De Marco, A., Cagliano, A. C., Mangano, G., & Scorrano, F. (2014). Current trends in smart city initiatives: Some stylised facts. *Cities*, 38, pp. 25–36.

Perkmann, M., Tartari, V., Mckelvey, M., Autio, E., Broström, A., D'este, P., & Krabel, S. (2013). Academic engagement and commercialisation: A review of the literature on university-industry relations. *Research Policy*, 42(2), pp. 423–442.

Perry, J. L., & Rainey, H. G. (1988). The public-private distinction in organization theory: A critique and research strategy. *Academy of Management Review*, 13(2), pp. 182–201.

Sandulli, F. D., Ferraris, A., & Bresciani, S. (2017). How to select the right public partner in smart city projects. *R&D Management*, 47(4), pp. 607–619.

Scuotto, V., Ferraris, A., & Bresciani, S. (2016). Internet of things: Applications and challenges in smart cities: A case study of IBM smart city projects. *Business Process Management Journal*, 22(2), pp. 357–367.

The role of international educational and science programs for sustainable development (systemic approach)

Ineza Gagnidze[1]

Keywords: *sustainable development; education; science; Entrepreneurial University; system*

1 Weak and strong sustainability

Since the 1960s, natural and social scientists have highlighted a series of sustainable development issues and recommended integrated policy action and commensurate means of implementation, such as technology, finance, capacity building and trade. From that time to the present, the research of the problems of sustainable development has been expanded in many organizations.

> The desire of developed countries to hold leading positions in the future converges the desire of less developed countries to develop their economies. This mutual interest facilitates the unification of these countries under a common goal, which is called the overcoming of the crisis and sustainable development.
> (Gagnidze, 2014, p. 6)

> Sustainability is a set of global problems that affect all humans equally and can be solved through appropriate application of science and technology. It hides the fact that some humans are more responsible for environmental degradation, some humans have benefitted more from environmental degradation, some have had more opportunities to participate in decision making about socio-environmental issues, and some have suffered more from their consequences.
> (Feinstein and Kirgchgasler, 2014, p. 11)

Sustainable development is development that meets the needs of the present without compromising the ability of future generations to meet their own needs.

It's known that sustainable development means the unification of three different spheres (sometimes four spheres are indicated: environment, society, culture and economy) in one system. The experience of studying sustainable development in separate countries has a short history. Weak sustainability and strong sustainability models (see Figure 30.1 "a" and "b") are actively discussed currently. The classical model of sustainable development treats economics, environment and

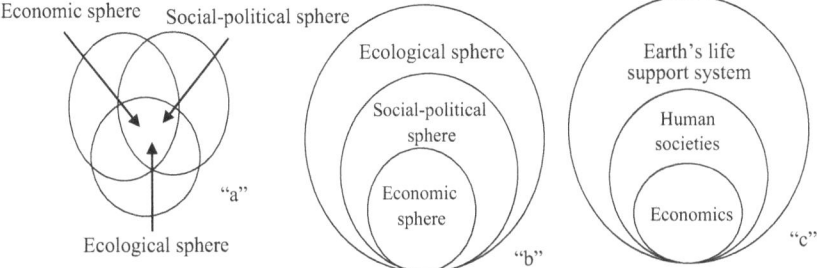

Figure 30.1 Weak sustainability ("a") and strong sustainability ("b") and the hierarchical model of sustainability ("c")

Source: Redrawn from Fischer et al. (2007) ("c"); Authors' elaboration

society essentially as three separate spheres, with only some overlap in the model; see Figure 30.1a.

Strong sustainability is an alternative model of sustainable development in which economy is embedded in society; society and economy are embedded in the ecological sphere. Society and economy are interdependent with ecology and cannot function without a healthy ecological sphere (Figure 30.1b). In its content it is like the strong sustainability and reflects the similar dependence of the approach on the hierarchical model of sustainability, redrawn from Fischer et al. (2007) (see Figure 30.1c).

There are other opinions expressed in the scientific paper, namely Sumi (2007 p. 75) argues that

> it is necessary to clarify the structure of environmental issues . . . the environment consists of three systems – the natural system, the social system, and the human system – and their mutual interaction. Environmental problems can then be defined as perturbations of these interactions; to solve such problems we must find a way to restore these interactions. Research activity should be organized in accordance with this structure. It is particularly important to note interactions with the human system. Environmental issues have a time horizon. To take action we need the agreement and support of society.

It's evident that achieving a strong sustainability policy in each country requires the elaboration of consistent and effective thought-out policy. This is connected with great difficulties.

2 The model of possible systemic effects of the international educational and scientific links

One of the most important supporting factors in the formation of sustainable economy of any country is to build and operate an effective education and science system. However, it is noteworthy that the efficiency of one education and science system, even of a very big country, can't solve the global problems, especially sustainable development problems.

> The difficulty of solving this problem is determined by the fact that sustainable development is connected to social-cultural sphere. This makes it difficult to establish worldwide universal conclusions. The education and science is exclusion under this context. We think they are the shortest and the most effective ways for reaching sustainable development simultaneously in many countries.
>
> (Gagnidze, 2014, p. 2)

Education changes the behavior of individuals and at the same time influences the three components necessary for sustainable development. Here, large-scale international education and science programs play a great role. In the European Union significant funds are allocated for the implementation of such programs and projects, namely Erasmus+ and Horizon2020.

In order to spend the mentioned funds efficiently, we believe it is necessary to take into consideration the formation requirements of sustainable economy during the implementation of the programs. In this regard, it is recommended to:

- Achieve a strong sustainability;
- Increase the number of participants majoring in sustainable development in the Erasmus+ and Horizon2020 programs;
- Elaborate new education programs considering the sustainable development;
- Support the consideration-implementation of the approach of sustainability science;
- Implement the education and science programs so it is mutually beneficial that the technology is transferred and the expertise is spread from the developed countries to the developing countries, thus effectively expanding the scope in solving general problems; develop the Entrepreneurial University Model, which plays the main role in achieving the systemic synergistic effect for more effectiveness.

We think that the effective implementation of the above-mentioned approach will be possible when the education and science policy is designed so that it obtains the synergistic effects characteristic for the system. Taking into consideration the above mentioned, we suppose it would be reasonable if the government of developing country introduces a multi-level policy of education and science and uses a cluster-based approach to manage them. We wrote earlier regarding these levels, namely (see Figure 30.2):

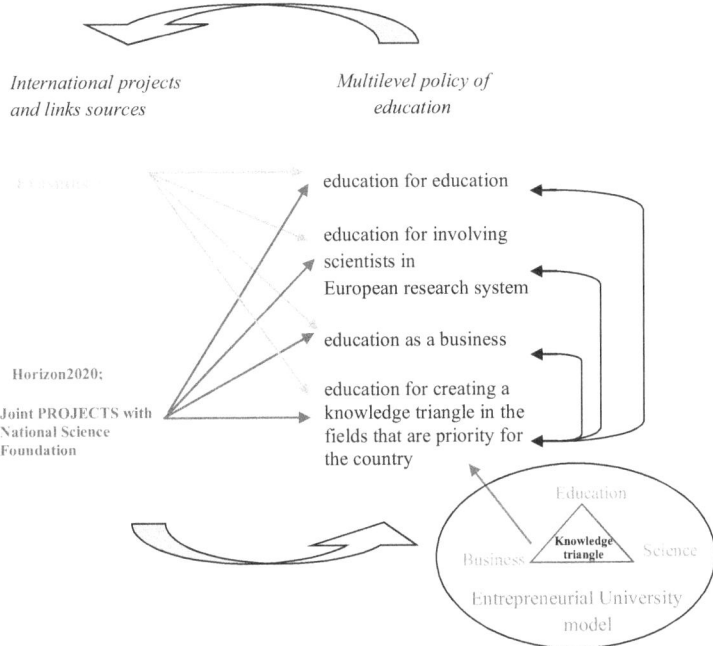

Figure 30.2 Mutually beneficial effective circular system

- *Education for education* (working on improving the training and retraining system for teachers in secondary schools, higher education and vocational institutions and improvement of curriculums);
- *Education for involving scientists of developing countries in international research systems* (development and implementation of different activities for promoting scientists' participation in international and joint scientific and scholarship programs);
- *Education as a business*;
- *Education for creating a knowledge triangle in priority fields of the country* (special attention should be paid to carrying out various activities for making science and education meet international standards in the fields where the country has competitive advantage).

(Gagnidze and Maisuradze, 2016, p. 39)

The model shows possible systemic synergistic effects of the international educational and scientific links. Such systemic approach is directly connected to the innovative development of the country. It supports to develop the

Entrepreneurial University model, which is in the core of the sub-system on Figure 30.2.

> "Sustainability" implies that the critical activities of a higher education institution are ecologically sound, socially just and economically viable, and that they will continue to be so for future generations. A truly sustainable college or university would emphasize these concepts in its curriculum and research, preparing students to contribute as working citizens to an environmentally healthy and equitable society. The institution would function as a sustainable community, embodying responsible consumption of energy, water, and food, and supporting sustainable development in its local community and region.
>
> (Association of University Leaders for a Sustainable Future, 2015)

3   Education and science for sustainable development

> Education for Sustainable Development (ESD) approaches and initiatives are diverse, and vary all across the world. To be effective, ESD processes must take into account the different sustainable development needs of the region, in addition to particular environmental contexts, cultural diversity and socio-economic, political and educational systems.
>
> (UNESCO, 2013)

The role of education and science in solving the problem of sustainable development is critical. The system of education and science in developing countries fails to meet modern challenges. We think it's desirable to start working on a different model. For this purpose we have developed a new approach. It connects the clusters and alternative design pattern of the education system for sustainable development.

> The university's contribution to innovation in economic and social development is the heart of the entrepreneurial university concept. Academic entrepreneurship transcends simple knowledge capitalization as the university interacts with innovative actors from other institutional spheres to promote regional growth. The interactions form a university – industry – government triple helix. The transformation of the university from a secondary to a primary institutional sphere is a key element in the triple helix.
>
> (Etzkowitz and Zhou, 2008)

Based on all the above mentioned we can make several conclusions:

- With the help of the alternative design pattern for the education system, the management of education and science becomes systemic, complex and effective;
- Consideration of sustainable development by international educational and science programs through the alternative design pattern for educational systems will be quickly implemented. In this case, the alternative design pattern will be an effective intermediate supporter.
- This alternative design pattern for educational systems contributes to the formation of innovative system of developing countries.

- The alternative design pattern for the education system serves as a guide for other clusters as technology transfer, introduction of innovations; obtaining and generation of new knowledge are easier in these fields. We consider that in this way it accelerates fast and innovative recovery of developing economy.
- Education and science, within the framework of the Entrepreneurial University Model may play a universal role to cope with the problems of sustainable development.

Note

1 IvaneJavakhishvili Tbilisi State University.

Bibliography

Association of University Leaders for a Sustainable Future (2015), Retrieved 21.9.2018 from http://ulsf.org/about/

Environment, Sustainable Development and the University in Africa, Module 1, Retrieved from www.unep.org/training/downloads/toolkit/4.0%20-%20 Module%201.pdf. p. 33.

Etzkowitz, H. and Zhou, Ch. (2008). Introduction to special issue Building the entrepreneurial university: A global perspective. *Science and Public Policy*, 35(9), 627–635, Retrieved from www.ingentaconnect.com/content/beech/spp

Feinstein, N. W. and Kirgchgasler, K. L. (2014, July 8). Sustainability in Science Education? How the Next Generation Science Standards Approach Sustainability, and Why It Matters, Retrieved from http://dces.wisc.edu/wp-content/uploads/ sites/30/2013/08/sustainability-in-science-ed.pdf

Fischer, J., Manning, A. D., Steffen, W., Rose, D. B., Daniell, K., Felton, A., . . . & MacDonald, B. (2007). Mind the sustainability gap. *Trends in Ecology & Evolution*, 22(12), 621–624.

Gagnidze, I. (2014). The Role of International Educational Programs for Sustainable Development, Proceedings, Business Systems Laboratory 2nd International Symposium 'Systems Thinking for a Sustainable Economy. Advancement in Economic and Managerial Theory and Practice', January 23–24, Universitas Mercatorum, Rome, Italy.

Gagnidze, I. and Maisuradze, N. (2016). Systemic effects of international educational and scientific links: Proposals for the development of educational and scientific national system in Georgia. *International Journal of Markets and Business Systems*, 2(1), 25–44.

Global Sustainable Development Report (2013, September). Building the Common Future We Want, Executive Summary, Retrieved from http://sustainabledevelopment.un.org/content/documents/975GSDR%20Executive%20Summary.pdf

National Journeys, towards Education for Sustainable Development, United Nations Decade of Education for Sustainable Development 2005–2014, United Nations Educational, Scientific and Cultural Organization (2013), Retrieved from http:// unesdoc.unesco.org/images/0022/002210/221008e.pdf

Sumi, A. (2007). On several issues regarding efforts toward a sustainable society. *Integrated Research System for Sustainability, Science and Springer*, 2(1), 67 76.

Formation of an environment suitable for the improvement of the education level of human resources and production of highly qualified staff in developing countries

Irina Gogorishvili[1]

Keywords: *sustainable economy; global crisis; the world organization of education; knowledge economy; Internet of Things*

1   Introduction

In order to overcome a global economic crisis on the basis of the sustainable development of the world economy, it is necessary to introduce strict economy and control over the use of resources. Meanwhile, this process will not take place without aggressive opposition of national interests, as developing countries do not own national investment capital required for the production of innovative technologies and products. Even some developed countries have insufficient amount of such capital.

The study aims to devise an economic policy that stimulates the enhancement of knowledge (education) of human resources and production of highly qualified workforces in developing countries.

Considering this goal, we seek to accomplish a few tasks including some of the most important ones below:

- To corroborate the necessity of the production of educated and highly qualified labor resources in developing small open economies (SOEs);
- To develop the areas of the economic policy for the formation of scientific and educational systems focused on knowledge economy; and
- To identify the chances of preventing obstacles related to the development and execution of an effective economic policy.

Our study is based on works by foreign and national scientists regarding knowledge economy and scientific and educational systems, and strategic activity of international organizations. Theoretical bases of the study include the Product Life Cycle, Economies of Scale and Gravitational theories. We based our work on the logical analysis and non-empirical research methods.

2 The problem of developing human resources in developing countries

Implementation of innovative technologies and production of innovative technical products (which ensures overcoming the global crisis) take place on the basis of the strict economy of investment capital; as for human capital, demand for it has long exceeded supply, especially in developing countries. Meanwhile, the lack of qualified workforces and the dissatisfactory development level of scientific and educational systems, that hinder development and execution of an effective economic policy in SOEs, should be regarded as a problem. The lack of educated people is the main reason for the inefficient use of investment capital and natural resources. To what extent is it possible to solve problems related to the lack of education and professional competence in developing countries? There are international financial-economic organizations that should play a special role in the field of production of completely new innovation-oriented projects. With the help of these organizations a new type of economic order should be created and established which is acceptable to countries and regional integrations. Clearly, developed and developing countries should be playing the main role in this process through the development of international scientific and technical cooperation.

The idea of the development of labor resources (human capital) is closely linked with ecology issues of society. In terms of socio-ecology, not only the application of scientifically approved research methods gains in importance, but the negative aspects of interdependence within human society and the outcomes of their impact on the environment, which are yet poorly studied (there is even less information available about risk insurance), are also noteworthy. High level of development of the education system is the most effective solution to socio-ecological problems. Important prerequisites for sustainable economic development of developing countries (that determine their place in the global system of division of labor) should correspond to the: developed labor resources and national education system as well as the level of integration (engagement) of this system into the areas of regional education.

In this respect, we should differentiate two (short- and long-term) periods of time. Due to a wide range of obstacles occurring in the processes of economic development in developing countries, risk prevention is put on the agenda of international organizations. Risks are, above all, related to the orientation of the economic policy over short time periods.

3 The creation of an international education organization determines the opportunities for sustainable development of developing countries

The process of the creation of a new global system of division of labor leads to sustainable development of the world economy. It is noteworthy that this system is presently being formed and its formation will be based on the implementation of basic innovations. Theoretical bases of the new global system of division of labor include scientific works of a number of scientists specializing in the Product Life Cycle and Scale Effect theories: Raymond Vernon, J. S. Haffbauer, Paul Krugman, James M. Lutz, Robert T. Green, Ian H. Giddy, Stephan B. Linder, Frank Graham, Elhanan Helpman, Henryk Kierzkowski, Michael Porter and AnnaLee Saxenian among others. A new generation of scientists such as Pfeiffer (2015); Braun (2010); Sotoudeh, Peissl, Gudowsky, Jacobi (2011); Spök (2006) have begun to develop theoretical questions of IoT (Internet of Things).

As a result of efficient use of colossal funds invested in the creation of knowledge economy and education systems (which will only be possible through joint effort of a coordinating and controlling organization), plenty of financial and material resources will be saved in the world economy. Education policy should become an essential constituent of the global economic policy (global economic order).

Successful integration with educational systems should be implemented slowly and in a detailing way – says scientist Hannah Augur (2016). International organizations such as: World Bank, WTO, WLO, JATA, etc. should concentrate their activities on this process. Their actions and aims should sustain significant revolutionary changes; at the same time processes and events they engage to promote new technological products are absolutely insufficient because of the following reasons:

- In spite of challenges in high technological cycles, with the most important slogan-people changing from IT product users to producers that ensure the realization of professional skills and aims of each member of the world society; worldwide processes proceed very slowly because of problems regarding utilization of investment capital;
- It requires a worldwide effort to transfer the work force of different specializations from IT soft product users to their producers, and no single Trans National Corporation (TNC) is capable of doing it because of two reasons: first, producers of soft products are not interested in starting large-scale campaigns of educating in soft products and the second reason is – in spite of major success and capital growth – IT soft producer companies are unable to provide investment capital for this process;
- The fourth industrial revolution that has started and will proceed in the sixth technological cycle (wave) may cause deep changes in world society. Differentiation processes may bring us to emergence of new countries and nations according to economy levels, and may cause some instability;
- Industry 4.0 process will create links between Information Technology (IT) systems and necessity of their (products) production. There will be leading as well as second and third degree roles in these processes and their performance will be differentiated in developed and developing countries too.

Due to the mentioned reasons, international organizations of education and science should form along with the existing economic-financial, labor, truism, healthcare, air transport and other organizations. Its functions will include promotion, spreading and propagation of knowledge on IT soft product creation. Besides, as long-term aims it will be necessary to have changes in educational sphere orientation and to create international classifiers of specialties (with regard to changes to rhythm).

In January 2014, International Organization of Standardization (ISO) issued a new series of standards ISO55000(2). ISO is oriented on the creation of a common basis for administration of assets.

The main purpose of programs on effective management of assets in the whole chain of value creation is the increase of asset values throughout their life cycle for all stakeholders. In the initial stage of the new technology cycle, along with development of skilled labor features and quality indicators, demand on work resources will be strictly differentiated. The labor market will have excessive demand on specialties engaged in IoT program product creation. Undoubtedly those constituents of labor forces will have permanent demand, will be in an asset position and thus for managing such assets internationally (for this is the only means to rationally use them); TNC activities and soft products that they create (e.g. IBM Maximo Asset Management and IBM Watson IoT) will not be enough.

The main purpose of educational policy in developed and developing countries should be formation of personalized educational system on every level (primary, secondary, special and higher). This system should be oriented on the whole life cycle of labor resources. Setting up specialties, learning courses and curriculums on the basis of cognitional technologies systems will give us the possibility to create a personalized educational system. Under these conditions we will be able to maximally increase the chance of effective usage of labor resources. TNCs have started working in this direction but their economic aims are somehow different from aims of society development, so to save time, it is necessary that the government stands as the initiator of any changes in this direction on the market.

The possibility of receiving personalized education will improve results of continuity of the educational process for students, professors and administrative workers. It is definitely difficult to achieve especially in developing countries, because of insufficient investment capital in the area of education. In spite of the situation we can control processes by means of regional integrations and international organizations (if they are equipped by innovative instruments of educational policy). There is a growing difference between knowledge provided by educational systems and skills demanded by the current global market. Educational systems (with insignificant exceptions) are in collapse in developing countries. These countries cannot cope with the situation independently, but if we cooperate on an international scale regarding changes in national educational systems, then world society will have a scale effect and rationality (and efficiency) of using labor resources will grow.

4   Conclusions

As a result of economical use of vast resources invested in creation of the knowledge economy and educational system (that will be possible only by common coordinating and controlling organizations' efforts) a large amount of financial and material resources will be saved in the world economy. Educational policy should become an essential part of global economic policy (global order).

Integration of IT and IoT with educational establishments (schools and universities) is within TNC interest and it supports its implementation. Authorities of developed countries show interest towards new IT soft programs and IoT, but it is fractioned and inconsistent. As for developing countries we only see interest from TNC.

Creation of a personalized educational system should be oriented to the whole life cycle of labor resources. Composing specialties, learning courses and curriculums on the basis of cognitional technologies systems will give us the possibility of creating personalized educational systems.

Speeding up the process of digital technologies utilization in educational systems is hindered by the absence of unified standards and rigid qualification classifiers.

A population aging process is going on in the world. It is irreversible and inevitable even without fast integration of innovative tools with every sphere of social life. We will face: deficiency of labor resources, international and regional economic conflicts, insufficient investment capital and other complex problems difficult to solve.

Note

1 Ivane Javakhishvili Tbilisi State University, Georgia.

Bibliography

Augur, H. (2016). Getting into data science: A guide for students and parents. Retrieved from http://dataconomy.com/author/hannah-augur/

Braun, E. (12/2010). The changing role of technology in society. Retrieved from http://epub.oeaw.ac.at/ita/ita-manuscript/ita_10_03.pdf

ISO55000 standards for asset management. Retrieved from www.assetmanagement-standards.com/

Lutz, J.M., Green, R.T. (1983). The Product Life Cycle and Export Position of the United States. *Journal of International Business Studies*. Winter, pp. 77–93.

Mager, A. (11/2013). In search of ideology: Socio-cultural dimensions of Google and alternative search engines. Retrieved from http://epub.oeaw.ac.at/ita/ita-manuscript/ita_13_02.pdf

Pfeiffer, S. (2015). Effects of industry 4.0 on vocational education and training. Retrieved from http://epub.oeaw.ac.at/ita/ita-manuscript/ita_15_04.pdf

Sotoudeh, M., Peissl, W., Gudowsky, N., Jacobi, A. (12/2011). Long-term planning for sustainable development: CIVISTI method for futures studies with strong participative elements. Retrieved from http://epub.oeaw.ac.at/ita/ita-manuscript/ita_11_03.pdf

Spök, A. (12/2006). From farming to "Pharming": Risks and policy challenges of third generation GM crops. Retrieved from http://epub.oeaw.ac.at/ita/ita-manuscript/ita_06_06.pdf

Importance of education in Georgia in terms of sustainable development of tourism

Marina Metreveli[1]

Keywords: *natural resources; protected areas; eco-tourism; sustainable development of tourism; environmental education*

1   Introduction

Progress of the society and all-round improvement of living standards are peculiarly related to ecological substantiation of economic development. The ultimate goal of use of nature is satisfaction of the social, economic and environmental requirements of mankind. Under the modern conditions, the individual and public requirements can be met if the economy is developed in environmentally rational way, implying sustainable development of the economy and the sectors thereof (Metreveli, 2011).

The necessity of consideration of the environment and development in the common light of politics is widely recognized worldwide since the establishment of "sustainable development" in capacity of the guiding principle of social development. In many countries, consideration of environmental aspects in relation with economic and social factors is one of the primary institutional challenges.

In general, integration of policy is an uninterrupted process allowing the consideration of the intersecting issues and preventing possible discord between the sectors. Integration of environmental policy ensures consideration of the environmental issues upon development of the policy of all related sectors. It is a practical approach unlike the reactive approach when environmental policy responds to the negative outcomes of socio-economic policy and practice instead of being integrated upon early development thereof. Integration of environmental policy shall result in general improvement of economic policy and implementation thereof. Consideration of all environmental issues upon development of economic policy might appear complicated, though the main trend shall be directed to sustainable development, implying healthy environment and social welfare along with economic development.

The paper aims at analysis of the current state of sustainable development of tourism and environmental education (EE) in Georgia; as well as development of the incentive mechanisms for their development.

The methods of the study are: a systemic approach to the evaluation of the outcomes of sustainable development of tourism and environmental education, and induction, deduction and statistical methods.

The study aims at the intensification of the state approach to the environmental education for all age groups and social layers of the population of Georgia and facilitation of the sustainable development of tourism. The unique nature of the paper lies in the fact that it provides the recommendations related to the issue, envisaging the top order undertakings of the government and the model of the system approach to the sustainable development of tourism and environmental education.

2   Sustainable development of tourism in Georgia

Famous tourism researchers Dredge and Jenkins note the controversial attitude of the civil society subjects towards tourism components, entailing conflicts of interests. "Government, business and community actors involved in the activities associated with tourism planning and policy will all respond differently to these settings, communications and relationships" (Dredge and Jenkins, 2011, p 34). Conflict-related fluctuations at this stage of tourism development (when the government sets a high pace for economic development to be achieved) are further enhanced. Taking the opinions of these scientists into account, mitigation of contradictions entailed with conflicts of interests requires emphasizing the priority of sustainable development to reduce the desired pace of development.

Eco-tourism (especially in developed countries), as the primary direction of sustainable tourism, shall be recognized as the primary economic means ensuring alternative income sources for the local population as on protected areas as well as on adjacent territories. Participation of the local population in modern environmental concept in ethical as well as in strategic terms, is recognized as a priority for better support of environmental protection. One of the basic principles of sustainable development of eco-tourism is provision of engagement of the local population therein. Often, locals are artificially delivered from this process (Gogelia, 2014).

There are mainly two types of local participation in environmental activities (Metreveli, 2011: (i) political, implying the right of locals to participate in decision-making on protected areas; (ii) economic, implying benefits of locals from the protected area or compensation from restriction of usage of the territory. Ross and Wall (1999), in their studies, underline the link between eco-tourism and sustainability – environmental protection and sustainable development.

Eco-tourism is a segment of sustainable tourism implying visits to relatively calm natural landscapes, including protected areas. Despite the absence of accurate data, approximately 15–20% of international tourism is attributed to eco-tourism. Growth pace of eco-tourism and other nature-based tourist activities shall be the highest compared to other tourist segments and may be equal to approximately 15% annually (Ceballos-Lascurian, 2000).

The National Park program is mainly focused on development of visitor service. This comprehensive network of protected areas covers East, Central, South and West Georgia. At present, there are 68 protected areas countrywide, most of which are capable to be subject to sustainable development through environmental and entertainment events. These parks and eeserves cover 500,000 hectares, which constitute 8% of the whole territory of Georgia (Kikodze & Gokhelashvili, 2006).

Eco-tourism can't independently ensure development of natural monuments in the region. It shall be considered as an important component of the traditional economy or sustainable development of these territories, providing additional incomes. Thus, travel business facilitates development of the local economy, and local participation ensures correct administration of environmental activity and formation of attractive and unique tourism production. Thus, it is important to define the percentage of the total number of tourists traveling to the protected areas as eco-tourists or those just visiting protected areas.

Table 32.1 provides that number of tourists visiting Georgia during 2007–2015 increased 5,6 times and to the protected areas – 67 times. Simultaneously, tourism-gained income increased 5,04 times in the country and 46,17 times in protected areas.

Table 32.1 Total number of tourists traveling to Georgia and comparison of due incomes with incomes from tourists traveling to protected areas according to years 2007–2015

Year	Total number of tourists	Number of tourists on protected areas	Number of tourists on protected areas/total number of tourists	Total income of tourism	Income from tourism to protected areas	Income from tourism to protected areas/total income from tourism
2007	1,051,749	7,714	0,73%	383,745	34,427	8,97%
2008	1,290,108	12,226	0,95%	446,645	55,686	12,47%
2009	1,500,049	68,761	4,58%	475,889	80,452	16,91%
2010	2,031,717	126,209	6,21%	659,245	107,095	16,25%
2011	2,822,363	303,686	10,76%	954,908	420,189	44,00%
2012	4,428,221	298,910	6,75%	1,410,902	736,464	52,20%
2013	5,392,303	355,681	6,60%	1,719,700	993,107	57,75%
2014	5,515,559	420,166	7,62%	1,787,140	1,320,793	72,91%
2015	5,901,094	518,218	8,78%	1,935,915	1,589,756	82,12%

Sources: The table compiled by the author is based on the statistic data from: www.apa.gov.ge/ge/angariSebi and www.gnta.ge and Geostat.ge

3   Trends and elaboration of the main directions of EE in Georgia

Ecological education is the multi-disciplinary training forming environmental awareness, values and skills allowing the society to participate in the maintenance and improvement of the environment. The training and equipment of future generations with respective environmental knowledge, skills and sustainable environmental attitudes is an urgent task. Environmental education facilitates the population to identify the environmental and related sustainable development problems, allowing problem analysis and solution (Annual reports of APA – Agency of Protected Areas, 2015).

The population of the agricultural-economic settlements has the least education, especially those residing near the National Parks (for the population of all age and social affiliations). It requires early provision of environmental information so that these individuals can play an important role in the protection and management of these territories, being the holders of the national heritage.

Correspondingly, one of the main directions of the APA is EE for local population. The APA, in view of public awareness, holds: meetings, lectures, workshops and trainings according to the green calendar, eco-lessons in the field, green actions, eco-tours and other environmental activities.

Figure 32.1 provides the trend of engagement of the population residing near protected areas in EE during 2006–2016. Within 2006–2016, the number of beneficiaries has increased averagely with 39%, namely in 2007 – with 65,08% compared to 2006; in 2008 – with 12,43%, in 2009–2010 with 22,29%, in 2010–2011 with 158%, in 2011–2012 with 46,33%, in 2012–2013 reduction with 9,83%, in 2013–2014 with 2,28% and in 2014–2015 with 29,96%.

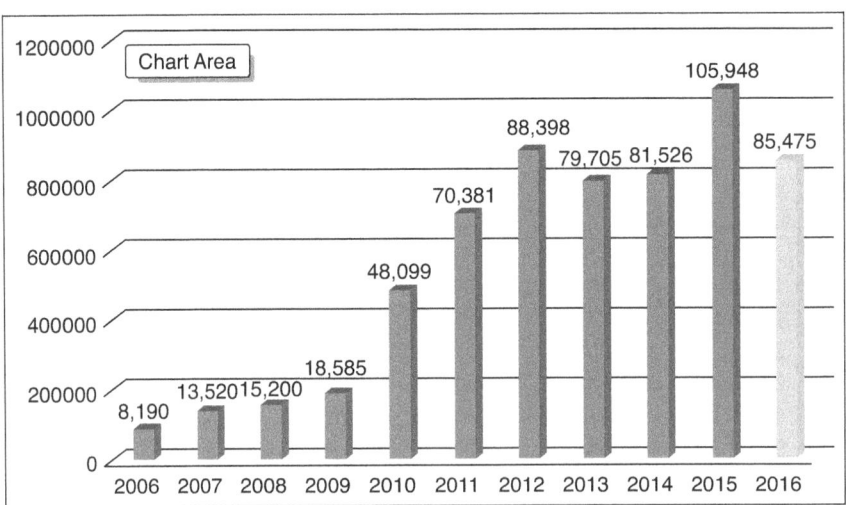

Figure 32.1 Dynamics of the number of beneficiaries engaged in EE for 2006–2016

Source: LEPL Agency of Protected Areas (2016), www.apa.gov.ge/uploads/other/3/3166.pdf

Because most of the population travels to protected areas, they shall be the carriers of systemic and diverse information about protected areas, sustainable development of tourism and environmental issues. Pre-schools, schools and universities countrywide launched special subjects on environmental protection but the material resources and academic methods are not enough for trainings and education on the environment and sustainable development, evidenced with the scarce financing allocated by the LEPL EIEC under the State Budget 2017 (100000 GEL).

As a result of the hereof Article, we might outline the following top-order undertakings of the promotion policy of sustainable development of tourism and environmental education in Georgia:

"Tourism Strategy 2025" shall necessarily envisage activities such as

- Advertisement and campaigning activities to attract tourists, develop eco-tourism and promote protected areas;
- Undertakings to improve environmental behavior of certain groups and increase public awareness on ecology;
- Creation of favorable conditions for recreation in natural and historical-cultural milieu, healthcare and tourism to increase the eco-tourism potential of Georgia.

Systematization of "environment and sustainable development" requires elaboration of and public awareness on main goals (to support among all age groups of Georgian population)

- Concept of sustainable development;
- Interrelation of sustainable tourism and environment and importance of eco-tourism;
- Global environ and respective threats – air, oil and water pollution, erosion resource insufficiency, wastes, climate change, types and eco-system, environmental degradation issues (desertification, forest loss, climate change, bio-diversity loss);
- Renewal energy;
- Environmental instruments;
- Legal instruments – international conventions, declarations and agreements, environmental legal basis.

The above-mentioned factors can be represented in the capacity of the system approach model of interrelation between environmental education, sustainable development of tourism and state policy in Georgia (Figure 32.2). Inculcation of these issues from childhood is the fundamental factor facilitating the development of environmental protection and sustainable development of tourism in Georgia.

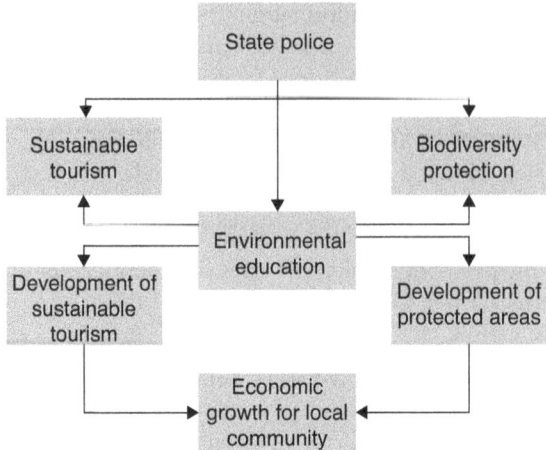

Figure 32.2 System approach model towards sustainable development of tourism and environmental protection

Source: The table compiled by the author

4   Conclusions

The survey revealed that travel to Georgian protected areas is as popular amongst domestic as foreign tourists. Travelers mainly constitute eco-tourists interested in environmental protection.

Facilitation of Development of Sustainable Tourism in Georgia requires consideration of EE and training enactment mechanisms under the strategic documents and respective legislative acts on tourism development.

Implementation of such approaches (mechanisms) will facilitate the present and future generations to obtain environmental protection and sustainable development awareness and to be equipped with due skills and the capacity of practical application of these skills (such as: participation in decision-making, sustainable management of natural resources, timely and proper response to local problems, etc.).

Sustainability of universal fundamental values conditions diversity of moral confidence towards the environment serving as the basis and outcome of eco-tourism. Sustainable tourism shall become the main qualitative characteristic of the tourist market of Georgia to ensure efficiency and a high level of success of the field.

Note

1 Georgian Technical University.

Bibliography

Ceballos-Lascurain, H. (1996). *Tourism, Ecotourism and Protected Areas.* Gland, Switzerland: IUCN.

Dredge, D., and Jenkins, J. (Eds.). (2011). *Stories of Practice: Tourism Policy and Planning.* Aldershot: Ashgate. pp. 37–55.

Gogelia, M. (2014). *Ecotourism Development Strategy's Forms.* Tbilisi: Universali (in Georgian).

http://apa.gov.ge/en/saagento

http://moe.gov.ge/res/images/file-manager/strategiuli-dokumentebi/strategiebi-gegmebi/garemosdacviTi-ganaTleba-mdgradi-ganviTarebisaTvis-saqarTvelos-erovnuli-strategia-da-samoqmedo-gegma-2012%E2%80%932014.pdf

Kajaia, G. (2003). *Basics of Applied Ecology.* Tbilisi: Universali (in Georgian).

Kikodze, A., & Gokhelashvili, R. (2006). *Kolkheti National Park: Field Guide.* Tbilisi: APA (in Georgian).

Metreveli, M. (2007). *Customs of Local Inhabitants and Rules of Relations with Them (Vashlovani and Lagodekhi Protected Areas) Guide-Book for Tourists.* Tbilisi: GAT.

Metreveli, M. (2011). *Environment and Ecotourism Management.* Tbilisi: Favoriti Printi (in Georgian).

National Biodiversity Strategy and Action Plan of Georgia 2014–2020 (2014). Tbilisi.

Ross, S., & Wall, G. (1999). Ecotourism: Towards congruence between theory and practice. *Tourism Management*, 20, pp. 123–132.

Sustainability of the Anthropocene
A cybernetic concept

Markus Schwaninger[1]

Keywords: *sustainability; anthropocene; viability; recursive organization*

There is a tight nexus between the two topics Anthropocene and Sustainability. At this stage, the salient feature of the Anthropocene is its non-sustainability. Not only are humans running amok. Humanity as a whole is ruining nature, and at a progressive rate. Finding sustainable ways is not an option: It is the only way of avoiding disaster. The purpose of my contribution is to offer a concept by which we can find such a path toward sustainability. The claims of people as to where sustainability "must come from" varies greatly. Psychologists often maintain that the problem "ticks in our head", i.e., its solution is in essence an issue at the level of the individual. On the other hand, many politicians claim that sustainability is fundamentally a matter of regulation at the global level. A cybernetic concept is presented, by which the quest for a sustainable world is grounded at all recursive levels, from individual to organizations . . . to the world. This would be an effective, multi-level solution providing requisite variety for coping with a burning, multi-level problem.

Note

1 University of St. Gallen.

Bibliography

Schwaninger, M. (2015). "Organizing for sustainability: A cybernetic concept for sustainable renewal", *Kybernetes* 44(6/7): 935–954.

Websites

http://wosc2017rome.asvsa.org/index.php

The trade-off between Italian universities' education plans and students' awareness regarding sustainability

Marzia Ventura,[1] *Rocco Reina*[2] *and Walter Vesperi*[3]

Keywords: *CSR; Higher Educational Program; Italian Universities*

1 Introduction

It's very interesting to think about the possibility that Corporate Social Responsibility (CSR) can become a new way of thinking and acting in the global community. But, in order to do this, it's important to work unified CSR's perception and obtain an impact on the social and economic culture of a territory. Under these conditions, the universities could have a key role.

They have the responsibility to train and lead the new generation of students and future citizens. If the universities insert in their didactic plans courses related to CSR, these will be useful to help younger generations in understanding the issue of CSR and their awareness regarding its positive impact on the served community and the context So, the object of the paper is to know how Italian universities have been operating, on the assumption that if the universities pay more attention to the practices of corporate social responsibility this will generate for students greater interest and awareness of the culture of ethics and social responsibility and improvement of the quality of life. The attention is focused on the learning programs present in the Italian medium universities and all teaching activity programs linked to the theme of CSR with a web analysis. The main objective of this paper is to investigate if the Italian universities have adequate educational programs able to answer the demand for new management skills in CSR.

In this way it will be possible to analyze the impact of training in respect to social and economic contexts. The outcome of this study can assist researchers, managers and institutions to better understand the effects of the culture on the context and therefore its quality of life.

2 Theoretical framework

According to the managerial literature (Table 34.1), companies who look beyond their economic targets and pay particular attention to social and environmental issues are rewarded by the stakeholders.

Table 34.1 Reference authors

Waddock and Graves (1997); Balbanis et al. (1998); Orlitzky (2001):	The authors showed a positive link between CSR and financial results
Owen and Scherer (1993); Paul et al. (1997); Maignan et al. (1999); Sen and Bhattacharya (2001)	The authors showed in their research that CSR has a positive influence on customers
McGuire et al. (1988); Wright et al. (1995); Greening and Turban (2000); Albinger and Freeman (2000); Backhaus et al. (2002); Ray (2006):	The authors showed in their research that the social involvement of companies and therefore their social responsibility influences employer attractiveness
Cable and Judge (1994); Chatman (1991); Judge and Cable (1997); Schneider (1987):	The authors showed in their research that people looking for a job prefer organizations that share common values with them

Source: Our elaboration, 2016

Following these arguments CSR is a medium to create loyalty and to attract customers, employees and investors that are all socially responsible (Freeman, 2006).

3   Results

The used methodology – through on the web analysis – is organized in three steps: (i) Check the universities' offers; (ii) analyze the training courses and specific programs related to CRS; (iii) analyze the student populations (e.g. attends the University of Catanzaro), in order to know their sensibility to the topic.

The first part was conducted in medium-sized Italian universities (n°31) and totally involved n°96 degree courses and n°2 master courses (training courses and specific programs related to Ethics-Sustainability and CSR keywords).

The first result of the survey starts with the analysis of the academic defined offer in order to find courses and lessons specifically linked with the issue of CSR, as well as the specific teaching programs adopted. The research on the universities reveals two different aspects: less attention is linked to the topics of Corporate Social Responsibility and Sustainability in respect to the topic Ethics (see Figure 34.1).

The analysis begins by considering the different degree courses and focusing only on those that have activated or inserted a module in the material created specifically on three main areas: Sustainability, Ethics and Responsibility. So, we can see that the concentration of those issues was a part of courses related to "Medicine", "Sociology", "Human Resource Management" and "Engineers".

The second part of the analysis began in the academic year 2015/16. The selected questionnaire, in general terms, measured on a Likert's scale (of 7 points – very

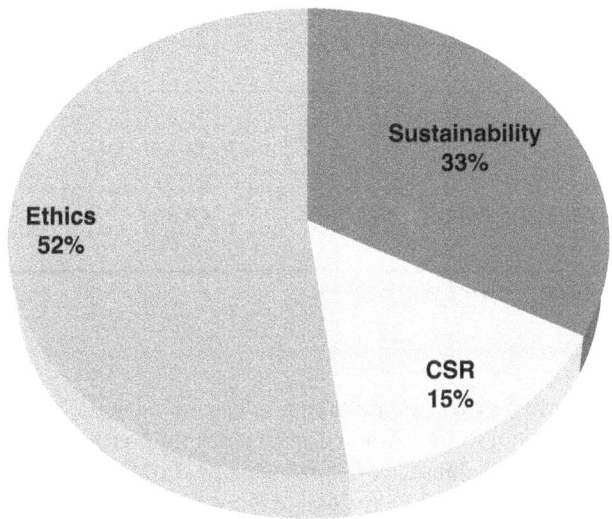

Figure 34.1 Training courses in the degree courses (%) (n°31 university)
Source: Our elaboration of universities' web sites, 2016

Table 34.2 General section

Nationality	1,5% Erasmus
	98,5% Italian
Sex	68,8% women
	30,2% men
Age	65,1% between 20–24 years
	21,1% between 25–29 years
	5,2 > 30 years
Current employment	66,9% studied only
	27,9% study and work
Member of environmental organization	97,7% not members
Member of ONG	93,8% not members

Source: Our elaboration, 2016

important/less important) the ranking of importance of CSR for the students during the job research process. The questionnaire's reliability is given by a Cronbach Alfa with a score of 0.963, indicating excellent reliability. The number of students that successfully completed the questionnaire was n°45, whose profiles are presented in Table 34.2.

Table 34.3 Five aspects

Aspects to arrange of importance	%	Importance average
System of rewards: pay, extras pay, etc.	71,55	2,40
Value related to their	53,37	2,93
Corporate Image	51,61	2,84
Corporate Social responsibility	49,56	2,99
Organizational structure	35,78	3,43

Source: Our elaboration, 2016

This attention can only create a different understanding in the students so as to consider relevant within various work environments, the issues concerning ethics, responsibility and sustainability.

The other result was that students would consider CSR as an attribute when seeking employment and, more specifically, 74.5% of the sample considered a company's social responsibility during their search for employment, but for only 38.5% of them was it very important to work in a socially responsible company.

In this regard, 23% of respondents would be willing to work for a company that is CSR, 27% would not be willing to work for a company that is CSR and for 50% it depended on the business location.

However, in order of importance from a number ranking of business attributes that influence the attractiveness of a company, CSR comes after the other aspects such as the reward system, company image and related values. Thus, as shown below, the CSR is the fourth aspect (out of five analyzed) that would evaluate job applicants during the job search process.

It has been observed that the student in giving less importance to the attributes related to the system of rewards and to the image of the enterprise prefers the CSR. Regarding activities that are perceived as socially responsible practices, the students consider that all the practices outlined in the questionnaire are closely related to CSR, although with varying importance. Thus, practices that are perceived as more related to CSR are those related to the social and environmental dimensions of CSR, labor relations and human rights.

According to the data in Table 34.4 "n°8 on n°19 dimensions of identified CSR", are those in which interviewees show greater interest (average value between 1 and 3).

The data reveal how students give less importance to parameters such as human rights and relations, as compared to parameters such as climate change, production and services and the governance. The results also show that students under 20 years (1.2%) are those who generally give less value to the dimensions of CSR.

Table 34.4 The eight dimensions of identified CSR

Parameters	Importance average
1 Human and labor rights: protection of fundamental human rights and labor rights	2,24
2 Industrial relations: health and safety, profits sharing, retirement	2,61
3 Diversity: employment for the disabled, women and minority recruitments	2,68
4 Operations and environmental management: environmental management systems for hazardous waste, recycling, pollution prevention	2,68
5 Products and service beneficial to the environment: no use of chemicals that affect ozone as well as agricultural chemicals	2,74
6 Climate change: the use of clean energy and eliminate climate change concerns	2,75
7 Production and services	2,84
8 Governance: elite and transparency	2,84

Source: European Commission, 2001

4 Conclusion

This paper aims to highlight if CSR is important for job seekers in their job search process, through the analysis of the most-valued CSR practices and which of them companies prefer to implement.

The first research question focused on the extent to which ethics, CSR and sustainability are recommended and incorporated in the universities' curricula. In the undergraduate courses teaching ethics assumes a higher percentage (52%) compared to sustainability (33%) and CSR (only 15%).

The results show that most of the topics are included; 56% of the time was spent in freshman-level courses and during the sophomore year. At this stage of work in progress, some suggestions are possible to consider. Most likely the theme of CSR appears to be strongly linked to cultural aspects; for these reasons time is a key variable on which to act for better job prospects and work.

The students consider CSR as an important attribute of attraction during the initial process of entering in the labor market, but after attributes like economic aspects and corporate image. The practices that are perceived as more related to CSR are the social and environmental related; within these practices, labor relations and human rights are less interesting in respect to economic and corporate governance. Surely, in order to understand the actions specifically taken by the universities, it would be interesting to make some interviews with the aim to better understand the problem and to propose these insights in order to be useful for the activation of pathways learning on ethical and sustainable aspects of work. In fact, the Education Program is a basic condition to develop the behavior mentalities.

Notes

1 University of Catanzaro.
2 University of Catanzaro.
3 University of Messina.

Bibliography

AACSB International, adopted (2003) revised (2008): Eligibility Procedures and Accreditation Standards for Business Accreditation, p. 62. Retrieved 21.9.2018 from https://www.aacsb.edu/accreditation/standards/2003-business

Albinger, H.S. and Freeman, S.J. (2000): "Corporate social performance and attractiveness as an employer to different job seeking populations". *Journal of Business Ethics*, n. 28, pp. 243–253.

Backhaus, K.B., Stone, B.A. and Heiner, K. (2002): "Exploring the relationship between corporate social performance and employer attractiveness". *Business and Society*, vol. 41, n. 3, pp. 292–319.

Balbanis, G., Phillips, H.C. and Lyall, J. (1998): "Corporate social responsibility and economic performance in the top British companies: Are they linked?" *European Business Review*, vol. 98, n. 1, pp. 25–44.

Cable, D.M. and Judge, T.A. (1994): "Pay preferences and job search decisions: A person organization fit perspective". *Personnel Psychology*, n. 47, pp. 317–349.

Chatman, J.A. (1991): "Matching people and organizations: Selection and socialization in public accounting firms". *Administrative Science Quarterly*, n. 36, pp. 459–484.

European Commission (2001): Green paper: Promoting a European framework for CSR. Retrieved 21.9.2018 from http://ec.europa.eu/transparency/regdoc/rep/1/2001/EN/1-2001-366-EN-1-0.Pdf

Freeman, B. (2006): "Substance sells: Aligning corporate reputation and corporate responsibility". *Public Relations Quarterly*, n. 51, pp. 12–19.

Greening, D.W. and Turban, D.B. (2000): "Corporate social performance as a competitive advantage in attracting a quality workforce". *Business y Society*, n. 39, pp. 254–280.

Judge, T.A. and Cable, D.M. (1997): "Applicant personality, organizational culture, and organization attraction". *Personnel Psychology*, n. 50, pp. 359–394.

Maignan, I., Ferrell, O.C. and Hult, G.T. (1999): "Corporate citizenship, cultural antecedents and business benefits". *Journal of the Academy of Marketing Science*, n. 27, pp. 455–469.

McGuire, J.B., Sundgren, A. and Schneeweis, T. (1988): "Corporate social responsibility and firm financial performance". *Academy of Management Journal*, n. 31, pp. 854–872.

Orlitzky, M. (2001): "Does firm size confound the relationship between corporate social performance and firm financial performance?" *Journal of Business Ethics*, vol. 33, n. 2, pp. 167–180.

Owen, C. and Scherer, R. (1993): "Social responsibility and market share". *Review of Business*, n. 15, pp. 11–17.

Paul, K., Zalka, L.M., Downes, M., Perry, S. and Friday, S. (1997): "U.S. consumer sensitivity to corporate social performance, development of a scale". *Business and Society*, n. 36, pp. 408–419.

Ray, J.R., Jr. (2006): Investigating relationships between corporate social responsibility orientation and employer attractiveness. Doctoral Thesis. The George Washington University.

Schneider, B. (1987): "The people make the place". *Personnel Psychology*, n. 40, pp. 437–454.

Sen, S. and Bhattacharya, C.B. (2001): "Does doing good always lead to doing better? Consumer reactions to corporate social responsibility". *Journal of Marketing Research*, n. 38, pp. 225–243.

Waddock, S.A. and Graves, S.B. (1997): "The corporate social performance-financial performance link". *Strategic Management Journal*, vol. 18, n. 4, pp. 303–319.

Wright, P., Ferris, S.P., Hiller, J.S. and Kroll, M. (1995): "Competitiveness through management of diversity, effects on stock price valuation". *Academy of Management Journal*, n. 38, pp. 272–284.

Systems analytics for change

Louis Klein[1]

Keywords: *systems analytics; systemic change; systemic inquiry; social complexity*

Facing the systemic challenges of the Anthropocene we need to acknowledge that the required change is either systemic or it is not. If challenges finally are articulated as being systemic, solutions need to be of the same kind, namely systemic.

What we witness is a shift in the public discourse. Explicit systemic and cybernetic approaches have been around for more than 50 years. It was not earlier than the financial crisis that the public discourse became familiar with the term "systemic".

Especially ecologists push now for systemic solutions and systemic change; however, we see little progress in systemic solution practices beyond modeling and scenario planning. And the generation of the so-called millennials and digital natives is amused to see the cumbersome layouts and petty aesthetics of the systemic modeling tools on the market. It is certainly not a mere question of a nice brush-up to reintroduce systems approaches and cybernetic models into the public discourse or political and economic decision making.

Systems analytics (Klein, 2016b) seem to provide a new label for the well-known systems approaches and cybernetic models; however, in times of the so-called digital transformation new technological possibilities come into sight. As we learnt from the net 2.0 or industry 4.0 we can expect more than modeling and illustration and we can expect more than a mere technological perspective. So in a combination of discourse praxis analysis and action research it will be necessary to learn more about the conditions for the possibility to explore and map the systemic properties and dynamics of social systems, respectively socio-technical systems as the means to navigate any kind of systemic change (Bosch & Nguyen, 2015).

Systems analytics need to go hand in hand with systemic inquiry (Burns, 2012; Collen, 2003; Churchman, 1968). This points at the fundamentals of the epistemological turn in social sciences and social systems research (Klein & Weiland, 2014; Klein, 2005). The number one critical success factor for systemic change in the Anthropocene is the ability to meet and handle social complexity (Klein, 2016a, 2016b; Letiche et al., 2011; Schwaninger, 2008; Thévenot, 2001). What do dynamic sensitivity models of political and cultural issues look like? What does a systemic stakeholder analysis (Raue & Klein, 2016; Klein, 2016b) look like? What can critical narrative inquiry (Jorgensen & Largacha-Martinez, 2014; Klein & Weiland, 2014) contribute to explore culture? And finally how can contemporary computing capabilities be used to model and process social complexity?

Successful systemic change is a question of exploring realms of possibilities to move, following a next practice approach (Klein, 2013; Bowers, 2011), from a current practice to a better practice. We cannot jump to solutions and we cannot jump to a best practice. Systemic change (Klein, 2016a; Collen, 2003; Midgley, 2000) unfolds step by step carefully observing context and emergence, dynamics and conditions. The rest is systemic project management. Systems analytics need to become the contemporary agent we need to support systemic change practices. And they will.

Note

1 European School of Governance.

Bibliography

Bosch, O., & Nguyen, N. C. (2015). *Systems Thinking for EVERYONE: The Journey from Theory to Making an Impact*. Canberra: Think2Impact.

Bowers, T. D. (2011). Towards a Framework for Multiparadigm Multimethodologies. *Systems Research and Behavioral Science*, 28(5), 537–552.

Burns, D. (2012). Participatory Systemic Inquiry. *IDS Bulletin*, 43(3), 88–100.

Churchman, C. W. (1968). *The Systems Approach*. New York, NY: Delacorte Press.

Collen, A. (2003). *Systemic Change through Praxis and Inquiry*. New Brunswick, NJ: Transaction Publishers.

Jorgensen, K. M., & Largacha-Martinez, C. (2014). *Critical Narrative Inquiry: Storytelling, Sustainability and Power*. New York, NY: Nova Science Publishers.

Klein, L. (2005). Systemic Inquiry: Exploring Organisations. *Kybernetes: Heinz von Förster in Memoriam: Part II*, 34(3/4), 439–447.

Klein, L. (2013). Notes on an Ecology of Paradigms. *Systems Research and Behavioral Science*, 30(6), 773–779.

Klein, L. (2016a). Towards a Practice of Systemic Change: Acknowledging Social Complexity in Project Management. *Systems Research and Behavioral Science*, 33(6). https://doi.org/10.1002/sres.2428

Klein, L. (2016b). Understanding Social Systems Research. In M. Nemiche & M. Essaaidi (Eds.), *Advances in Complex Societal, Environmental and Engineered Systems*. Springer.

Klein, L., & Weiland, C. A. P. (2014). Critical Systemic Inquiry: Ethics, Sustainability and Action. In K. M. Jorgensen & C. Largacha-Martinez (Eds.), *Critical Narrative Inquiry: Storytelling, Sustainability and Power* (pp. 145–158). New York, NY: Nova Science Publishers.

Letiche, H., Lissack, M., & Schultz, R. (2011). *Coherence in the Midst of Complexity: Advances in Social Complexity Theory*. New York, NY: Palgrave.

Midgley, G. (2000). *Systemic Intervention: Philosophy, Methodology, and Practice*. New York: Kluwer Academic/Plenum Publishers.

Raue, S., & Klein, L. (2016). Systemic Risk Management: A Practice Approach to the Systemic Management of Project Risk. In C.-N. Bodea, A. Purnus, M. Huemann, & M. Hajdu (Eds.), *Managing Project Risks for Competitive Advantage in Changing Business Environments* (pp. 70–85). Hershey, PA: IGI Global.

Schwaninger, M. (2008). *Intelligent Organizations: Powerful Models for Systemic Management* (2nd ed. 2009). Berlin: Springer.

Thévenot, L. (2001). Organized complexity conventions of coordination and the composition of economic arrangements. *European Journal of Social Theory*, 4(4), 405–425.

Co-learning for sustainability
Partnership between engineering students and green micro-entrepreneurs in Colombia

María Catalina Ramírez,[1] *Julia Díaz*[2]
and Andrés Acero[3]

Keywords: *sustainability; Triple Helix Model; green business; learning; micro-enterprises*

In the last three decades, profound analyses of the social and environmental responsibility of organizations have emerged as a need to thrive structural changes towards sustainable development. As suggested by the triple helix model, higher education institutions and research centers were not the exception. Universities have a crucial role in sustainable development because they have the duty to shape future professionals with the relevant skills to contribute to the viable well-being of society. In Colombia, as in other developing countries, this discussion should move educational systems and organizations to design mechanisms and methodologies that contribute to the synergy between companies, public institutions and social education organizations. Given the above, research will be focused on understanding how the universities can generate positive impact in the interaction of students with the environment.

Furthermore, social responsibility has become necessary in organizational strategic management and modern modes of production. In consequence, those responsibilities require an integral alignment of the organizations by minimizing the use of materials, reducing energy consumption, promoting environmental services and adapting to new patterns of consumption. This has been, consequently, a concern of the team of Ingenieros sin Fronteras Colombia (EWB Colombia, by their acronym in English). This interdisciplinary group of students and professors, from the Universidad de los Andes and the Corporación Universitaria Minuto de Dios, seeks to develop and apply innovative entrepreneurial business models that promote viable well-being, going hand in hand with the community. Therefore, the generation of this sustainable and innovative business model must integrate social and environmental sustainability since it plays a key role in the transformation process of the current economy. As an alternative, the implementation of green business strategies and models should strive for the preservation of the environment, as natural capital that supports the existence of territories, and should contribute to regional development through mitigation of social and environmental impact, inside and outside the boundaries of the organization.

In the previous context, the university plays a fundamental role since its formative role is the setting for socialization and inclusion of a sustainability approach in future managers, workers, government officials, entrepreneurs and researchers. From the point of view of social projection, our proposal tries to show how the academy has the opportunity to integrate the role of education for sustainable development by implementing and developing projects with high social impact or by promoting the creation or transformation of green business. This perspective allows universities to bridge theory and practice and encourages students to face real environments during their training.

Specifically, the present study explored the following characteristics of the partnership green business-universities:

- The required conditions for the consolidation of green business entrepreneurs in the emerging economies and in developing countries, as in the Colombian case.
- Exploring the triple helix model and the relationship between universities and the productive sector, focused on the micro and small enterprises.
- The consolidation of possible university social responsibility (USR) exercises in order to build a synergy to allow joint learning among college students and entrepreneurs.
- The design of a training process in order to gather professors and students about the role of engineering in promoting sustainable development of vulnerable communities.

During this research, "green businesses" will be a concept under construction. Indeed, although participant entrepreneurs had the intention to be green in practice, their work with the students was to build strategies for "greening" their businesses. As a primary aspect of the context and according to the needs of micro-entrepreneurs, it is more common to learn to be green than be born green, and it was the focus of the partnership.

Finally, conclusions are discussed related with URS plans for institutional development, as a process that goes through several stages, going from a discursive to a more active phase. Due to the high value in the contribution of small entrepreneurs to the university, the approach of a strategy of USR integrated the managerial, regulatory and transformative approaches benefiting the regional development and the potential of replication in several contexts.

Notes

1 Universidad de los Andes.
2 Universidad de los Andes.
3 Universidad de los Andes.

Bibliography

Bárcena, A., Prado, A., Cimoli, M., & Pérez, R. (2011). *Experiencias exitosas en innovación, inserción internacional e inclusión social: una mirada desde las PYMES*. CEPAL, LC/L, 3371.

Cámara de Comercio. (2006). *Descripción de la provincia del Guavio.* Recuperado el 05 de 11 de 2013, de http://aulas.alianzaporelguavio.net/pluginfile.php/89/mod_resource/content/1/Descripcion%20guavio.pdf

Cámara de Comercio de Bogotá. (2009). *Centro de Pensamiento en Estrategias Competitivas,* Universidad el Rosario.

Delgado, M., Cabrera, E., & Ortiz, N. (2008). *Informe sobre el estado de la biodiversidad en Colombia 2006–2007.* Bogotá.

Eisenhardt, K. (1989). Building theories from case study research. *Academy of Management Review,* 14(4), 532–550.

Espinosa, A., & Walker, J. (2011). *A complexity approach to sustainability.* Hull: Hull University Business School.

Gibbons, M., Limoges, C., Nowotny, H., Schwartzman, S., Scott, P., & Trow, M. (1997). *La nueva producción del conocimiento. La dinámica de la ciencia y la investigación en las sociedades contemporáneas.* Barcelona: Pomares-Corredor.

Glaser, B., & Strauss, A. (1967). *The discovery grounded theory: Strategies for qualitative inquiry.* London, England: Wiedenfeld and Nicholson.

Hernández, A., Sanabria, R., & Saavedra, J. (2007). *Los desafíos actuales de las empresas en Colombia: Serie Universidad, Ciencia, y Desarrollo,* 6(2). Bogotá: Centro de Estudios Empresariales, Universidad del Rosario.

Ingenieros Sin Fronteras Colombia. (2013). Informe Fase Cero, *Negocios Verdes.* Bogotá. Retrieved 21.9.2018 from https://isfcolombia.uniandes.edu.co/

Millenium Ecosystem Assesment Panel. (2005). *Ecosystems and human wellbeing: Synthesis.* Washington, DC: Island Press.

Ministerio de Ambiente y Desarrollo Sostenible. (2014). *Plan Nacional de Negocios Verdes.* Retrieved 21.9.2018 from http://www.minambiente.gov.co/index.php/component/content/article/1385-plantilla-negocios-verdes-y-sostenibles-40

Navas, L. (2014). *Análisis del aprendizaje organizacional en el programa de fortalecimiento de negocios verdes comunitarios en la región del Guavio.* Master Thesis on Industrial Engineering. Universidad de los Andes, Bogotá.

Pacheco, J. F., Ramírez, M. C., & González, M. A. (2013). Transformación de Unidades Productivas Tradicionales en Negocios Verdes. *Revista Digital Desarrollo Regional,* 1.

Quezada, R. A. G. (2011). La RSU como desafío para la gestión estratégica de la Educación Superior: el caso de España. *Revista de educación,* (355), 109–133.

Yin, R. K. (2003). *Case study research: Design and methods.* Beverly Hills, CA: Sage Publications.

Actors, smart technologies and governance in sustainable cities

Francesco Bifulco,[1] Cristina Caterina Amitrano,[2] Marco Tregua[3] and Anna D'Auria[4]

Keywords: *sustainability; governance; actors; technology; sustainable cities*

1 Overview

The innovation of urban contexts is even more a hot topic in the international debate (Cocchia 2014; Roberts et al. 2016). New technologies are currently considered as an unavoidable element in managing modern cities (Camagni et al. 1998; Ishida and Isbister 2000; Wolfram 2012), since they support spreading sophisticated services to improve the quality of life of citizens (Bifulco et al. 2014, 2016).

When scanning contributions, it emerges city managers are aiming at sustainability (Enquist et al. 2007); apart from the elements linked to the "triple-bottom-line" (Rogers and Ryan 2001), the way towards a "sustainable city" (Jenks and Dempsey 2005; Egger 2006; Tregua et al. 2015) is strictly related to the participation of citizens in management and governance of the urban contexts (Bingham et al. 2005; Saviano and Iorio 2010; Schaffers et al. 2011; Flint and Raco 2012); indeed actors – citizens, organizations, local agencies, etc. – could represent, together with ICTs, the way to create, realize and spread knowledge and services. For this reason, the relationship among actors and the relationship between them and local governance is often evaluated as coproduction (Pestoff 2014). Hence, citizens are both creators and users (Pestoff 2014); additionally they are part of a "smart community" (Nam and Pardo 2011), since they participate in managing the city, by interacting with all stakeholders and positively affecting the definition of city services.

2 Aim and methodology

The main aim of our research stands on the clarification of the elements shaping the definitions of sustainable cities; in this way we expect to grasp some more meanings about the approach to sustainable cities provided by both scholars and organizations dealing with projects involving sustainable goals for metropolitan and urban contexts.

In order to achieve the aforementioned aim, we focus on the evidences around the world provided by the "Sustainable Cities Index 2016", as proposed by

Arcadis, since it is the most updated among the available rankings. Additionally, the data are unbiased, since they are not provided by city or local agencies themselves, so the confidence and quality of information proposed is grounded. We would analyze the role played by actors, technology-based innovation, and governance in the cities listed in Arcadis index. The choice of the cases is aligned with the methodological suggestion by Yin (1994), since the index is proposing the Top 100 Sustainable Cities all over the world. Finally, since our approach is a qualitative one, the authors would discuss together how to interpret the content of the reports shaping the index, in order to decrease the subjectivity of a stand-alone researcher.

3   Findings

The cases discussed have been chosen from the "Sustainable Cities Index 2016" on the basis of their consonance with the three topics under investigation, namely actors, technologies and governance. From a general perspective most of the cities presented in the "Index" have some common elements: almost all of them are big cities and the activities they performed are mainly based on the environmental sphere as the reduction of pollution and the management of waste and water, or the improvement of the transport system with a sustainable approach.

As it concerns actors, the evidences show the roles of the different entities involved in urban sustainability projects, namely government and local offices, industry and business, universities and research centres, and citizens.

Furthermore, the main part of interventions realized in different fields are based on the employment of new technologies, with a major focus on environment and mobility that is revealed by many cases.

Finally, some key issues related to governance emerges with long-term goals, especially in relation to energy, education, wage level and environment, and with the attention addressed to both internal and external synergies to be created in order to achieve sustainable actions. Resilience is another key approach envisioned by cities aiming at being more sustainable and governance is implementing actions to make easier the reactions to external changes and factors, both planned and unplanned ones.

4   Conclusions, limitations and further research

The results of our research led to a better understanding of how cities are performing their activities and devoting their efforts towards sustainability (Roberts et al. 2016); thoroughly, cities are innovating the way public services are being provided, in order to challenge the new needs, namely the ones oriented to sustainability, with specific reference to energy policies, waste management, improvements in transport networks and even by involving people in changing their habits. As a confirmation of this last statement, actors' involvement emerged as pivotal in driving changes in contexts like cities, since different resources are carried by each actor, favouring the needed set to perform activities in innovative ways and support the role of public governance. Indeed, local agencies and other entities are linked as actors that should be involved by local governance (Saviano and Iorio 2010) and they have to contribute to governance decision, too. Finally, the results achievable by cities as presented above should take place thanks to the support of technology (Wolfram 2012) in two different perspectives, namely as a support to activities to be deployed and as a way to favour resource integration among actors through platforms.

To sum up, actors, governance and technologies can be considered as three intertwined pillars to drive cities towards sustainability; the first evidences provided in this research should be enhanced and complemented as long as reports from other cities will be available and even by adding insights through interviews or direct relationships with representatives from local agencies.

Notes

1 University of Naples Federico II.
2 University of Naples Federico II.
3 University of Naples Federico II.
4 University of Naples Federico II.

Bibliography

Bifulco, F., Tregua, M., & Amitrano, C.C. (2014). Living labs for smart innovation: A user-centric approach. In L. Freund & W. Cellary (Eds.), *Advances in the Human Side of Service Engineering* (pp. 282–294). Boca Raton, FL: CRC Press.

Bifulco, F., Tregua, M., Amitrano, C.C., & D'Auria, A. (2016). ICT and sustainability in smart cities management. *International Journal of Public Sector Management*, 29(2), 132–147.

Bingham, L.B., Nabatchi, T., & O'Leary, R. (2005). The new governance: Practices and processes for stakeholder and citizen participation in the work of government. *Public Administration Review*, 65(5), 547–558.

Camagni, R., Capello, R., & Nijkamp, P. (1998). Towards sustainable city policy: An economy environment technology nexus. *Ecological Economics*, 24(1), 103–108.

Cocchia, A. (2014). Smart and digital city: A systematic literature review. In R.P. Dameri & C. Rosenthal-Sabroux (Eds.), *Smart City: How to Create Public and Economic Value with High Technology in Urban Space* (pp. 13–43). Berlin: Springer.

Egger, S. (2006). Determining a sustainable city model. *Environmental Modelling & Software*, 21(9), 1235–1246.

Enquist, B., Edvardsson, B., & Sebhatu, S. (2007). Values-based service quality for sustainable business. *Managing Service Quality*, 17(4), 385–403.

Flint, J., & Raco, M. (2012). *The Future of Sustainable Cities: Critical Reflections*. Bristol: Policy Press.

Ishida, T., & Isbister, K. (2000). *Digital Cities: Technologies, Experiences, and Future Perspectives*. New York, NY: Springer.

Jenks, M., & Dempsey, N. (2005). *Future Forms and Design for Sustainable Cities*. New York, NY: Routledge.

Nam, T., & Pardo, T.A. (2011). Smart city as urban innovation: Focusing on management, policy, and context. In E. Estevez & M. Janssen (Eds.), *Proceedings of the*

5th International Conference on Theory and Practice of Electronic Governance (pp. 185–194). New York, NY: ACM.

Pestoff, V. (2014). Collective action and the sustainability of coproduction. *Public Management Review*, 16(3), 383–401.

Roberts, P., Sykes, H., & Granger, R. (Eds.). (2016). *Urban Regeneration*. Los Angeles, CA: Sage.

Rogers, M., & Ryan, R. (2001). The triple bottom line for sustainable community development. *Local Environment*, 6, 279–289.

Saviano, M., & Iorio, G. (2010). How far from participatory governance? A survey on e-democracy in Italian municipalities. *Journal of Management*, 1(2), 1–18.

Schaffers, H., Komninos, N., Pallot, M., Trousse, B., Nilsson, M., & Oliveira, A. (2011). Smart cities and the future internet: Towards cooperation frameworks for open innovation. In J.J. Domingue et al. (Eds.), *The Future Internet* (pp. 431–446). Berlin: Springer.

Tregua, M., D'Auria, A., & Bifulco, F. (2015). Comparing research streams on smart city and sustainable city. *China-USA Business Review*, 14(4), 203–215.

Wolfram, M. (2012). Deconstructing smart cities: An intertextual reading of concepts and practices for integrated urban and ICT development. In M. Schrenk, V.V. Popovich, P. Zeile & P. Elisei (Eds.), *Re-Mixing the City: Towards Sustainability and Resilience?* (pp. 171–181). Proceedings REAL CORP 2012, May, Schwechat, Austria.

Yin, R. (1994). *Case Study Research: Design and Methods* (2nd ed.). Newbury Park, CA: Sage.

Websites

www.arcadis.com

The organizational space in health

The mApp as a sustainable knowledge creation process

Rocco Reina,[1] *Marcello Martinez,*[2]
Primiano Di Nauta[3] *and Biagio Merola*[4]

Keywords: *organizational space in health; knowledge management; medicine mobile application*

1 Framework

In healthcare systems, knowledge management is a challenging issue for all involved stakeholders. On one side, patients are becoming more educated, empowered, informed and involved in decision-making processes related to their health. On the other side, professionals are engaging in designing new "forms" of knowledge structures for healthcare (Martinez and Galdiero, 2015; Reina, 2015). The greatest opportunity to strengthen and support this improvement process basically depends on the increasing availability of information systems and web-based infrastructures and platforms. For example, internet and virtual spaces can potentially favour new learning processes and management practices by disseminating knowledge and making codified knowledge retrievable (Hsiu-Mei Huang and Shu-Sheng Liaw, 2004). So, information technology earns an important role for sharing knowledge and information among people in and out of organizations. The most updated technologies (e.g. mobile applications, cloud computing and so on) allow the emergence of new space for knowledge creation, by facilitating new relationships (Gottscalk, 2005). Particularly, the use of technological innovations like mobile applications can positively impact on contexts and organizations by favouring data and knowledge transfer processes. Therefore, the concept of space earns a wide meaning, as the relational, organizational, but also social and virtual places where new learning processes arise and knowledge creation and transfer processes can be established (Nonaka et al., 2005).

In spite of the underlined contributions (Noennig and Jannack, 2013), very little is already known about non-conventional places – virtual and social places – able to allow the management of knowledge creation processes. Anyhow, these factors have interesting and desirable implications on healthcare systems. For example, *mHealth* is an emerging and rapidly developing framework for healthcare, able to contribute to the improvement of its quality and efficiency. In fact, the widespread use of smartphones and 3G/4G networks increased the use of

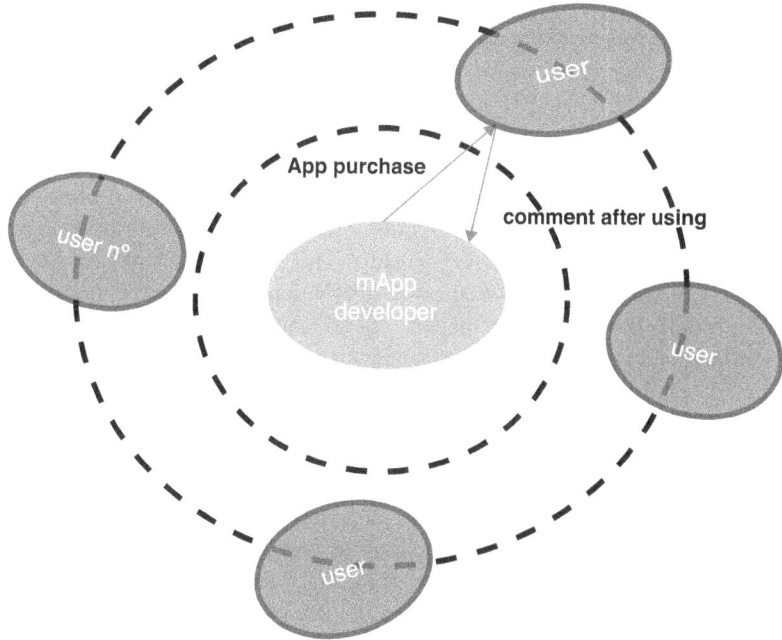

Figure 38.1 Organizational bullseye system
Source: Reina (2015)

mobile applications that provide healthcare services (European Commission, 2014). On this basis, this work proposes an analysis of the relation between knowledge creation space and mobile applications in healthcare in order to determine if and how new technologies influence the quality and effectiveness of knowledge, by improving patients' care. In fact, all the people who use and download the mApp have an active role in the creation of new knowledge and meanwhile can satisfy their own health needs. The work wants to analyze this relation with an empirical search finalized to explore the potential use of *mApp* as knowledge creation space for patients' sustainable care. In Figure 38.1, the relationship model is shown in order to present the different actors involved and the conjectured organizational space in Health.

On this basis, the paper proposes to highlight how in complex environments, like health systems, the use of medicine mobile applications can create new learning processes which, in turn, can create new knowledge space. In this direction, this work aims to better understand, both in theoretical and practical perspectives, if the possibility to create new spaces of knowledge in the context of healthcare exists, thanks to the use of the mobile applications. The results obtained can be utilized for further reflections of practical relevance.

2　Objective and methodology

Over the last few years new organizational theories emerged that focused on instruments able to manage the knowledge in organizations and so to make them more flexible and learning (Reina, 2015). In this way, the increasing and frequent use of the new technologies, linked to the use of the web, satisfied specific expectations related to the new emerging information needs, relationships and effective communication.

The development of these technologies to support healthcare professionals caused a new demand for innovative tools – the mobile applications – as new technological systems of transmission, creation and sharing information. On the other hand it is clear that this situation is largely dependent on the impact that new technologies have on doctor-patient relationships and therefore on transfer and knowledge creation. Knowledge management in healthcare is difficult for all parties involved. In fact, patients are becoming more educated and empowered as well informed and involved in decision-making processes; on the other hand, professionals are engaged in new processes of knowledge creation and transfer among different actors in healthcare.

This paper proposes to highlight how in complex environments, like health systems, the use of medicine mobile applications produces new learning processes which, in turn, can create new knowledge space. In order to do this, the paper used a deductive methodology: the desk and on-the-job phase (Eisenhard, 1989; Yin, 1994). (i) The phenomenon was explored and its characteristic and peculiarities was underlined through the study of specific literature; (ii) a general classification of words related to the Medicine category in the most used systems like iOS and Android (Sole24ore, December 2013); (iii) a taxonomy of the available mApps on smartphones and tablets by IOS and Android in the Italian context; (iv) focus on mApps for a fee by classifying the information found; (v) finally, a definition of specific databases. The process of research pursued has to converge in order to provide new answers to new emerging information needs in healthcare and to help and improve the approach to health of citizen/patient through specific and dedicated apps.

3　Result

This research presents the results of the exploratory and descriptive studies in order to show the general dynamics of the phenomenon (Eisenhard, 1989; Yin, 1994). The Juniper Report estimates about 44 million health apps were downloaded in 2012, while 142 million health apps were downloaded in 2016.

Since 2011 the Juniper Report stated that hardware peripherals that attach to smartphones would "greatly extend the capabilities" of health apps so, it was also foreseen that by 2017, 3.4 billion people worldwide would own a smartphone and half of them would be using mHealth apps. In fact, the evolution of ICT applications has inevitably led to a new clinical way to manage information flows (Buccoliero et al., 2005). The same Sole24Ore on Special Health Report (2013) was dedicated to the Medicine category of apps, as a particular tool able to connect and integrate data in order to obtain specific health information. So, the recognition of different keywords presented in this special report become our first starting point.

Table 38.1 Keywords

1 Psychology	2 Heart	3 Sports
4 Diabetes	5 Kids	6 Flu
7 Heart attack	8 Obesity	9 Cancer
10 Antioxidants	11 Women	12 Skin
13 Diet	14 DNA	15 Cancer
16 Genetics	17 Psyche	18 Drugs
19 Ictus		

Source: Sole24Ore, 2013

Table 38.2 Fee apps on IOS and Android

	Medicine Category		Other Category	
	2014	2016	2014	2016
Psychology	1	4	34	42
Heart	3	3	12	26
Diabetes	70	23	125	29
Flu	3	13	2	20
Heart attack	5	4	0	2
Obesity	0	0	5	13
Cancer	6	7	9	17
Antioxidants	0	0	0	0
Diet	2	1	110	51
Drugs	20	35	5	16
Genetics	0	11	0	37
Ictus	0	5	6	9
Kids	4	0	0	172
Women	2	14	13	136
DNA	10	20	21	149
Psyche	16	1	90	3
Sports	0	3	306	146
Cancer	4	3	72	2
Skin	2	4	548	50
Total	148	151	1358	920

Source: Reina (2015)

We adopted these initial considerations in order to follow the process of a simple user in the research of knowledge on own health, through specific apps present on web. So, we have two types of mApps, the first concerns apps which can be freely downloaded from the web (including those in the Medicine category), while the second regards apps for specific medicine uses, as the Guidelines UE Commission declared (MEDDEV 2.1\6, 2012). In Tables 38.2 and 38.3 the

Table 38.3 Fee apps on IOS and Android

	Medicine Category		Other Category	
	2014	2016	2014	2016
Psychology	0	2	20	26
Heart	1	1	43	16
Diabetes	2	30	53	21
Flu	0	8	17	15
Heart attack	4	28	35	22
Obesity	0	8	28	16
Cancer	1	18	14	13
Antioxidants	0	0	51	21
Diet	0	0	26	22
Drugs	7	17	42	4
Genetics	2	6	27	21
Ictus	3	14	33	20
Kids	0	0	9	5
Women	0	0	10	3
DNA	0	0	35	34
Psyche	0	2	30	28
Sports	0	0	4	7
Cancer	3	28	40	8
Skin	0	0	136	65
Total	19	114	653	367

Source: Reina (2015)

number of fee apps available for the Medicine category and other categories in the years 2014 and 2016 are given. This research was carried out using as a tool the smartphone with the IOS and Android operating systems.

The total number of mobile applications (fee and free) on smartphones in the Android operating system was in 2014 n°4525, in 2016 n°4560; in the IOS operating system was n°6009 in 2014 and n°6700 in 2016; therefore, in this primary step, our work analyzed 19 apps in 2014 and 114 apps in 2016 related to the fee app in the "Medicine" category in Android, and 148 in 2014 and 151 in 2016 in IOS system.

Table 38.4 identifies for different selected keywords the number of comments made by the users, after their experience.

The total number of comments (see Figures 38.2 and 38.3) is the difference between Android n°74 and IoS n°132. Not all the comments made by the users can be considered as development of new knowledge; in fact it's possible to identify technical comments and non-technical or quality comments.

Table 38.4 Number of comments

Keywords	Medicine category	Medicine category	Number of user comments	
	Android 2016	IoS 2016	Android 2016	IoS 2016
Psychology	2	4	2	4
Heart	1	3	0	0
Diabetes	30	23	4	3
Flu	8	13	4	14
Heart attack	28	4	6	0
Obesity	8	0	0	0
Cancer	18	7	0	4
Antioxidants	0	0	0	0
Diet	0	1	0	0
Drugs	17	35	57	72
Genetics	6	11	0	12
Ictus	17	5	1	0
Kids	0	0	0	0
Women	0	14	0	11
DNA	0	20	0	1
Psyche	2	1	0	10
Sports	0	3	0	9
Cancer	28	3	0	0
Skin	0	4	0	0
Total			74	132

Source: Reina (2015)

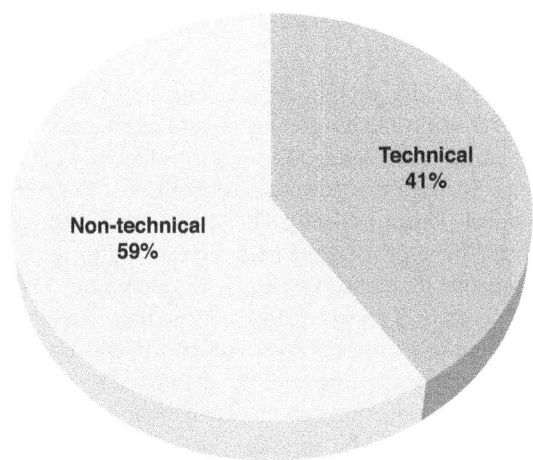

Figure 38.2 Shares of technical and non-technical comments in Android
Source: Reina (2015)

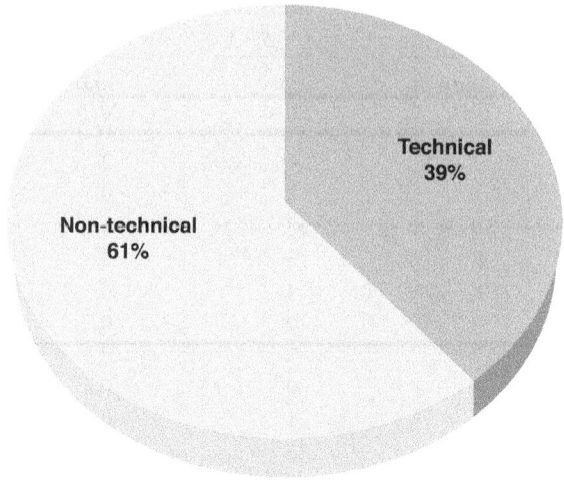

Figure 38.3 Shares of technical and non-technical comments in IoS
Source: Reina (2015)

On these observations, it's possible to define that only the first group of comments permit to create and to disseminate new knowledge. The possible comments found were: "*it's necessary to improve punctuation*", "*it's a real alternative to veterinary hand book*", "*it's possible to require major updates*", "*it's difficult to insert some data*", "*it contains only the basic information and it isn't a real help*", "*it's rich of useful information*", "*it's a useful data base able to define the substances present in food*", "*There are continuous updates and new features are developed*".

4   Conclusion

This paper provides a reflection on how the use of medical applications is able to create new processes of creation and transfer knowledge among doctor/patient actors. Mobile health applications are designed to interact directly with users in order to allow them to better manage their own health, with or without the presence of a healthcare professional (Parsons, 2011). This makes medicine apps ideal tools for supporting healthcare by improving the participation and the capacity for the self-management of patients. The observations derived from our research show that feedback from the developers go to enhance trust in users\patients who download the application, becoming a new way to attract customers. The user through the feedback is aware of being connected with the developer at anytime and this helps to increase the level of participation and involvement and to make him feel always attended. In this way, through the apps the relationship between people (users) becomes more direct and personalized. Some aspects remain unclear; in fact the use and the spreading of new technological systems give rise to doubts concerning safety, privacy and reliability of information itself. It would be important to investigate in depth the real motivations which drive patients to use the device. The perceptions of users themselves should not be disregarded in the overall process; they have an important role in applying the information regarding the product they are interested in and evaluating whether it meets their needs.

Notes

1 Università Magna Graecia.
2 Seconda Università degli Studi di Napoli.
3 Università degli Studi di Foggia.
4 Università degli Studi di Foggia.

Bibliography

Buccoliero, L., Caccia, C., Nasi, G. (2005), *E he@alth. Percorsi di implementazione dei Sistemi informative in sanità*. Milano: McGraw Hill.

Di Nauta, P., Merola, B., Caputo, F., Evangelista, F. (2015), "Reflections on the Role of University to Face the Challenges of Knowledge Society for the Local Economic Development", *Journal of the Knowledge Economy*, Springer Science Business Media New York, ISSN 1868-7873 (Electronic version), DOI: 10.1007/s13132-015-0333-9.

Eisenhard, K.M. (1989), "Building Theories from Case Study Research", *Academy of Management Review*, Vol. 14, No. 4, pp. 532–550.

European Commission (2014), "Green Paper on mHealth", open until 10 July 2014. Retrieved 21.9.2018 from https://ec.europa.eu/digital-single-market/en/news/deadline-public-consultation-green-paper-mhealth-coming-soon

Gottscalk, P. (2005), "Strategic Knowledge Management Technology", Idea Group.

Guidelines UE Commission declared (MEDDEV 2.1\6, 2012), available: http://ec.europa.eu/growth/sectors/medical-devices/guidance_it.

Hsiu-Mei Huang, Shu-Sheng Liaw (2004), "The Framework of Knowledge Creation for Online Learning Environments", *Canadian Journal of Learning and Technology*, Vol. 30, No. 1, Winter/hiver.

Martinez, M., Galdiero, C. (2015), "Public Private Partnership in Italian Health Care Management: An Organizational Maturity Assessment Model", *GSTF Journal of Nursing and Health Care (JNHC)*, Vol. 2, No. 1, March, DOI: 10.5176/2345-718X_2.1.61.

Noennig, J. R., Jannack, A. (2013, June), "Garage labs: micro-incubators for scientific entrepreneurship". In *IFKAD–8th International Forum for Knowledge Asset Dynamics: Smart Growth: Organizations, Cities and Communities*, Zagreb, Croatia.

Nonaka, I., von Krogh, G., Voelpel, S., Streb, C. (2005), "Knowledge Creation 10 Years After: A Review and Appraisal", EGOS Colloquium.

Parsons, T. (2011), "Consumer Electronics Can Help Improve Patient Health", available: www.pewinternet.org.

Reina, R. (2015), *Organizzare i sistemi informative per la sanità – Analisi ed esperienze*. Milano: Giuffrè Editore.

Yin, R. (1994), *Case Study Research: Design and Methods* (2nd ed.). Thousand Oaks, CA: Sage Publishing.

Understanding public goods and education from a sustainable development goals perspective for the 21st century

From economics to humankind

Jose-Rodolfo Hernandez-Carrion[1]
and Rafael Soler-Muñoz[2]

Keywords: *public goods; education; economic theory; sustainable development goals; knowledge*

Public economic goods from an economic perspective have two defining features: non-rivalry and non-excludability. Air quality, knowledge or education can be examples of public goods. We want to question if one person's breathing of fresh air does not reduce air quality for others. Public goods are studied by Economic Theory to explain a very rare kind of situation which is possible to connect this perspective to education and knowledge, and we would like to understand this from a system science approach.

Public economics has recently introduced the new concept of global public goods as a new category of public goods whose provision is central for promoting the well-being of individuals in today's globalized world. The extent to which a good is perceived as "public" does not depend as much on its inherent characteristics as on prevailing social values within a given society about what should be provided by non-market mechanisms; that's why we see an opportunity for having a new view for the 2015 challenge of the United Nations agenda for the Sustainable Development Goals (SDG).

The concept of "public goods" is confusing because it confounds three analytically distinct concepts: excludability, rivalry and public finance. Pure public goods are of limited relevance as an explanation of government spending. The broader policy community uses the term in ways that invoke different means of both "public" and "good" than economists favour. The vision of the next stage in the advancement of civilization can be formulated by reexamination of the attitudes and assumptions that currently underlie approaches to social, legal and economic development. Sustainable development is an idea originated from the economic field connected to money and distribution. Today a new perspective of education as a "public good" should be the catalyst for this potential sustainable development according to SDG (2017).

Notes

1 Group of Economics and Complexity, University of Valencia (Spain).
2 Group of Economics and Complexity, University of Valencia (Spain).

Bibliography

Bollier, D. (2007). The growth of the commons paradigm. In *Understanding knowledge as a commons*. Cambridge: MIT Press. Retrieved October 2011, from http://goo.gl/HygdY

Buchanan, J.M. (1968). *The demand and supply of public goods*. Chicago: Rand McNally & Company.

Drucker, P.F. (1969). The knowledge society. In *The age of discontinuity: Guidelines to our changing society*, 263–381. San Francisco: Harper and Row.

Ferrer-Figueras, L. (1997). *Del paradigma mecanicista de la ciencia al paradigma sistémico*. Valencia: Universitat de València.

Gonzalez-Rodriguez, D., & Hernandez-Carrion, J. R. (2014). A bacterial-based algorithm to simulate complex adaptive systems. *Lecture Notes in Artificial Intelligence (LNAI)*, 8575, 250–259. https://doi.org/10.1007/978-3-319-08864-8_24

Gonzalez-Rodriguez, D., & Kostakis, V. (2015). Information literacy and peer-to-peer infrastructures: An autopoietic perspective. *Telematics and Informatics*, 32, 586–593. https://doi.org/10.1016/j.tele.2015.01.001

Hardin, G. (1968). The tragedy of the commons. *Science*, 162, 1243–1248.

Harnad, S. (1991). Post-Gutenberg galaxy: The fourth revolution in the means of production of knowledge. *Public-Access Computer Systems Review*, 2(1), 39–53.

Head, J.G. (1974). *Public goods and public welfare*. Durham: Duke University Press.

Hernández-Carrión, J.R. (2016). Supporting Decent Work and Economic Growth (Goal 8 of the United Nations 2030 Agenda): Recommended Library Actions. Retrieved October 2016, from www.ifla.org/files/assets/hq/topics/libraries-development/documents/sdg-8-and-libraires-en.pdf

Ostrom, E., & Hess, C. (2007). A framework for analyzing the knowledge commons. In *Understanding knowledge as a commons*. Cambridge: MIT Press.

Ostrom, V., & Ostrom, E. (1999). Public goods and public choices. 75–105, Polycentricity and Local Public Economies. Readings from the Workshop in Political Theory and Policy Analysis. Retrieved October 2016, from http://johannes.lecture.ub.ac.id/files/2012/02/Public-Goods-and-Public-Choices.pdf

Sábada, I. (2008). *Propiedad intelectual: ¿Bienes públicos o mercancías privadas?* Madrid: Los libros de la catarata.

Shaw, J.S. (2010). Education: A bad public good? *The Independent Review*, 15(2), 241–256. Retrieved October 2016, from www.independent.org/pdf/tir/tir_15_02_05_shaw.pdf

Learning for the future

Operationalizing competences in ESD

Francesca Farioli[1] *and Michela Mayer*[2]

Keywords: *transformative learning; ESD competences; sustainable education; system perspective*

The urgent priority for our planet is to understand and activate a transition towards sustainability. This is not merely a question of reducing greenhouse gas emissions, or even an economic transition, but rather a complete transition of our worldview, our customs and institutions, our behaviour into a mode that takes sustainability into account.

Much knowledge and publicity about sustainability has been produced at so many levels; however this has led to so little action.

Many reasons have been cited for this lack of success but one that has been overlooked is the lack of awareness of conviction, reflexiveness and behavioural change: critical elements of transformative learning (Wiek et al. 2012).

To infuse change to allow us to move to a sustainability trajectory requires a social transformation involving a variety of actors providing guidance and leadership in formal, non-formal and informal learning, and for that to happen, a corresponding enhancement in the competences of educators, leaders and decision makers is needed (Sterling 2001).

In 2011 the UNECE has developed a set of core competences in ESD, which however remains a largely theoretical tool that has not been tested against real educational contexts (UNECE 2011).

The Erasmus plus project 'A Rounder Sense of Purpose' (RSP) – Integrating ESD educator competences into educator training – coordinated by University of Gloucestershire with the participation of the Italian Association for Sustainability Science as the reference Partner for Italy, aims to operationalize the UNECE Competences in ESD into an accreditation scheme which could be used to assess acquisition of ESD competences by educators and therefore contributing to the transformation needed.

The approach followed by RSP aims to 'distil' and 'reduce' the number of competences and at the same time to re-word them in a way that assessment of acquisition could be feasible. This 'distillation' process has been carried out through dedicated group and pairs discussion using the UNECE document as a base and comparing UNECE competences sets with other SD and ESD competences sets (Roorda 2016; Wiek et al. 2015). The expected outcome of this activity was to reduce the number of competences and to express them in accordance with the largely accepted definitions of 'competences', which focus on the combination of knowledge, skills, attitudes and values appropriate to the context and enable successful task performance and problem solving (OECD DeSeCo Project 2003; Baartman et al. 2007). Moreover the need for developing 'key competences' for life-long learning has been highlighted in the recommendation of the European Parliament and of the Council, as those competences "which all individuals need for personal fulfilment and development, active citizenship, social inclusion and employment" (EU 2006).

The competences 'distillation process' has been carried out by the RSP project through different means, e.g. collecting the opinions of teachers and experts from the different countries involved, and discussing, through several 'retroactive' cycles, a re-organized and re-formulated framework of competences.

A provisional result of this process is the matrix of twelve competences showed in Table 40.1. The Table presents three columns corresponding to those of the UNECE model and four rows which are defined as Integration, Involvement, Practice and Reflection. Each of the twelve competences is indicated by a name and by a description formulated as concrete, observable, actions put in place by the educator.

Together, the four rows could indicate the cyclic process of competences development of an educator: (i) starting from an integrated approach to the environmental and social transformation needed in the local educational context, (ii) adding to this their personal involvement and commitment, (iii) combining the two in their practical work as an educator, (iv) evaluating the process and the results of their work, and finally linking all this to assume responsibility and take decision, even in situations of uncertainty.

For each of the twelve RSP competences a number of learning objectives have been formulated as a more detailed description of the competences to be developed. The model will be validated and improved through a number of experimentations and pilot testing which are planned to take place during the project period, across different contexts and using various modalities: by focus groups involving teachers, environmental educators and scientists; by implementation of specific training modules within a number of higher education and vocational training courses.

The debate over the model, which has been carried out so far in different working environment settings and different countries, confirms the authors' premise that the proposed ESD competences could be seen as general competences for educators, useful and usable for whatever discipline, in a vision of transformative social learning (Lotz-Sisitka et al. 2015).

The model, which is expected to be delivered at the end of the project, promises to be appreciated by all people engaged in overcoming the disciplinary boundaries

Table 40.1 RSP competence framework

Thinking holistically	Envisioning change	Achieving transformation
Integration:		
Systems competence	**Futures competence**	**Participation competence**
The educator helps learners to develop an understanding of the world as an interconnected whole and look for connections across human and natural worlds and consider the consequences of our actions.	The educator uses a range of techniques to help learners explore alternative possibilities for the future and to use these to consider how our behaviours might need to change.	The educator contributes towards changes in education that will help sustainable development and encourages their learners to do the same.
Involvement:		
Attentiveness competence	**Empathy competence**	**Engagement competence**
The educator alerts learners to fundamentally unsustainable aspects of our society and the way it is developing and conveys the urgent need for change.	The educator is considerate of the emotional impact of the learning process on their learners and develops their self-awareness.	The educator works flexibly and responsively with others, remaining aware of their personal beliefs and values, and encourages their learners to do the same.
Practice:		
Transdisciplinarity competence	**Innovation competence**	**Action competence**
The educator acts collaboratively both within and outside of their own discipline, role, perspectives and values and encourages their learners to do the same.	The educator takes an innovative and creative approach using real-world contexts wherever possible.	The educator focuses on the development of learners' critical thinking skills and helps them to take considered actions in their own context.
Reflection:		
Evaluation competence	**Responsibility competence**	**Decisiveness competence**
The educator helps learners to critically evaluate the relevance and reliability of assertions, sources, models and theories.	The educator acts transparently and accepts personal responsibility for their work and encourages their learners to do the same.	The educator acts in a cautious and timely manner even in situations of uncertainty and encourages their learners to do the same.

in favour of transversal, critical and 'transgressive' knowledge-production process. The expected implementation and experimentation will allow to verify its consistency, usability and efficacy for the development of Sustainable Educators Competences.

Acknowledgements

The project 'A Rounder Sense of Purpose' – Integrating ESD educator competences into educator training – is supported by Erasmus plus EU Programme-key action: Cooperation for innovation and the exchange of good practices action type: Strategic partnerships for higher education. The project is coordinated by University of Gloucestershire, project partners are Frederick University (CY); Kutato Tanarok Orszagos Szovetsege (HU); Italian Association for Sustainability Science (IT); Duurzame PABO (NL); Tallinn University (EE).

Notes

1 Italian Association for Sustainability Science (IASS).
2 Italian Association for Sustainability Science (IASS).

Bibliography

Baartman, L.K.J., Bastiaens, T.J., Kirschner, P.A., Van der Vleuten, C.P.M. (2007) "Evaluation assessment quality in competence-based education: A qualitative comparison of two frameworks", *Educational Research and Reviews*, 2, 114–129.

EU (2006) European Parliament and of the Council of the EU, Recommendation of the European Parliament and of the Council of 18 December 2006 on key competences for lifelong learning (2006/962/EC).

Lotz-Sisitka, H., Wals, A. J., Kronlid, D., Mc Garry, D. (2015) "Transformative, transgressive social learning: Rethinking higher education pedagogy in times of systemic global dysfunction", *Current Opinion in Environmental Sustainability*, 16, 73–80.

OECD (2003) "DeSeCo project", in D.S. Rychen and L.H. Salganik (eds.). *Key Competencies for a Successful Life and a Well-Functioning Society*, Hogrefe & Huber Göttingen, Germany.

Roorda, N. (2016) "The seven competences of a sustainable professional: The RESFIA+D model for HRM, education and training", in C. Machado and J. P. Davim (eds.). *Management for Sustainable Development*, River Publishers, Aalborg, Denmark.

Sterling, S. (2001) *Sustainable Education: Re-Visioning Learning and Change, Green Books for the Schumacher Society*. Schumacher UK, CREATE Environment Centre, Seaton Road, Bristol, BS1 6XN, England

UNECE (2011) Learning for the Future: Competences in Education for Sustainable Development, UN.

Wiek, A., Bernstein, M.J., Foley, R. W., Cohen, M., Forrest, N., Kuzdas, C., Kay B., Withycombe Keeler, L. (2015) "Operationalising competencies in higher education for Sustainable Development", in Barth et al. (eds.). *Handbook of Higher Education for Sustainable Development*, Routledge, London, pp. 241–260.

Wiek, A., Farioli, F., Fukushi, K., Yarime, M. (2012) "Bridging the gap between science and society editorial special feature", *Sustainability Science Journal*, 7(Supplement 1), Springer.

Edu-care
Towards an ethos of holistic care

Hernán López Garay[1]

Keywords: *higher education; ontological fragmentation; enlightenment; freedom; learning environments*

1 Introduction

Universities, one of the oldest institutions of higher education in the Western culture, are facing a crucial challenge: To continue reinforcing a view of the world that is increasingly making life unsustainable in planet earth or to make a significant contribution to the development of an ability to unfold and sustain an ethos of holistic care i.e., care of the human and natural planet as a whole.

In this paper we will develop an onto-historical understanding of the current crisis that will allow us to argue that the role of higher education institutions (HEI) should be *edu-care*, i.e. the education of a new breed of agents of change able to cultivate a new paradigm centered on care. This will demand from HEI to embark in what we may call a new project of Enlightenment – a project centered not on the search of infinite perfection through reason but through care.

At Universidad de Ibagué, in Colombia, we are beginning to unfold and test some ideas of what such an *edu-caring* means.

2 How we become what we are: ontological fragmentation

The project of Enlightenment has failed (MacIntyre, 1984). We live in a world more and more fragmented at all levels: reason separated from morality, man from nature, body from soul, and in general man divided from himself, other men and nature.

We experience this fragmentation of the world in many different ways: as a complexity beyond science and the humanities (which in turn are very fragmented), as an unsustainable world threatened by the increasing danger of global warming, exhaustion of natural resources, potential serious conflicts all over the world, increasing poverty, and a general loss of capability to make holistic sense of life.

Whence did all this fragmentation arise? What are its conditions of possibility? As we will explain, there is a deeper fragmentation, an ontological fragmentation which starts in Ancient Greece with the crack opened in the unity of beings and their ground. The ensuing emergence of an unstable relation between the supra-sensory and the sensory ends up at the present in the dissolution of both. In this latter historical condition everything including human beings becomes entities lacking intrinsic meaning (which previously they derived from the supra-sensible and in the later stage, before the dissolution, they tried to derive it from within). These empty shells become thus mere 'resources' to be optimized and disposed of with maximal efficiency.

What are the conditions of possibility of the transformation of every human being in mere resource to be disposed of with maximal efficiency? Heidegger's (1977) answer is that these conditions lie in the essence of technology, that he calls en-framing. Now, en-framing is a particular setting up of things character-ized by things left without the attentive caring for its thingly essence. En-framing does not protect or care for the thing as the thing (Heidegger, 1949, p. 2, emphasis added).

> En-framing is a particular way of revealing the world, such that everything, not only technological artifacts, is understood and acted upon as mere resources ready-to-be-used. To be more precise, it belongs to the essence of technology to unfold a world-system of things-ready-to-be-used. Each com-ponent of this system, including human beings, must have this instantaneous disposability within the larger system.

But contrary to what one may imagine, the danger with technology, thus understood, is not that technological things might get out of hand. The prime danger is losing our freedom. Certainly, the freedom Heidegger is talking about is not that one of remaining in a position of control of nature. The problem is that the technological way of revealing could become imperialist, "driving out every other possibility of revealing" by overwriting our defining capacity for world disclosure. It is the freedom to allow beings to come into un-concealment in a more authentic mode. Heidegger (1959, p. 56) puts it this way: The greater danger is that man comes to deny and throw away his own special nature – that he is a meditative being. Therefore, the issue is the saving of man's essential nature. How could we face this major epochal challenge? Inspired by Heidegger (1977) we will say that the saving power lies in disclosing the essence of ancient technology, *tékhne*.

3   On the essence of ancient technology (*tékhne*)

Let us think of reality not as made of objects or processes but as grounded on the void, the un-expressible infinite represented by a blank space, and infinite canvas. Such a void/canvas is Being, the ground of any particular being. Let us say the fundamental event is the act of Distinction. Drawing a Distinction creates the distinguished (distinction), the not-distinguished and the boundary between the two. In this sense one can say Distinction is a bringing-forth.

In Ancient Greece, *tékhne* was a particular way of bringing-forth different from *physis*. *Physis* was considered *poiesis* in the highest sense for "what presences by means of physis has the bursting open belonging to bringing-forth, e.g., the bursting of a blossom into bloom, in itself" (Heidegger, 1977, p. 5). In contrast, what is brought forth by *tékhne*, which may be considered the *poiesis* of the fine arts, "has the bursting open belonging to bringing-forth not in itself, but in another . . . in the craftsman or artist" (ibid.).

The ancient artisan illustrates the particular respectful, *careful*, mediating bringing-forth nature of the bringing-forth of *tékhne* (Rojcewicz, 2006):

> techne involves "*caring* for beings and letting them grow." In other words, it involves letting beings come into their essence, letting their essence come forth in them, letting the essence come to actual existence in beings. The bringing forth of the actual things is thus a matter of *care*, of letting or abetting the essence. . . . Techne does not amount to ordering things, making them submit to human will; on the contrary, it is a submitting of oneself to the essence of things, putting oneself in service to that essence.
>
> (p. 44, emphasis added)

Amply defined, this *caring* for beings is a capacity for world disclosure imbued with respect, with pious devotion, like the ancient artisan abets the essence into revealing itself *nurturing* the essence forth.

4   Higher education and the fostering of the growth of the saving power: edu-care

The role of higher education may lie, then, in helping us recover our freedom (that is, our human essence). One may envision higher education institutions embarking in the development of an ethos where our capacity for care-full world disclosure is cultivated. In this sense universities will rehearse and exemplify the rest of society what this new paradigm could mean. It will amount for HEI to undertake what we may call a new project of Enlightenment – a project not centered primarily on reason but on care.

Practices/disciplines constitutive of an ethos of care

Based on the previous conceptual framework, we are carrying out an action research project at Universidad de Ibagué (UNIBAGUE), Colombia, where a set of practices – intended to foster the unfolding of an ethos of holistic care – are being developed. The project makes part of a larger project which aims at the reform of the two first semesters common to all careers (López-Garay and Reyes-Alvarado, 2017).

P1.   *The practice of not filling but emptying and letting be*

Current education reinforces enframing by developing competent individuals for calculative thinking, disciplinary fragmentation of reality, the design of systems aimed at the total control of nature and other human beings, and in general an ability to unfold reality as mere resources always ready to be used.

In this regard, what is most required is the development of emptying and letting-be practices rather than practices that continue to fill in the enframing mentality. Whereas we are referring here to general exercises, to dispose a state of mind, the following practices are more specific (Senge et al., 2008).

P2.   *The practice of meditative-systems thinking rather than calculative thinking*

Meditative thinking is the type of thinking related to the poietical mode of revealing of *tekhne*, and thus to care, since essential care is poiesis.

Meditative thinking means to attune with the situation, to open herself to the seed/essence and foster its growth.

P3.   *The practice of releasement toward things*

When we use technological devices we not only use these machineries at our own convenience; we also let ourselves be challenged by them, so as to develop new devices that would be more suitable for a certain project or more accurate in the carrying out of certain research. Now, by being so extremely useful, at the same time these devices are 'shackling' us. It is a matter of a different comportment towards them; it is a different disposition to which Heidegger gives the name "releasement toward things" (Pezze, 2006, p. 7).

P4.   *The practice of dialogue*

It is in the dialogue (a 'swaying' of people's thinking) that something existing, but otherwise not unfolding, is first revealed. In the dialogue our receptiveness opens up and we become more prepared to wait. The tendency of affirmation weakens and the truth of what occurs finds its way to us. During a conversation 'something else' is allowed to be; it regains its time and space in our existence. It is created through the dialogue, like a symphony (Bohm, 1984).

P5.   *The practice of teaching-learning*

Teaching is letting learn. The teacher must be an exemplary learner, capable of teaching his or her students to learn, by being capable of learning-in-public, actively responding to the emerging demands of each unique educational situation (Thomson, 2001, p. 259).

5   Conclusion

At UNIBAGUE we are practicing these practices and developing them further in several space-time learning environments (López-Garay and Reyes-Alvarado, 2017). One of them, the introductory course to systems thinking for freshmen, is in fact a learning laboratory to develop micro-communities imbued with the aforementioned caring-developing practices. A larger learning-practicing community, formed by the first year curriculum for all careers is our next step in the developing of the caring-ethos project. Some of the emergent properties we are expecting to see (as the caring practices take root) in our university community are: The gradual de-fragmentation of this community. In fact, in the ontologically reductive terms of *enframing*, "we cannot fully understand the being of an entity – be it a book, cup, rose, or, even oneself – because they can not burst really fully in the open. They burst just as mere resources and that is all" (Thomson, 2001, p. 258).

Hence the practice of careful disclosing the essential in all things helps to shatter the encapsulation of the sciences because it reveals things in their full richness which cannot be trapped in one single discipline. And shatters as well the capsule of the students as the caring practices help students realize that entities are more than mere resources and hence they can be understood otherwise, i.e., one can disclose other *non-enframed worlds*.

Note

1 University of Ibagué.

Bibliography

Bohm, D. (1984). *On Dialogue*. Retrieved from http://sprott.physics.wisc.edu/Chaos-Complexity/dialogue.pdf

Brown, S. (1969). *Laws of Form*. London: Allen & Unwin, hardcover.

Heidegger, M. (1949). The Danger: A Bremen Lecture. Retrieved from https://es.scribd.com/doc/106181555/Heidegger-Die-Gefahr

Heidegger, M. (1959). *Discourse on Thinking: A Translation of Gelassenheit*. New York: Harper & Row, Publishers.

Heidegger, M. (1977). The Question Concerning Technology. In M. Heidegger (Ed.), *The Question Concerning Technology and Other Essays*, translated by W. Lovitt (pp. 3–36). New York: Garland Publishing.

Hodge, S. (2015). *Martin Heidegger Challenge to Education*. New York: Springer Publishing.

López-Garay, H. & Reyes-Alvarado, A. (2017). Learning and Becoming Trans-Disciplinary in Multiple Contexts: The Current Challenge for Engineering Education (A Case Study). Research in Engineering Education Symposium. July 6 to July 8, 2017. Universidad de Los Andes, Colombia.

MacIntyre, A. (1984). *After Virtue: A Study in Moral Theory* (2nd Ed.). IN: University of Notre Dame Press.

Pezze, B. (2006). Heidegger on Gelassenheit. *Minerva, Internet Journal of Philosophy*, 10, 94–122. Retrieved from www.minerva.mic.ul.ie//vol10/Heidegger.html 1/30

Rojcewicz, R. (2006). *The Gods and Technology: A Reading of Heidegger*. New York: State University of New York Press.

Senge, P., Scharmer, O., Jaworski, J. & Sue-Flowers, B. (2008). *Presence: Human purpose and the Field of the Future*. New York: Doubleday.

Thomson, I. (2001). Heidegger on Ontological Education, Or: How We Become What We Are. *Inquiry*, 44, 243–268.

Thomson, I. (2011). *Heidegger, Art, and Postmodernity*. Cambridge: Cambridge University Press.

Reflecting upon the role of smart grids in linking smartness and sustainability

Francesco Caputo,[1] *Barbora Buhnova*[2]
and Leonard Walletzky[3]

Keywords: *smartness; sustainability; smart grid domain; systems thinking*

1   Introduction

In a social and economic world affected by the increasing scarcity of resources, level of competitiveness, and variety, the studies on sustainability are attracting the interest of numerous researches and research streams (Cox & Blake 1991; Bretschger 2005; Porter 2011). The sustainability as science interested in the definition of the 'better' balance among social, economic, and environmental dynamics proposes a radical change in perspective in the way to approach and manage the world we live in (Komiyama & Takeuchi 2006; Lang et al. 2012; Barile et al. 2013a, 2013b).

Over time, different contributions have analyzed the key variable of sustainability (Clark & Dickson 2003; Kajikawa 2008; Wiek et al. 2012). In such perspectives, several researchers have studied what the impact of sustainability on companies' and consumers' decisions (Málovics et al. 2008; Pickett-Baker & Ozaki 2008; Young et al. 2010), and in which way it is possible to link the companies' economic aims with a wider perspective inclusive of the interests of future generations (Gladwin et al. 1995; Wade-Benzoni 1999; Perrini 2006; Barile et al. 2015). Despite the amazing advancements in knowledge offered by all these contributions, a shared conceptual framework on the strategies required to build common approaches to sustainability still appears to be missing (Boström 2012).

In order to bridge this gap, a possible contribution is offered by the evolutions proposed by the computer science as research domain interested in the identification of more effective and effective solutions by acting on a better use of available information and resources (Omondi et al. 2007; Caputo et al. 2016a, 2016c). By analyzing recent advancements in knowledge offered by computer science it is possible to identify their strong correlation with the principles, aims, and interests of sustainability sciences (Lazer et al. 2009; Swart et al. 2002). From a more general viewpoint, it is possible to note that computer science's willingness to build a smart world is addressing the society towards more sustainable lifestyles by tracing a wider conceptual framework in which the boundaries among the disciplinary domains dissolve themselves and a common pathway based on the weaving between sustainability and technology is emerging (Elzen et al. 2004). As a consequence of this development, a more interconnected and information-based society is emerging and new interpretative approaches are requesting from us to face new challenges (Castells 2011; Caputo & Walletzký 2017).

With the aim to contribute to the amazing debate on the interconnections among sustainability and technology, the paper focuses the attention on the topic of smart grids as an example of technological innovation useful in supporting the achievement of sustainability aims (Nidumolu et al. 2009). In such a line, smart grids are analyzed as systems able to connect several actors by supporting more efficient information sharing directly to ensure a sustainable use of the available resources (Giordano et al. 2011; Momoh 2012).

In order to enlarge previous contributions offered on the topic of smart grids, the paper builds upon the conceptual framework offered by systems thinking (Von Bertalanffy 1968; Emery 1981; Bogdanov 1989; Golinelli 2010; Barile & Saviano 2011; Barile 2013; Basile & Caputo 2017) and cybernetics (Wiener 1948, 1961; Ashby 1956; Beer 1960). In this perspective, the utility of smart grids in supporting the survival of a community by acting on a dynamic and fast adaptation of providing systems to the users' needs is analysed, and the implications for collectivity in the light of sustainability are discussed.

2   Research pathway

By adopting the interpretative lens offered by systems thinking and cybernetics the paper proposes a multi-disciplinary literature review in order to define a possible conceptual framework for the topic of smart grids in the light of sustainability science. The conceptual framework herein is verified with a specific aim to analyze in depth the link between technologies and sustainability and the role of smart grids as possible bridges between smartness and sustainability. Finally, conceptual reflections herein are discussed in order to enrich previous knowledge about the relationship between smartness and sustainability with the aim to define possible implications and indications for managerial and organizational studies.

The followed research pathway offers several stimuli with reference to the ways in which technologies support the definition and satisfaction of collective needs through the building of innovative and sustainable pathways. According to this, smart technologies have considered a 'mediator factor' useful in support of the 'translation' of sustainability strategies defined by governance, industries, and universities in forming effective paths the domain of economy, environment, and society. In such a line, the smart technologies are considered the key element of a Smartness Cycle for Sustainability (SmaCySu) as shown in Figure 42.1.

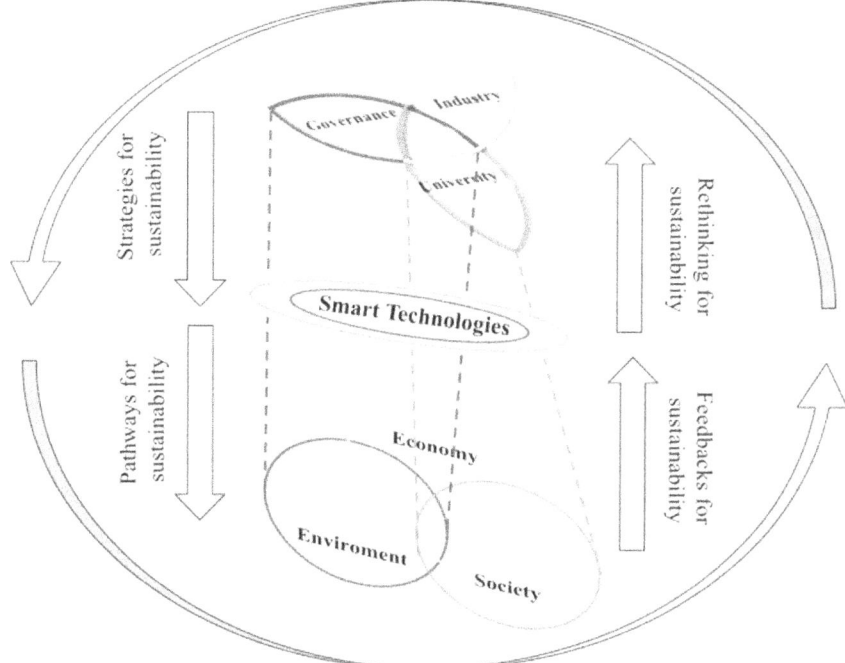

SMARTNESS CYCLE FOR SUSTAINABILITY

Figure 42.1 The Smartness Cycle for Sustainability (SmaCySu)
Source: Authors' elaboration

3   Theoretical and practical implications

The paper proposes relevant evidences of the application of systems thinking and cybernetics principles as consequences of the communication between smartness and sustainability. In such a line, a multi-disciplinary conceptual framework is proposed as a useful starting point for future research and studies.

At the same time, the paper offers several indications for decision makers and practitioners interested in better understanding the principles of sustainability and their role in the emerging smart world (Di Fatta et al. 2016).

By following the reflections herein, relevant guidelines to face the emerging challenges of smartness and sustainability can be derived both for decision makers and practitioners.

4   Conclusions and future directions for research

The vibrant scenario in which we live cannot be explained and understood by using conceptual models developed with references to old rules and balances (Caputo et al. 2016b; Di Nauta et al. 2015; Saviano et al. 2016). New pathways and perspectives must be defined to face the challenges of increasing variety in an emerging world on an emerging world affected by an.

In such a line, the topics of smartness and sustainability well synthesize most of the elements able to influence the future evolutions of our society and our world. With the aim to offer a contribution to the wide-reaching debate on the link between smartness and sustainability, the paper proposes some conceptual reflections with the aim to underline advantages and opportunities of smart approaches to achieve the sustainability aims under the shared conceptual umbrella of systems thinking and cybernetics studies.

Notes

1 Masaryk University.
2 Masaryk University.
3 Masaryk University.

Bibliography

Ashby, W. R. (1956). *An introduction to cybernetics: An introduction to cybernetics*. London: Chapman & Hall.

Barile, S. (2013). *Contributions to theoretical and practical advances in management: A Viable Systems Approach (VSA)*. Roma: ARACNE.

Barile, S., Carrubbo, L., Iandolo, F., Caputo, F. (2013a). From 'EGO' to 'ECO' in B2B relationships. *Journal of Business Market Management*, 6(4), 228–253.

Barile, S., Saviano, M., Polese, F., Di Nauta, P. (2013b). Il rapporto impresa-territorio tra efficienza locale, efficacia di contesto e sostenibilità ambientale. *Sinergie rivista di studi e ricerche*, 91(1), 25–49.

Barile, S., Saviano, M. (2011). Foundations of systems thinking: The structure-system paradigm. In Vv.Aa., *Contributions to theoretical and practical advances in management: A Viable Systems Approach (VSA)* (pp. 1–24). Avellino: International Printing.

Barile, S., Saviano, M., Iandolo, F., Caputo, F. (2015). La dinamica della sostenibilità tra vortici e correnti. In XXXVII Convegno Nazionale AIDEA Sviluppo, sostenibilità e competitività delle aziende: il contributo degli economisti aziendali, Università Cattolica del Sacro Cuore, Piacenza, 10–12 September.

Basile, G., Caputo, F. (2017). Theories and challenges for systems thinking in practice. *Journal of Organisational Transformation & Social Change*, 14(1), 1–3.

Beer, S. (1960). *Cybernetics and management*. New York: Wiley.

Bogdanov, A. A. (1989). *Tektology: Universal organizational science*. Moscow: Finance.

Boström, M. (2012). A missing pillar? Challenges in theorizing and practicing social sustainability: Introduction to the special issue. *Sustainability: Science, Practice, & Policy*, 8(1), 1–14.

Bretschger, L. (2005). Economics of technological change and the natural environment: How effective are innovations as a remedy for resource scarcity? *Ecological Economics*, 54(2), 148–163.

Caputo, F., Evangelista, F., Russo, G. (2016a). Information sharing and communication strategies: A stakeholder engagement view. In Vrontis, D., Weber, Y., Tsoukatos, E. (Eds.) *Innovation, entrepreneurship and digital ecosystems* (pp. 436–442). Cyprus: EuroMed Press.

Caputo, F., Formisano, V., Buronova, B., Walletzky, L. (2016b). Beyond the digital ecosystems view: Insights from Smart Communities. In Vrontis, D., Weber, Y., Tsoukatos, E. (Eds.) Innovation, entrepreneurship and digital ecosystems (pp. 443–454). Cyprus: EuroMed Press.

Caputo, F., Giudice, M. D., Evangelista, F., Russo, G. (2016c). Corporate disclosure and intellectual capital: The light side of information asymmetry. *International Journal of Managerial and Financial Accounting*, 8(1), 75–96.

Caputo, F., Walletzký, L. (2017). Investigating the users' approach to ICT platforms in the city management. *Systems*, 5(1), 1–15.

Castells, M. (2011). *The rise of the network society: The information age: Economy, society, and culture*. New York: John Wiley & Sons.

Clark, W. C., Dickson, N. M. (2003). Sustainability science: The emerging research program. *Proceedings of the National Academy of Sciences*, 100(14), 8059–8061.

Cox, T. H., & Blake, S. (1991). Managing cultural diversity: Implications for organizational competitiveness. *The Executive*, 45–56.

Di Fatta, D., Caputo, F., Evangelista, F., Dominici, G. (2016). Small world theory and the World Wide Web: Linking small world properties and website centrality. *International Journal of Markets and Business Systems*, 2(2), 126–140.

Di Nauta, P., Merola, B., Caputo, F., Evangelista, F. (2015). Reflections on the role of university to face the challenges of knowledge society for the local economic development. *Journal of Knowledge Economy*, 1–19. DOI:10.1007/s13132-015-0333-9

Elzen, B., Geels, F. W., Green, K. (Eds.). (2004). *System innovation and the transition to sustainability: Theory, evidence and policy*. London: Edward Elgar Publishing.

Emery, F. E. (Ed.). (1981). *Systems thinking: Selected readings*. New York: Penguin books.

Giordano, V., Gangale, F., Fulli, G., Jiménez, M. S., Onyeji, I., Colta, A., Maschio, I. (2011). smart grid projects in Europe: Lessons learned and current developments. JRC Reference Reports, Publications Office of the European Union.

Gladwin, T. N., Kennelly, J. J., Krause, T. S. (1995). Shifting paradigms for sustainable development: Implications for management theory and research. *Academy of Management Review*, 20(4), 874–907.

Golinelli, G. M. (2010). *Viable Systems Approach (VSA): Governing business dynamics*. Padova: Cedam.

Kajikawa, Y. (2008). Research core and framework of sustainability science. *Sustainability Science*, 3(2), 215–239.

Komiyama, H., Takeuchi, K. (2006). Sustainability science: Building a new discipline. *Sustainability Science*, 1(1), 1–6.

Lang, D. J., Wiek, A., Bergmann, M., Stauffacher, M., Martens, P., Moll, P., Thomas, C. J. (2012). Transdisciplinary research in sustainability science: Practice, principles, and challenges. *Sustainability Science*, 7(1), 25–43.

Lazer, D., Pentland, A. S., Adamic, L., Aral, S., Barabasi, A. L., Brewer, D., Jebara, T. (2009). Life in the network: The coming age of computational social science. *Science* (New York, NY), 323(5915), 721.

Málovics, G., Csigéné, N. N., Kraus, S. (2008). The role of corporate social responsibility in strong sustainability. *The Journal of Socio-Economics*, 37(3), 907–918.

Momoh, J. (2012). *Smart grid: Fundamentals of design and analysis*. New York: John Wiley & Sons.

Nidumolu, R., Prahalad, C. K., Rangaswami, M. R. (2009). Why sustainability is now the key driver of innovation. *Harvard Business Review*, 87(9), 56–64.

Omondi, A., Premkumar, B., Gelenbe, E., Mitrani, I., Whittle, P., Bailly, F., Walk, H. (2007). *Advances in computer science and engineering: Texts*. London: Imperial College Press.

Perrini, F. (2006). SMEs and CSR theory: Evidence and implications from an Italian perspective. *Journal of Business Ethics*, 67(3), 305–316.

Pickett-Baker, J., Ozaki, R. (2008). Pro-environmental products: Marketing influence on consumer purchase decision. *Journal of Consumer Marketing*, 25(5), 281–293.

Porter, M. E. (2011). *Competitive advantage of nations: Creating and sustaining superior performance*. New York: Simon and Schuster.

Saviano, M., Caputo, F., Formisano, V., Walletzký, L. (2016). From theory to practice in systems studies: A focus on Smart Cities. In Caputo, F. (Ed.) The 4rd international symposium advances in business management: 'Towards systemic approach' (pp. 35–40). Avellino: Business Systems. E-book series.

Swart, R., Raskin, P., & Robinson, J. (2002). Critical challenges for sustainability science. *Science*, 297(5589), 1994–1995.

Von Bertalanffy, L. (1968). *General systems theory*. New York: George Braziller.

Wade-Benzoni, K. A. (1999). Thinking about the future an intergenerational perspective on the conflict and compatibility between economic and environmental interests. *American Behavioral Scientist*, 42(8), 1393–1405.

Wiek, A., Ness, B., Schweizer-Ries, P., Brand, F. S., Farioli, F. (2012). From complex systems analysis to transformational change: A comparative appraisal of sustainability science projects. *Sustainability Science*, 7(1), 5–24.

Wiener, N. (1948). *Cybernetics*. Paris: Hermann.

Wiener, N. (1961). *Cybernetics or control and communication in the animal and the machine*. Cambridge: MIT Press.

Young, W., Hwang, K., McDonald, S., Oates, C. J. (2010). Sustainable consumption: Green consumer behaviour when purchasing products. *Sustainable Development*, 18(1), 20–31.

An analysis of FDI impact in GDP growth

Case of Albania

Remzi Sulo[1] and Aurel Koroci[2]

Keywords: *FDI; GDP growth; Albania; international trade*

Foreign Direct Investment (FDI) is seen as the fundamental part for an open and successful international economic system and a major mechanism for development. The entrance of FDI would normally be considered as a prerequisite for the success of the introduction of foreign capital. This paper aims to gain a better understanding of FDI impact in economic development in a country.

In modern economies, economic growth is attributed to three main factors: capital, labor, and technical and technological progress. Financial and industrial globalization is increasing substantially and is creating new opportunities for both industrialized and developing countries. The largest impact has been on developing countries because they now are able to attract foreign investors and foreign capital. FDI in developing countries brings economic development and enhances the international competitiveness of domestic enterprises. FDI, in particular in developing countries, is considered one of the most important factors of economic growth. Promoting FDI in economic growth in the host country has its own theoretical basis.

Since the beginning of the '90s of last century, FDI became a key factor in acceleration of economic globalization. The network of national economies is linked to international corporations and coordinates their activities and financial economy. Even the process of economic globalization itself has accelerated operations and direct investment of multinational corporations, mainly in developing countries of the world. FDI actually manifests some important features that relates to the country they come from, the direction in which they are invested in host countries, very good conditions they create for the development of international trade, etc.

FDI accounts for the largest and most important proportion of foreign capital in Albania, which undoubtedly plays an important role in Albania's economic development growth. FDI into Albania has been rising steadily since the early 2000s. Today, FDI stock has reached nearly 50% of the country's GDP. These investments are essentially in the oil, metal ore, infrastructure, construction and telecommunication sectors. Albania has set up reforms to boost FDI. The State has adopted a tax reform that is advantageous to foreign investors and aims at reducing corruption and administrative difficulties, which can be discouraging to investors. The long-winded procedures to obtain operating licenses in the trade, construction and tourism sectors have slowed down investment progress. In addition, investments continue to suffer from the lack of infrastructure and poorly defined property law. The program of structural reforms, carried out by Albania in the context of its agreement with the IMF and in order to accede to the EU, should encourage foreign investment, particularly in the energy sector. Since 2013, FDI flows to the country have exceeded USD 1 billion, a trend that should continue.

Although Albania remains one of the least developed countries in Europe, it has a strategic geographical position (with ports on both the Adriatic and the Ionian Sea); significant natural resources; cheap manpower; prospects of joining the European Union. Additionally, Albania is still a developing country which needs foreign investors to develop entire sections of its economy, a fact which provides interesting opportunities.

Empirical evidence from different countries suggests that FDI plays an important role in contributing to economic growth. However, most studies generally indicate that the effect of FDI on growth depends on other factors such as the degree of complimentarily and substitution between domestic investment and FDI, and other country-specific characteristics. From this point of view, the main objective of the study is to analyze the impact of foreign direct investment on economic growth in Albania.

This paper aims to gain a better understanding of the impact of Foreign Direct Investment (FDI) in GDP and to explore empirically the correlation between them. The stated issue is analyzed using time series data (2006–2015) of GDP growth and FDI.

The results of the study showed that there is a positive relation between FDI and GDP growth.

Notes

1 Albanian University.
2 Albanian University.

Bibliography

Aghion, P., D. Comin, and P. Howitt, (2006). When Does Domestic Saving Matter for Economic Growth? NBER Working Paper 12275.

Arrken Brian, J., and A. F. Harrison, (1999). Do Domestic Firms Benefit from Direct Foreign Investment: Evidence from Venezuela. *American Economic Review*, (89), 232–237.

Borensztein, E., J. De Gregorio, and J.-W. Lee, (1998). How Does Foreign Direct Investment Affect Economic Growth? *Journal of International Economics*, 45, 115–135.

De Mello Luiz, R., Jr., (1999). Foreign Direct Investment Led Growth: Evidence from Time Series and Panel Data. *Oxford Economic Papers*, (51), 133–151.

Durham, K. B., (2004). Absorptive Capacity and the Effects of Foreign Direct Investment and Equity Foreign Portfolio Investment on Economic Growth. *European Economic Review*, 48, 285–306.

Gao, T., (2005). Foreign Direct Investment and Growth under Economic Integration. *Journal of International Economics*, 67, 157–174.

Koroci, A., (2008). Is the China's FDI Excessive? Based on the Empirical Research of Crowding-Out (into) Effect. International Conference, Tirana, Albania, First Volume, p. 79.

Websites

http://ec.europa.eu/enlargement/pdf/key_documents/2 015/20151110_report_albania.pdf

Is a smarter planet also more sustainable?

Co-creating knowledge for sustainability

Sergio Barile,[1] *Francesca Farioli,*[2] *Fabio Orecchini*[3] *and Marialuisa Saviano*[4]

Keywords: *smartness; sustainability; service science; sustainability science; systems thinking; knowledge co-creation; boundary-crossing interaction*

1 Introduction and aims

Sustainability and Sustainable Development are becoming increasingly relevant in the global agenda of governments as well as businesses and civil society. They should be among the top priorities of what we would consider a really smarter planet.

The concept of *smartness* is widely used essentially to refer to digitalized processes and telematic interactions in several fields of social, environmental and economic human activities. The concept of *sustainability* is used to define successful human activities, processes and interactions from an integrated social-environmental-economic viewpoint. Although both the concepts imply multi- and inter-disciplinary views that involve economic, social and environmental sciences, the 'smartness' and 'sustainability' perspectives have different focus: the former is more focused on *socio-technical systems*, hence on human-technology interactions; the latter is more focused on *social-ecological systems* (Berkes et al., 2003; Ostrom, 2009a), hence on human-nature interactions.

By reflecting upon human-nature and human-technology interactions, i.e. relationships between socio-technical and social-ecological systems, in terms of contribution to sustainability, we wonder:

> *What is the relationship between smartness and sustainability?*
> *Is a smarter planet also more sustainable?*

With the aim of addressing these questions, by adopting a systems thinking view and co-creation logic, our essay outlines a possible boundary-crossing co-creation framework for Sustainable Development. Our interpretative methodology is built upon the roots of systems thinking (Barnard, 1938; Buckley, 1968; von Bertalanffy, 1968; Emery, 1969; Espejo, 1994; Basile and Caputo, 2017; Calabrese et al., 2017; Tronvoll et al., 2017). Systems thinking and, specifically, the *Viable Systems Approach* (*vSa*) (Golinelli, 2010; Barile, 2013; Barile et al.,

2012b) can be adopted as meta-level frameworks that provide general interpretation schemes to support understanding of complex phenomena like sustainability. Accordingly, our aim is to identify research domains that are engaged in knowledge co-creation efforts whose integration could accelerate progress toward sustainability (Di Nauta et al., 2015; Formisano et al., 2015; Caputo, 2017).

2   Discussion

Our findings suggest exploring the possible role of two among the worldwide-recognized communities of scientists, scholars and professionals that respectively envision a 'smarter' and 'more sustainable' planet: *Service Science* (Spohrer and Maglio, 2008), and *Sustainability Science* (Kates et al., 2001; Clark and Dickson, 2003; Komiyama and Takeuchi, 2006; Wiek et al., 2012b). The purpose is to seek commonalities that can highlight prospects of fruitful scientific collaboration for bridging smartness and sustainability perspectives (Golinelli et al., 2015; Polese et al., 2016). Both research streams are targeted to the challenge of developing inter- and trans-disciplinary bodies of knowledge which could contribute to the improvement of life conditions of individuals and organizations: the focus of the first is on making our planet *smarter*; the other on making it *more sustainable*. Both Service Science and Sustainability Science communities have to deal with the complexity of life phenomena and need a better understanding of reality through integration of disciplines. The governance of *social-technical-ecological* systems implies, in fact, complex trade-offs and systemic dynamics to deal with (Barile and Saviano, 2017; Caputo and Walletzky, 2017). Finally, both adopt a solution-oriented research approach and call for scientific collaborations among scholars as well as practitioners from various domains to co-create knowledge (Caputo and Evangelista, 2017; Caputo et al., 2016).

We believe that by recognizing and further exploring potential convergences between the two research streams, it will be possible to bridge the current gap between them. Such bridging elements could be integrated to develop a common framework to support knowledge co-design and co-creation processes through which *University-Industry-Government* collaboration, which is necessary to address the challenge of a smarter and more sustainable world, can be realized (Saviano, 2015; Carayannis et al., 2017).

A key element of such a knowledge co-creation context is effective *boundary-crossing interaction* that can allow the dialogue between disciplines (Barile et al., 2012a, 2012b; Barile et al., 2014a). Dealing with the complexity of sustainability and sustainable development requires, in fact, crossing of discipline borders through boundary interaction. The aim, in particular, is to favour the integration of the environmental and economic sciences' perspectives in a coherent framework.

This co-creation context can provide the knowledge necessary to support the required University-Industry-Government collaboration. A widely shared call for a functioning *Science-Policy* interface for SD was launched by the UN member States at Rio+20 (UN-DESA, 2014). On the other hand, there is wide consensus on the key role of a *Science-Industry* collaboration to address effective innovation through the 'Third mission' of Universities (Ranga and Etzkowitz, 2013). These double interfaces are necessary since co-creating harmonic social-ecological systems requires multi-actor collaboration, wide stakeholders involvement, information sharing, dialogue between different paradigms and sharing of sustainable development values.

In particular, an effective Science-Policy interface can help to (Cornell et al., 2013; Farioli, 2015; Saviano, 2015):

- Identify priorities and select targets.
- Integrate knowledge resources.
- Merge different perspectives.
- Provide decision makers with the necessary knowledge support.
- Make knowledge accessible.
- Enhance the educational impact of the process.
- Direct the emergent transformational change.

An effective Policy-Society interface can help to (Saviano, 2015):

- Favor multi-actor participation and collaboration by identifying spaces of consonance for effective interaction, balancing interests and solving emerging conflicts, and aligning actors efforts and integrating resources.
- Coordinate participated decision making processes.
- Connect the local and the national levels of government.
- Govern the variability of emergent dynamics and unexpected outcomes.

Coupled to effective science-industry and policy-industry interfaces, these roles can complete a possible boundary-crossing co-creation framework for Sustainable Development as showed in Figure 44.1.

In the proposed scheme, academia/university is expected to lead the knowledge generation process, defining the *possibilities* to work on; governments are expected to create the conditions (constraints, rules and incentives) for addressing sustainable development identifying the *necessities* to comply with; industry is expected to put in practice the possibilities, given the necessities, i.e. to develop feasible *solutions* to implement (Farioli et al., forthcoming).

A key aspect in the model is the role and function of actors that (Saviano, 2015):

- Are reciprocally engaged through interface processes.
- Dynamically take different perspectives.
- Dynamically play different *roles* going beyond their institutional *functions*.

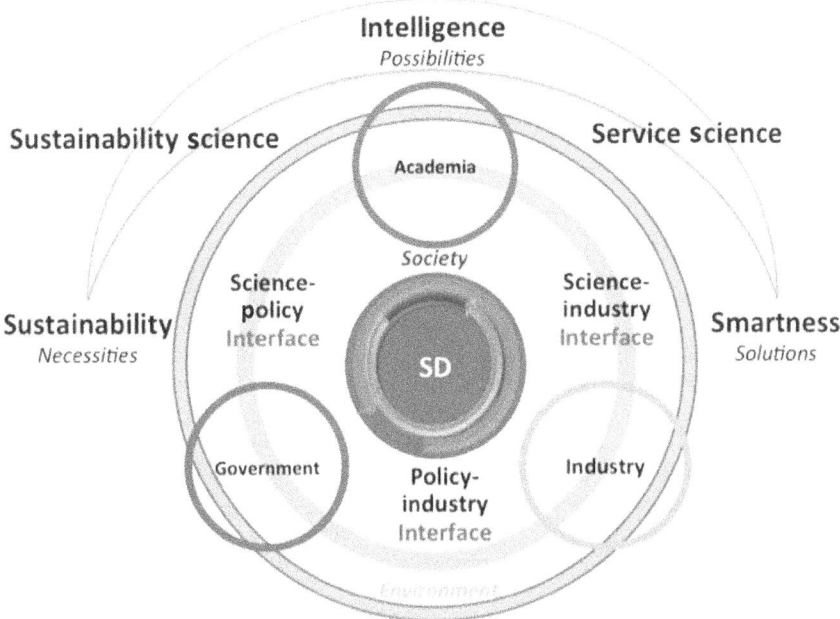

Figure 44.1 A boundary-crossing co-creation model for Sustainable Development (SD)

Sources: Elaboration from Saviano (2015); www.asvsa.org

3 Conclusions

In the light of our framework, it implicitly appears that a smarter planet does not necessarily imply a more sustainable planet: while sustainability, in terms of orienting behaviours, implies involvement of the values systems that guide choices and decisions, smartness acts at a more operative (structural and organizational) level regarding the way processes are organized and managed from a prevalent engineering and management problem-solving view.

Hence, a Smarter Planet does not necessarily mean a more Sustainable Planet.

In conclusion, this essay aims to launch a call for collaboration among scholars and practitioners from various domains to co-create knowledge for sustainability (Saviano et al., 2017a). It also provides insights for rethinking research, education, policy making and management approaches by putting sustainability at the centre of a shared science, policy and business agenda opening up new research pathways to collaboratively address the challenges of a *more intelligent* world.

Notes

1 Sapienza, University of Rome, Italy.
2 IASS, Italian Association for Sustainability Science, Italy.
3 University of Salerno and IASS, Italian Association for Sustainability Science, Italy (*corresponding author – msaviano@unisa.it*).
4 Marconi University, Rome, Italy and IASS, Italian Association for Sustainability Science, Italy.

Bibliography

Arnold, M. (2015). Fostering sustainability by linking co-creation and relationship management concepts. *Journal of Cleaner Production*, 140, 179–188.

Barile, S. (Ed.) (2013). *Contributions to Theoretical and Practical Advances in Management: A Viable Systems Approach (VSA)*. ARACNE, Roma.

Barile, S., Carrubbo, L., Iandolo, F., Caputo, F. (2013). From 'EGO' to 'ECO' in B2B relationships. *Journal of Business Market Management*, 6(4), 228–253.

Barile, S., Franco, G., Nota, G., Saviano, M. (2012a). Structure and dynamics of a 'T-shaped' knowledge: From individuals to cooperating communities of practice. *Service Science, Informs*, 4(2), 161–180.

Barile, S., Lusch, R., Reynoso, J., Saviano, M., Spohrer, J. (2016). Systems, networks, and eco-systems in service research. *Journal of Service Management*, 27(4), 652–674.

Barile, S., Pels, J., Polese, F., Saviano, M. (2012b). An introduction to the viable systems approach and its contribution to marketing. *Journal of Business Market Management*, 5(2), 54–78.

Barile, S., Saviano, M. (2011). Qualifying the concept of systems complexity. Various Authors (Eds.), *Contributions to Theoretical and Practical Advances in Management: A Viable Systems Approach (VSA)* (pp. 27–63). Associazione per la ricerca sui Sistemi Vitali International Printing, Avellino. www.asvsa.org.

Barile, S., Saviano, M. (2013). An introduction to a value co-creation model: Viability, syntropy and resonance in dyadic interaction. *Syntropy*, (2), 69–89.

Barile, S., Saviano, M. (2017). Complexity and Sustainability in Management: Insights from a systems perspective. In Barile, S., Pellicano, M., Polese, F. (Eds.), *Social Dynamics in a Systems Perspective*, New Economic Window Book Series, Springer International.

Barile, S., Saviano, M., Iandolo, F., Calabrese, M. (2014a). The viable systems approach and its contribution to the analysis of sustainable business behaviors. *Systems Research and Behavioral Science*, 31(6), 683–695.

Barile, S., Saviano, M., Iandolo, F., Caputo, F. (2015). La dinamica della sostenibilità tra vortici e correnti, *Proceedings XXXVII Convegno Nazionale AIDEA Sviluppo, sostenibilità e competitività delle aziende: il contributo degli economisti aziendali*, Bologna, Il Mulino.

Barile, S., Saviano, M., Simone, C. (2014b). Service economy, knowledge and the need for T-shaped innovators. *World Wide Web*, 18(4), 1177–1197. DOI: 10.1007/s11280-014-0305-1.

Barnard, C. (1938). *The Functions of the Executive*. Cambridge, MA: Harvard University Press.

Basile, G., Caputo, F. (2017). Theories and challenges for systems thinking in practice. *Journal of Organisational Transformation & Social Change*, 14(1), 1–3.

Berkes, F., Colding, J., Folke, C. (2003). *Navigating Social: Ecological Systems: Building Resilience for Complexity and Change.* Cambridge University Press, Cambridge, UK.

Bettencourt, L.M.A., Kaur, J. (2011). The evolution and structure of sustainability science.*Proceedings of the National Academy of Sciences*, 108, 19540–19545.

Brown, V.A., Harris, J.A., Russell, J.Y. (2010). *Tackling Wicked Problems: Through the Transdisciplinary Imagination.* Routledge, London.

Buckley, W. (1968). *Modern Systems Research for the Behavioural Scientist.* Aldine Transaction.

Calabrese, M., Iandolo, F., Caputo, F., Sarno, D. (2017). From mechanical to cognitive view: The changes of decision making in business environment. In Barile, S., Pellicano, M., Polese, F. (Eds.), *Social Dynamics in a System Perspective* (pp. 223–240). Springer, New York.

Caputo, F. (2016). A focus on company-stakeholder relationships in the light of the stakeholder engagement framework. In Vrontis, D., Weber, Y., Tsoukatos, E. (Eds.), *Innovation, Entrepreneurship and Digital Ecosystems* (pp. 455–470). EuroMed Press, Cipro.

Caputo, F. (2017). Reflecting upon knowledge management studies: Insights from systems thinking. *International Journal of Knowledge Management Studies*, 8(3/4), 177–190.

Caputo, F., Del Giudice, M., Evangelista, F., Russo, G. (2016). Corporate disclosure and intellectual capital: The light side of information asymmetry. *International Journal of Managerial and Financial Accounting*, 8(1), 75–96.

Caputo, F., Evangelista, F. (2017). Information sharing and cognitive involvement for sustainable workplaces. In Leon, R.D. (Ed.), *Managerial Strategies for Business Sustainability during Turbulent Times* (pp. 122–139). IGI Global, New York.

Caputo, F., Evangelista, F., Perko, I., Russo, G. (2017). The role of big data in value co-creation for the knowledge economy. In Vrontis, S., Weber, T., Tsoukatos, E. (Eds.), *Global and National Business Theories and Practice: Bridging the Past with the Future* (pp. 269–280). EuroMed Press, Cyprus.

Caputo, F., Evangelista, F., Russo, G. Buhnova, B. (2017). A systems view of companies' communication in online social environment. *Journal of Organizational Transformation & Social Change*, 14(1), 21–38.

Caputo, F., Perano, M., Mamuti, A. (2017). A macro-level view of tourism sector: Between smartness and sustainability. *Enlightening Tourism: A Pathmaking Journal*, 7(1), 36–61.

Caputo, F., Walletzky, L. (2017). Investigating the users' approach to ICT platforms in the city management. *Systems*, 5, 1. DOI: 10.3390/systems5010001.

Carayannis, E.G., Barth, T.D., Campbell, D.F. (2012). The Quintuple Helix innovation model: Global warming as a challenge and driver for innovation. *Journal of Innovation and Entrepreneurship*, 1(1), 1–12.

Carayannis, E.G., Caputo, F., Del Giudice, M. (2017). Technology transfer as driver of smart growth: A Quadruple/Quintuple innovation framework approach. In Vrontis, S., Weber, T., Tsoukatos, E. (Eds.), *Global and National Business Theories and Practice: Bridging the Past with the Future* (pp. 295–315). EuroMed Press, Cyprus.

Cash, D., Clark, W.C., Alcock, F., Dickson, N., Eckley, N. (2003). Knowledge systems for sustainable development. *PNAS*, 100, 8086–8091.

Christopher, W.F. (2010). From system science: A new way to structure and manage the company for sustainable success. *Service Science*, 2(1–2), 62–75.

Clark, W.C., Dickson, N.M. (2003). Sustainability science: The emerging research program. *PNAS*, 100, 8059–8061.

Cornell, S., Berkhout, F., Tuinstra, W., Tàbara, J.D., Jäger, J., Chabay, I., . . . Otto, I.M. (2013). Opening up knowledge systems for better responses to global environmental change. *Environmental Science & Policy*, 28, 60–70.

Del Giudice, M., Ahmad, A., Scuotto, V., Caputo, F. (2017). Influences of cognitive dimensions on the collaborative entry mode choice of small and medium-sized enterprises. *International Marketing Review*, 34(5), 652–673.

Del Giudice, M., Scuotto, V., Khan, Z., Caputo, F., Carayannis, E. (2017). Micro level actions by owners-managers in improving sustainability practices of culture and creative small and medium enterprises: UK-Italy comparison. *Journal of Organizational Behaviour*, 1–19. DOI: 10.1002/job.2237.

Di Nauta, P., Merola, B., Caputo, F., Evangelista, F. (2015). Reflections on the role of university to face the challenges of knowledge society for the local economic development. *Journal of Knowledge Economy*, 1–19.

Dovers, S. (1996). Sustainability: Demands on policy. *Journal of Public Policy*, 16, 303–318.

Elkington, J. (1997). *Cannibals with Forks: The Triple Bottom Line of 21st Century Business*. Capstone, London.

Elkington, J., Fennell, S. (1998). Can business leaders satisfy the triple bottom line? *Visions of Ethical Business*, 1, 34–36.

Emery, F. (Ed.) (1969). *Systems Thinking*. Penguin Books, Harmondsworth.

Espejo, R. (1994). What is systemic thinking. *System Dynamic Review*, 10(2/3), 199–212.

Espejo, R. (2014). Organizational transparency, inclusion and sustainability, *Refereed Proceedings of the Business Systems Laboratory: 2nd International Symposium*, January 23–24, Universitas Mercatorum, Rome, Italy.

Espejo, R., Harnden, R. (1989). *The Viable System Model: Interpretations and Applications of Stafford Beer's VSM*. Wiley, New York.

Farioli, F. (2015). Co-design and co-creation of knowledge for sustainability: An introduction, *5th International Conference on Sustainability Science (ICSSS)*, Tokyo.

Farioli, F., Barile, S., Saviano, M., Iandolo, F. (2018 in press). Re-reading sustainability through the Triple Helix model in the frame of a systems perspective. In Marsden, K. (Ed.), *The Sage Handbook of Nature*.

Folke, C., Carpenter, S., Elmqvist, T., Gunderson, L., Holling, C.S., Walker, B. (2002). Resilience and sustainable development: Building adaptive capacity in a world of transformations. *AMBIO: A Journal of the Human Environment*, 31(5), 437–440.

Formisano, V., Caputo, F., D'amore, R. (2015). Tratti evolutivi della società della conoscenza: il contributo degli studi sulle reti nella prospettiva sistemica. *Esperienze d'impresa*, 2, 73–94.

Frame, B. (2008). 'Wicked', 'messy' and 'clumsy': Long-term frameworks for sustainability. *Environment and Planning C: Government and Policy*, 26(6), 1113–1128.

Funtowicz, S.O., Ravetz, J.R. (1993). Science for the post-normal age. *Futures*, 25, 735–755.

Gibbons, M. (1994). *The New Production of Knowledge: The Dynamics of Science and Research in Contemporary Societies*. Sage, London.

Goerner, S.J., Lietaer, B., Ulanowicz, R.E. (2009). Quantifying economic sustainability: Implications for free-enterprise theory, policy and practice. *Ecological Economics*, 69(1), 76–81.

Golinelli, G.M. (2010). *Viable Systems Approach: Governing Business Dynamics.* Cedam Kluwer, Padova.

Golinelli, G.M. (2015). *Cultural Heritage and Value Creation: Towards New Pathways.* Springer International Publishing, New York.

Golinelli, G.M., Barile, S., Saviano, M., Farioli, F., Masaru, Y. (2015). Towards a common framework for knowledge co-creation: Opportunities of collaboration between Service Science and Sustainability Science. In Gummesson, E., Mele, C., Polese, F. (Eds.), *Service Dominant Logic, Network and Systems Theory and Service Science: Integrating Three Perspectives for a New Service Agenda.* Giannini, Napoli.

Golinelli, G.M., Volpe, L. (2012). *Consonanza, valore, sostenibilità: verso l'impresa sostenibile.* Cedam, Padova.

Hildreth, P.M., Kimble, C. (2002). The duality of knowledge. *Information Research*, 8(1), 1–18.

Ison, R., Schlindwein, S.L. (2015). Navigating through an 'ecological desert and a sociological hell': A cyber-systemic governance approach for the Anthropocene. *Kybernetes*, 44(6/7), 891–902.

IUCN (2006). The future of sustainability re-thinking environment and development in the twenty-first century. *Report of the IUCN Renowned Thinkers Meeting*, 29–31 January.

Kates, R.W., Clark, W.C., Corell, R., Hall, J.M., Jaeger, C.C., Lowe, I., McCarthy, J.J., Schellnhuber, H.J., Bolin, B., Dickson, N.M., Faucheux, S., Gallopin, G.C., Grubler, A., Huntley, B., Jager, J., Jodha, N.S., Kasperson, R.E., Mabogunje, A., Matson, P., Mooney, H., Moore, B., O'Riordan, T., Svedin, U. (2001). Environment and development: Sustainability science. *Science*, 292, 641–642.

Komiyama, H., Takeuchi, K. (2006). Sustainability science: Building a new discipline. *Sustainability Science*, 1(1), 1–6.

Lietaer, B., Ulanowicz, R.E., Goerner, S.J. (2009). Options for managing a systemic bank crisis. *Sapiens*, 2(1), 1–15.

Marsden, T., Farioli, F. (2015). Natural powers: From the bio-economy to the eco-economy and sustainable place-making. *Sustainability Science*, 10(2), 331–344.

Martens, P. (2006). Sustainability: Science or fiction? *Sustainability: Science, Practice, & Policy*, 2(1).

Meadows, H.D., Randers, J., Meadows, D. (1972). *The Limits to Growth: A Report for the Club of Rome's Project on the Predicament of Mankind.* New American Library, New York.

Nonaka, I., Takeuchi, H. (1995). *The Knowledge-Creating Company: How Japanese Companies Create the Dynamics of Innovation.* Oxford University Press, Cambridge.

O'Connor, M. (2006). The 'Four Spheres' framework for sustainability. *Ecological Complexity*, 3(4), 285–292.

Orecchini, F. (2007). A 'measurable' definition of sustainable development based on closed cycles of resources and it's application to energy systems. *Sustainability Science*, 2. DOI: 10.1007/s11625-007-0035-8.

Orecchini, F. (2011). Energy sustainability pillars. *International Journal of Hydrogen Energy*, 36, 7748–7749.

Orecchini, F., Naso, V. (2006). *La società no oil. Un nuovo sviluppo è possibile ma senza petrolio.* Orme Editori, Milano.

Orecchini, F., Naso, V. (2012). Energy systems in the era of energy vectors: A key to define, analyze and design energy systems beyond fossil fuels. In *Green Energy and Technology*. Springer, New York.

Orecchini, F., Santiangeli, A. (2011). Beyond smart grids: The need of intelligent energy networks for a higher global efficiency through energy vectors integration. *International Journal of Hydrogen Energy*, 36, 8126–8133.

Ostrom, E. (2009a). A general framework for analyzing sustainability of social-ecological systems. *Science*, 325(419). DOI: 10.1126/science.1172133.

Ostrom, E. (2009b). *Understanding Institutional Diversity*. Princeton University Press, Princeton, NJ.

Pearce, D.W., Atkinson, G.D., Dubourg, W.R. (1994). The economics of sustainable development. *Annual Review of Energy and the Environment*, 19, 457–474.

Polese, F., Caputo, F., Carrubbo, L., Sarno, D. (2016). The value (co)creation as peak of social pyramid, 26th Annual RESER Conference, 'What's ahead in Service Research: New Perspectives for Business and Society', Università di Napoli 'Federico II', Italy, 8–10 Settembre.

Rammel, C., Stagl, S., Wilfing, H. (2007). Managing complex adaptive systems: A co-evolutionary perspective on natural resource management. *Ecological Economics*, 63(1), 9–21.

Ranga, M., Etzkowitz, H. (2013). Triple Helix systems: An analytical framework for innovation policy and practice in the knowledge society. *Industry and Higher Education*, 27(4), 237–262.

Sala, S., Farioli, F., Zamagni, A. (2013). Life cycle based methods: Where are we in the context of sustainability science progress? *The International Journal of Life Cycle Assessment*, 18(9), 1653–1672.

Saviano, M. (2015). Multi-actor co-creation systems for progressing toward sustainability: Criticalities and challenges, *5th International Conference on Sustainability Science (ICSSS)*, 22–23 January, Tokyo.

Saviano, M., Barile, S., Caputo, F. (2017a). Re-affirming the need for systems thinking in social sciences: A viable systems view of smart city. In Vrontis, S., Weber, T., Tsoukatos, E. (Eds.), *Global and National Business Theories and Practice: Bridging the past with the Future* (pp. 1552–1567). EuroMed Press, Cyprus.

Saviano, M., Barile, S., Spohrer, J., Caputo, F. (2016). A service research contribution to the global challenge of sustainability. *Journal of Service Theory and Practice* (forthcoming), 27(5), 951–976. https://doi.org/10.1108/JSTP-10-2015-0228.

Saviano, M., Barile, S., Spohrer, J., Caputo, F. (2017b). A service research contribution to the global challenge of sustainability. *Journal of Service Theory and Practice*, 27(5), 951–976.

Saviano, M., Bassano, C., Calabrese, M. (2010). A VSA-SS approach to healthcare service systems the triple target of efficiency, effectiveness and sustainability. *Service Science, Informs*, 2(1–2), 41–61.

Saviano, M., Nenci, L., Caputo, F. (2017c). The financial gap for women in the MENA region: A systemic perspective. *Gender in Management: An International Journal*, 32(4), 203–217.

Scalia, M., Angelini, A., Farioli, F., Mattioli, G.F., Saviano, M. (2016). The chariots of Pharaoh at the red sea: The crises of capitalism and environment: A modest proposal towards sustainability. *Culture della sostenibilità*, 1, 3–63.

Schwaninger, M. (2015). Organizing for sustainability: A cybernetic concept for sustainable renewal. *Kybernetes*, 44(6/7), 935–954.

Shiroyama, H., Yarime, M., Matsuo, M., Schroeder, H., Scholz, R., Ulrich, A.E. (2012). Governance for sustainability: Knowledge integration and multi-actor dimensions in risk management. *Sustainability Science*, 7(1), 45–55.

Spangenberg, J.H. (2011). Sustainability science: A review, an analysis and some empirical lessons. *Environmental Conservation*, 38(3), 275–287.

Spohrer, J., Anderson, L., Pass, N., Ager, T. (2008). Service science and service dominant logic. *Otago Forum*, 2, 4–18.

Spohrer, J., Maglio, P.P. (2006). *The Emergence of Service Science: Toward Systematic Service Innovations to Accelerate Co-creation of Value*. IBM Almaden Research Center, New York.

Spohrer, J.C., Freund, L. (2014). Measuring T-shapes for ISSIP professional development. AHFE 2014, *2nd Conference on Human Side of Service Engineering*, 22–24 July, Kracow, Polland.

Spohrer, J.C., Gregory, M., Ren, G. (2010). The Cambridge-IBM SSME White Paper revisited. In Maglio, P.P., Kieliszewski, C.A., Spohrer, J.C. (Eds.), *Handbook of Service Science* (pp. 677–706). Springer, New York.

Spohrer, J.C., Maglio, P.P. (2008). The emergence of service science: Toward systematic service innovations to accelerate co-creation of value. *Production and Operations Management*, 17(3), 238–246.

Trencher, G., Yarime, M., McCormick, K., Doll, C., Kraines, S. (2014). Beyond the third mission: Exploring the emerging university function of co-creation for sustainability. *Science and Public Policy*, 41(2), 151–179.

Tronvoll, B., Barile, S., Caputo, F. (2017). A systems approach to understanding the philosophical foundation of marketing studies. In Barile, S., Pellicano, M., Polese, F. (Eds.), *Social Dynamics in a System Perspective* (pp. 1–18). Springer.

Ulanowicz, R.E., Goerner, S.J., Lietaer, B., Gomez, R. (2009). Quantifying sustainability: Resilience, efficiency and the return of information theory. *Ecological Complexity*, 6(1), 27–36.

von Bertalanffy, L. (1968). *General System Theory: Foundations, Development, Applications*. George Braziller, New York.

Vos, R.O. (2007). Perspective defining sustainability: A conceptual orientation. *Journal of Chemical Technology & Biotechnology: International Research in Process, Environmental & Clean Technology*, 82, 334–339.

WCED-World Commission on Environment and Development (1987). *Our Common Future*. Oxford University Press, Oxford.

Wiek, A., Farioli, F., Fukushi, K., Yarime, M. (2012a). Bridging the gap between science and society, 2012 editorial Special Feature *Sustainability Science Journal*, 7(Supplement 1). Springer.

Wiek, A., Harlow, J., Melnick, R., van der Leeuw, S., Fukushi, K., Takeuchi, K., Farioli, F., Yamba, F., Blake, A., Kutter, R. (2014). Sustainability science in action: A review of the state of the field through case studies on disaster recovery, bioenergy, and precautionary purchasing. *Sustainability Science Journal*, published on line 15 August 2014 Springer. DOI: 10.1007/s11625-014-0261-9.

Wiek, A., Ness, B., Schweizer-Ries, P., Brand Fridolin, S., Farioli, F. (2012b). From complex systems thinking to transformational change: A comparative study on the epistemological and methodological challenges in sustainability science projects. *Sustainability Science Journal*, 7(5). DOI: 10.1007/s11625-011-0148-y.

Wiek, A., Ness, B., Schweizer-Ries, P., Brand Fridolin, S., Farioli, F. (2013). Collaboration for transformation. *Sustainability Science Journal*, October. DOI: 10.1007/s11625-013-0231-7.

Wolfson, A., Tavor, D., Mark, S., Schermann, M., Krcmar, H. (2010). S3-Sustainability and Services Science: Novel perspective and challenge. *Service Science*, 2(4), 216–224.

Wolfson, A., Tavor, D., Mark, S., Schermann, M., Krcmar, H. (2011). Better place: A case study of the reciprocal relations between sustainability and service. *Service Science, Informs*, 3(2), 172–181.

Yarime, M., Trencher, G., Mino, T., Scholz, R.W., Olsson, L., Ness, B., Rotmans, J. (2012). Establishing sustainability science in higher education institutions: Towards an integration of academic development, institutionalization, and stakeholder collaborations. *Sustainability Science*, 7(1), 101–113.

Zuccari, F., Dell'Era, A., Orecchini, F., Santiangeli, A. (2015). The concept of energy traceability: Application to EV electricity charging by RES. *Energy Procedia*, 82, 637–644.

UN-DESA. (2014). *Prototype Global Sustainable Development Report*. UN, New York.

An integrated model of governance for sustainability

Massimo Scalia,[1] *Sergio Barile,*[2] *Marialuisa Saviano*[3] *and Francesca Farioli*[4]

Keywords: *Anthropocene; governance; sustainability; Triple-Helix Model; Triple Bottom Line Model*

1 Introduction and aims

The term "Anthropocene" qualifies a human epoch characterized by the apparent influence of humans on the Earth dynamics, dramatically changing the conditions of equilibrium at the biophysical and social level. Current governance systems do not seem to fit the needs of the new epoch. Scientists and professionals from all over the world are debating about the appropriate governance in the Anthropocene, adopting different disciplinary perspectives. With the purpose of highlighting the contribution of a systems perspective, this essay aims to promote discussion about the need of integrating the multiple dimensions that should be considered in the framework of governance for sustainability. We embrace a systems view of sustainability and sustainable development (Pearce, Atkinson and Dubourg, 1994; Clayton and Radcliff, 1996; Barile et al., 2014), also contributing to highlight cybersystemic possibilities (Espejo, 2014, 2015; Espinosa, 2015; Schwaninger, 2001, 2015; Ison and Schlindwein, 2015).

Among the various models that interpret socio-economic phenomena by integrating multiple dimensions, the Triple helix model seems to offer an important contribution (Etzkowitz, 1998; Etzkowitz and Leyesdorff, 2000). Although developed to sustain the thesis of the so-called "third mission" of universities in the governance of socio-economic innovation, the model has been used in several disciplinary domains to different aims. In the field of sustainability, it has been elaborated in several domains (Leydesdorff and Etzkowitz, 2003; Lombardi, 2012) also including further elements in a fashion of quadruple and even quintuple helix models (Carayannis, Barth and Campbell, 2012).

The use of the model in the field of enquiry of sustainability, however, does not seem to offer evidences or examples useful to understand its real contribution. Hence, our aim is to propose an elaborated version of the model providing an example that highlights its interpretative potential. More specifically, our purpose is to offer a possible evidence of the general reasoning related to the use of the model as a possible reference in the development of a governance model for sustainable development.

2   Discussion

The kinematic functioning of the Triple helix model, i.e. the movement of the three blades – *academy, industry* and *government* – recall a well-known theorem of mechanics, according to which every "act of motion" is a helical motion. More precisely, physicists state that if we consider a motion during a small interval of time, it is not distinguishable from a helical motion (i.e. the product of a translation along an axis for a rotation around the same axis).

The framework for which we provide a possible example has been developed within the Italian research stream of the Viable Systems Approach (Barile et al., 2015;Golinelli et al., 2015; Farioli et al., forthcoming; Saviano, 2015; Saviano and Caputo, 2017) and essentially proposes to integrate the Etzkowitz's Triple helix model with the Triple Bottom Line Model of Elkington (Elkington, 1997a; Elkington and Fennel, 1998). This integration allows applying the Triple helix in the field of sustainability providing a possible reference model for the governance of sustainable development in the Anthropocene. The aim is to overcome dominant dualistic thinking and scientism (Ison and Schlindwein, 2015) and the limits of a reductionist view, trying to provide an evolution model, that is trying to give a time-representation of the interaction between the agent and the moving blades.

The model developed by integrating the Triple helix and the Triple Bottom line frameworks (Figure 45.1) can be used to explain how sustainability can be reached when the three actors of the Triple helix, *academy, industry* and *government*, interact in a "virtuous dynamics" capable of changing the balance between the three spheres of sustainability: the *environmental*, the *social* and the *economic* ones.

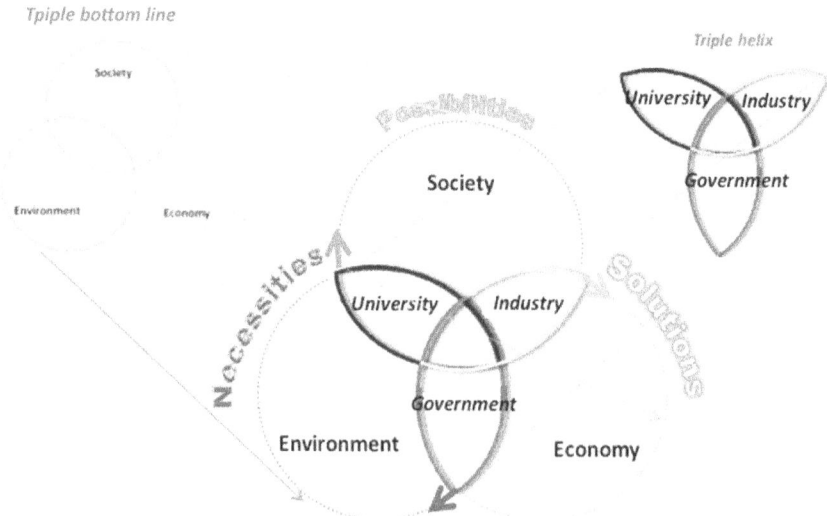

Figure 45.1 An integrated model of governance for sustainability

Sources: Barile and Saviano, www.asvsa.org and Barile et al. (2015)

The basic idea of the model is that the conditions of equilibrium between the three spheres of sustainability can be read as the outcome of the three actors' intertwined behaviours (Barile et al., 2015). Harmonic interaction between the three is the key for a balanced dynamic of development. This harmonic interaction implies that each category of actor plays a specific role, although hybridization among them occurs or is even required when a role is not adequately played. Basically, governments are expected to define the "necessities" by interpreting the needs that emerge from the environment viewed as a set of resources used for any kind of life process. Universities as representatives of the "science" but also the "education" areas are responsible for envisioning all the possibilities taking into account the above necessities. Finally, the industries are responsible for developing feasible solutions that can stand on the market, possibly with public support, and complying with the necessities and possibilities defined by governments and universities. Clearly, science, hence the universities, plays the key role for implementing sustainable development. However, only through a shared and harmonic view of necessities, possibilities and solutions, the development trajectories can be synergistically and sustainably pursued. To understand the integrated version of the model, it is fundamental to take into account that these helix processes include (although not explicitly) interaction with the stakeholders involved in the specific dynamics under focus. Stakeholders, especially society, influence in various ways behaviours and decisions on the main actors and can significantly contribute to create the context conditions for favouring the "helix" movement toward sustainability. In the proposed integrated model, society as well as environment and economy are included not only as passive outcomes but also as active influencers of the three actors' behaviours and decisions.

By exemplifying the use of the model in a context of governance for sustainability, it is possible to highlight many events that have been almost ignored, although they have had an impact by acting as a tract of motion of a blade. The case we refer to provides an example of the dynamics of governance for sustainability in terms of policies and actions emerging from interaction between the three actors directed to foster more conscious industry policies. In particular, it shows how the three blades, i.e. the three actors, have been working in the process that has led to the "Paris agreement" about the policies to adopt for addressing the challenge of climate change (United Nations, United Nations Treaty Collection, 2016). This agreement has important historical roots that cannot be ignored to understand the mechanisms that are at the basis of an effective governance approach for sustainability.

The example has been traced in previous works (Angelini et al., 2015, 2016; Scalia et al., 2016) and will be further developed in forthcoming ones as a relevant sustainability stream that recognizes climate change as "the greater threat of this century". Essentially, the pathway that has led to the "Paris agreement" is rooted in the years 2005–2006, when the world's major academies urgently informed major governments about the problems emerging from the current development model in terms of global warming, highlighting the damaging effects of the dominant industrial model in terms of sustainability, and calling for a "prompt action" against global warming by pressuring policy makers. As an outcome, the European governments in 2007, within the Council of Europe, promised to increase their commitment towards sustainable development and established the "20–20–20" targets. Subsequently, industry was required to develop and use more sustainable energy and production systems, to invest in renewable energies, with the support of the government. Apparently, the academies' engagement acted as a driving force that incepts the movement of the other two elements of the helix (governments and industry) generating a force field that creates the conditions of alignment for all. This is the key of the model.

3   Conclusions

By exemplifying the use of the Triple helix model in a context of governance for sustainability, the intent of this essay is to provide insights for developing a possible model of reference for integrating the roles of key actors and perspectives that must be involved in the action for sustainability.

The example provided underlines that harmonic interaction between the three spheres is the key for a balanced dynamic of development. Hence, a shared framework of reference for scientists, policy makers, industries and civil society can significantly help to organize action for coherently and synergistically promoting sustainable development.

Notes

1 CIRPS, Interuniversity Research Center for Sustainable Development – Sapienza University of Rome, Italy.
2 Sapienza, University of Rome, Italy.
3 University of Salerno, Italy (*corresponding author* – msaviano@unisa.it).
4 Italian Association for Sustainability Science (IASS), Italy.

Bibliography

Angelini, A., Farioli, F., Mattioli, G., Scalia, M. (2015). Le due crisi: crisi del capitalismo e crisi ambientale. Una soluzione sostenibile? Parte I. *Culture della sostenibiltà, 16.*

Angelini, A., Farioli, F., Mattioli, G., Scalia, M. (2016). Le due crisi: crisi del capitalismo e crisi ambientale. Una soluzione sostenibile? Parte II. *Culture della sostenibilità, 17.*

Barile, S., Saviano, M., Iandolo, F., Calabrese, M. (2014). The viable systems approach and its contribution to the analysis of sustainable business behaviors. *Systems Research and Behavioral Science, 31*(6), 683–695.

Barile, S., Saviano, M., Iandolo, F., Caputo, F. (2015). La dinamica della sostenibilità tra vortici e correnti. In XXXVII Convegno Nazionale AIDEA Sviluppo, sostenibilità e competitività delle aziende: il contributo degli economisti aziendali, Università Cattolica del Sacro Cuore, Piacenza, 10–11–12 september.

Carayannis, E.G., Barth, T.D., Campbell, D.F. (2012). The Quintuple Helix innovation model: Global warming as a challenge and driver for innovation. *Journal of Innovation and Entrepreneurship, 1*(1), 1–12.

Clark, W.C. (2007). Sustainability science: A room of its own. *Proceedings of the National Academy of Sciences, 104*(6), 1737.

Clayton, A.M.H., Radcliff, N.J. (1996). *Sustainability: A systems approach.* Earthscan Publishing Limited, London.

Daly, H. (1977). *Steady-state economics,* Island Press, Washington, DC. (1991). *Steady-state economics.* Second edition with new essays, Island Press, Washington, DC.

Dzisah, J., Etzkowitz, H. (2008). Triple helix circulation: The heart of innovation and development. *International Journal of Technology Management and Sustainable Development, 7*(2), 101–115.

Elkington, J. (1997). *Cannibals with forks: The triple bottom line of 21st century.* New Society Publishers, London.

Elkington, J., Fennell, S. (1998). Can business leaders satisfy the triple bottom line? *Visions of Ethical Business, 1,* 34–36.

Espejo, R. (2014). Organizational transparency, inclusion and sustainability. In Refereed Proceedings of the Business Systems Laboratory: 2nd International Symposium, January 23–24, Universitas Mercatorum, Rome, Italy.

Espejo, R. (2015). Good social cybernetics is a must in policy processes. *Kybernetes, 44*(6/7), 874–890.

Espejo, R., Schuhmann, W., Schwaninger, M., Bilello, U. (1996). *Organizational transformation and learning: A cybernetic approach to management.* John Wiley & Sons, Chichester.

Espinosa, A. (2015). Governance for sustainability: Learning from VSM practice. *Kybernetes, 44*(6/7), 955–969.

Etzkowitz, H. (1998). The triple helix as a model for innovation studies. *Science and Public Policy, 25*(3), 195–203.

Etzkowitz, H., Leyesdorff, L. (2000). The dynamics of innovation: From national systems and "mode 2" to a Triple Helix of university-industry-government relations. *Research Policy, 29*(2), 109–123.

Farioli, F., Barile, S., Saviano, M., Iandolo, F. (2016). *Re-reading sustainability through the Triple Helix model in the frame of a systems perspective.* Sage (forthcoming).

Georgescu-Rögen, N. (1971). *The entropy law and the economic process.* Harvard University Press, Cambridge, MA.

Georgescu-Rögen, N. (1979). Energy and economics myths: Institutional and analytical economic essays. *Southern Economic Journal, 46*(2), 655–657.

Golinelli, G.M., Barile, S., Saviano, M., Farioli, F., Masaru, Y. (2015). Towards a common framework for knowledge co-creation: Opportunities of collaboration between Service Science and Sustainability Science. In Gummesson, E., Mele, C., Polese, F. (Eds.), *Service dominant logic, network and systems theory and service science: Integrating three perspectives for a new service agenda,* Giannini, Napoli.

Ison, R., Schlindwein, S.L. (2015). Navigating through an "ecological desert and a sociological hell": A cyber-systemic governance approach for the Anthropocene. *Kybernetes, 44*(6/7), 891–902.

Leydesdorff, L., Etzkowitz, H. (2003). Can "the public'be considered as a fourth helix in university-industry-government relations? Report on the Fourth Triple Helix Conference, 2002. *Science and Public Policy, 30*(1), 55–61.

Lombardi, P. (2012). An advanced Triple-Helix network model for smart cities performance. In Ozge Yalciner Ercoskun (Ed.), *Green and ecological technologies for urban planning: Creating smart cities.* IGI Global.

230 Massimo Scalia et al.

Lovelock, J.E., Margulis, L. (1974). Atmospheric homeostasis by and for the biosphere: The Gaia hypothesis. *Tellus*, *26*, 2–10. doi: 10.1111/j.2153-3490.1974.tb01946.x.

Martens, P. (2006). Sustainability: Science or fiction? *Sustainability: Science, Practice, & Policy*, *2*(1).

Maturana, H., Varela, F., (1980). *Autopoiesis and cognition. Boston Studies in Philosophical Science*, Vol. 42. Reidel Publishing Company, Dordrecht, Holland.

Meadows, H.D., Randers, J., Meadows, D. (1972). *The limits to growth: A report for the club of Rome's project on the predicament of mankind*. New American Library, New York.

Nurse, K. (2006). Culture as the fourth pillar of sustainable development. *Small States: Economic Review and Basic Statistics*, *11*, 28–40.

O'Connor, M. (2006). The "Four Spheres" framework for sustainability. *Ecological Complexity*, *3*(4), 285–292.

Ostrom, E. (2009). *Understanding institutional diversity*. Princeton University Press, Princeton, NJ.

Paris Agreement. United Nations Treaty Collection. https://treaties.un.org/pages/ViewDetails.aspx?src=TREATY&mtdsg_no=XXVII-7-d&chapter=27&clang=_en

Pearce, D.W., Atkinson, G.D., Dubourg, W.R. (1994). The economics of sustainable development. *Annual Review of Energy and the Environment*, *19*, 457–474.

Pohl, C. (2008). From science to policy through transdisciplinary research. *Environmental Science & Policy*, *11*(1), 46–53.

Ranga, M., Etzkowitz, H. (2013). Triple Helix systems: An analytical framework for innovation policy and practice in the knowledge society. *Industry and Higher Education*, *27*(4), 237–262.

Sala, S., Farioli, F., Zamagni, A. (2013). Life cycle based methods: Where are we in the context of sustainability science progress? (Part I). *The International Journal of Life Cycle Assessment*, *18*(9), 1653–1672.

Saviano, M. (2015). Multi-actor co-creation systems for progressing toward sustainability: Criticalities and challenges. In *5th International Conference on Sustainability Science (ICSS)*, 22–23 January, Tokyo.

Saviano, M. (2016). Il valore culturale del patrimonio naturale nella promozione dello sviluppo sostenibile. *Sinergie, Italian Journal of Management*, *34*(99), 167–194.

Saviano, M., Caputo, F. (2017). Re-affirming the need for systems thinking in social sciences: A viable systems view of smart city. In *10th Annual Conference of the EuroMed Academy of Business*.

Scalia, M., Angelini, A., Farioli, F., Mattioli, G.F., Saviano, M. (2016). The chariots of Pharaoh at the red sea: The crises of capitalism and environment: A modest proposal towards sustainability. *Culture della sostenibilità*, *1*(2016 Numero Speciale), 3–63.

Scholz, R.W. (2011). *Environmental literacy in science and society: From knowledge to decisions*. Cambridge University Press, Cambridge.

Schwaninger, M. (2001). System theory and cybernetics: A solid basis for transdisciplinarity in management education and research. *Kybernetes*, *30*(9/10), 1209–1222.

Schwaninger, M. (2015). Organizing for sustainability: A cybernetic concept for sustainable renewal. *Kybernetes*, *44*(6/7), 935–954.

Seager, T.P., Melton, J., Eighmy, T.T. (2004). Working towards sustainable science and engineering: Introduction to the special issue on highway infrastructure. *Resources, Conservation & Recycling*, *42*(3), 205–207.

Spangenberg, J.H. (2011). Sustainability science: A review, an analysis and some empirical lessons. *Environmental Conservation*, 38(3), 275–287.

Vos, R.O. (2007). Perspective defining sustainability: A conceptual orientation. *Journal of Chemical Technology & Biotechnology: International Research in Process, Environmental & Clean Technology*, 82, 334–339.

WCED-World Commission on Environment and Development (1987). *Our common future.* Oxford University Press, Oxford.

Wiek, A., Farioli, F., Fukushi, K., Yarime, M. (2012). Bridging the gap between science and society. 2012 editorial Special Feature *Sustainability Science Journal*, 7(Supplement 1). Springer.

Wiek, A., Ness, B., Schweizer-Ries, P., Brand Fridolin, S., Farioli, F. (2013). Collaboration for transformation. *Sustainability Science Journal*, October. doi: 10.1007/s11625-013-0231-7.

Innovation between redundancy and vicariance

The rising need for a culture of variety

Bernardino Quattrociocchi,[1] *Cristina Simone,*[2]
Mario Calabrese,[3] *Francesca Iandolo*[4]
and Irene Fulco[5]

Keywords: *innovation; redundancy; vicariance; culture of variety*

Innovative processes (Figure 46.1), by their very definition, relate to phases of instability and transformation, characterized by the circumstances in which traditional schemes and models that belong to the cognitive endowment of the decision maker – and that can be defined as consolidated within its specific context of reference – appear ineffective, ending to be progressively not adequate, when reiterated in the attempt to be used to recover the balance of the organization.

The concept of complexity – understood as the inability of the decision maker to perceive a problem by analyzing and detailing with usual procedures and proceeding to the identification of related solutions according to the consolidated schemes – has the characteristic of consistency with scenarios that are typical of the innovative processes. In fact, innovation is necessarily achieved through the preparation and implementation of change. Change, which should not be seen just as a substitution or replacement, intervenes determining new methods, alternative routes, procedures that are different from the usual.

Having agreed that, in the interpretation of any innovative change, intervenes the use of a factor that can be defined as creative emphasis, as something that is necessary, but not sufficient, we proceed to point out that this factor indicates a mode, an emerging approach, which, different and new compared to the action of the past, cannot avoid borrowing the cognitive elements already present in the set of the organizational culture. What we want to substantiate is that the set of knowledge that is not usually used in the routine management processes, and is normally considered a burden, also in economic terms, compared to the productivity of the organization, becomes an element of wealth and opportunities in creative processes and organizational transformation. From these considerations, our contribution proposes the introduction of the concepts of redundancy and vicariance, and the need to define their role in fostering the notion of information variety introduced by the Viable Systems Approach, *vSa* (Barile, 2009).

Redundancy enables an organization to engage in experimentation and organizational change (Mansfield, 1961; Schumpeter, 1934; Nohria and Gulati, 1996; Cohen et al., 1972; Hambrick and Snow, 1977; March, 1981), relaxes

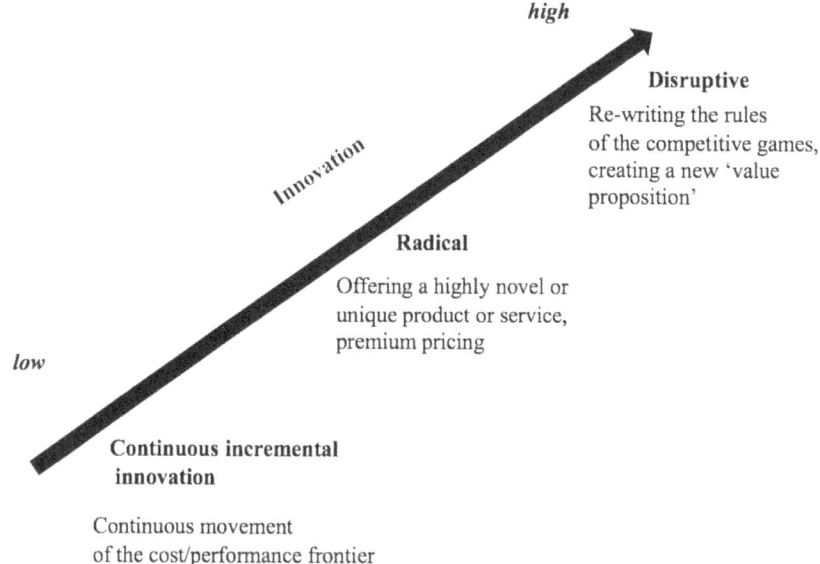

Disruptive

Re-writing the rules
of the competitive games,
creating a new 'value
proposition'

Innovation

Radical

Offering a highly novel or
unique product or service,
premium pricing

low

**Continuous incremental
innovation**

Continuous movement
of the cost/performance frontier

Figure 46.1 Innovation: a taxonomy based on the literature review

Source: Authors' elaboration based on Schumpeter (1934); Teece (1986, 2010); Abernathy and
Clark (1985); Christensen and Bower (1996); Tidd et al. (1997)

internal controls and creates funds that can be redirected toward projects with
uncertain outcomes, fostering an environment for innovation (Bromiley, 1991;
Damanpour, 1991; Greve, 2003), allows firms to adapt to complex competitive
landscapes (Levinthal, 1997) and provides organizations with the ability to be
proactive as well as defensive in adopting new technologies or designing new lines
of services (Nohria and Gulati, 1996; Nonaka and Takeuchi, 1997).

In addition, vicariance needs an approach that can be defined analogically, as
it deals with re-combining, resetting, re-shaping, the consolidated model. This
approach implies the recourse to lateral thinking (de Bono, 1969, 1970), and the
adoption of double loop learning models (Watzlawick et al., 1974; Argyris and
Schön, 1978; Senge, 1990). These concepts are related to the sensorial vicariance,
that entails breaking the rules, inventing and problem creating (Berthoz, 2013).

From the above, we can say that innovating in complex contexts can be considered as derivable from the 'redundancy' as a result of a principle of 'vicariance'.
The definitions of the two concepts introduce what we called the 'need for a
culture of variety', that is the need to make reference to the cognitive elements
that intervene in the definition of innovative paths. As we defined innovation
in complex contexts as the need to refer to elements that do not belong to the
cognitive endowment of the decision maker, we can make reference to the cultural dimension and its nature (rich or poor in variety), that deeply influences

the cognitive capability of a system (individual or organization) in promoting, accepting, impeding or refusing innovation. In literature, culture can be defined as the set of meanings, assumptions, cognitive frames, norms, symbols and values that the members of a group have in common or share; moreover, it is characterized by path dependence and is strongly related to history, both of individuals and of organizations (Jacques, 1952; Pettigrew, 1979; Siehl and Martin, 1984; Schein, 1985; Trice and Beyer, 1993). This dimension can be represented by the *vSa* notion of value categories (Barile, 2009), that are the strong beliefs that each system owns and that have a key role in the resolution of complex problems, like innovative processes.

Considering the cultural dimension in innovative processes allows to shift from a culture that is related to 'protocol' in favor of a culture that considers and includes 'variety'. In fact, if we consider these two cultures as representative of two polar frameworks, we can identify how value categories differently behave with relation to a variety of dimension, as represented in Figure 46.2.

The culture of protocol implies that any behavior is led by a sort of formulary, by manuals of right rules to follow in each situation according to specific methods, techniques and tools. The culture of variety, instead, contemplates that the challenge of innovation asks for an antithetical culture based on a necessary heterogeneity granted by the consideration of value categories. If we put in relation these cultures with the types of innovation, we can have, as represented in Figure 46.3:

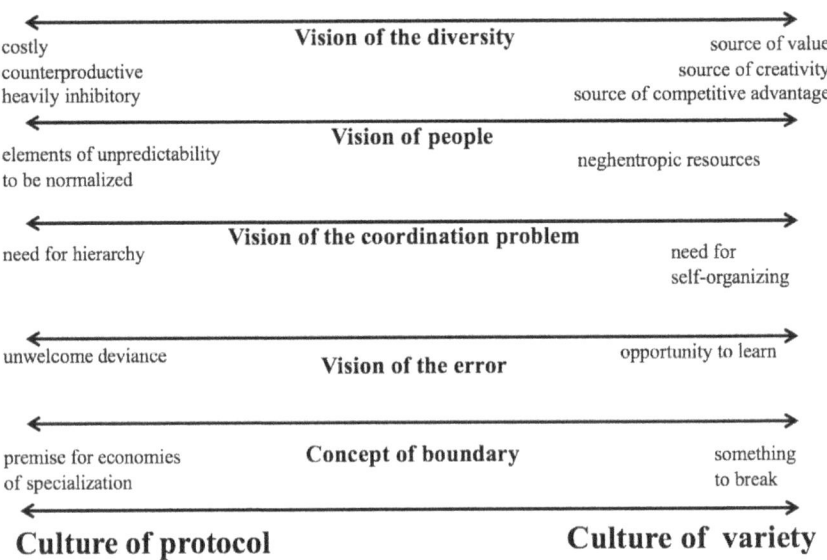

Figure 46.2 Value categories related to the information variety: a two polar framework

Source: Authors' elaboration

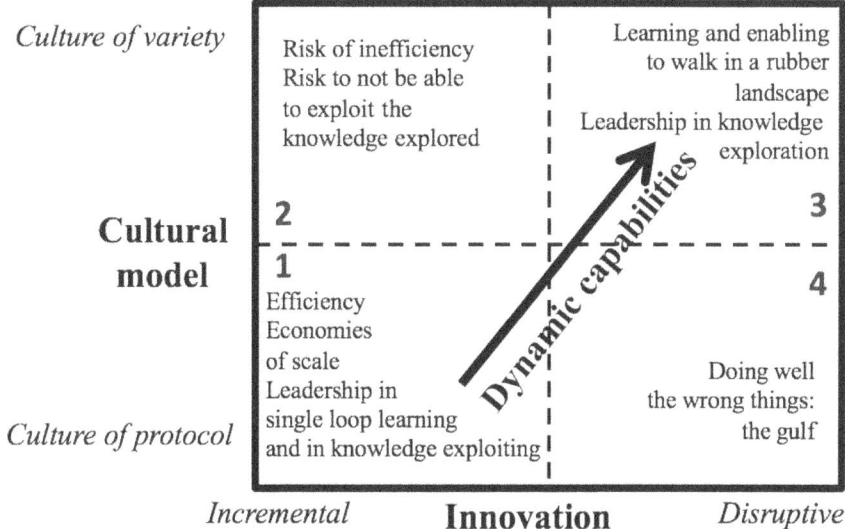

Figure 46.3 'Innovation–Cultural model' matrix
Source: Authors' elaboration

If we define incremental innovation as characterized by stability, predictability, linear dynamics and relations, well-known technological trajectories, low inter-dependences among socio-economic phenomena; and disruptive innovation as characterized by instability, unpredictability, nonlinear dynamics and relations, Schumpeterian shocks and high interdependences among socio-economic phenomena, we can conclude what follows.

The protocol culture is winning in an environment characterized by high stability, predictability and certainty, where the concepts of 'standard', 'repeatable and repetitive' and 'efficiency' are considered a categorical imperative. On the contrary, it is inefficacy in an environment characterized by the emergence of complexity where innovation capability is vital. The culture of variety, instead, promotes the economies of flexibility, creativity and integration of knowledge rather than the conquest of economies of scale. The principle of requisite information variety (Ashby, 1956) is not an abstract concept. Rather, it is a very concrete managerial principle which is fundamental in the *vSa*: if an organization is not capable to change by adapting itself in response to the external changes, it will not be able to survive. Here is a call for manager and researchers: to direct the development of business models. A shift to a more flexible, open and responsible view of organizations, less focused on the 'certainty' of the structure and technology, and open to the unpredictable outcomes of the human side of the system's dynamics, which is essentially emergent in nature. As well as organizations of the past have been marked by the culture of protocol, the organizations of the future will need to be based on the culture of variety. What will be the new balance between the values of protocol and the values of variety? In which way and to what extent will the command and control be permeated by self-organization? Will the self-organizing networks crumble the pyramids? Some and important elements of the culture of protocol will not be eliminated; the principles of top-down and of work standardization will not disappear. But the 'variety-based' organizations will be organizational systems in which the protocol organizational mechanisms will be subject to major stresses, due to the strong push towards multipolarity, self-organization, the search for a local adaptation. The top-down logic, the hierarchy principle, the hyperdetemination obsession will be integrated, diluted with other principles, which will have to find a compromise. Maybe sometimes collide, but that, over time, will have to learn to share with each other the realm of human organization.

Notes

1 Sapienza University of Rome.
2 Sapienza University of Rome.
3 Sapienza University of Rome.
4 Sapienza University of Rome.
5 Sapienza University of Rome.

Bibliography

Abernathy, W. J., & Clark, K. B. 1985. Innovation: Mapping the winds of creative destruction. *Research Policy*, 14(1), 3–22.

Argyris, C., & Schön, D. A. 1978. *Organizational learning, readings*. MA: Addison.

Ashby, W.R. 1956. *An introduction to cybernetics*. London: Chapman & Hall.

Barile, S. 2009. *Management Sistemico Vitale*. Torino: Giappichelli Editore.

Barile, S., & Saviano, M. 2013. Dynamic capabilities and T-shaped knowledge: A viable systems approach. In *Contributions to theoretical and practical advances in management: A Viable Systems Approach (VSA)*. Roma: ARACNE Editrice Srl, 39–59.

Barile, S., Saviano, M., Iandolo, F., & Calabrese, M. 2014. The viable systems approach and its contribution to the analysis of sustainable business behaviors. *Systems Research and Behavioral Science*, 31(6), 683–695.

Berthoz, A. 2013. *Vicariance (La): Le cerveau créateur de mondes*. Odile Jacob.

Bono, E. D. 1969. *The mechanism of mind*. Harmondsworth: Penguin Books.

Bono, E. D. 1970. Lateral thinking: Creativity step by step. In *Harper colophon books*. Harper & Row.

Bromiley, P. 1991. Testing a causal model of corporate risk taking and performance. *Academy of Management Journal*, 34(1), 37–59.

Christensen, C. M., & Bower, J. L. 1996. Customer power, strategic investment, and the failure of leading firms. *Strategic Management Journal*, 197–218.

Cohen, M. D., March, J. G., & Olsen, J. P. 1972. A garbage can model of organizational choice. *Administrative Science Quarterly*, 1–25.

Damanpour, F. 1991. Organizational innovation: A meta-analysis of effects of determinants and moderators. *Academy of Management Journal*, 34(3), 555–590.

Espejo, R., & Reyes, A. 2011. Organizational systems: Managing complexity with the viable system model. Heidelberg: *Springer Science & Business Media*.

Greve, H. R. 2003. *Organizational learning from performance feedback: A behavioral perspective on innovation and change*. Cambridge: Cambridge University Press.

Hambrick, D. C., & Snow, C. C. 1977, August. A contextual model of strategic decision making in organizations. *Academy of Management Proceedings*, 1977(1), 109–112. Academy of Management.

Jacques, E. 1952. *The changing culture of a factory*. New York: Dryden.

Levinthal, D. A. 1997. Adaptation on rugged landscapes. *Management Science*, 43(7), 934–950.

Mansfield, E. 1961. Technical change and the rate of imitation. *Econometrica: Journal of the Econometric Society*, 741–766.

March, J. G. 1981. Footnotes to organizational change. *Administrative Science Quarterly*, 563–577.

Nohria, N., & Gulati, R. 1996. Is slack good or bad for innovation. *Academy of Management Journal*, 5, 1245–1264.

Nohria, N., & Gulati, R. 1997. What is the optimum amount of organizational slack? A study of relationship between slack and innovation in multinational firms. *European Management Journal*, 15, 603–611.

Nonaka, I., & Takeuchi, H. 1997. A new organizational structure. In *Knowledge in Organisations*. Boston: Butterworth-Heinemann, 99–133.

Pettigrew, A. M. 1979. On studying organizational cultures. *Administrative Science Quarterly*, 24(4), 570–581.

Pondy, L. R. 1983. *Organizational symbolism* (Vol. 1). Jai Press.

Roger, C. C., & Ashby, W. R. 1970. Every good regulator of a system must be a model of that system. *International Journal of Systems Science*, 1, 89–97.

Schein, E. H. 1985. *Organisational culture and leadership: A dynamic view*. San Francisco: John Wiley and Sons.

Schumpeter, J. A. 1934. *The theory of economic development: An inquiry into profits, capital, credit, interest, and the business cycle* (Vol. 55). Transaction publishers.

Senge, P. 1990. *The fifth discipline*. New York: Currency Doubleday.

Siehl, C., & Martin, J. 1984. The role of symbolic management: How can managers effectively transmit organizational culture. *Leaders and Managers: International Perspectives on Managerial Behavior and Leadership*, 7, 227–239.

Teece, D. J. 1986. Profiting from technological innovation: Implications for integration, collaboration, licensing and public policy. *Research Policy*, 15(6), 285–305.

Teece, D. J. 2010. Business models, business strategy and innovation. *Long Range Planning*, 43(2), 172–194.

Tidd, J., Bessant, J. R., & Pavitt, K. 1997. *Managing innovation: Integrating technological, market and organizational change* (Vol. 4). Chichester: Wiley.

Trice, H. M., & Beyer, J. M. 1993. *The cultures of work organizations*. Prentice-Hall, Inc.

Watzlawick, P., Weakland, J. H., & Fisch, R. 1974. *Change: Principles of problem formation and problem resolution*. Norton.

The sustainability laboratory and Project Wadi Attir

Michael Ben-Eli[1]

Keywords: *sustainability; whole systems design; dry land agriculture; ecosystem restoration; Bedouin community; green technology*

1 General

Project Wadi Attir, The Sustainability Laboratory's flagship project, is a ground-breaking initiative with the Bedouin community in the Negev desert, and demonstrates an approach to sustainable dry land agriculture that leverages Bedouin traditional values, know-how and experience with modern-day science and cutting edge technologies.

Initiated by The Lab and the Hura Municipal Council, a local Bedouin township, the project showcases implementation of holistic sustainability principles developed by The Lab. It demonstrates an approach to sustainable development in an arid environment, valid and replicable locally as well as in other similar regions around the world.

2 Areas of sustainability innovation

Project Wadi Attir is designed to innovate in five key dimensions, reflecting The Lab's five core sustainability principles, and related to the Material, Economic, Life, Social and Spiritual domains:

> **The Material Domain:** The project's system of integrated green technologies and waste-to-resources approach maximizes the use of renewable resources, eliminates harmful emissions and aims for near-zero waste.
> **The Social Domain:** The project has created a unique coalition of individuals and groups representing all sectors of Israeli society, including government, civil society and private sector companies, and it features a cooperative organizational structure and cooperation between different Bedouin tribes and women in leadership roles.
> **The Economic Domain:** The project goes beyond "job creation" by creating new enterprises and empowering a community of entrepreneurs who take responsibility for their own future.

The Domain of Life: The project takes a humane and low-impact approach to raising farm animals, while orchestrating the enrichment of biodiversity on the project site.

The Spiritual Domain: The project is anchored in a value proposition, upheld by the community and articulated in its Declaration of Principles.

The integration of all these domains into a single vision and one comprehensive design is a key distinguishing factor in The Lab's whole systems approach to development. These innovations are expressed through the project's many functions and initiatives.

3   Key functions and initiatives

Herding and Dairy Initiative: The Herding and Dairy Initiative demonstrates a modern, economically viable model for animal husbandry that is consistent with traditional practices. It produces a variety of high quality cheeses, and is designed to utilize the full range of herding by-products, including dairy products, biogas fuel, manure for quality compost fertilizer and wool for weaving and crafts.

Medicinal Plants Initiative: The Medicinal Plants Initiative preserves, documents and showcases traditional Bedouin knowledge in natural healing remedies and body care utilizing desert herbs. It also established a high-quality brand of healing and cosmetic products, including creams, soaps, infusion teas and essential oils.

Indigenous Vegetables Initiative: The Indigenous Vegetables Initiative involves the cultivation of a variety of authentic, indigenous desert vegetables, in order to preserve heirloom desert varieties that are disappearing from use and contribute to better nutrition within the community. A women-led training program helps reintroduce the cultivation of indigenous vegetables on family-managed plots.

Ecosystem Restoration Initiative: The Project's Ecosystem Restoration Initiative incorporates an extensive soil enhancement, rainwater harvesting and biodiversity enrichment agenda, demonstrating an effective, low-impact approach to combating desertification.

Integrated Infrastructure of Green Technologies: The project 100-acre site is supported by an integrated infrastructure of green technologies. It includes a pioneering hybrid wind/solar energy system, a state-of-the-art drip irrigation system, a bio-gas production system, a wastewater treatment system and a composting facility.

The Visitor, Training and Education Center: The project's Visitor Center is designed to serve as an important eco-tourism destination, providing a source of income while introducing visitors to Bedouin society, tradition and culture. The center also offers technical training for surrounding communities, acting as a source of on-going empowerment, and it functions as an important regional education hub, serving primary and high schools from around the Negev.

4   Impacts

Consistent with its multifaceted characteristics, the project is:

- Contributing to the welfare and empowerment of the Bedouin community, replacing a model of apathy, helplessness and dependence with a model of proactive engagement, self-reliance, confidence and success.
- Contributing to the global sustainability agenda, exemplifying best practices in whole systems design and practical incorporation of sustainability principles.
- Changing the community's marginalized image in the eyes of the rest of society and cultivating fruitful partnerships with government and other sectors.
- Spreading The Lab's approach to sustainability and development by attracting attention, locally and internationally, through online resources and a growing number of visitors to the project's site.

Acknowledgements

Implementation of Project Wadi Attir has been supported by a consortium of Israeli Government Ministries, the JNF-USA, and by a group of some 45 foundations and individual donors.

Note

1 Founder, The Sustainability Laboratory.

Bibliography

De Gonzaga, S. (2013). Holistic Thinking at Work: The Sustainability Laboratory and Project Wadi Attir. An Interview with Dr. Michael Ben-Eli. *Centerpoint Now*. Special Issue on Sustainability, 32–34.

Websites

http://wadiattir.sustainabilitylabs.org

A path towards an evolutionary interpretation of the education service in Italian Universities

A systems variety perspective

Sergio Barile,[1] *Primiano Di Nauta*[2] *and Francesco Caputo*[3]

Keywords: *higher education; information variety; multi- and trans-disciplinarity; competences; capabilities*

In the last few years, the domain of Higher Education (HE) went through a revolution. People's expectations and requirements for education has changed, as well as the competences required by the market. As a matter of fact, the demand for Higher Education services deeply changed, highlighting serviced offered not always can grasp the market evolution. As consequence, the topic of higher education is progressively becoming more relevant for managerial researchers and practitioners opening multiple debates about the planning and managing of education service. Several contributions (Calandra Buonaura and Di Nauta 2004; Polese et al. 2016) have been provided with the aim to align consolidated education approaches to the increasing variety and variability of social and economic dynamics without producing the expected results.

The need for rethinking consolidated reductionist approaches extending the researchers' perspectives in the light of multi- and trans-disciplinary studies is almost evident (Barile et al. 2015).

With reference to this ongoing paradigmatic revolution, several elements can be considered relevant in the Italian Higher Education System (IHES):

- students and their families are paying more attention to innovative study programmes able to quickly grasp the ongoing market changes;
- work activities are becoming less structured and programmable with high relational contents and often they are supported by sophisticated technologies;
- individual awareness about the need for acquiring useful knowledge and soft skills is clearer;
- the emerging digital economy is requiring more flexibility;
- Higher Education Institutions (HEIs) are engaged in the so-called "third mission" for supporting companies, institutions, and practitioners in increasing their capability to produce value for the society;

- the consumption market has improved its general knowledge and as consequence it is less sensitive to consolidated companies' proposals, reducing companies' opportunities for survival;
- companies and economic organizations are perceiving the need for changing their approaches to the market and, for this reason, they are searching for professional profiles able to propose and promote innovative strategies.

Regarding the elements pointed out, the Italian HEIs often show a plastered situation as a result of national laws and regulations on one side, and the incapability for understanding the ongoing changes from the other side (Barile et al. 2015).

Nowadays, Italian HEIs seem to be mostly addressing qualifying competences fostering the relevant student capabilities while modern social and economic configurations are requiring a deployment of the educational offerings, inviting researchers and teaching professionals to retrain their contributions on the basis of emerging needs and technologies, readapting user functions.

According to previous managerial contributions, competencies-based models and approaches seem to be unable to support an effective understanding of changing social and economic dynamics. Teaching institutions are not interested in identifying and addressing the implications related to the overlapping of the multiple dimensions, pathways, and resources involved in the ongoing market and social changes, and therefore they cannot react in a short time to the multiple evolutions in the social and economic domains.

Building upon these preliminary reflections, the need emerges for rethinking higher education approaches by increasing the attention from the so-called competencies to abilities. In this sense, competencies are related to the application of coded knowledge for solving a defined and (usually) experienced problem, while the abilities grant students to decontextualize endowed capabilities for exploring, understanding, describing, and managing unknown and unexperienced problems (Barile et al. 2015).

The proposed change in perspective is a challenge for the multiple levels involved in the planning and management of higher education paths because it requires modifying the relationship between the market and education providers who have to reconsider the temporal dimension.

Accordingly, the real challenges for the educational services in Italian universities are related to the definition of paths able to combine the competencies required from the market for solving actual problems with the capabilities needed to support students in understanding and facing the emerging market dynamics for managing future trends in social and economic configurations.

In this direction, building upon the conceptual framework proposed by systems studies and, specifically, by the Viable Systems Approach (VSA) (Saviano et al. 2018; Badinelli et al. 2012; Barile et al. 2012), Italian universities should extend their focus including paths not only direct-to-transfer information units – as data that are processed and transformed during the educational path – but also to support students in defining useful interpretation schemes – in terms of paths through which information are organized for ensuring an effective understanding of external dynamics and processes – and in understanding the relevance of categorical values – as values and strong beliefs that define individual identity and subjective perception and definition of the reality.

Acting in this direction raises the possibility for changing traditional views and approaches in the education sector usually based on a reductionist and specialist representation of knowledge. The real advancement in knowledge distribution and teaching practice is related to the opportunities for defining education paths inspired by the multi- and trans-disciplinarity.

The proposed change in perspective offers the opportunity for supporting the emergence of conditions for an effective collaboration among the future researchers and practitioners for facing the variety and variability of emerging social and economic challenges. In such a vein, the contribution recognizes the validity of information sharing and cognitive alignment as relevant pillars for rethinking higher education approaches. The main idea is that higher education institutions should provide to the students not only knowledge and competences, but they should also teach them ways the endowed knowledge and competences can be combined with the resources of the multiple actors that compose their social and economic environment. Only in this direction will education become the starting point for rethinking the ways in which society perceives the world and reacts to its changes. Only by developing education approaches inspired and rooted in the aspiration of promoting the value of co-creation, is it possible to hope for a social and economic evolution, able to involve everyone in building conditions for social inclusivity and sustainable development.

Notes

1 Sapienza University of Rome.
2 University of Foggia.
3 University of Salerno.

Bibliography

Aguiari, R., & Di Nauta, P. (2012). Governing business dynamics in complex contexts. (ed.) Franco Angeli. *Mercati e Competitività, fascicolo* 1, 39–59.

Badinelli, R., Barile, S., Ng, I., Polese, F., Saviano, M., & Di Nauta, P. (2012). Viable service systems and decision making in service management. *Journal of Service Management*, 23(4), 498–526.

Barile, S. (2009). *Management sistemico vitale* (Vol. 1). Giappichelli, Torino.

Barile, S., Franco, G., Nota, G., & Saviano, M. (2012). Structure and dynamics of a "T-Shaped" knowledge: From individuals to cooperating communities of practice. *Service Science*, 4(2), 161–180.

Barile, S., Saviano, M., & Polese, F. (2014). Information asymmetry and co-creation in health care services. *Australasian Marketing Journal (AMJ)*, 22(3), 205–217.

Barile, S., Saviano, M., Polese, F., & Di Nauta, P. (2012). Reflections on service systems boundaries: A viable systems perspective: The case of the London Borough of Sutton. *European Management Journal*, 30(5), 451–465.

Barile, S., Saviano, M., & Simone, C. (2015). Service economy, knowledge, and the need for T-shaped innovators. *World Wide Web*, 18(4), 1177–1197.

Calabrese, M., Iandolo, F., Caputo, F., & Sarno, D. (2017). From mechanical to cognitive view: The changes of decision making in business environment. In Barile, S., Pellicano, M., & Polese, F. (eds.), *Social Dynamics in a System Perspective* (pp. 223–240). Springer, New York.

Calandra Buonaura, C., & Di Nauta, P. (2004). An approach to accreditation: The path of the Italian higher education. In Omar, P.L., Schade, A., & Scheele, J.P. (eds.), *Accreditation Models in Higher Education: Experiences and Perspectives, ENQA Workshop Reports 3, European Network for Quality Assurance in Higher Education 2004* (pp. 51–55). Multiprint, Helsinki, Finland.

Caputo, F. (2018). *Approccio sistemico e co – creazione di valore in sanità*. Nuove Culture, Roma.

Caputo, F., Carrubbo, L., & Sarno, D. (2018). The influence of cognitive dimensions on consumers-SMEs relationship: A sustainability oriented view. *Sustainability*, 10(9), 3238.

Caputo, F., Formisano, V., Buronova, B., & Walletzky, L. (2016). Beyond the digital ecosystems view: Insights from Smart Communities. In Vrontis, D., Weber, Y., & Tsoukatos, E. (eds.), *Innovation, Entrepreneurship and Digital Ecosystems* (pp. 443–454). EuroMed Press, Cyprus.

Caputo, F., Walletzky, L., & Štepánek, P. (2018). Towards a systems thinking based view for the governance of a smart city's ecosystem: A bridge to link Smart Technologies and Big Data. *Kybernetes*. https://doi.org/10.1108/K-07-2017-0274

Carayannis, E. G., & Campbell, D. F. (2006). *Knowledge Creation, Diffusion, and Use in Innovation Networks and Knoge Clusters: A Comparative Systems Approach across the United States, Europe, and Asia*. Greenwood Publishing Group.

Daňa, J., Caputo, F., & Ráček, J. (2018). Complex network analysis for knowledge management and organizational intelligence. *Journal of the Knowledge Economy*, 1–20. DOI: 10.1007/s13132-018-0553-x

Del Giudice, M., Khan, Z., De Silva, M. Scuotto, V., Caputo, F., & Carayannis, E. (2017). The micro-level actions undertaken by owner-managers in improving the sustainability practices of cultural and creative Small and Medium Enterprises: A UK-Italy Comparison. *Journal of Organizational Behaviour*, 38(9), 1396–1414.

Di Fatta, D., Caputo, F., & Dominici, G. (2018). A relational view of start-up firms inside an incubator: The case of the ARCA consortium. *European Journal of Innovation Management*, 21(4), 601–619.

Di Nauta, P., Merola, B., Caputo, F., & Evangelista, F. (2018). Reflections on the role of university to face the challenges of knowledge society for the local economic development. *Journal of the Knowledge Economy*, 9(1), 180–198.

Di Nauta, P., Omar, P.L., Schade, A., & Scheele, J.P. (eds.) (2004). Accreditation models in higher education: Experiences and perspectives. *ENQA Workshop Reports 3, European Network for Quality Assurance in Higher Education*, Multiprint, Helsinki, Finland.

Evangelista, F., Caputo, F., Russo, G., & Buhnova, B. (2016). Voluntary corporate disclosure in the Era of Social Media. In Caputo, F. (ed.), *The 4rd International Symposium Advances in Business Management: "Towards Systemic Approach"* (pp. 124–128). Business Systems. E-Book Series, Avellino.

Freeman, C. (1987). Technical innovation, diffusion, and long cycles of economic development. In *The Long-Wave Debate* (pp. 295–309). Springer, Berlin and Heidelberg.

Goldenberg, M., Kamoji, W., Orton, L., & Williamson, M. (2009). Social innovation in Canada: An update. *CPRN Research Report*, Canadian Policy Research Networks.

Iandolo, F., & Caputo, F. (2018). *La creazione di valore tra economia, impresa e sostenibilità*. Nuove Culture, Roma.

Ng, I., Badinelli, R., Polese, F., Nauta, P.D., Löbler, H., & Halliday, S. (2012). SD logic research directions and opportunities: The perspective of systems, complexity and engineering. *Marketing Theory*, 12(2), 213–217.

Nonaka, I., & Takeuchi, H. (1995). *The Knowledge-Creating Company*. Oxford University Press, Oxford.

Polese, F., Caputo, F., Carrubbo, L., & Sarno, D. (2016). The value (co)creation as peak of social pyramid. In Russo-Spena, T., & Mele, C. (eds.), *Proceedings 26th Annual Research Conference* (pp. 1232–1248). RESER, University of Naples "Federico II".

Saviano, M., Caputo, F., Mueller, J., & Belyaeva, Z. (2018). Competing through consonance: A stakeholder engagement view of corporate relational environment. *Sinergie, Italian Journal of Management*, 105, 63–82.

Scuotto, V., Caputo, F., Villasalero Diaz, M., & Del Giudice, M. (2017). A multiply buyer-supply relationship in the context of digital supply chain management of SMEs. *Production Planning & Control*, 28(16), 1378–1388.

Tronvoll, B., Barile, S., & Caputo, F. (2017). A systems approach to understanding the philosophical foundation of marketing studies. In Barile, S., Pellicano, M., & Polese, F. (eds.), *Social Dynamics in a System Perspective* (pp. 1–18). Springer, New York.

Theme III

Smartness and Big Data

Fabio Orecchini, Francesco Polese
and Igor Perko

How to measure the level of smartness of a destination?

Giuseppe Tardivo, Milena Viassone and Gabriele Santoro[1]

Keywords: *Smart City; smart destination; smartness*

1 Introduction and background

In the last years, the term "smart" has been used to describe everything that can be improved through the implementation and utilization of technology (Boes et al., 2015). This concept was firstly applied to cities that show a high growing rate (Nam and Pardo, 2011), and then has been applied to several other fields such as tourism destinations (Smart Destination) (Wang et al., 2013; Ferraris and Santoro, 2014; Neirotti et al., 2014). In such a context, a Smart City is an urban area where information and communication technologies (ICTs) allow the improvements of life of people, firms and other organizations operating in the city. The concept has gained a lot of attention in both politics and academia, and a large body of recent literature has been published (Nam and Pardo, 2011; Angelidou, 2014; Sandulli et al., 2017), analyzing various aspects.

Recently, the notion of Smart Destinations emerged, expanding from the Smart City concept (Zhu et al. 2014), and scholars essentially focus on the role of ICTs for the management and promotion of a destination (Guo et al. 2014; Wang et al. 2013).

In this guise, a destination can be defined smart when firms, bodies and tourists interact constantly in order to realize in a continuous way the following activities: (i) data collection about the activities within the destination; (ii) data analysis of tourist behavior and activities within the destination in order to improve the management of destination and tourist satisfaction; (iii) the identification of measures of feasibility (financial and technical) for the destination development in order to make it more sustainable and adaptable to tourists' needs and preferences.

Despite several attempts to describe Smart Destinations (Wang et al., 2013) or Smart Cities (Hollands, 2008; Scuotto et al., 2016), no study provides a measure of the level of smartness of a destination.

In order to bridge this gap, this paper aims at measuring the level of smartness of a destination on the base of five main dimensions: smart people, smart mobility, smart living, smart environment, smart governance and at applying this framework to a 2020 Smart City destination candidate: Turin.

Therefore, this paper provides important managerial implications because it suggests key guidelines to policymakers in choosing the right intervention on the inadequate areas in order to improve the smartness of the destination and create value for all the stakeholders.

2 Methodology

This paper aims at (i) developing a model for measuring the level of smartness of a destination on the base of five main dimensions: smart people, smart mobility, smart living, smart environment and smart governance; (ii) applying it to a Smart City candidate: Turin. This city has been chosen because it, applying for the challenge launched by the European Union in 2011 with the initiative Smart Cities & Communities, is a candidate to become a Smart City and is particularly dynamic in proposing initiatives to achieve this goal, such as TAPE (Turin Action Plan for Energy), the Torino Master Plan for Torino Smart City and the participation in several other campaigns. The practical application of the smartness measure can be addressed only to candidate Smart Cities because the process started in 2011 will finish in 2020.

The objective of the paper is achieved throughout a process consisting of three main phases: (i) a review of literature on the topic of smartness of a destination; (ii) the administration of a structured questionnaire to a sample of 150 people from Piedmont consisting of accommodation operators, tourist bodies, tourists, asking them to express a level of agreement/disagreement on different items describing the dimensions of smartness through a 5-point Likert scale. The composition of the sample is shown in Table 48.1, and the survey was carried out in April 2015.

The questionnaire is based on 27 items organized in the five dimensions already cited. These dimensions are evaluated by stakeholders through a 5-point Likert scale (Likert, 1932), asking them to express their assessment in a range from "Totally disagree" to "Totally agree". In order to get the global index of destination smartness it is important to evaluate the level of smartness perceived by every single interviewee. This corresponds to the average (SMARTj) of judgments provided by every interviewee to the 27 items (SMARTij). In a second time, it is important to compute the average of values of SMARTj for all the interviewees, which corresponds to a global level of "smartness" perceived.

The survey was uploaded on a web platform, and the link was sent to potential respondents.

The reliability of the instrument and its dimensions was tested through using Cronbach's Alpha. The overall survey showed a strong reliability, expressed by a high level of Cronbach's Alpha (0,827); this corresponds to a considerable level of internal consistence of the scale used with the sample of reference.

(iii) Finally, strategic paths of action are suggested to territorial bodies in order to further improve the level of smartness of a Smart City candidate.

3 Results

Results show a positive average level of smartness (SMARTij = 3,39) with several differences among the dimensions and subdimensions. In fact, the highest percentages with regard to good levels of agreement (levels 4 and 5) are shown for smart living, in particular for what concerns the ease of knowing (i) the location and the availability of the most various activities linked to the tourist sector as hotels, bars and restaurants, shops, news, etc. by web sites, applications; and (ii) the history and all that information about Turin as a destination and about the various tourist attractions by the web (e.g. interactive tourist offices) or specific applications. In particular, this city has launched the Torino Living Lab in order to promote, develop and test new innovative solutions in a real context. With Turin Living Lab, citizens, firms and Public Administration explore and try together products, technologies and innovative services in a specific area of the city with the aim to test functions and utility for final users and to evaluate the effects on life quality of its inhabitants. This lab consists in the experimentation of 32 projects and it is hosted in the Campidoglio Quarter, which becomes in this way the first Turin urban area devoted to innovation and Smart City principles.

Good performances are also registered by most Smart Mobility perspectives such as the possibility to track – through a smartphone – the time and the route of the main public transports and the availability of a QR Code used by various hotels/restaurants in order to provide additional information about the main available services.

The lowest levels of agreement (1 and 2) have been registered by the dimension Smart Environment and by a subdimension of Smart Mobility ("The wifi connection is functioning and it is easy to use it in tourist points"). In fact, unfortunately Turin is placed in the first position for the level of pollution in Italy, exceeding for 62 days the limit value of fine dust Pm10. These data confirm a certain inhomogeneity among the subdimensions of each single dimension because another item describing Smart Mobility – "The time and the route of the main public transports and the opening time/closure of the main tourist sites/restaurants are easily available on the own smartphone" – collects one of the highest agreements. Turin has recently launched several mobility projects such as the project PUMAS Planning Sustainable Regional-Urban Mobility in the Alpine Space; the online Carpooling hub, the aggregator of sites for the carpooling realized by CSI Piemonte in the European project OPTICITIES for the pilot site of Turin; Bike2Work, the project that incentives the use of the bikes to go to work.

Smart Governance is settled in a middle position because it does not show high levels of agreement nor high levels of disagreement: this is the picture of a town that is pointing toward the development of government transparency, involvement of the public in decision making, public and social service, but it is not still mature in this field.

Despite the positive index of smartness, Turin must face important challenges that involve in particular environment; anyway results of this paper allow to emphasize the main opportunities to exploit and main weaknesses to solve in order to reach this important objective.

4   Conclusions

In conclusion, this paper results in the creation of an index able to measure the level of smartness of a destination. Throughout its application, it shows how the level of smartness of Turin is positive and affected by five important dimensions: smart people, smart mobility, smart living, smart environment and smart governance supported by literature and by practice (Cohen, 2011). In particular, while Turin excels with regard to smart living, it shows important weaknesses for what regards smart environment. Therefore, despite the positive index of smartness of Turin, it should carry out further efforts, in particular in terms of reduction of pollution, recycling and the function of Wi-Fi.

This research has several implications for both academic and practitioners. First, it contributes to literature on Smart Cities and Smart Destinations, proposing a method for the assessment of the smartness of a destination since, to the best of our knowledge, no studies have done this so far. In addition, from a managerial point of view, this index represents a useful measure for Smart Cities candidates like Turin that allows organizations and tourist operators to have a projection of the level of smartness before 2020 (Portolese, 2013) and to draw different development strategies.

More generally, this paper contributes to the existent bulk of knowledge because it provides key guidelines to policymakers in choosing the right intervention on the inadequate areas in order to improve the smartness of the destination.

In this regard, this methodology can help in identifying the right strategies and adjustments. Moreover, after identifying the weak dimensions, policymakers can select potential public-private partnerships.

Note

1 Department of Management, University of Turin.

Bibliography

Angelidou, M. (2014). Smart city policies: A spatial approach. *Cities, 41*, S3–S11.
Boes, K., Buhalis, D., & Inversini, A. (2015). Conceptualising smart tourism destination dimensions. In *Information and communication technologies in tourism 2015* (pp. 391-403). Springer International Publishing.
Cohen, B. (2011). Smart cities wheel. Retrieved September 24, 2014, from www.boydcohen.com/smartcities.html
Ferraris, A., & Santoro, G. (2014). Come dovrebbero essere sviluppati i progetti di Social Innovation nelle Smart City? Un'analisi Comparativa. *Impresa Progetto-Electronic Journal of Management, 4*, 1–15.
Guo, Y., Liu, H., & Chai, Y. (2014). The embedding convergence of smart cities and tourism internet of things in China: An advance perspective. *Advances in Hospitality and Tourism Research, 2*(1), 54–69.

Hollands, R. G. (2008). Will the real smart city please stand up? Intelligent, progressive or entrepreneurial? *City*, *12*(3), 303–320.

Likert, R. (1932). A technique for the measurement of attitudes. *Archives of Psychology*, 22, 5–55.

Nam, T., & Pardo, T. A. (2011, June). Conceptualizing smart city with dimensions of technology, people, and institutions. In *Proceedings of the 12th Annual International Digital Government Research Conference: Digital Government Innovation in Challenging Times* (pp. 282–291). ACM.

Neirotti, P., De Marco, A., Cagliano, A. C., Mangano, G., & Scorrano, F. (2014). Current trends in smart city initiatives: Some stylised facts. *Cities*, *38*, 25–36.

Portolese (2013). Retrieved from www.onleco.com/sites/default/files/contenuti/news/Smart%20city%20e%20green%20revolution/Arch_%20Portolese_FondazioneTorinoSmartCity.pdf

Sandulli, F. D., Ferraris, A., & Bresciani, S. (2017). How to select the right public partner in smart city projects. *R&D Management*, *47*(4), 607–619.

Scuotto, V., Ferraris, A., & Bresciani, S. (2016). Internet of things: Applications and challenges in smart cities: A case study of IBM smart city projects. *Business Process Management Journal*, *22*(2), 357–367.

Wang, D., Li, X. R., & Li, Y. (2013). China's "smart tourism destination" initiative: A taste of the service-dominant logic. *Journal of Destination Marketing & Management*, *2*(2), 59–61.

Zhu, W., Zhang, L., & Li, N. (2014). Challenges, function changing of government and enterprises in Chinese smart tourism. In Z. Xiang & L. Tussyadiah (Eds.), *Information and communication technologies in tourism 2014*. Dublin: Springer.

Discovering young adult preferences
The social media analysis state of the art

Igor Perko[1]

Keywords: *young adult; social media; European Union; Big Data; predictive analytics; systems thinking*

1 Introduction

In the EU young adults are an important population group, since it is expected that they will change the current EU environment. Their opportunities, capabilities, rights, deliverables and threats are addressed in multiple EU documents (European Commission, 2016; European Commission, 2015), designed with the goal of their social integration – with special focus on the disadvantaged individuals.

Social media provide potent and heavily indexed communication channels. Their structure enables communication monitoring, analysis and management – algorithms decide who will receive every sent message. The social media environment hosts multiple players: individuals are provided with a wide range of intelligently managed tools for massive peer-to-peer communication. Organisations can advocate their product and services to the individuals. They can measure the reaching and responses of the individuals to their proposals, and adjust their behaviour accordingly. Regulators and social media providers have full access to the social media resources and use it according to their objectives.

We can analyse behaviour on multiple levels: trends on the entire population, on multiple levels of segmentation (nationality, religion, sex, age, education, occupation, etc.), on the level of a single individual, behaviour on an instance level, and lastly: by analysing the drivers for the actual behaviour. The question arises: Which of those do we want to influence? The answer is clear: We want to influence each of the members in the targeted group, by pressing the right buttons, directing their behaviour in the right situation. Let me give you two examples: First, we want our child to make the bed, after he or she gets up, and second, a politician wants to have enough votes to win the current election at the national level.

The complexity of finding a correct answer is enormous; thereby in this paper, we will limit ourselves to identifying the gap between the state of the art in existing data sources, data analysis methods and behavioural studies and the actual demands from the real life. For this, we will analyse published scientific and professional R&D results in the fields: young adults in EU, social media, Big Data and behavioural theories. The goal is to analyse the technology and methodology state of the art and to assess the gap between the real life and the resulting digital models.

2   The young adult social media preliminary model

In the literature review, a preliminary model of social media influences on young people is elaborated. It is based on the analysis of the publications, related to the issues of young adults in EU (European Commission, 2016; European Commission, 2015) and the research results on social media interactions with young people (Kann et al., 2016; Keating, Hendy, and Can, 2016; Lin et al., 2016; Loader, Vromen, and Xenos, 2016; Shensa, Sidani, Lin, Bowman, and Primack, 2016) and Big Data analysis methodologies (Waller and Fawcett, 2013; Perko and Ototsky, 2016).

In the model, it is asserted that the EU young adults are faced with multiple issues: the development of capabilities, ranging from initiative taking to communication skills; they are confronted with multiple deliverables, opportunities and threats; nevertheless, they request the fulfilment of certain rights, as for instance the right for privacy. We assume that in social media communication only a limited number of EU young adult issues are discussed, while the others cannot be referenced.

The other identified players in the social media environment are social media providers, organisations and regulators. All of these players are using Big Data tools to monitor the communication in the social media to predict the behaviour

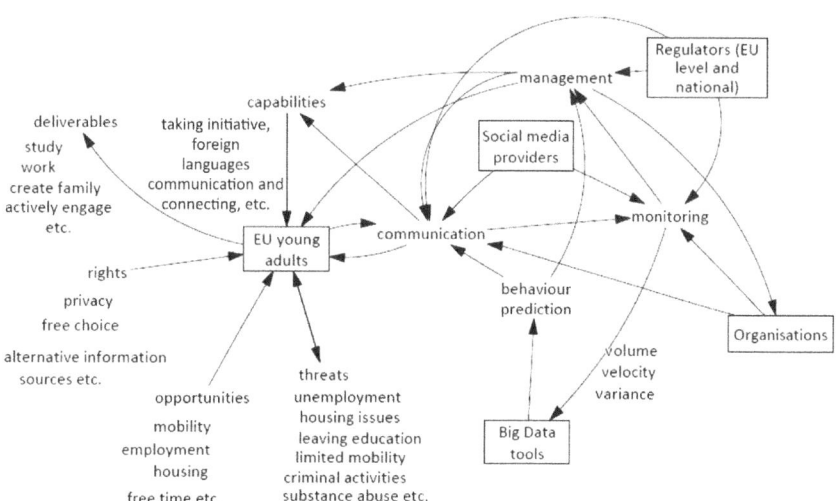

Figure 49.1 The young adults social media preliminary relations model

of young adults. The results are then used to manage the young adults directly, develop their capabilities, form expectations to fulfil deliverables or influence the communication flow.

The second assumption is that the other social media players do not address the EU young people's issues, but follow their own agenda. Their monitoring (direct, and via Big Data toolset) is therefore focused in filling the niche they decided to fill in.

We assume to find certain distinctions among the social media players:

- The organisations have limited monitoring scope – they can monitor only communication shared with them; therefore, they can manage only communication in their limited domain (with a limited number of young adults and on a limited number of issues). They tend to use only limited Big Data tools. In their limited scope they manage communication using multiple communication channels.
- The regulators can monitor full communication, but they are largely focused in on monitoring and analysing specific issues – for instance threats posted by the young adults. Even though the social media communications could provide direct insight in the young adult issues, the lack of mechanisms preventing the potential misuses of this information is the main cause of not monitoring, analysing this information. Therefore, its content is not integrated with the management mechanisms used by the regulators.
- The social media providers have full monitoring access, and are co-developing the Big Data toolset. They also actively manage the communication in social media by deciding which messages are delivered and how they are delivered. The question arises in whose behalf are they providing these services: is it for the young adults, to optimize their communication, is it for the organisations, to support their (largely economic) strategies, or is it for the regulators to provide insight?

Can these assumptions be verified with the social media analysis, and secondly, can the findings affect the processes in social media?

3   The research proposition

The relations between the young adults and the social media players are particularly interesting. The research could focus on three essential elements:

1   The importance of the young adult issues identified in the literature can be assessed by monitoring and mining communication, by using the same algorithms that are already used in the social media analysis (Waller and Fawcett, 2013; Shensa et al., 2016; Loader et al., 2016) by the social media players. We assume to find evidence on issues, addressed in the literature, identify new issues and above all assess the importance of the issues.

2   The direct social media communication management and the implications to the young adults in EU are used in preventing the threats and in building the communication skills. We do assume to find some support in the capabilities building, especially in the field of the communication skills development.

3   The indirect approach, by monitoring the young adult issues, and designing policies and strategies in addressing the most important issues, opening debates of privacy versus security. In the indirect back loop, we can assume that social media is only partially used to support other management activities.

To assess the closing loop's strength, the communication, executed by the regulators or the organisations, provides the young people support to resolve their issues and can be analysed. We assume to find little or no evidence of support in social media. The reasons for that can originate from:

1   The social media players are not monitoring or data mining the social media communication;

2   Since social media is not their native communication channel, they do not communicate through the social media;

3   They are using other means to execute their strategies and policies.

In the last step a critical comparison of the state of the art and the identified issues of young adults in EU is to identify the gap between the real life and the models presented and hopefully provide some direction for further research and for further development of the young adult-related policies.

Big Data and predictive analytics are the current driver for the understanding of complex systems. By examining properties of huge amounts of events, models are developed that forecast the situational-based behaviour. These models are then used to predict future behaviour in newly occurring instances, where currently only the properties are known.

We assume that the research results will provide evidence that the social media communication, monitoring, Big Data analysis and management tools are not used in such a way, that they could help understand or address the identified young adult issues. Therefore, there is a considerable gap between the current state of affairs and the desired state, where the mechanisms would be focused in supporting the holistic set of the young adult issues.

4   Conclusions

The proposed research results could provide the background information required to rethink the young adult support in EU.

First, it would provide a model to enhance the understanding of the young adult issues in EU.

Second, it would provide the regulators the method of assessing the direct feedback loop on the success of their policies.

Last, it would identify the gap between the young adults' real life and its current perception by the regulators and organisations. This way, social media players could adapt their strategies and policies to directly address the issues of young adults.

The final conclusion is a word of caution. The understanding of young adults' needs can easily be used for manipulation to achieve organisational goals (Espejo, 2000; Stokes, 2007). Before using the social media toolset, it is imperative to ensure that it is used to address the EU young people's issues, to enable them to live together, and to redesign the social media providers, the organisation and the regulator goals to address these issues and provide the value added.

Note

1 University of Maribor.

Bibliography

Espejo, R. (2000). Self-construction of desirable social systems. *Kybernetes*, *29*(7–8), 949–963. doi: 10.1108/03684920010342062

European_Commission. (2015). Find-er. Retrieved 2016, from http://ec-europa-finder.hosted.exlibrisgroup.com/

European_Commission. (2016). EU youth document library. Retrieved 2016, from http://ec.europa.eu/youth/library/index_en.htm

Kann, L., McManus, T., Harris, W. A., Shanklin, S. L., Flint, K. H., Hawkins, J., . . . Zaza, S. (2016). Youth risk behavior surveillance: United States, 2015. *Mmwr Surveillance Summaries*, *65*(6), 1–174.

Keating, R. T., Hendy, H. M., & Can, S. H. (2016). Demographic and psychosocial variables associated with good and bad perceptions of social media use. *Computers in Human Behavior*, *57*, 93–98. doi: 10.1016/j.chb.2015.12.002

Lin, L. Y., Sidani, J. E., Shensa, A., Radovic, A., Miller, E., Colditz, J. B., . . . Primack, B. A. (2016). Association between social media use and depression among US young adults. *Depression and Anxiety*, *33*(4), 323–331. doi: 10.1002/da.22466

Loader, B. D., Vromen, A., & Xenos, M. A. (2016). Performing for the young networked citizen? Celebrity politics, social networking and the political engagement of young people. *Media Culture & Society*, *38*(3), 400–419. doi: 10.1177/0163443715608261

Perko, I., & Ototsky, P. (2016). Big data for business ecosystem players. *Our economy*, *62*(2), 12–24.

Shensa, A., Sidani, J. E., Lin, L. Y., Bowman, N. D., & Primack, B. A. (2016). Social media use and perceived emotional support among US young adults. *Journal of Community Health*, *41*(3), 541–549. doi: 10.1007/s10900-015-0128-8

Stokes, P. A. (2007). From management science to sociology: Cybernetics, finalization and the possibility of a social science. *Kybernetes*, *36*(3–4), 420–436. doi: 10.1108/03684920710747048

Waller, M. A., & Fawcett, S. E. (2013). Data science, predictive analytics, and big data: A revolution that will transform supply chain design and management. *Journal of Business Logistics*, *34*(2), 77–84. doi: 10.1111/jbl.12010

Money saving or what?

Understanding the advantages of carpooling through Big Data analysis

Maria Vincenza Ciasullo,[1] *María Jimena Crespo Garrido,*[2] *Gennaro Maione,*[3] *Carlo Torre*[4] *and Orlando Troisi*[5]

Keywords: *Big Data analysis; Twitter Crawler; smart technology; carpooling*

1 Objective

The problems concerning the transport sector highlight the need to allocate increasingly material and immaterial resources for the development of high-tech devices able to improve greatly city liveability and citizens' transport conditions. In this regard, high attention has been paid to the development and subsequent implementation of smart technologies, able not only to collect data, but also to interpret them. Innovative transport systems, such as car sharing, bike sharing, boat sharing and especially carpooling, thanks to the implementation of more and more performing technologies (ICT), lead to greater economic progress, ensuring a strong reduction of costs and time.

Carpooling is a system of private cars shared among a group of people, thanks to the use of smart technologies. Specifically, carpooling is based on a virtual platform (developed on a website or app) that allows users to provide or be willing to travel by a car directed in a certain place, encouraging the exploitation of resources (free seats in a car) that otherwise would be wasted.

Over the years, several scholars have tried to identify the socio-demographic determinants of carpooling, in order to understand the real reasons for which people tend to resort to it. Specifically, the paper aims to facilitate the acquisition of a full awareness about the benefits of new forms of sustainable mobility, encouraged by an increasingly extensive and widespread use of technology (smartphone, tablet, laptop, etc.) (Ciasullo et al., 2016) for trips.

Furthermore, the work attempts to provide an empirical evidence of the real reasons inducing consumers to use alternative "smart" journey systems and, more specifically, carpooling. In this regard, the study seeks to highlight whether, in addition to money saving, the rising success of this phenomenon is linkable to other reasons, such as environmental protection, desire to socialize, curiosity, comfort in travelling, etc.

2   Methodology

The study has been conducted by means of a Big Data analysis, developed in three consecutive stages. First, taking into account the research goals, among different available alternatives, the authors have identified the IT tool considered most appropriate to the automated collection of the opinions expressed by the members of the Twitter community (De Maio et al., 2016). The choice to automate the data collection phase has not been accidental but dictated by the belief that it appears to be the most efficient way to get the highest possible number of feedbacks in a limited time span. Subsequently, the selected tool has been implemented within the chosen online community for four continuous months. Finally, the authors have appropriately analyzed the summary sheet provided by the IT tool (Twitter Crawler), interpreting results in an effort to define a kind of ranking of the reasons for which people resort to carpooling.

3   Results

The results of analysis show that, although the sample was particularly heterogeneous in terms of age, education, profession and socio-economic position, the respondents indicate resorting to carpooling for several reasons.

   In order to take into consideration only the hashtags most commonly associated with the word #carpooling, 673912 tweets have been automatically identified, selected, collected and categorized, allowing arriving at a classification which highlights the existence of various motives for which people state to use carpooling.

   Concretely, the IT tool has allowed grouping all the words emerged from the analysis on the basis of their semantic commonality. In other words, terms with similar meaning have been automatically recognized by the IT tool and have been combined in a single category. In this regard, however, it is worth pointing out that the authors' supervision has been necessary: after having gotten the output with all the words grouped by category, a check was performed to avoid possible bias related to the automation of the data collection process or to a possible wrong "interpretation" by the programme. After this screening, carried out the appropriate modification, the authors have identified seven categories of keywords used by Twitter users with regard to the hashtag #carpooling. The analysis shows that the main reason for which people declare to resort to carpooling seems to be the chance to save money. This finding is in line with other studies (Chen and Hsu, 2013), in which what emerges is a high attention of carpooling users towards money saving. Even, in many tweets, people write that the sole ground that pushes them to use carpooling is linkable to the possibility to reduce the price for travel, comparing to spending by using traditional means of transport. Not by chance, the term #moneysaving represents one of the most used hashtags in combination with #carpooling, demonstrating the decisive role played by the economic aspect in addressing the choices of who use carpooling (Shewmake, 2012). The fact that the analysis has shown that money saving represents the main reason why people use carpooling does not raise any wonder, given that several studies (Becker, 2013), carried out in different contexts, show that among all possible variables capable of orienting and, sometimes, conditioning consumer's behaviour, the price is one of the most decisive.

The second most used hashtag, #sustainability, suggests that people seem to be particularly concerned about the issue of environmental health. This result is consistent with the findings emerged from other researches (McKenzie-Mohr, 2013; Vlek and Steg, 2007), according to which the growing awareness about the need to ensure the respect for environment and public health is increasingly orienting market demand towards the adoption of sustainable behaviours and lifestyles. This consideration highlights the radical change that is affecting many consumers' purchasing behaviour (Osbaldiston and Schott, 2011), oriented not only to search for individual well-being but also to pursue common goals (Arbuthnott, 2009), thanks to the adoption of market principles based on the respect for public and environmental protection health (Amel et al., 2009).

At the third position as regards the number of hashtags posted by Twitter users, we find #traffic, which includes all issues related to urban congestion of vehicles. It is a theme particularly felt by people (Ma et al., 2016), mainly by commuters, that is, by those who for various reasons (study, work, etc.) travel daily a certain distance. In this regard, in fact, it seems that carpooling, especially for long-lasting trips, is becoming increasingly popular (Bento et al., 2013), with positive implications also for urban traffic (Dakroub et al., 2013). In fact, one of the most obvious benefits of this alternative travel system consists in the possibility of several people going in the same direction to use a single car, reducing the overall risk of possible traffic congestion (Zhang et al., 2015).

Likewise, the opportunity of #sharing seems to be particularly appreciated by carpooling users. In this regard, some studies (Matos et al., 2014) point out that many people resort to an alternative transportation system for the pleasure of sharing their time with others, broadening their circle of acquaintances and making new friends (Matos et al., 2014). Carpooling, therefore, embodies an innovative way of transporting, based not only on quantifiable data (such as, money saving or the reduction of the number of cars on the road), but also on the opportunity for users to get involved under a social and emotional profile.

The fifth position is occupied by #technology, underling the great relevance attached by consumers to technological component, not only in a transport context (Bazzan and Klügl, 2014) but also in any other service environment, such as education (Loia et al., 2016), work (Troisi et al., 2016), energy (Tang and Tan, 2013), etc. More specifically, what assumes importance, in addition to technology as such, is the perception that users have about usefulness (Renko and Druzijanic, 2014) and ease of use (Lu et al., 2014) of technological devices. Some studies (Borup et al., 2006; Jerram and Purdy, 2001), in fact, demonstrate that the higher users' expectations about technology usability is, the greater the contribution generated by using the device in terms of both effectiveness and efficiency is (Borup et al., 2006). This would explain why the hashtag #technology is so frequently used by those on Twitter who want to leave their comments about carpooling.

The #comfort, instead, ranks in sixth position, pointing out that, although carpooling offers the opportunity to travel and reach the desired destination cozily sitting in a car driven by another person, actually, as emerged from the analysis, there are at least five best reasons to resort to this alternative transportation system. At this point, it is worth highlighting that this result is not widely shared in literature, since the prevailing orientation (Gaur et al., 2009) believes comfort is the most decisive variable taken into account by travellers in their purchasing choices. However, the partially different finding revealed from the analysis could be justified by considering that the majority of carpooling users is represented by young people (Morency, 2007), who, being able to still dispose of much energy, do not care about the possibility to travel comfortably, turning their interest to other advantages, such as money saving, environment sustainability, traffic reduction, opportunity to make friends and technology usability.

In last place, finally, we find #curiosity. According to the authors, this is an unexpected result, given the central importance of this variable in orienting consumers' purchasing choices. In fact, several studies (Park et al., 2015; Foulds, 2014; Hill and McGinnis, 2007; Bennett, 2002) consider curiosity as a crucial factor in influencing consumer behaviour. Furthermore, according to other scholars (Yanamandram, 2009), curiosity, besides being very important, is also the first variable to be unconsciously considered by any consumer in the formation process of his/her purchasing intentions, without which it seems that the interest in a particular good or service, even, cannot be born. However, according to the authors, the result emerged from the analysis could be derived from the fact that, as typically occurs, people tend to express their thoughts about a specific service only after having used it. This would mean that all analyzed tweets have been actually posted after having benefited from all advantages of carpooling, a phase in which curiosity could be a feeling already satisfied and, therefore, no longer very considered.

4   Conclusion

The Big Data analysis highlights seven main reasons pushing people to use #carpooling: #money (1915 hashtags), #sustainability (1519 hashtags), #traffic (1286 hashtags), #sharing (683 hashtags), #technology (511 hashtags), #comfort (147 hashtags) and #curiosity (57 hashtags).

Therefore, it seems that, although money keeps on representing the main motive leading people to resort to carpooling (Bento et al., 2013), they are attentive also to other aspects of this smart journey system (Dakroub, et al., 2013; Le et al., 2001). This is an appreciable result, especially considering the current period of deep global crisis, in which, very often, the only variable taken into consideration by consumers seems to be money saving.

This trend is also confirmed in the context of urban transport, where it frequently occurs that, in the choices of different travel alternatives, consumers appear to be influenced mainly by the cost of the transport (De Grange et al., 2013). In the light of this consideration, carpooling actually seems to respond to multiple journey needs, taking shape as a valid alternative to the traditional way of travelling, in an era increasingly voted to a widespread use of technology.

5 Theoretical and practical implication

The paper could be understood as a useful tool for both practitioners (entrepreneurs, managers, etc.) and scholars (researchers, students, etc.), since it seeks to help consumers become aware of the opportunity of using smart technologies also in journey contexts to efficiency and timely respond to changing market needs. Furthermore, the work offers some interesting insights for future research on carpooling, suggesting deepening, on one hand, the primary role that nowadays smart technologies hold in any socio economic context (Loia et al., 2016; Troisi et al., 2016) and, on the other hand, the reasons pushing people to use more and more often alternative journey systems. The research presents two main limits: first, the analysis has been conducted by gathering feedbacks in a limited time span (only two months); moreover, in order to use a particularly large sample, the authors have resorted to a Big Data analysis by using an automated collection of Twitter people's opinions, but this way of operating has prevented the authors from going deeper in the analysis of consumers' thinking.

6 Research limits

The work presents two main limits: firstly, it has been realized by means of an analysis performed by collecting online users' opinions in a short period of time (about four months); furthermore, in the attempt to constitute a quite wide sample, an automated collection of tweets has been preferred to a deeper analysis of users' opinions. According to these limits, it could be appropriate conducing future researches in order to confront the findings arisen from this work with the results that could be derived by gathering users' thoughts through qualitative analysis.

Notes

1 University of Salerno.
2 University of Alcalá.
3 University of Salerno.
4 University of Salerno.
5 University of Salerno.

Bibliography

Amel, E. L., Manning, C. M. and Scott, B. A. (2009). "Mindfulness and sustainable behavior: Pondering attention and awareness as means for increasing green behavior", *Ecopsychology*, Vol. 1 No. 1, pp. 14–25.

Arbuthnott, K. D. (2009). "Education for sustainable development beyond attitude change", *International Journal of Sustainability in Higher Education*, Vol. 10 No. 2, pp. 152–163.

Bazzan, A. L. and Klügl, F. (2014). "A review on agent-based technology for traffic and transportation", *The Knowledge Engineering Review*, Vol. 29 No. 03, pp. 375–403.

Becker, G. S. (2013). *The Economic Approach to Human Behavior*, University of Chicago Press, Chicago.

Bennett, R. (2002). "Use of curiosity arousing websites for business-to-business internet marketing", *Quarterly Journal of Electronic Commerce*, Vol. 3, pp. 125–134.

Bento, A. M., Hughes, J. E. and Kaffine, D. (2013). "Carpooling and driver responses to fuel price changes: Evidence from traffic flows in Los Angeles", *Journal of Urban Economics*, Vol. 77, pp. 41–56.

Borup, M., Brown, N., Konrad, K. and Van Lente, H. (2006). "The sociology of expectations in science and technology", *Technology Analysis & Strategic Management*, Vol. 18 No. 3–4, pp. 285–298.

Chandy, R. K., Prabhu, J. C. and Antia, K. D. (2003). "What will the future bring? Dominance, technology expectations, and radical innovation", *Journal of Marketing*, Vol. 67 No. 3, pp. 1–18.

Chen, Y. T. and Hsu, C. H. (2013). "Improve the carpooling applications with using a social community based travel cost reduction mechanism", *International Journal of Social Science and Humanity*, Vol. 3 No. 2.

Ciasullo, M. V., Polese, F., Troisi, O. and Carrubbo, L. (2016). *How service innovation contributes to co-create value in service networks*. In International Conference on Exploring Services Science, Springer International Publishing, pp. 170–183, May.

Dakroub, O., Boukhater, C. M., Lahoud, F., Awad, M. and Artail, H. (2013). *An intelligent carpooling app for a green social solution to traffic and parking congestions*. In 16th International IEEE Conference on Intelligent Transportation Systems (ITSC 2013), pp. 2401–2408, October. IEEE.

De Grange, L., González, F., Muñoz, J. C. and Troncoso, R. (2013). "Aggregate estimation of the price elasticity of demand for public transport in integrated fare systems: The case of Transantiago", *Transport Policy*, Vol. 29, pp. 178–185.

De Maio, C., Fenza, G., Loia, V. and Parente, M. (2016). "Time aware knowledge extraction for microblog summarization on Twitter", *Information Fusion*, Vol. 28, pp. 60–74.

Dewan, K. K. and Ahmad, I. (2007). "Carpooling: A step to reduce congestion", *Engineering Letters*, Vol. 14 No. 1.

Ferreira, J., Trigo, P. and Filipe, P. (2009). "Collaborative carpooling system", *World Academy of Science, Engineering and Technology*, Vol. 54, pp. 721–725.

Foulds, M. (2014). "Consumer curiosity drives trends: Snacks & confectionary", *South African Food Review*, Vol. 41 No. 11, pp. 24–26.

Gaur, S. S., Madan, S. and Xu, Y. (2009). "Consumer comfort and its role in relationship marketing outcomes: An empirical investigation", *AP-Asia-Pacific Advances in Consumer Research*, Vol. 8.

Hill, M. E. and McGinnis, J. (2007). "The curiosity in marketing thinking", *Journal of Marketing Education*, Vol. 29 No. 1, pp. 52–62.

Jerram, J. C. and Purdy, S. C. (2001). "Technology, expectations, and adjustment to hearing loss: Predictors of hearing aid outcome", *Journal-American Academy of Audiology*, Vol. 12 No. 2, pp. 64–79.

Le, C. N., Thompson, F., Cresswell, A. M. and Dawes, S. (2001). *Carpooling on the superhighway: Information technology and interorganizational knowledge-sharing networks in the public sector*. In American Sociological Association Conference.

Loia, V., Maione, G., Tommasetti, A., Torre, C., Troisi, O. and Botti, A. (2016). Toward smart value co-education. In *Smart Education and E-Learning*, Springer International Publishing, pp. 61–71.

Lu, J., Lu, C., Yu, C. S. and Yao, J. E. (2014). "Exploring factors associated with wireless internet via mobile technology acceptance in Mainland China", *Communications of the IIMA*, Vol. 3 No. 1.

Ma, J., Smith, B. L. and Zhou, X. (2016). "Personalized real-time traffic information provision: Agent-based optimization model and solution framework", *Transportation Research Part C: Emerging Technologies*, Vol. 64, pp. 164–182.

Matos, M. L., Cruz, M., Guimarães, A. and Macedo, H. (2014). *A social network for carpooling*, in Proceedings of the 7th Euro American Conference on Telematics and Information Systems, ACM, April.

McKenzie-Mohr, D. (2013). *Fostering Sustainable Behavior: An Introduction to Community-Based Social Marketing*, New Society Publishers, Gabriola Island.

Morency, C. (2007). "The ambivalence of ridesharing", *Transportation*, Vol. 34 No. 2, pp. 239–253.

Osbaldiston, R. and Schott, J. P. (2011). "Environmental sustainability and behavioral science: Meta-analysis of proenvironmental behavior experiments", *Environment and Behavior*, Vol. 44.

Park, S. H., Mahony, D. F., Kim, Y. and Do Kim, Y. (2015). "Curiosity generating advertisements and their impact on sport consumer behavior", *Sport Management Review*, Vol. 18 No. 3, pp. 359–369.

Renko, S. and Druzijanic, M. (2014). "Perceived usefulness of innovative technology in retailing: Consumers' and retailers' point of view", *Journal of Retailing and Consumer Services*, Vol. 21 No. 5, pp. 836–843.

Selker, T. and Saphir, P. H. (2010). *TravelRole: A carpooling/physical social network creator*, in Proceedings the 2010 International Symposium on Collaborative Technologies and Systems, Chicago, IL, May.

Shewmake, S. (2012). "Can carpooling clear the road and clean the air? Evidence from the literature on the impact of HOV lanes on VMT and air pollution", *Journal of Planning Literature*, Vol. 27.

Tang, C. F. and Tan, E. C. (2013). "Exploring the nexus of electricity consumption, economic growth, energy prices and technology innovation in Malaysia", *Applied Energy*, Vol. 104, pp. 297–305.

Troisi, O., Carrubbo, L., Maione, G. and Torre, C. (2016). *The more, the merrier: Co-working as practical expression of value co-creation in sharing economy*, in XXVI RESER Conference, pp. 1130–1144.

Vlek, C. and Steg, L. (2007). "Human behavior and environmental sustainability: Problems, driving forces, and research topics", *Journal of Social Issues*, Vol. 63 No. 1, pp. 1–19.

Yanamandram, V. (2009). "Fostering curiosity about social responsibility and marketing ethics", *Australasian Accounting Business & Finance Journal*, Vol. 3, No. 3.

Zhang, Z., Wang, G., Cao, B. and Han, Y. (2015). *Data services for carpooling based on large-scale traffic data analysis*, in Services Computing (SCC), 2015 IEEE International Conference on IEEE, pp. 672–679, June.

A new method for asynchronous mapping, localization and control of vehicles

Damian Petrecki[1] *and Jerzy Józefczyk*[2]

Keywords: *Big Data; control of vehicles; environment awareness; autonomous cars; SLAM*

1 Introduction

Vehicles operated by drivers are still a dominant part of different transport facilities. For example, then a car follows an operator's instruction using mechanical executors, and it uses a lot of electronic systems to improve stability, safety, and comfort, or even to provide semi-autonomous driving possibilities. These systems are composed of many sensors and separated controllers connected by a common bus. In consequence, an enormous amount of data should be processed in the on-line mode which generates many technical drawbacks. For example, modern cars have kilometers of wires, several dozens of computers and a slow centralized computer network without the necessary level of security. Moreover, it is worth noting that every domain is handled separately, and there is no standard method to share and use high-level data between controllers. Finally, a car has data providing potential to perform better driving, but installed controllers are not designed to use them in a useful way (Johansson, Törngren and Nielsen, 2005).

The aim of this paper is to propose a method for gathering all signals from sensors, creating a map of an environment, improving a trajectory previously defined by a driver, and calculating the best control signals at the same time. Such methods and resulting special algorithms can be applied for supporting drivers and potentially for using by control systems of autonomous cars where drivers are not involved.

2 Asynchronous Mapping Localisation

The proposed method is referred to as *Asynchronous Mapping Localisation and Control* (AMLAC). It is based on the SLAM (Simultaneous Localisation And Mapping) concept which enables combining data measured by sensors with those expected according to a calculated vehicle's movement to finally create a map of an environment (e.g. Amitava and Fumitoshi, 2010; Davison, Reid, Molton and Stasse, 2007; Durrant-Whyte and Bailey, 2006; Schleicher, Bergasa, Ocaña, Barea

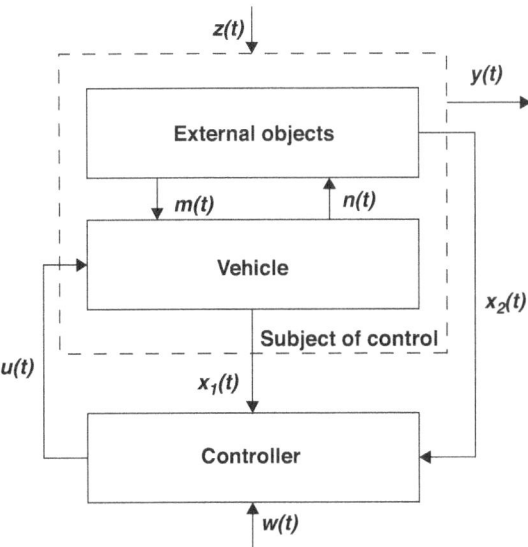

Figure 51.1 Control system structure

and López, 2010; Zhang and Singh, 2014). On the other hand, AMLAC method avoids some basic SLAM restrictions like external coordinates, discrete approach, and time synchronization between measurements and decisions. The important aspect of the method is the ability to model of all objects, which dynamically appear in a vehicle's environment, taking into account all relations between objects (including a vehicle itself). Consequently, a complex plant is defined.

As presented in Figure 51.1, a driver sends his (her) decision signal $w(t)$ to the controller which also reads vehicle's current state $x_1(t)$. $x_1(t)$ receives a measurement $x_2(t)$ from an environment, and finally determines a control signal $u(t)$ to fulfill a predicted output $y(t)$. Signals $m(t)$ and $n(t)$ represent a relation between a vehicle and external objects, even though they are not directly measured or used in the method. In contrast with classical approaches, AMLAC method is based on continuous time functions, which represent a probable position and rotation of classified objects in a vehicle environment, defined in the coordinate system which has its origin in the current position of a vehicle. Every identified external object is described using its individual parameters and the trajectory being a set of functions in the vehicle's coordinate system comprising both positions and rotations. The AMLAC method controller accepts sensor data, which describe control input devices, vehicle, and its environment. It sends a control signal to executors and to central data storage, which can also be used by external processes such as lights control to prevent blinding other drivers. The internal structure of the controller is shown in Figure 51.2.

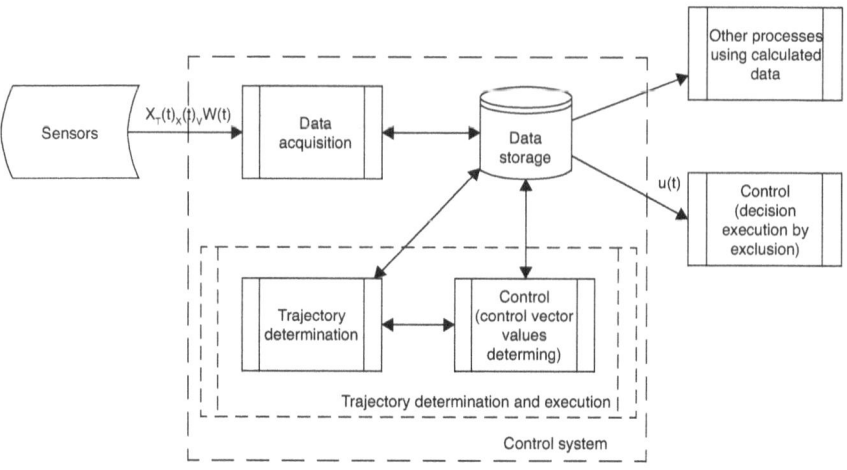

Figure 51.2 Controller model

3   Controller implementation

The preliminary implementation of the controller based on an actual vehicle and environment simulator has been elaborated. To meet AMLAC method postulates, the following sub-processes have been introduced in this implementation:

1   Function identifier – to identify continuous functions based on discreet values.
2   System scheduler and queues watchdogs – to manage in-system processes.
3   Drive recorder – to collect data for future analysis.
4   Simulator – to calculate expected localizations of all objects.
5   Static environment map generator – to transform coordinate systems.
6   Control vector computer – to generate control signals.
7   Trajectory determiner – to calculate best vehicle trajectory.
8   Sensor data analyzer – complex task of building environment map.

The implementation makes it possible to evaluate individual sub-processes as well as the AMLAC method as a whole. Some results of the evaluation are presented in the paper. The computation time and the resource usage (memory and CPU) are the basis of the assessment. The evaluation of AMLAC method is focused on the difference between the trajectory made by an algorithmically controlled vehicle and determined objectively in the simulated environment. The accuracy of the used knowledge about the environment, the quality of the calculated trajectory, and finally the result of the execution of the control signal are taken then into account.

The proposed *AMLAC* method is still being developed. The next important steps are detailing all processes, developing of the simulator and evaluating proposed solution.

Note

1 Wroclaw University of Science and Technology.
2 Wroclaw University of Science and Technology.

Bibliography

Amitava, C. and Fumitoshi, M. (2010). A Geese PSO Tuned Fuzzy Supervisor for EKF Based Solutions of Simultaneous Localization and Mapping (SLAM) Problems in Mobile Robots. *Expert Systems with Applications*, 37(8), 5542–5548.

Davison, A.J., Reid, I.D., Molton, N.D. and Stasse O. (2007). MonoSLAM: Real-Time Single Camera SLAM. *IEEE Transactions on Pattern Analysis & Machine Intelligence*, 29(6), 1052–1067.

Durrant-Whyte, H. and Bailey, T. (2006). Simultaneous Localization and Mapping: Part I. *IEEE Robotics & Automation Magazine*, 13(2), 99–110.

Johansson, K.H., Törngren, M. and Nielsen, L. (2005). Vehicle Applications of Controller Area Network. In *Handbook of Networked and Embedded Control Systems*, Basel: Birkhäuser, 741–765.

Schleicher, D., Bergasa, L.M., Ocaña, M., Barea, R. and López, E. (2010). Real-Time Hierarchical Stereo Visual SLAM in Large-Scale Environments. *Robotics and Autonomous Systems*, 16(3), 991–1002.

Zhang, J. and Singh, S. (2014). Robotics: Science and Systems X. In *LOAM: Lidar Odometry and Mapping in Real-time*, Berkeley: University of California Press.

Ubiquitous smart devices
Source of Big Data and a solution for Big Data

Bojan Žlahtič,[1] *Grega Žlahtič,*[2] *Peter Kokol*[3] *and Milan Zorman*[4]

Keywords: *smart devices; smart phones; tablets; data mining; knowledge discovery; Big Data; ubiquitous devices; cloud processing*

The amount of data that is being accumulated every day is enormous and for the human mind even hard to grasp. To put the growth of data into perspective, one could imagine that in order to store this amount of data into human's most valuable storage space, the brain, the data accumulated through the last four years would fill up all the brains of whole Slovenia. It is estimated that the human brain can hold an astonishing 0.0025 EB of data (Drachman, 2005) (Bartol et al., 2015). There are about 2 million people in Slovenia; this adds up to about 5000 EB of storage capacity. As the estimate of the daily collected data lies in the range of 2.5 EB (Siegel, 2016) of data, we can clearly see that we are talking about enormous amounts of data and that this data presents a big computational problem for any industry collecting data. Collection of data has become a relatively easy task, since thanks to the internet and the ubiquitous presence of smart devices, everyone has data at their disposal. In other words, we have more data than we can cope with. Even if we restrict ourselves to just one domain of data collection, the amount of data will most likely still be overwhelming. But the amount of data to be stored is only one problem we are facing; the other problem is even greater, the processing of such data. Usually we gather data in order to obtain knowledge from it, to find/discover something of importance, something that we can use to our benefit. If time was of no importance to the data processing and knowledge discovery, the only problem remaining would be finding the needle in the biggest data haystack there is, which is as indicated, in itself is a huge problem. Sadly, in modern society time and timing is one of the most important factors in data mining, so not only do we have to develop/find smarter ways to mine through the collected data, we also have to find new innovative ways to find wanted information/knowledge at the right time. The usual way to deal with data faster is to employ parallel computing, using one of the many tools available. The mentioned method is one of the most promising methods, until the arrival of quantum computers that is. But for now we are stuck using multiple computers to process data faster, which usually means that we have to have multiple computers at our disposal, or at least multiple processing hardware units. Of course everything can be done in the cloud, but still at the end it is done employing parallelisation. There is another solution for multiprocessing using the cloud, that could additionally speed up the process of mining data. There is one type of device that is persistently infiltrating everyone's lives and is one of the culprits for the enormous amount of data that is being accumulated, smart devices. Smart watches, phones, tablets, glasses, bracelets, televisions, cars, etc. All of the mentioned devices bear a great potential and could be exploited as a data mining and knowledge discovery tool.

Notes

1 University of Maribor.
2 University of Maribor.
3 University of Maribor.
4 University of Maribor.

Bibliography

Bartol, T. M., Jr., Bromer, C., Kinney, J., Chirillo, M. A., Bourne, J. N., Harris, K. M., & Sejnowski, T. J. (2015). Nanoconnectomic upper bound on the variability of synaptic plasticity. *Elife*, 4, e10778.
Drachman, D. A. (2005). Do we have brain to spare? *Neurology*, 2004–2005.
Siegel, E. (2016). *Predictive Analytics: The Power to Predict Who Will Click, Buy, Lie, or Die*. NJ: Wiley.

Co-creation in action as the acid test of smart service systems viability

Francesco Polese,[1] *Luca Carrubbo*[2]
and Debora Sarno[3]

Keywords: *smart service systems; value co-creation; structure/system dichotomy; system viability*

1 Introduction

The topic of Smart Service Systems (SSS), as a recent evolution of Service Systems (SS), can be a ground of studies on Service (Spohrer et al. 2008) and a multi-domain application of many theoretical foundations affecting the systems view. This paper aims to focus on the connection between structural/system traits of SSS and the inner value co-creation, trying to explain how the achievement of a co-creative experience is effectively supportive of the SSS viability. Two main scientific propositions are methodologically presented:

1   Are we able to distinguish between 'structural' and 'system' features of SSS as drivers for value co-creation processes?
2   Are we able to define and investigate effective value co-creation processes as the acid test for SSS viability and their ability to survive in the long run?

System studies (Von Bertalanffy 1968; Beer 1972; Von Foerster 1981; Checkland 1981; Espejo and Harnden 1989) and, in particular, the Viable Systems Approach (VSA) (Golinelli 2005; Barile and Polese 2010; Barile et al. 2012a, 2012b, 2014), are used as the interpretative lens to answer to these questions.

2 Working together for a smarter planet

SS researchers have investigated the potential evidences of service research 'on stage', referring to something really iterative, interactive, instrumented, interconnected, intelligent. In this direction, a new generation of SS capable of describing and analyzing occurring situations and making decisions based on the available data in a predictive or adaptive manner could be defined as Smart (Demirkan et al. 2011a, 2011b).

In SSS, service is seen as the final goal, rather than a normal throughput. SSS are a kind of human-centred SS, meaning that knowledge and capabilities are determined by people (Maglio et al. 2009). The intelligence of SSS derives not from intuition or chance, but from systemic methods of learning, service thinking, rational mode in actions, social responsibility and networked governance (Barile et al. 2013). These concepts have introduced the new view of the world as a 'smarter planet', an interconnected globe in which there is a growing attention to data measurement, networks development, enhanced learning and responsive adaptation processes (www.ibm.com/smarterplanet/us/en/). Applying smart context, smart practicalities, smart organizations, smart operations and smart outcomes to modern service 'events', we can enjoy several important changes in our lives (Napoletano and Carrubbo 2010).

3   Structural and system traits of SSS

Using the lens of VSA, the *structural traits* of SSS can be identified in:

- *Distinctive Resources*, as People, Products, Specialized Competences, ICT, etc. All actors are resources, and all service tools are considered useful instruments for business activities (Mele et al. 2010);
- *Relevant Information*, such as environmental external information, business internal information, etc.;
- *Static Relations*, as internal/external relationships (among sub/ supra-systems);
- SSS structure is characterized by a static condition. All elements are connected, possibly convergent, following consonant relationships and aimed to catch the same proposition of survival over time, in the long run;

   VSA can also support in pointing out the *SSS system traits*:

- *Resource integration*, as mode in actions or, products and processes enhancing. In SSS, all actors are motivated to develop harmonic interactions, offering and integrating resources needed for 'viable' service exchanges;
- *Information sharing and decision making*, by means of smart technologies and other solutions. IT-based SSS can support in assets and goals management while being capable of self-reconfiguration to satisfy stakeholders over time (Barile and Polese, 2010);
- *Dynamic Interactions*, such as internal and external interactions. SSS is not simply the sum of its parts; rather, system interactions form a higher-order construct that becomes the driver of success and sustainability Vargo and Lusch 2008);
- Dynamically, SSS differ from other types of socio-technical systems in that they depend on entities sharing capabilities to increase mutual satisfaction. SSS appear today as resource integrators, socially constructed and knowledge-based.

4   Looking for static pre-conditions and dynamic determinants featuring value co-creation in SSS

Value co-creation in SSS (Prahalad and Ramanswamy 2004; Ballantyne and Varey 2006) is difficult to measure and predict for its emergent style. The VSA lens can help in defining the structural traits, that is the static pre-conditions for co-creative processes:

- *Distinctive Resources for co-creation*: through participation, the user may broaden and transform his/her resources into skills that can be positively activated while resource integration takes place and ensure the potential compatibility among the actors who operate in the same context (Wieland et al. 2012);
- *Relevant Information for co-creation*: any SSS is considered as an open system that establishes relationships, not only with the sub-systems, which it contains and manages, but also with supra-systems in which it is included (Hall and Fagen 1956). As the world is becoming smarter, SSS must be people-centric, information-driven, and e-oriented to adapt and mutually satisfy any participant involved within the same service eco-system;
- *Static Relations for co-creation*: the sense of responsibility of actors and the intensity of their participation determine the development of positive commitment and a spontaneous participation to the generation of value (Barile et al. 2012a, 2012b) and definitively allow a co-creative situation;
- *Resource Integration for co-creation*: consumers obtain value from its use, processing or consumption and by comparing it with other entities interested in the building process (Katzan 2011);
- *Information Sharing and Decision Making for co-creation*: the variety and variability of information about possible connections between SSS promote new forms of cooperation, interpreted as interactions between cognitively aligned actors;
- *Dynamic Interactions for co-creation*: value is co-determined by providers and users at the time of purchasing, through a personal 'consumption' process favoured by constant interaction with other parts of the SSS in which users operate (Vargo and Lusch 2016).

5   Discussion

Value creation takes place when a potential resource becomes an effective specific benefit. Then, value co-creation follows a dynamic flow by means of the interaction among different SSS possessing critical resources. The trigger is the desire to reach collective mutual satisfaction, the active contribution is multiple, the integration is the highest and complementarity is fundamental.

Since 'harmony' between actors can be understood as a fusion of listening skills, consideration, dialogue, recognition and respect in intra- and inter-systemic relationships, we can verify how system consonance/resonance qualifies competitiveness in business (Polese et al. 2009) and then how SSS viability is encouraged by value co-creation. Direct consequences in practice are detailed below:

- *Fitting.* Fitting, as an adaptive set of actions, can converge on different levels of the SSS structure, with different levels of depth; it depends on a combination of factors, regarding the strategies of the system (decisional area) and the constraints coming from the outside;
- *Engagement.* The 'involvement' follows the service-centred view of exchange and indicates participation in co-production (Vargo and Lusch 2008). The term 'engagement' is very often used by SS scholars to indicate the active, equal and reciprocal participation of users and providers in the co-creation of value (Polese and Carrubbo 2016);
- *Innovation.* Innovation may result as an experimental process during which continuous learning obtainable by doing, by using, by failing, by interacting is fostered. Value co-creation in this sense implies the active multi-actor contribution.

Since the SS paradigm is today used to model each economic, managerial, organizational, industrial or computer system, the idea of a Smarter Planet can be easily pursued by focusing on the design of SSS, that evolves by adapting to the changing conditions and thus reducing the mismatch and loss of resources (Demirkan et al. 2011a). The governance of SSS should direct the system towards a final goal, transforming static structural relationships into dynamic interactions with other entities.

According to the VSA perspective, SSS are capable of simultaneously optimizing the use of resources and improving the quality of the provided service. Viability appears as the final result of all service operations, obtained by merging the contributions coming from all system parts. It represents an essential and necessary prerequisite to operate in the contemporary context, and smartness is a trait of proactive behaviour; then viable organizations (or viable systems, VS) are cybernetic, cognitive, autopoietic, aligned. Finally, it is important to notice that although VS are always smart, not all SSS are viable (Badinelli et al. 2012). In this sense, the smartness strictly connects to the system viability.

Value co-creation is the main leverage for competitiveness today and for the possibility to survive over time in markets. Value co-creation is the evidence of SSS capacity to maintain their own market-share during the time and then foster their presence, by adapting to external changes that can occur. Value co-creation is a sort of acid test of SSS viability as it is frameworked by VSA and system studies.

Notes

1 University of Salerno.
2 University of Salerno.
3 University of Foggia.

Bibliography

Badinelli, R., Barile, S., Ng, I., Polese, F., Saviano, M., & Di Nauta, P. (2012). Viable service systems and decision making in service management. *Journal of Service Management, 23*(4), 498–526. https://doi.org/10.1108/09564231211260396

Ballantyne, D. & Varey, R.J. (2006). Creating value-in-use through marketing interaction: The exchange logic of relating, communicating and knowing. *Marketing Theory, 6*(3), 335–348.

Barile, S., Carrubbo, L., Iandolo, F., & Caputo, F. (2013). From 'EGO' to 'ECO' in B2B relationships. *Journal of Business Market Management, 6*(4), 228–253.

Barile, S., Pels, J., Polese, F., & Saviano, M. (2012a). An introduction to the viable systems approach and its contribution to marketing. *Journal of Business Market Management, 5*(2), 54–78.

Barile, S., Polese, F., & Carrubbo, L. (2012b). Il Cambiamento quale Fattore Strategico per la Sopravvivenza delle Organizzazioni Imprenditoriali., In S. Barile, F., Polese, & M. Saviano (Eds.), *Immaginare l'innovazione* (pp. 2–32). Torino: e-book Giappichelli Editore.

Barile, S. & Polese, F. (2010). Smart service systems and viable service systems: Applying systems theory to service science. *Service Science, 2*(1–2), 21–40. https://doi.org/10.1287/serv.2.1_2.21

Barile, S., Polese, F., Saviano, M., Pels, J., & Carrubbo, L. (2014). The contribution of VSA and SDL perspectives to strategic thinking in emerging economies. *Managing Service Quality, 24*(6), 565–591.

Beer, S. (1972). *Brain of the Firm.* London: The Penguin Press.

Checkland, P. (1981). *Systems Thinking, Systems Practice.* Chichester: John Wiley & Sons Ltd.

Demirkan, H., Spohrer, J., & Krishna, V. (Eds.) (2011a). *Service Systems Implementation.* New York: Springer.

Demirkan, H., Spohrer, J., & Krishna, V. (Eds.) (2011b). *The Science of Service Systems.* New York: Springer.

Espejo, R. & Harnden, R.J. (1989). *The Viable System Model.* London: John Wiley.

Golinelli, G.M. (2005). *L'approccio sistemico al governo dell'impresa. L'impresa sistema vitale. I* (3 ed.). Padova: Cedam.

Hall, A.D. & Fagen, R.E. (1956). Definition of system, general systems. *Yearbook of the Society for the Advancement of General Systems Theory, 1*, 18–28.

Katzan, H. (2011). Foundations of service science concepts and facilities. *Journal of Service Science (JSS), 1*(1), 1–22. https://doi.org/10.19030/jss.v1i1.4297

Maglio, P.P., Vargo, S.L., Caswell, N., & Spohrer, J. (2009). The service system is the basic abstraction of service science. *Information Systems and E-Business Management, 7*(4), 395–406.

Mele, C., Pels, J., & Polese, F. (2010). A brief review of systems theories and their managerial applications. *Service Science, 2*(1–2), 126–135.

Napoletano, P. & Carrubbo, L. (2010). Becoming smarter: Towards a new generation of service systems. *Impresa, Ambiente, Management, 4*(3), 415–438.

Polese, F. & Carrubbo, L. (2016). Eco-sistemi di Servizio in Sanità, Collana Studi e Ricerche Aziendali. *Ed Giappichelli*, 65.

Polese, F., Carrubbo, L., & Russo, G. (2009). Managing business relationships: Between service culture and a viable system approach. *Esperienze d'Impresa*, 2.

Prahalad, C.K. & Ramanswamy, V. (2004). *The Future of Competition: Co-Creating Unique Value with Customers.* Cambridge, MA: Harvard University Press.

Spohrer, J., Anderson, L., Pass, N., & Ager, T. (2008). Service science e service dominant logic. *Otago Forum*, *2*, 4–18.

Vargo, S.L. & Lusch, R.F. (2008). Service-dominant logic: Continuing the evolution. *Journal of the Academy of Marketing Science*, *36*(1), 1–10. https://doi.org/10.1007/s11747-007-0069-6

Vargo, S.L. & Lusch, R.F. (2016). Institutions and axioms: An extension and update of service-dominant logic. *Journal of the Academy of Marketing Science*, *44*(1), 5–23. https://doi.org/10.1007/s11747-015-0456-3

Von Bertalanffy, L. (1968). *General System Theory: Foundations, Development, Applications.* New York: George Braziller.

Von Foerster, H. (1981). *Observing Systems.* Seaside: InterSystems Pubblication.

Wieland, H., Polese, F., Vargo, S., & Lusch, R. (2012). Toward a service (eco)systems perspective on value creation. *International Journal of Service Science, Management, Engineering and Technology*, *3*(3), 12–25.

The ICT adoption in the European Museum sector

Building a positioning model

Claudio Nigro,[1] *Primiano Di Nauta,*[2]
Enrica Iannuzzi[3] *and Miriam Petracca*[4]

Keywords: *ICT; tourism sector; museum organization; type of governance; museums size*

1 Introduction

'Innovation' is an important path for all companies aiming at being competitive, whatever the reference sector or the territorial system. Over time, numerous scholars have focused on innovation trends implemented by companies, looking at the inclination level of its adoption as one of the discriminating factors for the success of these business actors. Particularly, in the international debate regarding innovation in the tourism sector, it has been paid a lot of attention to the role played by the 'Information and Communication Technology' (hereafter, ICT) as a key driver to revolutionize the worldwide tourism (Poon 1988; Sundbo et al. 2007; Hall and Allan 2008; Hjalager 2010; Aldebert et al. 2011; Meneses and Teixeira 2011; Camisón and Monfort-Mir 2012).

This is confirmed by several empirical evidences, in which the ICT would produce important effects/advantages on tourism that, at a macro level, impact on: tourists, for the acceleration/automation of the interface between the artistic-cultural heritage and tourism; consumers, able to identify, personalize, and buy tourist products; companies, able to develop, manage, and deliver their offer worldwide; the competitiveness of tourism organizations and destinations (among others, Buhalis and Amaranggana, 2013; Carrubbo et al. 2012).

Many empirical evidences confirm this trend in a micro-level analysis, i.e. the single tourism organization aimed at contributing to the development of the 'tourism system'.

This work focuses on the meso level, exclusively focusing on the inclination of a particular category of business, a museums' sample, to highlight ICT as a strategic resource for its competition.

2 The brief reconstruction of the literature

Innovation is a very topical issue in the economic, managerial, political, and social debate. Particularly, with regard to the tourism sector, the economic-managerial studies about the approach to innovation were developed within the governance dynamics of the artistic-cultural Italian heritage (Zan 1999; Nigro et al. 2011), expecially for some aspects related to the managerialization process – as the opportunity to make concrete the promotion of 'culture' and 'territory' – and, then, to the role of 'professionals of culture' aimed at value co-creation (Rullani 2004; Montella 2009; Franch 2010; Golinelli 2012; Nigro et al. 2015, 2016a, 2016b).

In this scenario, we think it's useful to mention some interesting contributions: La Rocca (2014), who identified three main classes of ICT aimed at promoting a better quality of the tourism experience: *Information-centered*, focused on the number of application that tourists can use to visit the chosen destination; *Tourist-centered*, to strengthen and make unique and long-lasting the tourist experience; *Tourist-engaging*, in which tourists play the role of 'urban sensors' through specific tools (social media, Big Data, and so on); with regard to the museum sector, Ioannidis et al. (2014) proposed an analysis on how ICT could add value to collections of museums and cultural sites, and enhance access and communication between stakeholders and users/visitors. Camarero et al. (2010) identified three types of innovation in museums: *Technological innovation*; *Innovation in value creation*; *Organizational innovation*.

From the analysis of the literature, the 'size' and 'nature of governance' (public, or private) emerges as recurrent factors able to orient the attitude of innovation in museum organizations.

3 Purposes of the research

In line with the recurring themes emerging from the literature review, this contribution focuses on the inclusion of ICT as a strategic resource for tourism organizations, with particular attention to the museums. Particularly, the work analyzes the museum decision makers propensity to invest in ICT based on two factors: the dimension of the museum; the type of governance (public or private).

The focus of the work has been dictated by the peculiar characteristics of museums. On one side, these organizations can be interpreted as non-profit, for which the social objectives (education, conservation, housing, etc.) are of main interest. On the other side, museums can be interpreted in the light of for-profits, as they pursue, among others, both business objectives and financial ones. Furthermore, museums are able to adapt to the changing context, related to several factors: size, ownership structure, organizational structure, funding sources, and so on.

In this scenario the contribution consists in a quali-quantitative field research aiming at: *proposing a positioning model able to examine the bent of museum management to include ICT in their own assets; exploring the attitude of museums to invest in ICT due to museum dimension and type of governance.*

4 Methodology

Coherent with the aim of this contribution, the research group analyzes the adoption of ICT for the artistic-cultural heritage promotion in a museum sample identified as follows: the top ten museums, according to the number of visitors, of the ten major European cities with a clear cultural tourism vocation. The cities have been identified taking into account those urban centers that in the year 2014 recorded up to 100,000 museum visits. As a result, the sample is composed of the biggest museums of Paris, Barcelona, London, Amsterdam, Rome, Berlin, Vienna, Stockholm, Lisbon, and Helsinki. For each of the selected museums, the research verifies, based on secondary data, the presence/absence of the ICT implementation (Tourism-engaging, Tourist-centered, and Information-centered), through the following analysis: descriptive, factor, cluster, and discriminant.

Particularly, the research group developed two different analytical moments: two factorial analyses and two cluster analyses in order to build a positioning map that allowed to rank the museum organizations depending on their willingness to invest in ICT; then, two discriminant analyses in order to explore the propensity of museum decision makers in investing in ICT based on the 'dimension/size of the museum'[5] and the 'type of governance' (public or private).

5   Findings

Referring the reader to the previous works conducted by the research group (Nigro et al. 2016a, 2016b), it's useful to highlight that the cluster analysis responds to a service-driven logic. This is to say that it is possible to group the museums on the basis of two different categories of infrastructure services: *Site-uncentered* services that can be provided without requiring the physical presence of the visitor (e.g., Virtual Guide and Interactive Maps items); *Site-centered* services, able to be purchased and consumed with the physical presence of the visitor (e.g., Augmented Reality, Free Wi-Fi, and QR code items).

As a follow-up of the obtained results, the research group defined a positioning map using the latent variables *Site-uncentered* and *Site-centered* services. Each latent variable describes a different level of the museum organization in acquiring technologies, which are an expression of an underlying strategy.

Figure 54.1 shows four groups of museums with a specific level of propensity to invest in ICT:

>   *Low level*: museums that do not have yet an appropriate implementation level of ICT;
>   *Moderate level*: museums that are implementing new technologies, particularly the *site-uncentered* ones;
>   *Good level*: museums that are highly bent to innovation, especially for those technologies attracting visitors (ICT *site-centered*);
>   *Advanced level*: museums able to implement a range of technology solutions for the visitors' benefits, both inside and outside the structure.

The second stage of the research refers to the possibility to investigate whether or not the size and the type of governance of museums discriminate the adoption of ICT. Among other aspects, the discriminant analysis shows that: only the bigger museums invest in most sophisticated and expensive innovation services (i.e. Big Data), while both big and small museums have the same tendency in investing in

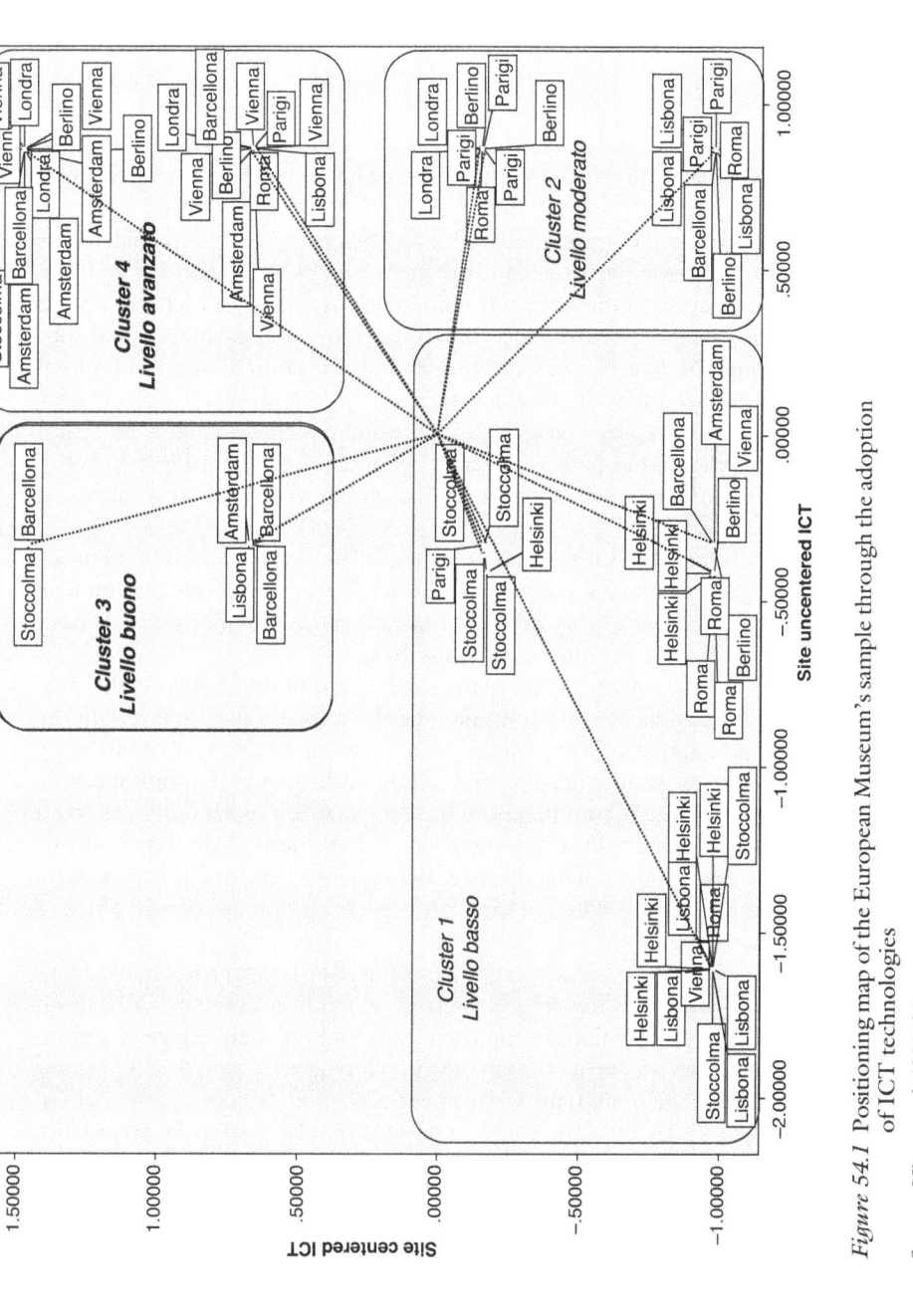

Figure 54.1 Positioning map of the European Museum's sample through the adoption
of ICT technologies

Source: Nigro et al. (2016b)

Virtual Guide and Interactive Maps; the public museums have a moderate positive orientation towards sophisticated ICT (such as Big Data Analytics), despite the high costs that they require, while among the private museums it is possible to observe a tendency to adopt the Augmented Reality.

6 Conclusions

This paper aims at providing a contribution to the debate on the adoption of ICT solutions in organizations committed to the promotion of the artistic-cultural heritage.

In particular: on the epistemological level, this study can be considered as a first exploratory attempt to define two latent variables (*site-centered* and *site-uncentered* services); on the managerial level these variables are useful to discuss the different strategic paths of museums. In fact, the strategic paths could support these organizations to access new market segments or to enhance the appeal of those already served by offering new services.

Remaining on the epistemological level, the findings arising from work return a link to the relationship (although causal) between the propensity to invest in ICT and 'antecedents', such as museum 'size' and 'governance assets', anything but intuitive.

On the basis of the results, it can be argued there is no relationship between the types of governance and the propensity to adopt more performing technology solutions. This may mean that the ICT strategic options, implemented in both public and private sector, would not be dissimilar.

Different results emerge with regard to the relationship between the propensity to the adoption of ICT solutions and the museum size, in line with the basic principles aimed to consider the measurement parameters of efficiency underlying any investment decision (e.g., ROI). Albeit an ICT solution can be considered one of many (although important) strategic levers adoptable by a museum, the investment in such leverage implies the need to intervene on two fundamental aspects: the criteria of choice between new investment and its mode of coverage; the need to redesign the 'core' operational processes with a strong work reorganization.

Not by chance, in fact, the research results showed little inclination to the adoption and use of Big Data (except for the large and public museums). This result could be due, in our opinion, to the complexity and the high degree of sophistication that represents such a technology, which requires a high level of professionalism and, at the same time, a consistent allocation of human and financial resources. This is why Big Data seems to represent, today, a strategic prerogative of museums with a leadership position in the market.

This contribution is at a first stage of development. In the next steps, the research group intends to verify the completeness of the variables used to analyze the dynamics of the phenomenon and increase the sample to enhance the level of its representativeness.

Notes

1 University of Foggia.
2 University of Foggia.
3 University of Foggia.
4 Giustino Fortunato University.
5 The average value of the tourist flow (1420221,22) of the museums sample has been adopted as a classification criteria aiming at dividing museums according to the size variable. Museums below the average flow of visitors are small; museums above the average flow of visitors are big.

Bibliography

Aldebert, B., Dang, R. J., & Longhi, C. (2011). Innovation in the tourism industry: The case of Tourism@. *Tourism Management, 32*(5), 1204–1213.

Buhalis, D., & Amaranggana, A. (2013). *Smart tourism destinations: Information and communication technologies in tourism 2014.* Proceedings of the International Conference in Dublin, Ireland, January 21–24, 2014, Springer International Publishing, 553–564.

Camarero, C., Garrido, M. J., & Vicente, E. (2010). Components of art exhibition brand equity for internal and external visitors. *Tourism Management, 31*(4), 495–504.

Camisón, C., & Monfort-Mir, V. M. (2012). Measuring innovation in tourism from the Schumpeterian and the dynamic-capabilities perspectives. *Tourism Management, 33*(4), 776–789.

Carrubbo, L., Moretta Tartaglione, A., Di Nauta, P., & Bilotta, A. (2012). A service science view of a sustainable destination management. *China-USA Business Review, 11*(11), 1017–1035.

Franch, M. (2010). Le frontiere manageriali per la valorizzazione della cultura e dell'arte. *Sinergie Journal of Management, 82*, 95–107.

Golinelli, G. M. (2012). *Patrimonio culturale e creazione di valore. Verso nuovi percorsi.* Padova: Cedam.

Hall, M. C., & Allan, W. (2008). *Tourism and innovation.* Abingdon, UK: Routledge.

Hjalager, A. M. (2010). A review of innovation research in tourism. *Tourism Management, 31*(1), 1–12.

Ioannidis, Y., Toli, E., El Raheb, K., & Boile, M. (2014, November). *Using ICT in cultural heritage, bless or mess? Stakeholders' and practitioners' view through the eCultValue project.* In Euro-Mediterranean Conference (pp. 811–818), Springer International Publishing.

La Rocca, R. A. (2014). The role of tourism in planning the Smart City: Tema. *Journal of Land Use, Mobility and Environment, 7*(3), 269–284.

Meneses, O.A., & Teixeira, A. A. (2011). The innovative behaviour of tourism firms. *Economics and Management Research Projects: An International Journal, 11*, 25–35.

Montella, M. (2009). *Valore e valorizzazione del patrimonio culturale storico.* Milano: Mondadori.

Nigro, C., Iannuzzi, E., & Carolillo, G. (2011). Comunicazione e strutturazione di un quadro istituzionale. Riflessioni sulla recente crisi del sistema finanziario. *Sinergie Italian Journal of Management, 89*, 109–130.

Nigro, C., Iannuzzi, E., Petracca, M., & Montagano, V. (2015). *Isomorfismo e decoupling nelle dinamiche di governance dei musei statali italiani.* Referred Electronic Conference Proceeding del XXVII Convegno annuale di Sinergie, 945–964.

Nigro, C., Iannuzzi, E., Petracca, M., & Montagano, V. (2016a). *L'adozione dell'ICT in un campione di musei Europei: verso una mappa di posizionamento.* Sinergie Referred Electronic Conference Proceedings, 945–964.

Nigro, C., Iannuzzi, E., Petracca, M., & Montagano, V. (2016b). *A preparatory research on ICT adoption by a sample of European museums: Toward a positioning model.* EURAM Conference Proceedings.

Poon, A. (1988). Innovation and the future of Caribbean tourism. *Tourism Management, 93,* 213–220.

Rullani, E. (2004). *Economia della conoscenza. Creatività e valore nel capitalismo delle reti.* Roma: Carocci.

Sundbo, J., Orfila-Sintes, F., & Sørensen, F. (2007). The innovative behaviour of tourism firms: Comparative studies of Denmark and Spain. *Research Policy, 361,* 88–106.

Zan, L. (ed.). (1999). *Conservazione e innovazione nei musei italiani. Management e processi di cambiamento.* Milano: Etas.

Towards management of construction site Big Data

Andrej Tibaut[1] and Damjan Zazula[2]

Keywords: *automated data collection; Big Data management; construction site monitoring; building information modelling; knowledge management*

1 Introduction to management of construction site Big Data

Model-based engineering approach in the AECO (Architecture, Engineering, Construction, Operation) industry ranges from modelling of labour productivity (Thomas et al., 1990), to model-based design and engineering (Rebolj, Tibaut, Čuš-Babič, Magdič, & Podbreznik, 2008), information-based modelling (Suermann, Issa, & Suermann, 2009), and to the critical thoughts about building information modelling approach (Turk, 2016). Information-based modelling allows for more complex construction projects, therefore the risk that they don't progress on schedule has become big concern. Monitoring of complex transdisciplinary activities on a construction site is a pressing problem that challenges practitioners and researchers to achieve the shortest project duration and minimum project cost (Omar & Nehdi, 2016).

A construction site is an engineering phenomenon where many mechanistic workflows interoperate in a predicted manner (Beardsworth, Keil, Bresnen, & Bryman, 1988). The predicted behaviour is result of the model-based view that engineers use when they design buildings and infrastructures. Unplanned events during the construction phase may lead to undesired consequences like safety risks (Keng & Razak, 2014), risk of delays, and schedule overruns (Arashpour, Wakefield, Lee, Chan, & Hosseini, 2015). To minimize disruptive effects of the events, continuous monitoring of all scheduled activities on the construction site must be ensured. The monitoring process involves manual data collection obtained by direct human observation and/or automated data collection (i.e. measurements [Paolo Rocchi, 2016]), which results in extensive data records. The empirical Big Data collected during the monitoring process is used for inspection of construction works as they proceed.

There are a growing number of sources of Big Data in construction projects because several technologies have been developed for automated data collection. These technologies include laser scanning of construction sites (Wang, Cho, &

Kim, 2015), GPS-based location tracking (Jiang, Lin, Qiang, & Fan, 2015), the use of RFID technology (Kuipers, Tomé, Pinheiro, Nunes, & Heitor, 2014) or bar codes (Navon & Sacks, 2007), augmented reality (Wang et al., 2013), and the applications of video cameras (Yang, Park, Vela, & Golparvar-Fard, 2015)

Use of video cameras is one of the cost efficient approaches to Big Data collection on construction sites (Brilakis, Park, & Jog, 2011; Rodriguez-Gonzalvez, Gonzalez-Aguilera, Lopez-Jimenez, & Picon-Cabrera, 2014).

In this work, the authors propose a transdisciplinary approach that orchestrates domains of project management, building information modelling, knowledge engineering and building construction based on a shared building information model and construction schedule and facilitated through the new ICT framework. The approach was verified with a field experiment on a construction site where the domains' Big Data (site images) was automatically collected and processed. The Big Data collection was limited to images of the externally visible building elements. Our approach included applied research with science-practice relationships.

2   Methodology and results

The research method was designed to include both a laboratory (lab) and a field experiment. The lab experiment was designed to be a replication of the field experiment. In the lab, the equipment was tested and validated. The field experiment took place on a real construction site setting where the Vinarium lookout tower (see Figure 55.2b) was built (reinforced concrete building) and assembled (steel columns assembly). The technical framework for collection of Big Data on the construction site (Figure 55.1), designed in our research, is based on the conceptual model, which contains three subsystems: material flow monitoring, image-based tracking, and personal on-site observations (Podbreznik & Rebolj, 2005; Rebolj, Čuš Babič, Magdič, Podbreznik, & Pšunder, 2008).

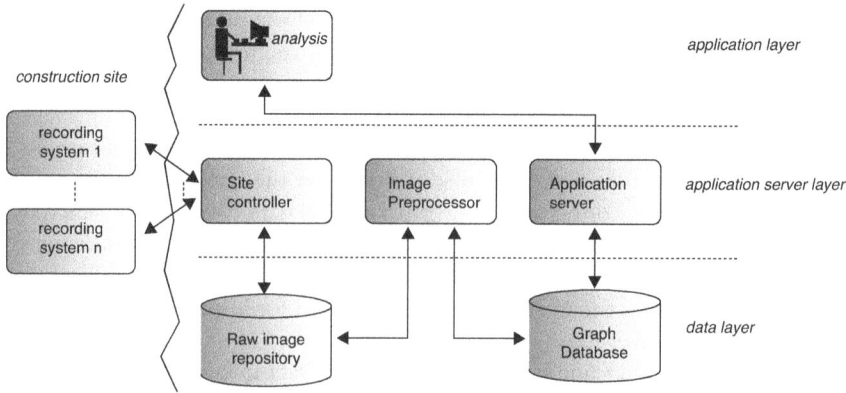

Figure 55.1 Technical framework for collection of Big Data on construction site

For the field experiment an innovative image and video recording device (Zazula, Sedej, Cigale, & Munda, 2013) was used to record images as part of the technical framework.

At the beginning of the experiment, a single recording device was used with the aim to manually increase the inclination angle as the tower grows. During the work the second recording device was installed at the same vertical position (30 meters away from the tower). The two recording devices were vertically aligned to independently cover the upper and lower part of the facility, respectively. The two resulting separate images were buffered locally on the device's storage. A computer, which was part of the recording device, pushed recorded images over the internet to the site controller at regular intervals. The two images were then stored in the raw image repository. The image pre-processor pulled images from the repository and processed them to produce a merged image from corresponding ones. Figure 55.2a is a Building Information Modeling (BIM) model, and the right is merged from two corresponding images. In our experiment 55000 images were recorded during the 199 project days. A user could observe and analyze images through the web application.

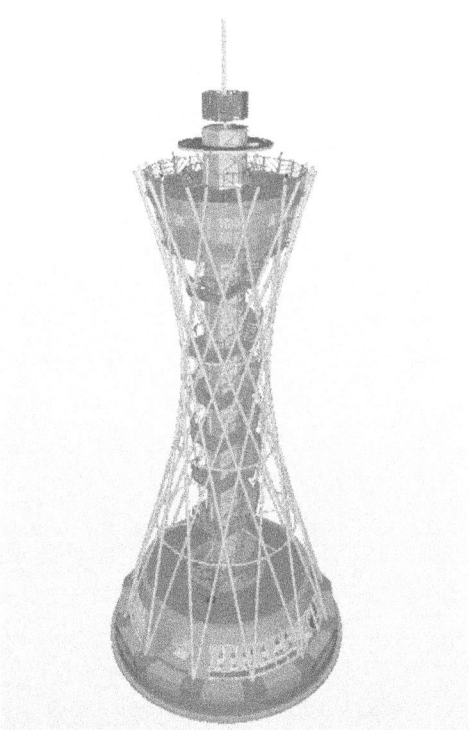

Figure 55.2 (a) BIM model of the "Vinarium lookout tower" and (b) corresponding construction site

Figure 55.2 (Continued)

3   Main conclusions

The research has resulted in a prototype framework that automatically collects Big Data during the monitoring process and efficiently inspects construction work as it proceeds. This leads to new forms of collaboration and legal framings such as access to data but also allows precise monitoring of complex construction schedule design on-site.

Research outcomes are of interest for both construction practitioners (i.e. investors, consultants) and transdisciplinary research domains (i.e. system integration, BIM, computer vision, Big Data management, knowledge management, construction scheduling, sustainability in building lifecycle), which call for applied research that includes participation of many stakeholders.

Notes

1 University of Maribor Faculty of Civil Engineering, Transportation Engineering and Architecture, Maribor, Slovenia.
2 University of Maribor Faculty of Electrical Engineering and Computer Science, Maribor, Slovenia.

Bibliography

Arashpour, M., Wakefield, R., Lee, E. W. M., Chan, R., & Hosseini, M. R. (2015). Analysis of interacting uncertainties in on-site and off-site activities: Implications for hybrid construction. *International Journal of Project Management*, *34*(7), 1393–1402. http://doi.org/10.1016/j.ijproman.2016.02.004

Beardsworth, A. D., Keil, E. T., Bresnen, M., & Bryman, A. (1988). Management, transience and subcontracting: The case of the construction site. *Journal of Management Studies*, *25*(6), 603–625. http://doi.org/10.1111/j.1467-6486.1988.tb00049.x

Brilakis, I., Park, M.-W., & Jog, G. (2011). Automated vision tracking of project related entities. *Advanced Engineering Informatics*, *25*(4), 713–724. http://doi.org/10.1016/j.aei.2011.01.003

Jiang, H., Lin, P., Qiang, M., & Fan, Q. (2015). A labor consumption measurement system based on real-time tracking technology for dam construction site. *Automation in Construction*, *52*, 1–15.

Keng, T. C., & Razak, N. A. (2014). Case studies on the safety management at construction site. *Journal of Sustainability Science and Management*, *9*(2), 90–108.

Kuipers, M., Tomé, A., Pinheiro, T., Nunes, M., & Heitor, T. (2014). Building space-use analysis system: A multi location/multi sensor platform. *Automation in Construction*, *47*, 10–23. http://doi.org/10.1016/j.autcon.2014.07.001

Navon, R., & Sacks, R. (2007). Assessing research issues in Automated Project Performance Control (APPC). *Automation in Construction*, *16*(4), 474–484.

Omar, T., & Nehdi, M. L. (2016). Data acquisition technologies for construction progress tracking. *Automation in Construction*, *70*, 143–155.

Paolo Rocchi. (2016). What information to measure? How to measure it? *Kybernetes*, *45*(5), 718–731. http://doi.org/10.1108/K-06-2015-0161

Podbreznik, P., & Rebolj, D. (2005). Automatic comparison of site images and the 4D model of the buiding. In R. J. Scherer, P. Katranuschkov, & S.-E. Schapke (Eds.), *CIB W78 22nd conference on information technology in construction* (pp. 235–239). Dresden, Germany: Institute for Construction Informatics and Technische Universitat and Dresden.

Rebolj, D., Čuš Babič, N., Magdič, A., Podbreznik, P., & Pšunder, M. (2008). Automated construction activity monitoring system. *Advanced Engineering Informatics*, *22*(4), 493–503.

Rebolj, D., Tibaut, A., Čuš-Babič, N., Magdič, A., & Podbreznik, P. (2008). Development and application of a road product model. *Automation in Construction*, *17*(6), 719–728. http://doi.org/10.1016/j.autcon.2007.12.004

Rodriguez-Gonzalvez, P., Gonzalez-Aguilera, D., Lopez-Jimenez, G., & Picon-Cabrera, I. (2014). Image-based modeling of built environment from an unmanned aerial system. *Automation in Construction*, *48*, 44–52.

Suermann, P. C., Issa, R., & Suermann, P. C. (2009). Evaluating industry perceptions of Building Information Modeling (BIM) impact on construction. *Journal of Information Technology in Construction*, *14*(December 2007), 574–594.

Thomas, H. R., Maloney, W. F., Horner, R. M. W., Smith, G. R., Handa, V. K., & Sanders, S. R. (1990). Modeling construction labor productivity. *Journal of Construction Engineering and Management*, *116*(4), 705. http://doi.org/10.1061/(ASCE)0733-9364(1990)116:4(705)

Turk, Z. (2016). Ten questions concerning building information modelling. *Building and Environment*, 1–11. http://doi.org/10.1016/j.buildenv.2016.08.001

Wang, C., Cho, Y. K., & Kim, C. (2015). Automatic BIM component extraction from point clouds of existing buildings for sustainability applications. *Automation in Construction*, *56*, 1–13.

Wang, X., Love, P. E. D., Kim, M. J., Park, C.-S., Sing, C.-P., & Hou, L. (2013). A conceptual framework for integrating building information modeling with augmented reality. *Automation in Construction*, *34*, 37–44.

Yang, J., Park, M.-W., Vela, P. A., & Golparvar-Fard, M. (2015). Construction performance monitoring via still images, time-lapse photos, and video streams: Now, tomorrow, and the future. *Advanced Engineering Informatics*, *29*(2), 211–224.

Zazula, D., Sedej, G., Cigale, B., & Munda, J. (2013). Računalniška naprava in postopek za oddaljeno upravljanje video toka in visokoločljivostnih slik iz digitalnega fotoaparata: odločba o podelitvi patenta št. 23859 A, št. prijave P-201100290, datum vložitve prijave 3. 8. 2011, datum objave prijave: 28. 2. Ljubljana: Urad Republike Slovenije za intelektualno lastnino.

How smart technologies and Big Data affect systems' lives? Conceptual reflections on the smart city ecosystem

Francesco Caputo,[1] Leonard Walletzky[2] and Petr Stepanek[3]

Keywords: *smart technologies; Big Data, smart city; complex adaptive systems; network analysis*

1 Introduction

The progress in knowledge offered by Computer Science and Information and Communication Technology (ICT) is changing the world in which we live (Bijker & Law 1994; Rip et al. 1995; Klein et al. 2012; Barile et al. 2015; Del Giudice et al. 2016; Evangelista et al. 2016). As underlined by several researchers and research streams (Danneels 2004, 2006; Kostoff et al. 2004; Latzer 2009; Del Giudice et al. 2012), the technology is 'disrupting' traditional market balances and rules by supporting the emergence of a more efficient, effective, and sustainable world. In such a context, the emerging view of the world is addressed by an increasing relevance of information and technology in supporting a better understanding of social and economic dynamics (Castells 1997; Bassellier & Blaize Horner Reich 2001; Turban et al. 2008; Di Nauta et al. 2015). As consequence, decision makers are paying more attention to the acquisition and processing of data as a strategic pathway on which to act to enlarge the existing knowledge about the emerging rules of the market (Bhatt 2001; Caputo et al. 2016a, 2016b).

Reflecting on this trend, some relevant research questions appear with reference to the utility of this increasing attention to data management in solving social and economic dynamics and in better understanding social and economic problems (Michael & Miller 2013; Chen & Zhang 2014). In such a line, different contributions have been offered with reference to the way in which the management of data can change our daily lives (Saha & Mukherjee 2003). More specifically, several authors have pointed out the risks and opportunities of so-called Big Data (McAfee et al. 2012; Mayer-Schönberger & Cukier 2013), and the possible future developments of a society strongly based on information sharing (Beniger 2009; Castells 2011). Despite the increasing attention on these topics, a clear conceptual framework to explain the impact of Big Data and technologies on our daily life seems to be still missing (Holmes 2005).

In order to bridge this gap, the paper focuses on the domain of Smart City as a relevant example of 'contamination' among users, suppliers, resources, information, and technologies, used to improve the quality of citizen lives and the effectiveness of citizen management (Nam & Pardo 2011; Jin et al. 2014; Scuotto et al. 2016). In such a context, the paper aims to investigate in which way the Smart City represents an evidence of disruptive technology to directly change the ways in which citizens perceive and interact within the citizen ecosystem (Dickinson et al. 2012; Zygiaris 2013; Caputo et al. 2016c; Polese et al. 2016; Dominici et al. 2017). More specifically, by analyzing the Smart City in terms of a complex adaptive system (Buckley 1968; Holland 1992), the paper aims to identify in which ways smart technologies and Big Data impact on the system's ability to adapt itself to the challenging environment in order to better satisfy markets' needs and expectations (Barile et al. 2014; Saviano et al. 2016).

2   Research pathway and methodology

By adopting the interpretative lens offered by Systems Thinking (Von Bertalanffy 1968; Beer 1985; Maturana & Varela 1991; Espejo 1994; Jackson 2003; Golinelli 2010; Barile 2013), the paper investigates the domain of Smart City with the purpose of defining a wider conceptual framework inclusive of the different dimensions on which a Smart City is based.

Afterwards, the Smart City as a complex adaptive system is analyzed to underline opportunities and risks of strategies and pathways based on the environmental changes. Finally, the conceptual framework herein is analyzed by using Network Analysis (Bode 1945; Haythornthwaite 1996; Anderson & Vongpanitlerd 2006; Borgatti et al. 2009) in order to explain in which ways the different dimensions of a Smart City can dynamically change their relations by defining different systems' configurations to directly satisfy the need of different actors involved in the city ecosystem.

Thanks to the adoption of the interpretative lens offered by Systems thinking, the relevant role of smart technologies in supporting the interaction among the different elements involved in a system by ensuring a fast reciprocal adaptation over the time (Streitz et al. 2005; Di Fatta et al. 2016) and the key role of Big Data as pathways to ensure the building of a strong feedback process able to increase the alignment between the linked systems are formalized as shown in Figure 56.1.

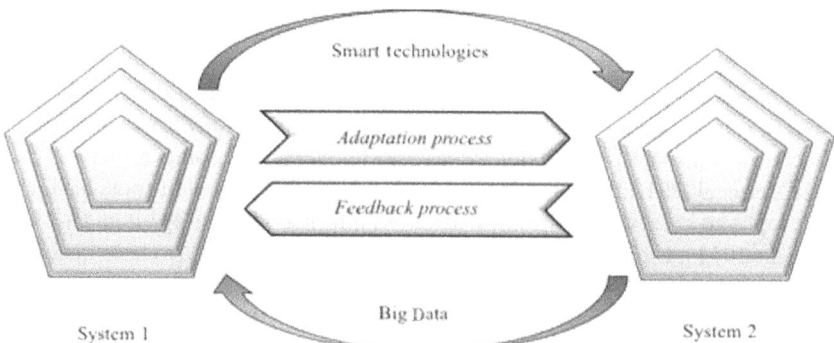

Figure 56.1 The role of smart technologies and Big Data in systems' linking

Source: Authors' elaboration

3 Theoretical and practical implications

The paper enlarges previous knowledge in domains of smart technologies and Big Data by showing their evidences with reference to the topic of the Smart City. At same time, the paper contributes to the debate on the Smart City by highlighting its systems nature and its relevance in ensuring an efficient management of emerging social and economic challenges thanks to the smart technologies and Big Data.

From a practical point of view, the paper defines a wider conceptual framework useful for practitioners and decision makers interested in a better understanding of principles and rules of the Smart City. This helps to define more effective managerial models able to act on the 'modularity' of the Smart City to satisfy needs and expectations of several actors by using the same resources.

4 Conclusions and future directions for research

In a world affected by an increasing attention to the topics of technologies and data, a relevant challenge to face refers to the ways in which smart technologies and Big Data can build more efficient, effective, and sustainable managerial models to satisfy the market's needs and expectations.

In such a context, the domain of the Smart City offers interesting stimuli of reflections because it shows in which ways, by acting on the same structure, it is possible to dynamically combine resources, people, technologies, infrastructures, and information to build different value propositions to satisfy the need of several actors involved in the citizen ecosystem. The reflections herein, with reference to the ways in which the Smart City can modify the links among its 'parts' to ensure the emergence of different systems, directly satisfy different actors, represent a starting point on which to base future studies, to investigate the dynamics of this 'adaptation processes' in order to enrich existing knowledge about the role of smart technologies and Big Data in the emerging society and to define possible guidelines to build more performant managerial models for the Smart City.

Notes

1 Masaryk University.
2 Masaryk University.
3 Masaryk University.

Bibliography

Anderson, B. & Vongpanitlerd, S. (2006). *Network analysis and synthesis.* New York: Dover.
Barile, S. (2013). *Contributions to theoretical and practical advances in management: A Viable Systems Approach (VSA).* Roma: Aracne.
Barile, S., Saviano, M., & Caputo, F. (2014). *A systems view of customer satisfaction.* In National Conference "Excellence in quality, statistical quality control and customer satisfaction", University Campus "Luigi Einaudi", Turin University, September 18–19.
Barile, S., Saviano, M., & Caputo, F. (2015). How are markets changing? The emergence of consumers market systems. In Dominici, G. (Ed.), *3rd international*

symposium advances in business management: "Towards systemic approach" (pp. 203–207). Business systems. Avellino: E-book series.

Bassellier, G. & Blaize Horner Reich, I. B. (2001). Information technology competence of business managers: A definition and research model. *Journal of Management Information Systems, 17*(4), 159–182.

Beer, S. (1985). *Diagnosing the system for organizations.* New York: John Wiley & Sons Inc.

Beniger, J. (2009). *The control revolution: Technological and economic origins of the information society.* Cambridge: Harvard University Press.

Bhatt, G. D. (2001). Knowledge management in organizations: Examining the interaction between technologies, techniques, and people. *Journal of Knowledge Management, 5*(1), 68–75.

Bijker, W. & Law, J. (1994). *Shaping technology/building society: Studies in sociotechnical change.* Cambridge: MIT Press.

Bode, H.W. (1945). *Network analysis and feedback amplifier design.* New York: Van Nostrand Company.

Borgatti, S. P., Mehra, A., Brass, D. J., & Labianca, G. (2009). Network analysis in the social sciences. *Science, 323*(5916), 892–895.

Buckley, W. (1968). *Society as a complex adaptive system: Essays in social theory.* New York: Taylor & Francis.

Caputo, F., Del Giudice, M., Evangelista, F., & Russo, G. (2016a). Corporate disclosure and intellectual capital: The light side of information asymmetry. *International Journal of Managerial and Financial Accounting, 8*(1), 75–96.

Caputo, F., Evangelista, F., & Russo, G. (2016b). Information sharing and communication strategies: A stakeholder engagement view. In Vrontis, D., Weber, Y., & Tsoukatos, E. (Eds.), *Innovation, entrepreneurship and digital ecosystems* (pp. 436–442). Cipro: EuroMed Press.

Caputo, F., Formisano, V., Buronova, B., & Walletzky, L. (2016c). Beyond the digital ecosystems view: Insights from smart communities. In Vrontis, D., Weber, Y., & Tsoukatos, E. (Eds.), *Innovation, entrepreneurship and digital ecosystems* (pp. 443–454). Cyprus: EuroMed Press.

Castells, M. (1997). *The information age: Economy, society and culture, Vol. 2: The power of identity.* London: Blackwell.

Castells, M. (2011). *The rise of the network society: The information age: Economy, society, and culture.* New York: John Wiley & Sons.

Chen, C. P. & Zhang, C. Y. (2014). Data-intensive applications, challenges, techniques and technologies: A survey on big data. *Information Sciences, 275,* 314–347.

Danneels, E. (2004). Disruptive technology reconsidered: A critique and research agenda. *Journal of Product Innovation Management, 21*(4), 246–258.

Danneels, E. (2006). Dialogue on the effects of disruptive technology on firms and industries. *Journal of Product Innovation Management, 23*(1), 2–4.

Del Giudice, M., Caputo, F., & Evangelista, F. (2016). How decision systems changing? The contribution of social media to the management decisional liquefaction. *Journal of Decision Systems, 25*(3), 214–226.

Del Giudice, M., Carayannis, E. G., & Della Peruta, M. R. (2012). *Cross-cultural knowledge management and open innovation diplomacy: Definition of terms.* In Giudice, M. D., Peruta, M. R. D., & Carayannis, E. G. (Eds.), *Cross-cultural knowledge management* (pp. 117–135). New York: Springer.

Dickinson, J. L., Shirk, J., Bonter, D., Bonney, R., Crain, R. L., Martin, J., & Purcell, K. (2012). The current state of citizen science as a tool for ecological research and public engagement. *Frontiers in Ecology and the Environment*, *10*(6), 291–297.

Di Fatta, D., Caputo, F., Evangelista, F., & Dominici G. (2016). Small world theory and the World Wide Web: Linking small world properties and website centrality. *International Journal of Markets and Business Systems*, *2*(2). DOI: 10.1504/ IJMABS.2016.080237.

Di Nauta, P., Merola, B., Caputo, F., & Evangelista, F. (2015). Reflections on the role of university to face the challenges of knowledge society for the local economic development. *Journal of Knowledge Economy*, 1–19. DOI: 10.1007/ s13132-015-0333-9.

Dominici, G., Yolles, M., & Caputo, F. (2017). Decoding the dynamics of value cocreation in consumer tribes: An agency theory approach. *Cybernetics and Systems*, *48*(2), 84–101.

Espejo, R. (1994). What is systemic thinking? *System Dynamics Review*, *10*(2–3), 199–212.

Evangelista, F., Caputo, F., Russo, G., & Buhnova B. (2016). Voluntary corporate disclosure in the era of social media. In Caputo, F. (Ed.), *The 4rd international symposium advances in business management: "Towards systemic approach"* (pp. 124–128). Business systems. Avellino: E-book series.

Golinelli, G. M. (2010). *Viable Systems Approach (VSA): Governing business dynamics*. Padova: Cedam.

Haythornthwaite, C. (1996). Social network analysis: An approach and technique for the study of information exchange. *Library & Information Science Research*, *18*(4), 323–342.

Holland, J. H. (1992). Complex adaptive systems. *Daedalus*, 17–30.

Holmes, D. (2005). *Communication theory: Media, technology and society*. London: Sage.

Jackson, M. C. (2003). *Systems thinking: Creative holism for managers*. Chichester: Wiley.

Jin, J., Gubbi, J., Marusic, S., & Palaniswami, M. (2014). An information framework for creating a smart city through Internet of Things. *IEEE Internet of Things Journal*, *1*(2), 112–121.

Klein, J. T., Grossenbacher-Mansuy, W., Häberli, R., Bill, A., Scholz, R. W., & Welti, M. (Eds.). (2012). *Transdisciplinarity: Joint problem solving among science, technology, and society: An effective way for managing complexity*. London: Birkhäuser.

Kostoff, R. N., Boylan, R., & Simons, G. R. (2004). Disruptive technology roadmaps. *Technological Forecasting and Social Change*, *71*(1), 141–159.

Latzer, M. (2009). Information and communication technology innovations: Radical and disruptive? *New Media & Society*, *11*(4), 599–619.

Maturana, H. R. & Varela, F. J. (1991). *Autopoiesis and cognition: The realization of the living* (Vol. 42). New York: Springer Science & Business Media.

Mayer-Schönberger, V. & Cukier, K. (2013). *Big data: A revolution that will transform how we live, work, and think*. Boston: Houghton Mifflin Harcourt.

McAfee, A., Brynjolfsson, E., Davenport, T. H., Patil, D. J., & Barton, D. (2012). Big data: The management revolution. *Harvard Business Review*, *90*(10), 61–67.

Michael, K., & Miller, K. W. (2013). Big data: New opportunities and new challenges. *Computer*, *46*(6), 22–24.

Nam, T., & Pardo, T. A. (2011). *Conceptualizing smart city with dimensions of technology, people, and institutions.* In Proceedings of the 12th Annual International Digital Government Research Conference: Digital Government Innovation in Challenging Times (pp. 282–291). ACM, June.

Polese, F., Caputo, F., Carrubbo, L., & Sarno, D. (2016). *The value (co)creation as peak of social pyramid.* In 26th Annual RESER Conference, "What's ahead in service research: New perspectives for business and society", University of Naples "Federico II", Italy, September 8–10.

Rip, A., Misa, T. J., & Schot, J. (Eds.). (1995). *Managing technology in society.* New York and London: Pinter Publishers.

Saha, D. & Mukherjee, A. (2003). Pervasive computing: A paradigm for the 21st century. *Computer, 36*(3), 25–31.

Saviano, M., Caputo, F., Formisano, V., & Walletzký, L. (2016). From theory to practice in systems studies: A focus on smart cities. In Caputo, F. (Ed.), *The 4rd international symposium advances in business management: "Towards systemic approach"* (pp. 35–40). Business systems. Avellino: E-book series.

Scuotto, V., Ferraris, A., & Bresciani, S. (2016). Internet of Things: Applications and challenges in smart cities: A case study of IBM smart city projects. *Business Process Management Journal, 22*(2), 357–367.

Streitz, N. A., Rocker, C., Prante, T., van Alphen, D., Stenzel, R., & Magerkurth, C. (2005). Designing smart artifacts for smart environments. *Computer, 38*(3), 41–49.

Turban, E., Leidner, D., McLean, E., & Wetherbe, J. (2008). *Information technology for management.* New York: John Wiley & Sons.

Von Bertalanffy, L. (1968). *General systems theory.* New York: George Braziller.

Zygiaris, S. (2013). Smart city reference model: Assisting planners to conceptualize the building of smart city innovation ecosystems. *Journal of the Knowledge Economy, 4*(2), 217–231.

Beyond Big Data

From *smart* to *wise* knowledge management

Elias G. Carayannis,[1] *Manlio Del Giudice,*[2]
Marialuisa Saviano[3] *and Francesco Caputo*[4]

Keywords: *Big Data; knowledge management; systems thinking; smartness; wisdom*

1   Introduction and aims

Ikujiro Nonaka and Hirotaka Takeuchi are widely recognized as among the most impactful scholars in the field of knowledge management. Not surprisingly for us, one of their recent works, published in *Harvard Business Review* in 2011, was listed as a contribution about 'Ethics'. It was titled "The Big Idea: The Wise Leader" and the incipit sentence was as follows:

> In an era when discontinuity is the only constant, the ability to lead wisely has nearly vanished. All the knowledge in the world did not prevent the collapse of the global financial system three years ago or stop institutions like Lehman Brothers and Washington Mutual from failing.

Hence, what Nonaka and Takeuchi considered 'big' was the idea of embedding leaders with a wisdom that was clearly long lacking. The impact of their message has been probably lower than that of their previous works. In fact, in the subsequent years, attention has been increasingly put on the challenge of managing ever bigger amounts of data; accordingly, the topic of 'Big Data' is attracting increasing interest.

Just one year later, in 2012, Andrew McAfee and Erik Brynjolfsson, discussing Big Data as a 'Management Revolution' and envisioning 'Smart leaders across industries' dealing with this revolution, argued that: "as with any other major change in business, the challenges of becoming a Big Data-enabled organization can be enormous and require hands-on – or in some cases hands-off – leadership" (McAfee et al., 2012: 62). We would put the emphasis on the 'hands-off' word, as we recognize and deeply agree not only on the revolutionary impact of Big Data on management but mainly on the necessity to change the 'decision-making culture' of organizations (McAfee et al., 2012: 61). Hence, the discussion shifts again to different requirements, like a change in 'culture' and in the management 'style', that do not currently appear adequately embedded into the knowledge management framework, especially considering a major trend like that of 'Big Data'.

Undoubtedly, data management represents one of the most relevant challenges both for decision makers and researchers interested in the governance of social and economic dynamics (Yew Wong & Aspinwall, 2005; Espejo & Reyes, 2011; Barile & Saviano, 2011; Chen et al., 2013). Fast progress in Computer Science is increasing opportunities to acquire and manage big amounts of data in a really short time. Of course, although requiring appropriate techniques and tools (Maimon & Rokach, 2005; Carayannis & Campbell, 2006a; Carayannis et al., 2006b; Hey et al., 2009; Del Giudice et al., 2016), almost any disciplinary domain has an interest in the use of data that are at the basis of information and knowledge management (Beer, 1979; Espejo, 1996; Carayannis, 1999, 2010; Del Giudice et al., 2012; Di Nauta et al., 2015; Caputo et al. 2016; Caputo & Walletzký, 2017). However, it seems that this increasing focus on producing new data (i.e. the means) is attracting more interest than solving the problems (the goals) (Linoff & Berry, 2011): new data are produced to explain other data, giving rise to a circular dynamic that may not help solve the problems (Boyd & Crawford, 2012). The capability of humans to solve problems appears constantly improved despite the evidence of big failures in the most relevant human issues. This situation is well synthesized by Schumacher who, analyzing the 'problem of production' in the modern era, argued that

> the changes of the last twenty-five years, both in the quantity and in the quality of man's industrial processes, have produced an entirely new situation – a situation resulting not from our failures but from what we thought were our greatest successes.
>
> (Schumacher, 2011: 6)

This is the reality of our ever-smarter knowledge society. In this context of reflection, we wonder:

> *Given the complex nature of the main issues to face on our planet, are ever-bigger amounts of data and ever-smarter technologies what we only need? Would managing Big Data really help to better solve complex humans' problems?*

Acknowledging the increasing relevance of managing Big Data, this paper does not neglect their importance as well as the value of 'smart' technologies to efficiently manage them. However, given the emblematic evidence highlighted by Nonaka and Takeuchi about the uselessness of all the knowledge available in the world to prevent the last collapse of the global financial system, the aim of this work is to stimulate reflections upon the kind of knowledge that is required to first prevent and then solve the big issues of humanity, i.e. complex issues for which decision makers seem to lack the support of useful information (Barile, 2009b).

2 Discussion

By adopting a systems thinking view through the interpretative lens of the Viable Systems Approach (Golinelli, 2010; Barile et al., 2013, 2016), this paper states the necessity to shift focus from the objective and quantitative management of data to the subjective and qualitative understanding of problems, when dealing with complex issues. On this basis, the paper investigates the notions of *bigness*, *smartness* and *wisdom* in the context of knowledge management, outlining the key elements of a general interpretation framework.

The pathway of knowledge creation under focus is viewed as a process through which decision makers find solutions to new problems, highlighting that more information (hence more data) are useful to solve new (never experienced) problems, only in the case in which the decision maker has the capability to correctly manage and interpret them (Barile, 2009a; Saviano & Caputo, 2012; Barile et al., 2014; Calabrese et al., 2017). A very attractive and surprisingly simple curve can show this dynamic of knowledge as a process of progressive shifts from conditions of chaos to complexity to complication to certainty: the knowledge '4Cs' curve (Figure 57.1) (Barile, 2009b). Any decision maker as well as individual can easily experience this kind of cognitive dynamic when facing a problematic situation. The curve shows that when facing conditions of complexity/chaos, incoming units of information (data collected) do not help to solve the problem and increase the variety generating entropy and chaos. This process keeps on until something happens (abduction) that supports the effective understanding of the problem by directing towards a possible interpretation (induction), hence explanation and/or solution (deduction).

The model suggests that in conditions of chaos or complexity, collecting information is useless. Information and data are useful only when an interpretation scheme exists that has proven to be effective (Barile & Saviano, 2017).

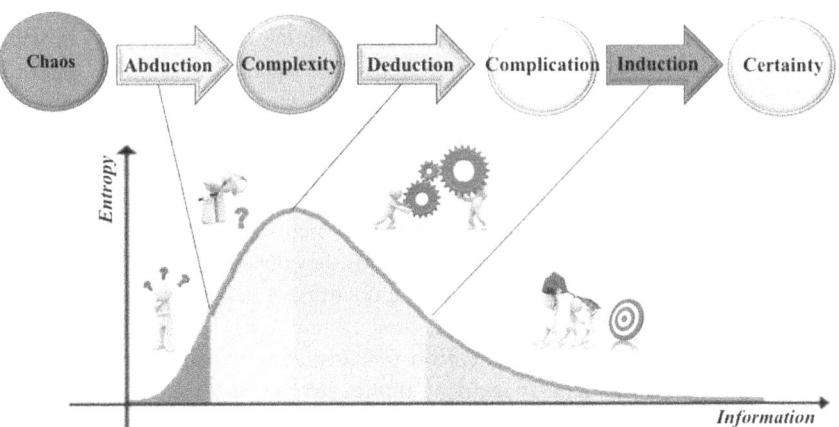

Figure 57.1 The 4Cs knowledge curve

Source: Authors' elaboration on Barile (2009b), www.asvsa.org

According to the interpretative framework of the Viable Systems Approach (Golinelli, 2010; Barile et al., 2013, 2016), a decision in the condition of chaos/complexity is made on the basis of what the decision maker 'feels' to do: essentially, he/she decides guided by his/her emotions and strong beliefs not by the information collected and processed as in problem-solving contexts. Therefore, it is fundamental to distinguish between *problem solving* and *decision making*: the former can be faced simply relying on information processing, as solutions exist and are of consolidated effectiveness; the latter instead requires making decisions without the support of information or consolidated interpretation schemes. According to the Viable Systems Approach, viable systems' dynamics of interaction can be interpreted in terms of cognitive processes that imply variations of the variety possessed. Viable systems, in fact, are represented through their 'Information Variety' that is a multi-dimensional construct made not only of Information Units, but also of Interpretation Schemes and Categorical Values (Barile, 2009b). More specifically (Barile, 2009b):

- *Information units* represent the 'structural' composition of knowledge in terms of the total amount of data held by the system.
- *Interpretation schemes* represent the system's organization of information variety.
- *Categorical values* represent the system's values and strong beliefs that affect the personality and identity of the system over the time.

By distinguishing between these three dimensions, the Viable Systems Approach highlights the deep differences that characterize the cognitive process in different problematic contexts, indicating the variables relevant for decision.

Building upon this interpretation, the traditional knowledge management framework can be integrated by highlighting the action of different approaches that go beyond the scope of data and information management. These approaches highlight the knowledge potential related not so much to the 'quantity' of data objectively available (information units) but to the 'quality' of decision makers' cognitive capabilities (interpretation schemes) and, what is more interesting, to his/her values systems and strong beliefs (categorical values). Thus, the knowledge creation potential is much more linked to the individual's capabilities than to the availability of data however 'Big' they are.

In line with this interpretation, the subjective dimension of knowledge management processes becomes more apparent and justifies that the potential of knowledge management is not linked only to the quantity of data available and to the technological power of processing data.

Introducing this subjective dimension of knowledge management allows us to enrich the knowledge management framework by distinguishing different approaches on the basis of the decision makers' knowledge potential (resources, competences, capabilities, experiences, and values) expressed through the dimensions of the information variety model. Accordingly, by intertwining the problematic context conditions (certainty, complication, complexity/chaos) and the dimensions of the information variety, it is possible to distinguish between situations in which the keys to effective knowledge are (Figure 57.2):

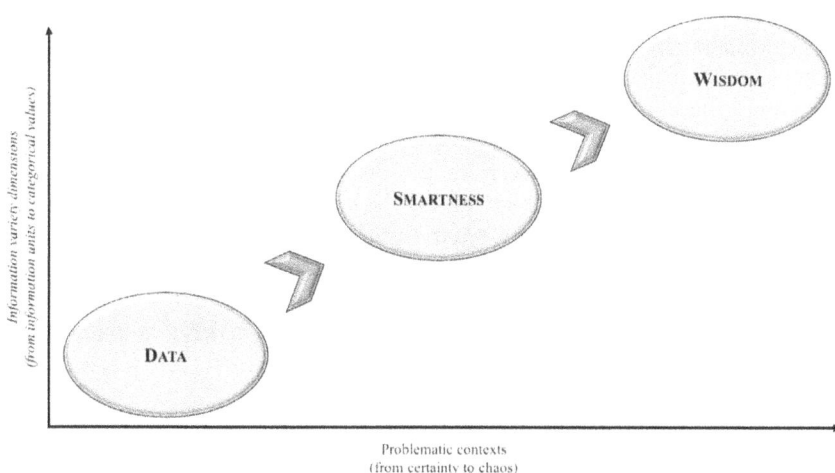

Figure 57.2 Problematic contexts and key leverages of knowledge management
Source: Authors' elaboration

Table 57.1 A *vSA*-based taxonomy of knowledge management approaches

Information variety dimension KM approaches	Information units	Interpretation schemes	Categorical values
Data processing (Bigness)	⬭	⬭	⬭
Problem solving (Smartness)	⬭	⬭	⬭
Decision making (Wisdom)	⬭	⬭	⬭

Source: Authors' elaboration on Barile (2009b)

1   Availability and processing capacity of *data*;
2   The *smartness* of the decision maker and technology;
3   The *wisdom* of the decision maker.

   In Table 57.1, different approaches to knowledge management can be further distinguished on the basis of the degree of relevance of each of the three dimensions of Information Variety to act upon when facing problem solving (certainty/complication) or decision-making (complexity/chaos) problematic contexts.

3   Conclusions

In conclusion, by essentially suggesting to shift focus from the quantitative management of data to the qualitative understanding of problems, the paper underlines the necessity to put attention on the decision maker's capabilities of performing 'smart' or even 'wise' uses of information (Spohrer et al., 2017).

Relevant implications derive from the proposed interpretation both at the theoretical and practical level highlighting the importance of the decision maker's knowledge potential not only related to the amount of information accessible, to the processing power of available technologies and to the humans' technical competences (on which the highest attention is currently put), but also to humans' personal endowment of capabilities, experiences, values, beliefs that uniquely and more deeply qualify his/her identity and personality.

Notes

1 The George Washington University School of Business, Washington, United States, caraye@gwu.edu.
2 Link Campus University, Rome, Italy, m.delgiudice@unilink.it.
3 University of Salerno, Italy, msaviano@unisa.it (*corresponding author*).
4 Masaryk University, Czech Republic, fcaputo@mail.muni.cz.

Bibliography

Barile, S. (2009a). *The dynamic of informative varieties in the processes of decision making.* In The 3rd International Conference on Knowledge Generation, Communication and Management, Orlando, FL, July.

Barile, S. (2009b). *Management sistemico vitale. Decidere in contesti complessi. Parte Prima.* Torino: Giappichelli.

Barile, S., Carrubbo, L., Iandolo, F., & Caputo, F. (2013). From "EGO" to "ECO" in B2B relationships. *Journal of Business Market Management*, 6(4), 228–253.

Barile, S., Lusch, R., Reynoso, J., Saviano, M., & Spohrer, J. (2016). Systems, networks, and ecosystems in service research. *Journal of Service Management*, 27(4), 652–674.

Barile, S. & Saviano, M. (2011). Foundations of systems thinking: The structure-system paradigm. In Vv. Aa., *Contributions to theoretical and practical advances in management: A Viable Systems Approach (VSA)* (pp. 1–24). Avellino: International Printing.

Barile, S., Saviano, M., & Caputo, F. (2014). *A systems view of customer satisfaction.* National Conference "Excellence in quality, statistical quality control and customer satisfaction", University Campus "Luigi Einaudi", University of Turin, September 18–19.

Barile, S. & Saviano, M. (2017). Complexity and sustainability in management: Insights from a systems perspective. In Barile, S., Pellicano, M., & Polese, F. (Eds.), *Social dynamics in a system perspective.* New York: Springer.

Beer, S. (1979). *The heart of enterprise.* New York: John Wiley & Sons.

Boyd, D. & Crawford, K. (2012). Critical questions for big data: Provocations for a cultural, technological, and scholarly phenomenon. *Information, Communication & Society*, 15(5), 662–679.

Calabrese, M., Iandolo, F., Caputo, F., & Sarno, D. (2017). From mechanical to cognitive view: The changes of decision making in business environment. In Barile, S., Pellicano, M., & Polese, F. (Eds.), *Social dynamics in a system perspective*. New York: Springer.

Caputo, F., Formisano, V., Buronova, B., & Walletzky, L. (2016). Beyond the digital ecosystems view: Insights from smart communities. In Vrontis, D., Weber, Y., & Tsoukatos, E. (Eds.), *Innovation, entrepreneurship and digital ecosystems* (pp. 443–454). Cyprus: EuroMed Press.

Caputo, F. & Walletzký, L. (2017). Investigating the users' approach to ICT platforms in the city management. *Systems, 5*(1), 1–15.

Carayannis, E. G. (1999). Fostering synergies between information technology and managerial and organizational cognition: The role of knowledge management. *Technovation, 19*(4), 219–231.

Carayannis, E. G. (2010). *Innovation, technology, and knowledge management*. New York: Springer.

Carayannis, E. G. & Campbell, D. F. (2006a). *Knowledge creation, diffusion, and use in innovation networks and knowledge clusters: A comparative systems approach across the United States, Europe, and Asia*. London: Greenwood Publishing Group.

Carayannis, E. G., Popescu, D., Sipp, C., & Stewart, M. (2006b). Technological learning for entrepreneurial development (TL4ED) in the knowledge economy (KE): Case studies and lessons learned. *Technovation, 26*(4), 419–443.

Chen, J., Chen, Y., Du, X., Li, C., Lu, J., Zhao, S., & Zhou, X. (2013). Big data challenge: A data management perspective. *Frontiers of Computer Science, 7*(2), 157–164.

Del Giudice, M., Caputo, F., & Evangelista, F. (2016). How are decision systems changing? The contribution of social media to the management of decisional lique-faction. *Journal of Decision Systems, 25*(3), 214–226.

Del Giudice, M., Carayannis, E. G., & Della Peruta, M. R. (2012). Culture and cooperative strategies: Knowledge management perspectives. In Carayannis, E. G., Del Giudice, M., & Della Peruta, M. R. (Eds.), *Cross-cultural knowledge management* (pp. 49–62). New York: Springer.

Di Nauta, P., Merola, B., Caputo, F., & Evangelista, F. (2015). Reflections on the role of university to face the challenges of knowledge society for the local economic development. *Journal of Knowledge Economy*, 1–19. DOI:10.1007/s13132-015-0333-9.

Espejo, R. (1996). Requirements for effective participation in self-constructed organizations. *European Management Journal, 14*(4), 414–422.

Espejo, R. & Reyes, A. (2011). *Organizational systems: Managing complexity with the viable system model*. New York: Springer Science & Business Media.

Golinelli, G. M. (2010). *Viable Systems Approach (VSA): Governing business dynamics*. Padova: Cedam.

Hey, T., Tansley, S., & Tolle, K. M. (2009). *The fourth paradigm: Data-intensive scientific discovery* (Vol. 1). Redmond, WA: Microsoft Research.

Linoff, G. S. & Berry, M. J. (2011). *Data mining techniques: For marketing, sales, and customer relationship management*. New York: John Wiley & Sons.

Maimon, O. & Rokach, L. (Eds.). (2005). *Data mining and knowledge discovery handbook* (Vol. 2). New York: Springer.

McAfee, A., Brynjolfsson, E., Davenport, T. H., Patil, D. J., & Barton, D. (2012). Big data: The management revolution. *Harvard Business Review, 90*(10), 61–67.

Nonaka, I. & Takeuchi, H. (2011). The wise leader. *Harvard Business Review, 89*(5), 58–67.

Saviano, M. & Caputo F. (2012). *Le scelte manageriali tra sistemi, conoscenza e vitalità*. In Management senza confini. Gli studi di management: tradizione e paradigmi emergenti, XXXV Convegno annuale AIDEA, University of Salerno, October, 4–5, pp. 1–21.

Schumacher, E. F. (2011). *Small is beautiful: A study of economics as if people mattered*. New York: Random House.

Spohrer, J., Bassano, C., Piciocchi, P., & Siddike, M. A. K. (2017). What makes a system smart? Wise? In Freund, L. E. (Ed.), *Advances in the human side of service engineering* (pp. 23–34). New York: Springer International Publishing.

Yew Wong, K. & Aspinwall, E. (2005). An empirical study of the important factors for knowledge-management adoption in the SME sector. *Journal of Knowledge Management, 9*(3), 64–82.

Theme IV

Democracy, transparency and social dynamics

Amanda Gregory, Daniele Bourcier,
Raul Espejo and Zoraida Mendiwelso-Bendek

Government transparency

Reality or mirage?

Noémia Bessa Vilela,[1] *Zan Jan Oplotnik*[2] *and Paulo Morais*[3]

Keywords: *governance; democracy; freedom of information; transparency; human rights*

1 Introduction

All citizens are entitled to access governmental information as they can only participate in politics if properly informed. The right to information is enshrined in a number of international agreements, including Article 10 of the Universal Declaration of Human Rights. The right of citizens to access public authorities' information is called freedom of information. This information plays a crucial role in informing the population so that they can make accurate political choices.

The World Bank and the European Union are some of the international organizations that have been actively promoting government transparency. Several countries have been adopting measures to promote transparency, namely widening the access to information and adopting accessto information legislation (Relly and Sabharwal, 2009).

One of the main reasons presented to justify transparency in governments is to curtail corruption, as openness can stop misappropriations and conflicts of interest when public money spending is displayed before the eyes of the citizens.

2 Transparency

Transparency is then used to keep governments honest (Kierkegaard, 2009). However, the success of this effort is largely dependent on the social, political and cultural environment. Anticorruption measures may be achieved by administrative reform, law enforcement or social change, when tolerant perceptions of corruption are tackled and changed (Bertot et al., 2010). In either case, the focus on information is foreseen as being essential.

Access to information can be either proactively promoted by governments, with different kinds of mechanisms of disclosure and publication, or they may simply release information on request by civil organizations, media and citizens. Some countries have launched internetbased initiatives, like egovernments or public disclosure portals for making information available to the public. In

Latin America, transparency portals, typically an initiative of the local minister of finance, are an example of how public information, in this case public financial information and procurement processes, is released (Solana, 2004; Matheus et al., 2010). The internet has greatly reduced the cost of collecting, distributing and accessing government information (Roberts, 2006), making it the master tool in this framework. In the past years, a new trend has emerged in what concerns the increasing access to government records and greater focus on the proactive disclosure of information .

Research suggests that there is a positive effect of the implementation of egovernments in downgrading corruption (Andersen, 2009).

However, transparency is not a consensual concept and neither has it been receiving consistent attention from academics, although it is a hot issue for political agents and media (Meijer et al., 2012). The importance of widely distributed and accessible government information is undeniable (Bertot et al., 2010); nonetheless several drawbacks have been pointed out to the total release of information. Bannister and Connolly (2011) list some problems that can rise from transparency in public administration, such as cost provision and the risk of data misinterpretation or inadvertent release. Other concerns are national security and personal privacy, as the unreasoned disclosure of information can jeopardize sensible information or basic human rights (Coglianese, 2009; Piotrowski and Van Ryzin, 2007).

Most of academic research has focused on government-led initiatives to promote transparency, but civil organizations and citizens are also in the forefront when it comes to making public information available to the public. Several initiatives around the world, on a solo responsibility or under the umbrella of international projects, have been collecting and presenting public data.

3   OpenSpending

OpenSpending is one of these projects, supported by the Open Knowledge Foundation, presenting datasets, monitoring and storing the financial operations of governments. Several sites under the broad name of "Where does my money go?" have been launched in countries like the UK, Japan and Macedonia.

Although these last projects have not been widely studied, the literature addressing transparency and disclosure of public information has identified some gaps to be filled in by academic research that can apply to both government and citizen initiatives, for instance, the assessment of the level and type of governmental transparency that is appropriate (Piotrowski and Van Ryzin, 2007). Other questions are yet to be answered, such as: which sectors are becoming more transparent and does transparency change the behavior of civil servants and public organizations (Meijer et al., 2012).

Challenges for academic research addressing transparency and openness in public administration have been identified: a monitoring is needed in order to identify possible negative side effects (Meijer et al., 2012); government information must survive in an accessible format and location to provide for long-term transparency (Jaeger and Bertot, 2010). On the other side, the mere disposal of information is not enough, as much of public documents suffer from complex language and other technicalities that can prevent an informed use of the information. Thus, sites displaying information must be adapted in order to guarantee usability, functionality and accessibility; there must be different levels of information, as there are diverse groups of individuals with different skills.

Notes

1 University of Maribor.
2 University of Maribor.
3 University Portucalense Infante.

Bibliography

Andersen, T. B. (2009). E-Government as an anti-corruption strategy. *Information Economics and Policy*, *21*(3), 201–210.

Anderson, J. E., & Marcouiller, D. (2002). Insecurity and the pattern of trade: An empirical investigation. *Review of Economics and Statistics*, *84*(2), 342–352.

Bannister, F., & Connolly, R. (2011). The trouble with transparency: A critical review of openness in e-government. *Policy & Internet*, *3*(1), 1–30.

Bertot, J. C., Jaeger, P. T., & Grimes, J. M. (2010). Using ICTs to create a culture of transparency: E-government and social media as openness and anti-corruption tools for societies. *Government Information Quarterly*, *27*(3), 264–271.

Coglianese, C. (2009). The transparency president? The Obama administration and open government. *Governance*, *22*(4), 529–544.

Jaeger, P. T., & Bertot, J. C. (2010). Transparency and technological change: Ensuring equal and sustained public access to government information. *Government Information Quarterly*, *27*(4), 371–376.

Kierkegaard, S. (2009). Open access to public documents: More secrecy, less transparency! *Computer Law & Security Review*, *25*(1), 3–27.

Matheus, R., Ribeiro, M. M., Vaz, J. C., & de Souza, C. A. (2010, October). Using internet to promote the transparency and fight corruption: Latin American transparency portals. *Proceedings of the 4th International Conference on Theory and Practice of Electronic Governance* (pp. 391–392). ACM.

Meijer, K., Smit, C., Beukeboom, L. W., & Schilthuizen, M. (2012). Native insects on non-native plants in the Netherlands: Curiosities or common practice. *Entomologische Berichten*, *72*(6), 288–293.

Piotrowski, S. J., & Van Ryzin, G. G. (2007). Citizen attitudes toward transparency in local government. *The American Review of Public Administration*, *37*(3), 306–323.

Relly, J. E., & Sabharwal, M. (2009). Perceptions of transparency of government policymaking: A cross-national study. *Government Information Quarterly*, *26*(1), 1

Roberts, A. (2006). *Blacked out: Government secrecy in the information age*. Cambridge, England: Cambridge University Press.

Solana, M. (2004). Transparency portals: delivering public financial information to citizens in Latin America. *Thinking out loud V-Innovative Case Studies on Participatory Instruments*. Retreived 22.9.2018 from https://dl.acm.org/citation.cfm?id=1930411

Post-agreement in Colombia
A linear pretext for a circular causation process

Germán Bula Escobar[1]

Keywords: *post-agreement; historic narrative; integrity; honor codes; moral revolutions; scapegoat archetype; bodyguard syndrome; active citizenship; appreciative processes; ICT; networks*

In 2016, the Colombian government reached a momentous agreement with the FARC guerrillas, the fate of which lies in post-agreement policies and actions that are in the eye of the hurricane, although the effectiveness of the State and government corruption continue to be society's underlying and core concerns. This paper puts forward "walking on two legs"; connecting peace-building with strengthening the integrity and effectiveness of the State is paramount and would create great synergy.

Post-conflict or post-agreement as an opening for various historic explanations

The term "post-conflict" would inaccurately suggest that signing the agreements entails the appearance of a conflict-free country. Colombia's war is without heroes and its peace is without greatness. All of this given the extreme complexity of the Colombian case, to which are tied both the degradation of the conflict and the particularities of its peace negotiations.

The post-agreement: a situation that requires and fosters dismantling corruption

The conjuncture is feasible to move the country forward, even if the guerrillas involved in the negotiations cannot be necessarily standard-bearers for change. Reviewing the changes aspired to shows that this is a case of old challenges with poor progress. Those that compromise the State run into corruption, whose dissolution requires not only structural and strategic but cultural changes.

How did people change their approach to drug trafficking and to paramilitarism?

Colombian society has been changing its "codes of honor" and today rejects both phenomena. These moral revolutions have been polycentric and have answered to no acronym or manifesto, nor to any particular leader.

The insufficiency of the usual democratic controls and the "scapegoat" archetype

Beyond the cultural change that indeed took place, the deep structures that supported, originated and financed drug trafficking and paramilitarism, and their political influence, remained almost unaltered. The role of legislature, executive, media and judiciary was insufficient. The scapegoat archetype that picks one or a few guilty parties among several, to exculpate all others in a de facto way, can be seen as a manifestation of the political archetype.

Is a "moral revolution" regarding peace and integrity possible in Colombia?

This article responds affirmatively, on condition of learning from the Colombians' own experience and using a systemic approach.

Citizenship, ICT and networks: a crucial role in the moral revolution

All these revolutions are interrelated: it is impossible to triumph in developing integrity and overcoming fear in a dissolute society that doesn't reject drug trafficking and paramilitarism; hard regimes – backed by a culture of fear – ultimately favored corruption; during armed conflicts corruption tends to play the part of "backdrop", and it emerges with strength as a "figure" only if societies experience a growing climate of peace. Equipping citizens with the milestones of their own unveiled and rationalized experience in the two "ongoing" moral revolutions must allow for the development of transformative forces from the perspective of creating environments where millions of human beings have the capacity to amplify the dynamics of change and exercise power by creating meanings that will contribute to the emergence of revolutions of integrity and peace; stimulating horizontal relationships between citizens' groups, movements and networks, appreciative processes and dialogue at every level of Colombian society. ICT coverage and social networks are improving, and they have had an important role in recent political events. Nevertheless, knowing to what extent these were used in opposition to the peace process is impossible. Potentially, this ICT progress will favor the challenges of strengthening integrity and moving towards education for peace and reconciliation.

Note

1 German Bula Escobar Consultores.

Bibliography

Appiah, K. A. (2010). *The honor code: How moral revolutions happen*. New York: W. W. Norton & Company Inc.

Beer, S. (1995). *Platform for change*. Chichester: Wiley.

El País de España. (2016). Pepe Mujica critica el modelo "gerencial" de Santos para alcanzar la paz. [Online] *Pulzo*. Available from www.pulzo.com/nacion/criticas-pepe-mujica-proceso-paz-colombia/PP142387 [Accessed 10 January 2017].

El Tiempo. (2016). "The Army has the morality and ethics to care for the FARC". Army Commander General Alberto Mejía said since August 2016. August 25.

El Tiempo. (2017). "Paz territorial y regional. Llevamos varias décadas pensando y construyendo región frente a la superación del conflicto". Available from www.eltiempo.com/opinion/columnistas/francisco-de-roux/paz-territorial-y-regional-francisco-de-roux-84326.

El Tiempo Express. (2016). "La del plebiscito fue la mayor abstención en 22 años. El 62,59 por ciento de los colombianos habilitados no votó". [Online] *El Tiempo Journal*. Available from www.eltiempo.com/politica/proceso-de-paz/abstencion-en-el-plebiscito-por-la-paz/16716874 [Accessed 10 January 2017].

Espejo, R. (2008). Editorial. *International Journal of Applied Systemic Studies*. Vol. 2, Nos. 1/2.

Espejo, R., Bula, G., & Zarama, R. (2001). "Auditing as the Dissolution of Corruption", *Systemic Practice and Action Research*, Vol. 14, No. 2, 139–156.

Espejo, R., and Dominici, G. (2016). "Cybernetics of value cocreation for product development. *Systems Research and Behavioral Science. Syst. Res*. Published online in Wiley Online Library. Available from wileyonlinelibrary.com. DOI: 10.1002/sres.2392

Espejo, R., and Reyes, A. (2011). *Organizational systems: Managing complexity with the viable system model*. London: Springer.

Government of Colombia. (2014). *Open government partnership*. [online] Available from www.opengovpartnership.org/country/colombia [Accessed 10 January 2017].

Hardin, G. (1968, December 13). "The Tragedy of the Commons", *Science*, Vol. 162, No. 3859.

IBA Global Insight. (2016, August/September). "Global Leaders Chair of the OECD Working Group on Bribery". In this interview, Drago Kos says: "Nowadays, we speak less and less about fighting corruption: We speak more and more about enhancing integrity". Available from www.ibanet.org.

Mead, M. [Online] *Brainy Quote*. Available from www.brainyquote.com/quotes/authors/m/margaret_mead.html [Accessed 10 January 2017].

Pecaut, D. (2001). *Guerra contra la Sociedad*. Colombia: Espasa.

Redacción Llano SIE7EDÍAS. (2016). "Por preacuerdo con la Fiscalía se mueve el caso del juez de tierras". [Online] *El Tiempo Journal*. Available from www.eltiempo.com/

colombia/llano-7-dias/corrupcion-en-restitucion-de-tierras/16688718 [Accessed 10 January 2017].

Semana (2017). *10 fórmulas para vencer la corrupción*. Availab le from www.semana.com/nacion/articulo/10-formulas-para-vencer-la-corrupcion/522734

Taylor, L. (2016, October 31). "In defense of Netflix's 'Narcos' and its depiction of Colombia". *The City Paper Bogotá*, Retreived 22.9.2018 from https://thecitypaperbogota.com/opinion/defense-netflix-narcos/15255

Von Foerster, H. (1996). *Las semillas de la cibernética*. Barcelona: Gedisa.

Managing the commons – corporate governance perspective

Jernej Belak[1] and Andreja Primec[2]

Keywords: *managing the commons; profit vs. non-profit organizations; natural resources; water*

1 Introduction

There is no doubt that the commons constitute one of the most awkward issues on the national (local) and international (global), political and civil society level. Inside the EU institutions, numerous discussions are held, such as the discourse of the commons in respect of the TTIP, evaluation of EU mechanisms to encourage innovation for medicines, etc.

Civil society suggests that Internet, copyright, health and climate as shared resources have to be established as part of the EU knowledge policy.

It is important to recognise that commons are necessary for everyone's wellbeing and even more, in the long run, for humankind's existence. Consequently, governing the commons is the essential question. Not only to manage these vital and limited resources efficiently, but also to achieve environmental and social sustainability by promoting participation and equitable access for people, being much more than ordinary individual consumers. Therefore, the efforts for the best possible way of managing the commons does not concern just state and local governments and their institutions, organisations, etc., but also individuals and business entities (corporations). As already Aristotle doubted the common consciousness of humans, people care much more for private ownership than for the common, it is necessary to recognise the commons as precious value.

How to implant the awareness of their collective planetary right to all the participants in this circle? It is true that the right to free access to the commons belongs to everyone, but the system of exploring those shared sources has to be legally settled, "while freedom and government have been considered opposites, the former generally is not safe without the latter" (Dietze, 1973). How to protect the commons with democratic tools – like legal rules? Which form of legal rule is appropriate for this purpose? Should the classic law or modern codes (of ethics for example) be chosen? How to influence the behaviour of corporations, which carry a substantial share of general social responsibility on commons exploitation?

These are basic research questions explored in this paper. For the purpose of our contribution, we will present the importance of the individual values in the process of influencing the values of corporations. Furthermore, we will argue that in order to be able to protect the right for commons, the appropriate legislation should be adopted at the national as well as international levels. The argumentation thinking in this way will lead us to the important question of business ethics and the impact of corporations and their ethical behaviour to the problem of managing the commons in general. Therefore, we will also address the questions and topics as to who is responsible for the management of commons in corporations, the role of corporate governance institutions in managing the commons and the quality of corporate governance in respect to managing the commons.

In a context of managing the commons we cannot avoid the question of how and in what institutional form the commons are managed best. Therefore, the first part of the paper focuses on the issues concerning the legal form of a company, in accordance with the principles ensuring the best governance as well as protection of the commons. The second part of the paper, however, deals with an example of the practice from Slovenia that is well on the way to developing a model of managing water as a public good, whereby the right to water is everybody's right and is protected by the highest legal act, i.e. the Constitution of the Republic of Slovenia.

2 Defining the problem

Scientists, researchers and academics from the field of corporate governance and management are also dealing with the issue of managing the commons. The most topical questions are primarily those on how and in what formal and institutional way managing the commons should be carried out and what starting points are required for managing the commons in accordance with the principles of specific and selected institutional forms. The first issue encountered in the considerations about the institutional and legal form of managing the commons is the selection and justification for the choice between companies as profit-oriented organisations and non-economic organisations aimed to perform a non-profit-making activity.

The major principles underlying the governance of companies operating under normal market conditions are becoming increasingly important also in Slovenia. The process, from the development of the vision, mission, goals to the determination and implementation of strategies as well as their modification and supervision is an integral part, despite certain differences in details, of ever more frequently used national and international models and concepts of management in our territory (e.g. Belak Ja. et al., 2003; Belak, Ja., 2010; Bleicher, 1994; Kralj, 2001; Rüegg-Stürm, 2002; Pučko, 2006; Tavčar, 2006; Thompson, 2001).

A general conclusion drawn from the above-indicated management concepts and models may be that a company is (also) an interest organisation and as a result, the coordination of the stakeholders in a company is key to its short-term success as well as to its long-term existence and development. The achievement of the latter depends mainly on how successful the managers (owners) of a company are in developing the vision and policy of the company (mission, purpose of existence and operation of the company and defining the main orientations in the operation of the company), with which all key officers identify themselves. The same applies for, the citizens, in the case of managing the commons.

Due to the increasingly frequent requests for greater efficiency of the public sector, in some developed democracies (primarily the USA, the UK, New Zealand and Australia), a new mindset started to be created in the second half of the 1980s concerning the management in the public sector under a common name of New Public Management, supposed to make the public service more "businesslike" by using private sector management models (Pečar, 2001).

The issues concerning the transfer of contemporary management models and concepts to the public sector are still topical (Chait, Ryan, & Taylor, 2005). This is confirmed by the conclusions of the authors who also focused on the management in the public sector after initially analysing the general management models (e.g. Mintzberg, 2000; Tavčar, 2005; Thompson, 2001). In spite of the time distance, Drucker's claim still holds for the public sector, namely that due to the complexity formed by such organisations, adjustments are required in the development of their policies and strategies depending on particular circumstances (1993, in Pečar, 2001). This is further highlighted by other authors. Tavčar (2006) mentions the need for criticism and selectivity when moving the general management ideas and concepts into the non-profit-making organisations, whereas the necessary consideration, taking into account the so-called situation factors in the transfer of management ideas into specific environments is also stated by Kropfberger (1998), while Kralj (2001, 2003) draws attention to the necessary accommodation of the general model of conducting the policy of a company to non-profit-making organisations.

Such specific circumstance applies to certain utility companies, where economic and non-economic activities are interlinked. Basically, these are organisations, companies, formally operating as companies governed by private law, although often their founders and majority (or even sole) owner is the local community or the state. Their core (in many companies even sole) activity is provision of services of public interest. In terms of the structure of the participants and the coordination of their interests, these companies constitute comprehensive complex systems, resulting in their governance and management being less predictable if compared to ordinary companies and the use of established and standardised approaches to management often impossible, especially under changing circumstances (Pučko, 2006).

Furthermore, the difficulty of overcoming such complexity is reflected in the recent developments in Slovenia regarding the management of organisations and companies of the so-called "public sector", as well as some others with the majority state ownership. Irrespective of the documents adopted in the recent period by the Slovenian Sovereign Holding, it is possible to agree, following a more detailed analysis, with Pučko's conclusion, namely that what is particularly noticeable is a lack of the process of developing a company's vision and policy (Pučko, 2006), or of strategic awareness as this is called by Thompson (2001). As a result, this causes frequent controversy in the public on whether the state is at all capable of acting as a responsible owner, as it often seems that the state lacks knowledge to perform the management functions in companies owned or co-owned by the state or that it lacks clear interests and objectives in relation to these companies (Jeras & Belak, 2014).

As regards these companies, it should be noted that difficulties in their management are also reflected in their performance. When seeking the possibilities of remedying the situation over the long run, frequently a change in the ownership structure or privatisation tries to be used as an appropriate solution. This should result in the companies with "non-optimal" ownership being transformed into the "normal" companies with known ownership, thus changing the governance arrangements. Typically, the difficulty resulting from that transformation is that the key stakeholders in a company are against it, namely both internal (mainly the employees) and external (users of services), since it is impossible to say whether in the event of managing the commons, all (interested) users would have the right to use the commons.

3 Conclusion

In view of the above, our opinion is that the decision of the Slovenian legislator was correct, because due to the significance of this natural resource, its management should under no condition be left to private companies. The primary purpose of private companies is to generate profit and that is absolutely out of the question as regards the management and exploitation of water resources. By enshrining the right to drinking water in its Constitution, Slovenia demonstrated its attitude towards this good and laid the foundation, at the same time, for equitable and efficient management of water resources. In our view, the position that the water regulatory framework belongs primarily to the state is correct; however, the ownership over water resources does not belong to the state and water still remains a public good. Moreover, water services may solely be provided by public companies in the form of services of general economic interest, which, naturally, means pursuing other values rather than making money. Management of public companies and ownership of the stakes in these companies, on the other hand, is exclusively in the hands of the state or local communities.

Notes

1 University of Maribor.
2 University of Maribor.

Bibliography

Aristoteles (350 B.C.E.). *Politics*, Book 2, Part III. http://classics.mit.edu/Aristotle/politics.2.two.html (19.1.2017)
Badelt, Christph. *Handbuch der Nonprofit Organisation: Strukturen und Management*. Stuttgart: Schäffer-Poeschl, 1997.

Belak, Janko. *Politika podjetja in strateški management.* Izv. Zbirka management in razvoj, 9. knjiga. Maribor; Gubno: MER Evrocenter, January, 2002.

Belak Janko (2010). *Integral Management MER Model.* Maribor: MER. (in Slo).

Bleicher, Knot (1994). Normatives Management: Politik, Verfassung und Philosophie des Unternehmens. Frankfurt: Campus Verlag.

Bovaird, Tony in Elke Löffler. *Public Management and Governance.* London: Routledge, 2003.

Brezovnik, Boštjan. "Aktualna vprašanja ureditve javnega podjetja v Sloveniji." *Lex Localis – Revija za lokalno samoupravo* 7.2 (2009): 177–195.

Chait, Richard P., William P. Ryan in Barbara E. Taylor. *Governance as Leadership: Reframing the Work of Nonprofit Boards.* Hoboken, NJ: John Wiley, 2005.

Dietze, Gottfried. *Two concepts of the rule of law.* Indianopolis: Liberty Fund, 1973.

Drucker, Peter F. (1993). *The Practice of Management,* Harper and Brothers, New York.

Duh, Mojca. *Upravljanje podjetja in strateški management.* Ljubljana: IUS Software, GV založba, 2015.

Fung, Archon. "Varietes of Participation in Complex Governance." 2006. *Harvard University: Public Administration Review.* 31 December 2015. http://unpan1. un.org/intradoc/groups/public/documents/un-dpadm/unpan039946.pdf.

Hauc, Aleš. "Načela korporativnega upravljanja OECD – Uvod." *OECD.* Ljubljana: SOCIUS, 2009. 3–5.

Jelovac, Dejan. "Odisejada krmarjev neprofitnega sektorja." *Jadranje po nemirnih vodah menedžmenta nevladnih organizacij.* Ured. Dejan, Jelovac. Ljubljana. Koper: Radio Študent, Študentska organizacija Univerze v Ljubljani, Visoka šola za management v Kopru, 2002. 11–27.

Jeras, Roman in Jernej Belak. "Etika in upravljanje v podjetjih v večinski državni lasti." *Finance* 222/2014 (2014).

Kovač, Bogomir. "Ko Butale postanejo Tepanje." *Mladina* (2015): 25.

Kovač, Polona. "Javna uprava v znamenju ljudi." *Teorija in praksa* 37.2 (2000): 279–293.

Kralj, Janko. *Management: temelji managementa, odločanje in ostale naloge managerjev.* Koper: Visoka šola za management Koper, 2003.

———. *Temelji managementa in naloge managerjev.* Koper: Visoka šola za management Koper, 2001.

Kropfberger, Dietrich. "Sistem managementa kot kritični dejavnik uspešnosti v razvoju podjetij – z vidika teorije kontingence." Belak, Ja. in soavtorji. *Razvoj podjetja in razvojni management.* Izv. Zbirka management in razvoj, 1. knjiga. Gubno: MER Evrocenter, 1998. 87–93.

Mintzberg, Henry in Jacques Bourgeault. *Managing Publicly.* Toronto: The Institute of Public Administration of Canada, 2000.

Mulej, Matjaž. "Wast ist – was soll der Übergangsprozeß." Belak, Janko in Štefan Kajzer. *Unternehmen im bergangsprozeß zur Marktwirthschaft.* Wien, 1994. 7–17.

OdSUKND. "Odlok o strategiji upravljanja kapitalskih naložb države." *Uradni list RS* 53/2015 (2015).

Pečar, Zdravko (2001). *Management v javnem sektorju. Ljubljana*: Univerza v Ljubljani – Visoka upravna šola (in Slo).

Peklar, F. Leonardo in Gorazd Podbevšek. "Načela korporativnega upravljanja OECD – Predgovor." *OECD. Načela korporativnega upravljanja OECD.* Ljubljana: SOCIUS, 2008. 3–4.

Pučko, Danijel. *Strateško upravljanje.* Ljubljana: Ekonomska fakulteta., 2006.

Rüeg-Stürm, Johannes (2002). Das Neue St. Galler Management-Modell. Grundkategorien einer integrierte Managementlehre: *Der HSG-Ansatz*. Bern: Haupt. (in Ger).

Rus, Andrej. "Upravljanje podjetjih v državni lasti: Slovenska reforma v luči vstopa v OECD." *Teorija in praksa* 1.44 (2011): 25–44.

SDH. "Merila za merjenje uspešnosti poslovanja družb s kapitalsko naložbo države." 2015. 27 February 2016. www.sdh.si/doc/Upravljanje/MERILA%20 ZA%20MERJENJE%20USPE%C5%A0NOSTI%20DRU%C5%BDB%20S%20 KAPITALSKO%20NALO%C5%BDBO%20DR%C5%BDAVE.pdf.

———. "SDH – Vizija, poslanstvo, strateške usmeritve in vrednote." 2014. *Slovenski državni holding.* 9 February 2015. www.sdh.si/sl-si/o-druzbi/poslanstvo%2c-vizija.

Tavčar, Mitja. *Management in organizacija – Sinteza konceptov organizacije kot instrumenta in kot skupnosti interesov.* Koper: Univerza na Primorskem, Fakulteta za management, 2006.

Thompson, J. L. (2001). *Understanding corporate strategy.* Cengage Learning EMEA.

VRS. "Vlada Republike Slovenije." 2009. *Politika upravljanja podjetij vdržavni lasti.* 26 October 2015. www.mf.gov.si/fileadmin/mf.gov.si/pageuploads/ Finan%C4%8Dno_premo%C5%BEenje_in_poro%C5%A1tva/Kapitalske_ nalo%C5%BEbe_RS/Politika_upravljanja_podjetij_v_dr%C5%BEavni_lasti/ sprejeta_politika_upravljanja_podjetij.pdf.

———. "Vlada Republike Slovenije." 2012. *Sporočilo za javnost.* 26 October 2015. www.vlada.si/fileadmin/dokumenti/si/Sporocila_za_javnost/sporocilo_AUKN. doc.

ZBS-1-UPB1. "Zakon o Banki Slovenije." *Uradni list RS* 2006.

ZGD-1. "Zakon o gospodarskih družbah." *Uradni list RS* 42/2006, s spremembami in dopolnitvami.

ZGGLRS. "Zakon o gospodarjenju z gozdovi v lasti Republike Slovenije." *Uradni list RS* 9/16 2016.

ZGJS. "Zakon o gospodarskih javnih službah." *Uradni list RS* 1993.

ZPIZ-2. "Zakon o pokojninskem in invalidskem zavarovanju." *Uradni list RS* 2012.

ZZ. "Zakon o zavodih." *Uradni list RS* 1991.

An agenda for social transparency
Making sense of Big Data

Raul Espejo[1]

Keywords: *democracy; transparency; complexity; communicative action; Big Data*

Brexit and the American elections last November provide examples of the art of lies in advanced democracies. It can be argued that we encounter this problem with referenda and elections in all democracies; effective interactions between citizens, experts and policy-makers are a major challenge. Representative and participative democracies need further development to be effective. We find that there is a significant distinction between the "emotional truth" emerging in citizens' minds and the "real truth" as constructed by solid debates supported by experts, think tanks and political parties and also by the serious press. This distinction touches key aspects of communications in a complex world, today dominated by Big Data, which in practice implies that there are data overload for citizens and politicians. For both it is increasingly difficult to distinguish lies from truths. For the former, Big Data may support conflating ungrounded trends of immigration, with deciding whether or not to be part of the European Union in the UK. Politicians, also overwhelmed by data – in an uncertain world – may construct and impose their truths influenced by ideology, weak expert advice and short-term political interests. The challenge is reducing the gap between sound evidence and emotional constructions. It may be argued that it is a social responsibility, similar to that of having a Justice System, to create aiding procedures to contextualise fairly what is heard through the media and social networks. In advanced democracies, for social issues whether of global or local relevance, it is irresponsible not to challenge the arguments advanced by those forming public opinion with the filter of authenticity, legitimacy and truthfulness (Habermas, 1979).

Some may argue that the huge complexity of social processes makes unavoidable poor data management. However, this is not necessarily the case. Complexity management tools, such as variety engineering (Beer, 1979, 1985, Espejo & Reyes, 2011), should expose in daily conversations the damage produced by those

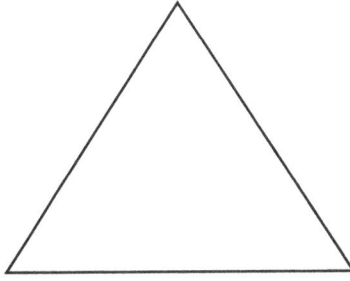

Three roots of communicative action

Truth
- Scientific methods and technology
- Efficiency: "Are we doing things right?"

Legitimacy
- Norms and interpersonal
  relations
- "Is this right and fair?"

Authenticity
- Integrity and identity expressed in
  words and actions (consistency/values)
- "Is this what I/we intend?"
  "Is it good for me/us?"
  "Am I/are we truthful?"

Figure 61.1 Three roots of communicative action
Source: Wene & Espejo, 1999

charismatic demagogues that give evidence lacking in authenticity, legitimacy and truthfulness.

Not only it is necessary to keep open checks and balances between multiple viewpoints to bridge gaps between emotional and real truths, but also it is necessary to count with the moral guidance of experts regulating on-going dialogues, offering judgements about precisely the authenticity, legitimacy and truthfulness of those constructing social opinions. These judgements of the dialogues constructing "real truths" – enmeshed in moral mazes –should be distributed throughout society; they are necessary at multiple levels from the local to the global. This proposal may appear as utopia; however, I propose that its realisation is necessary for mature democracies. This proposed utopia is an invitation to move in the direction of more transparent societies (Wene & Espejo, 1999).

Note

1 World Organisation of Systems and Cybernetics.

Bibliography

Beer, S. (1979). *The Heart of Enterprise*. Chichester: Wiley.

Beer, S. (1985). *Diagnosing the System for Organizations*. Chichester: Wiley.

Espejo, R. & Reyes, A. (2011). *Organizational Systems: Managing Complexity with the Viable System Model*. Heidelberg and New York: Springer Verlag.

Habermas, J. (1979). *Communication and the Evolution of Society*. Boston: Beacon Press.

Wene, C. & Espejo, R. (1999, June). A Meaning for Transparency in Decision Processes. In K. Andersson (Ed.), *Proceeding of Conference on Values in Decisions on Risk*. Stockholm: Valdor.

Websites

http://wosc2017rome.asvsa.org/index.php

http://wosc.co

Systemic performance for human/citizen/managers
Pragmatizing systems thinking through forum theatre

Tom Scholte[1]

Keywords: *Critical Systems Heuristics; boundary critique; enactive management; Forum Theatre*

1 Introduction: citizen-managers and the need for a didactic pragmatizing systems tool

Over the course of several papers detailing the main features of his Critical Systems Heuristics, Werner Ulrich puts forth a "new understanding of management" in which "competent management has something to do with competent citizenship" (Ulrich 2000, p 12). The particular competencies he identifies that "professionals, like citizens, need to develop" are "reflective" in nature (Ulrich 1998). Similar themes can be traced through the work of Osvaldo Garcia de la Cerda and Maria Soledad Saavedra Ulloa on their "Enactive Management" training programme. They remind us that

> one of the starting points required to recognize and understand the managerial dilemma is the fact that the manager is not only an instrument, a resource, a well-trained element, or at best a professional, but also – and foremost – he is a human being.
>
> (Cerda 2009)

The "human re-engineering" they have developed is designed to "create enactive meta-observers able to open new possibilities of action in different organizational contexts and domains." This is achieved by providing "embodied learning tools" enabling these (human) managers "to observe the observers they are, the distinctions they make when they interact with others, and, as a result, the reality they produce" (Cerda 2009).

For Ulrich, the "critical kernel of the systems idea" entails the manner in which the "boundary judgements" we use to demarcate a "system of concern" from its environment impact upon the "facts" produced by our observations of said system and the "values" produced by our evaluations of those "facts" (Ulrich 1998). As a result, Ulrich rates the ability to "critique" such boundary judgements, including one's own, as the key reflective competency required by the citizen-managers of the reinvigorated civil society he envisions; and he laments the lack of a suitable "didactic tool" with which to train all of its members in this "critical competency." In fact, a highly effective practice fit for just this purpose, as well as the embodied explication of a host of other systems perspectives including those of de la Cerda and Saavedra Ulloa, is already operational: Forum Theatre. This chapter will rigorously bring forward the systemic awareness embedded in its methodology (but, until now, not yet fully realized) in a manner designed to pave the way for its robust deployment across the various fields of applied social science including, of course, all forms of management involving human activity systems.

2   Forum Theatre

Inspired by the critical pedagogy of Paulo Freire, "Forum Theatre" was originally developed by Brazilian director, activist and, later, city councillor, Augusto Boal as part of his programme of the "Theatre of the Oppressed." It has been developed further by, among others, Vancouver's David Diamond, whose "Theatre for Living" has moved away from the binary language and model of "oppressor/ oppressed" and now "approaches community-based cultural work from a systems-based perspective; understanding that a community is a complexly integrated, living organism" (theatreforliving.com). The Theatre for Living website describes this component of their work as follows:

> In Forum Theatre, we show the audience the play all the way through once – the play builds to a crisis, and stops, offering no solutions. The play is then performed a second time, where audience members can then stop the action and enter the stage themselves, by replacing characters with whom they identify and try to solve problems or issues inside the story. The rest of the cast stays in character and improvises. . . . The theatre becomes a creative laboratory where we can try ways to transform ourselves, our communities, and the world.
>
> *(theatreforliving.com)*

In addition to fully produced "main-stage" shows developed over several weeks, involving paid casts and professional designers and, often, performed in "soft seat" theatres, Diamond has evolved his one-week "Power Play" process through which he will take members of a community into which he has been invited to explore a particular issue of concern through an intensive workshop process culminating in a short Forum piece to be performed before other community members in school gymnasiums, community centres and other "non-professional" spaces. A substantial percentage of his work takes place within indigenous communities across western Canada and has tackled such "wicked problems" as homelessness, addiction, and family violence.

Diamond has introduced key elements of cyber-systemic theory into the work including Niklas Luhmann's notion of social autopoiesis. In Diamond's words, "[b]ecause Theatre for Living approaches the community as a living organism . . .

when plays are created, they are made to help us investigate ways to change the behaviours that create the structure, not only the structure itself" (Diamond 2007). While Diamond's use of cyber-systemic theory as a framing device is certainly illuminating, the specific and deliberate application of cyber-systemic modes of analysis have not yet been employed within his work in the development of the strategies with which participants might respond to the exceedingly complex problems portrayed. This is the critical next step that this chapter explores.

3 Systemic case study of a Forum Theatre event

In February/March 2017, the author was a creator/performer in Theatre for Living's *šx̌ʷʔamət (home); a Forum Theatre piece exploring issues* surrounding reconciliation between Canada's indigenous and non-indigenous peoples performed at the Firehall Arts Centre in Vancouver. The author's reflections upon this experience, parsed through a more deliberately robust set of systemic heuristics than were employed in its creation and performance, will provide the case study for this chapter's proposal for the ongoing evolution of a particular systemic strain of this artistic/social practice.

In their Embodied Management training programmes, de la Cerda and Saavedra Ulloa have already recognized and utilized the effectiveness of the "inclusion of artistic and ludic techniques [including elements of the Theatre of the Oppressed] as an educational methodology" to enable human/citizen-managers to develop "skills for observation, self observation, and design of new interaction" (Cerda 2009). The first of the three main sections of this chapter will analyze the author's experience portraying Doug, a non-indigenous character with deeply internalized racist assumptions, and improvising with various audience members who attempted to shift the character's behaviour in both successful and unsuccessful ways. These interventions will be analyzed through the lens of an "ontological tool" known as CLEHES, developed by de la Cerda and Saavedra Ulloa representing the "idea of the human being as a system, as a dynamic unit composed by six interactive and criss-crossing elements" within which human/citizen-managers can make new self-reflective distinctions opening up new and more positive "dispositions to act" (ibid.). Through their categories of Body (Cuerpo), Language, Emotion, History, Eros, and Silence, this analysis will unpack the cognitive, affective, and haptic distinctions of particular improvisatory feedback loops and the behavioural responses they engendered.

The second major section will apply Ulrich's Critical Systems Heuristics regarding Boundary Judgements to the perspective of the character of Doug as developed by the author, as well as a number of the play's other characters, in order to demonstrate the analytic power of the combination of the Forum actor's work in constructing a character's distinct onto-epistemology and Ulrich's tools of Boundary Critique in illuminating, among other systems perspectives, Jay Forrester's assertion that the mental models of the participants in a human activity system are, perhaps, the single most powerful determinant of that system's behaviour.

The final main section, followed by a summary conclusion, will examine some of the ways in which the insights of the previous two sections can be didactically pragmatized for audience members of the Forum event itself. One of Diamond's modifications of Boal's original methods creates space for a greater analytical role for the moderator of the evening (known as the Joker). Whatever the degree of relative "success" or "failure" with which a given audience intervention was met, Boal would move the proceedings swiftly on to the next opportunity. Diamond, on the other hand, will lead the audience and actors through a brief analysis of the event before continuing on with the action of the play. In the more systems-inflected mode of Forum Theatre proposed here, these periods of analysis open up space for educational discussions and applications of specific systems heuristics that audience members might employ in order to make effective interventions in the play's action. There is an inherent danger, however, that the Joker ends up occupying the place of "privileged observer" endowed with authority, who describes and explains reality in the only correct and binding (for everyone) way. . . . The privileged observer instructs those who stray, corrects their mistakes and guarantees the final convergence of all observations (Matuszek 2015). So, in the self-reflexive spirit of second-order cybernetics, it is proposed that the heuristics of Boundary Critique be demonstrated for the audience by turning them upon the play itself and addressing the value-laden implications of the the-atre's company's choices as to what elements of the human activity system under examination to dramatize and which to leave "off-stage." Given the traditional employment of Forum Theatre as a tool of social critique and advocacy, this becomes an ethical imperative in light of Ulrich's admonition that "critique must be grounded, otherwise it is empty" while "on the other hand, systems thinking without critique amounts to a covert use of boundary judgements, the normative implications of which are not made a subject of discussion; its claims to systemic understanding and comprehensiveness merely cover its partiality" (Ulrich n.d.). Again, *x̌ʷʔamət (home) will provide the case study for the employment of this self-reflexive critique.*

In its entirety, this chapter sets out to make a convincing case that a version of Forum Theatre specifically adapted to include self-reflexive systems heuristics could well be the powerful pragmatic and didactic tool capable of communicating the systems perspectives of Ulrich, de la Cerda and Saavedra Ulloa, Forrester, and a host of others, to the human/citizen/managers that need them most.

Note

1 University of British Columbia.

Bibliography

Based in Vancouver, Canada, Theatre for Living (Headlines Theatre) Has Grown from Boal's Theatre of the Oppressed: It Articulates a True Voice of the Living Community, Directed by David Diamond. (n.d.). Retrieved April 10, 2017, from www.theatreforliving.com/

Cerda, O. G. (2009). Human Re-Engineering for Action: An Enactive Educational Management Program. *Kybernetes*, *38*(7/8), 1332–1343. doi:10.1108/03684920910977005

Diamond, D (2007). *Theatre for Living: The Art and Science of Community-Based Dialogue*. Oxford, UK: Trafford Publishing. doi:10.1007/978-0-585-34651-9_9

Matuszek, K. C. (2015). Ontology, Reality and Construction in Luhmann's Theory. *Constructivist Foundations*, *10*(2), 203–210. Retrieved from www.univie.ac.at/constructivism/journal/10/2

Ulrich, W. (1998). *Systems Thinking as if People Mattered: Critical Systems Thinking for Citizens and Managers*. Working Paper No. 23, Lincoln School of Management, University of Lincolnshire & Humberside.

Ulrich, W. (2000). Reflective Practice in the Civil Society: The Contribution of Critically Systemic Thinking. *Reflective Practice*, *1*(2), 247–268. doi:10.1080/713693151

Ulrich, W. (n.d.). Critical Systems Thinking for Citizens. *Critical Systems Thinking*, 165–178.

Workers' self-organization
The case of Empresas Recuperadas in Argentina and Uruguay

Nathalie Colasanti,[1] *Rocco Frondizi*[2]
and Marco Meneguzzo[3]

Keywords: *Argentina; Empresas Recuperadas; factories; Uruguay; workers' self-management*

The occupation and self-management of factories by workers is not uncommon in history, with examples during the years following the Russian revolution in countries such as Hungary, Germany and Italy, later during the Spanish civil war, and again after 1968 in France, Cuba, Chile and other countries.

The most notable example in recent years is the one of Empresas Recuperadas por sus Trabajadores (ERTs), i.e. factories which are abandoned or dismissed by their owners and are occupied and managed collectively by their employees. This specific phenomenon was first registered in Argentine after the 2001 crisis, as is now present in other Hispano-American countries as well as European ones.

The process starts when the factory owner decides to close it, either because the activity is not profitable, or because he/she wants to move it elsewhere. Often the owner ensures that production cannot continue by transferring machines to other factories or by having them destroyed or not replaced when they stop working (Ruggeri, 2014b).

Workers then can decide to occupy the factory in order to preserve their employment and their salaries, which leads to self-managing the factory through assemblies and councils.

ERTs often have strong ties with the local communities, from which they can gather support and find a potential market for their products.

This paper will be organized as follows. In the first section, we will review the literature on the topic of ERTs, to understand how they work and what issues they face. Secondly, we will select two countries, Argentina and Uruguay, where the ERT phenomenon is especially strong, and where it has also come to the attention of politicians, who have regulated it with specific laws; the social, economic and politic context of these countries will be explained. Thirdly, we will compare several case studies of ERTs in both Argentina and Uruguay. The methodology is based on case studies as this is the most appropriate technique for external observers with no influence on the phenomenon; our cases contain detailed interviews with workers of ERTs, with the questions being determined by the specificities of each factory. We will try to highlight how ERTs are managed, how they compete within the capitalist production system and what their relationships are with local communities, as well as with national government. Each case study will represent a detailed description of workers' self-management and the techniques used to organize production.

The objective of the paper is to shed light on how ERTs work and what the impact and consequences of workers' self-management are, as well as to understand the cases in which occupying a dismissed factory is actually a better choice for workers than that of looking for another job. We also want to identify the main issues with self-managed factories, that workers have to face both in their families and communities and with respect to the government.

The main limitation of this research is the difficulty of creating a predictable model to regulate the life of ERTs, on the one hand because of possible shifts in politics which lead to more or less favourable governments, and on the other hand because of the ever-evolving nature of factories that are forced to compete with capitalistic counterparts while their mode of production is not aligned with capitalist theory.

Notes

1 University of Rome "Tor Vergata".
2 University of Rome "Tor Vergata".
3 University of Rome "Tor Vergata".

Bibliography

Ben-Ner, A. (1988). Comparative Empirical Observations on Worker-Owned and Capitalist Firms. *International Journal of Industrial Organization*, 6(1), 7–31.
Birchall, J., and Hammond Ketilson, L. (2009). *Resilience of the Cooperative Business Model in Times of Crisis*, Geneva, ILO.
Coraggio, J.L. (2008). *Economía social, acción pública y política (hay vida después del neoliberalismo*. Buenos Aires, Ediciones Ciccus.
Dagnino, R. (2008). *Um debate sobre a Tecnociência: neutralidade da ciência e determinismo tecnológico*. Campinas, Editora da Unicamp.
Dow, G.K. (2003). *Governing the Firm: Workers' Control in Theory and Practice*. Cambridge, Cambridge University Press.
Gómez Solórzano, M.A., and Pacheco Reyes, C. (2014). *Trabajo informal, economía solidaria y autogestión*. Buenos Aires, Ediciones Continente.
Patrón, S. (2004). Fábricas y empresas recuperadas por sus trabajadores. Una historia, una guía. Buenos Aires: Lavaca.
Ruggeri, A. (ed.). (2014a). *Informe del cuarto relevamiento de empresas recuperadas en la Argentina, 2014. Las empresas recuperadas en el período 2010–2013*. Programa Facultad Abierta, Facultad de Filosofía y Letras, Universidad de Buenos Aires.
Ruggeri, A. (2014b). *Qué son las empresas recuperadas?* Buenos Aires, Ediciones Continente.
Ruggeri, A., Novaes, H.T., and Sardá de Faria, M. (2014). *Crisis y autogestión en el siglo XXI*. Buenos Aires, Ediciones Continente.
Vieta, M., and Depedri, S. (2015). Le imprese recuperate in Italia. In Borzaga, C. (ed.), *Economia Cooperativa. Rilevanza, evoluzione e nuove frontiere della cooperazione italiana*. Euricse.

Community self-organization

Democratic confederalism and its application in Rojava

Nathalie Colasanti[1] and Marco Meneguzzo[2]

Keywords: *community self-organization; Democratic Confederalism; Kurdish movement; Rojava*

1 Introduction

The Kurdish question has been going on since the first half of the XX century, with Kurds trying to obtain national independence only to be faced with repressive responses.

In 1978, Öcalan and other Kurdish activists founded the PKK, Kurdistan's Workers Party, to fight against the Turkish state and obtain national liberation and independence; today, however, the PKK is listed as a terrorist organization by the European Union, Turkey, the US and NATO. The armed warfare between Kurds and Turkey went on until the late '90s, when Öcalan was arrested and imprisoned in solitary confinement.

Following the change in external conditions, such as the collapse of the Soviet Union, his time in jail led to the creation of a new socio-political model, both thanks to his theoretical studies and in order to provide a new alternative for the Kurdish movement, given the failure of the struggle for independence. This model, called "democratic confederalism", does not seek national independence, but the possibility for Kurds, and those of other ethnicities as well, to build democratic autonomy. The civil war in Syria, which led to the destabilization of Assad's power, has represented an opportunity for Syrian Kurds to implement their model in the region of Rojava, the area of the country that is thought to be part of the wider region of Kurdistan. Of course, this would not have been possible without the preparatory work carried out by Kurdish activists from the '70s onwards, which created the grounds for the present revolution.

2 Democratic confederalism in Rojava

Democratic confederalism is now applied in the three cantons of Rojava, and it is an example of community self-organization to obtain a collective goal, which is social change and the possibility to manage the area in a democratic way and autonomously from the Syrian nation state. The model is based on the three pillars of grassroots democracy, ecology and women's emancipation. It aims to build a society made of small self-managed communes, each with its own assembly where all members can determine the needs and duties of the community, which in turn are coordinated by general assemblies. Democratic confederalism strongly echoes the theory of libertarian municipalism built by Bookchin in the 1950s.

Politics become a key part of everyday life, as all men and women alike are involved in democratic decision-making to manage their community; the political model is organically experienced by communities, with little to no distinction between politics and other areas of life.

The model also includes economic guidelines, such as the establishment of a cooperative economy, where the goods to be produced are determined first and foremost by the needs of each community, and then by the chance to be exported in exchange for other supplies. Part of the economy should also be open to external investments, although this should happen under the strict supervision of local communities and assemblies.

The main limitation in the study of this phenomenon is that, although democratic autonomy is being applied, Syria is currently facing a civil war, which makes it much more complicated for communities to actually abide by all provisions of the model, such as ecology. At the same time, shifting political equilibria between Turkey, Russia, the European Union and the United States mean that, depending on whose opinion is prevailing, the Kurdish experiment can either be viewed as a terrorist menace or as a possibly positive new force in the area.

Notes

1 Università degli Studi di Roma "Tor Vergata".
2 Università degli Studi di Roma "Tor Vergata".

Bibliography

Akkaya, A.H., and Jongerden, J. (2011). The PKK in the 2000s: Continuity through breaks? In M. Casier and J. Jongerden (Eds.), *Nationalism and politics in Turkey: Political Islam, Kemalism and the Kurdish issue*. New York, NY: Routledge.

Graeber, D. (2016). [Foreword]. In M. Knapp, A. Flach, and E. Ayboga. (2016), *Revolution in Rojava: Democratic autonomy and women's liberation in Syrian Kurdistan*. London, UK: Pluto Press.

Gunes, C. (2012). *The Kurdish national movement in Turkey from protest to resistance*. New York, NY: Routledge.

Jongerden, J., and Akkaya, A.H. (2012). The Kurdistan Workers' Party and a new left in Turkey: Analysis of the revolutionary movement in Turkey through the PKK's memorial text on Haki Karer. *European Journal of Turkish Studies* (online), *14*.

Knapp, M., Flach, A., and Ayboga, E. (2016). *Revolution in Rojava: Democratic autonomy and women's liberation in Syrian Kurdistan*. London, UK: Pluto Press.

Knapp, M., and Jongerden, J. (2014). Communal democracy: The social contract and confederalism in Rojava. *Comparative Islamic Studies*, *10*(1), 87–109.

Küçük, B., and Özselçük, C. (2016). The Rojava experience: Possibilities and challenges of building a democratic life. *The South Atlantic Quarterly*, *115*(1), 184–196.

McDowall, D. (2007). *A modern history of the Kurds.* London, UK: I.B. Tauris.

Öcalan, A. (2010). *Democratic confederalism.* Neuss: Mesopotamian Publishers.

Özcan, A.K. (2006). *Turkey's Kurds: A theoretical analysis of the PKK and Abdullah Öcalan.* New York, NY: Routledge.

Tokar, B. (2008). On Bookchin's social ecology and its contributions to social move ments. *Capitalism Nature Socialism, 19*(1), 51–66.

Yarkın, G. (2015). The ideological transformation of the PKK regarding the political economy of the Kurdish region in Turkey. *Kurdish Studies, 3*(1), 26–46.

Subject-subject management approach

Case: "Civilian Strategy" regional development from below

Peter Ototsky and Sergey Manenkov[1]

Keywords: *civilian strategy; complexity management; innovations; digital ecosystems*

1 Introduction

The work describes a paradigm shift in the management from the subject-object approach to the subject-subject, in which each person, an employee, is considered, not as a function, but as the subject who wants to be realized (Ackoff and Emery, 1972). The subject-object approach is useful for describing the factory. The subject-subject approach is useful when considering the territories, industry sectors and digital ecosystems (Digital Business Ecosystem, 2007) in which each participant is an entrepreneur. Simple performers more and more are replaced by robots – go "below the API" (Kosner, 2015). The paper describes the project "Civilian Strategy". The project was held by public associations during election campaigns: to intensify the political activity, to gather information about the mood, problems and aspirations of the citizens. "Civilian Strategy" was a platform for dialogue between society, government and business. The project provides the ground for the simulation of control systems and self-management of society. It raised the questions – "how to include the people in management as subjects?" (not professionals, but ordinary people); "how to start the process of innovation from below?" The path from the problems and aspirations of each of the participants to the regional development strategy has been proposed and implemented. During the sessions participants primarily needed to realize the strategies of their own lives and feel themselves as citizens and residents of the city (Mulej, 2013; Dudchenko, 1996).

Management paradigm shift

Cybernetics combined the models of engineering, biology and social sciences, showed the universal model of management and communication. However, cybernetics looked at a certain angle at all three areas of knowledge, from the paradigm of its time – the XX century.

For example when a mechanical watch was invented, people expanded the model of the watch to the whole world – "God started the rotating spheres a mechanism of the Universe". Ancient people used the model of the tribal community to interpret the world – the myths, the Pantheon of gods. Today in the era of information technology, we believe that the DNA is a program code. There is a viewpoint that Plato's division between the world of ideas and the material world was a projection of the division of people into masters and slaves (Losev, 1989). Masters are carriers of meanings, but they do nothing. Slaves do everything, but don't know why.

The XX century is the industrial age of factories. A machine is a tool that processes the input raw materials, resources into products, goods and services. The factory is a large machine, in which people are considered as a human resources in one line with the money and the ore. The Soviet Union in some way was a factory increased to the level of the country.

The same paradigm is popular in big business:

- Goal setting and rules come from the top,
- People are treated as functions,
- Information technologies are used as conveyors to control and set the tempo of work. For example, the operator is given 15 minutes to process the order; if delayed he/she will be penalized.

Big business is inhuman, so managers have to hold corporate events in order to add humanity. Employers spend half of their life at work together but they need to be taken into a different context to see other colleagues as persons not their functions.

Today people are the bottleneck of development. Technology has outpaced the ability of the person. Effective interaction between networks of humans and computers, the disclosure of creative abilities of people, learning and the reproduction of knowledge is the key to development (Beer, 1981; Espejo and Reyes, 2011). It's important to get out above the technical paradigm of management (consumer view on people). Today the subject-subject view is becoming more popular and useful. Even in biology, we can see it. For example, the immune system is not seen as the security department of the body, but as the self-reproducing populations of different species of leukocytes.

On this view, it is clear that the person wants to realize his/her potential, to find his path, to change the world by his activities, not just to carry out the functions written by somebody in the instructions.

Jacobs (1961) criticized the traditional approach to the design of cities, when architects plan the life of the city in the form of functions (activities of the citizens). The author offers a higher level of abstraction approach – planning the generators and attractors of variety in the city. People with different experiences, outlooks, desires, which are collected in one place – this is the soil on which the business grows up; and the business provides all the necessary functions and quickly adapts to needs. Such a great variety of activities that exist in real life cannot be planned in advance by the architect. The variety of people attracts the business, and the variety of activities presented by business attracts more people (positive feedback).

This is a way how the successful areas bloom; the life there is in full swing with the people living active lives. The areas go into decline in a similar way. During competition there are few winners – businesses that copy themselves everywhere. The variety of activities drops down and the area becomes less attractive for the people. When the traffic of people goes away from the area, all the business networks go into decline.

Successful areas differ from the synthetic designed "ideal" areas (=reduced models). People are bored and fade. The same thing can be seen not only in geographical areas but also in large corporations. The architects planned every function, task and detail. People in such schemes are performers and the consumers must follow instructions written in the concepts.

Jacobs (1961) wrote that small cities grow up around the big business that contains the entire necessary infrastructure (kitchens, gyms, laundries, etc.). Small business cannot survive with little variety. Small business thrives in the big cities in crowded places around the attractors of variety.

In a sense, today we all live in a big city, the "Internet". Today it is much easier to raise millions of people together all over the world. A new type of business (meta-business) sells entrance into the ecosystem = access to variety. For example, Uber sells for passengers the access to a variety of drivers, and for drivers it sells access to a variety of the passengers (Kosner, 2015). Another digital company Duolingo went even further – it equates production and consumption. On the one hand the company provides the free service of teaching foreign languages in a game form. On the other hand when millions of people translate educational texts, they perform commercial translation that feeds the Doulingo company (Luis von Ahn, 2011).

Such meta-businesses are fed by a variety of users and variety of tasks. They cannot be imagined outside the "big city" of the Internet. They provide a super small business type = human to human (H2H). Everyone here is an entrepreneur working its own interests. Providers of the ecosystem take the profit of the managers, coordinators, owners of the medium companies. Compared to the traditional business for buyers (passengers), the price is lower, and for sellers (drivers) wages are getting higher. The management is robotized; it has the ability to treat a fundamentally greater amount of variety. Digital managers coordinate the activities of millions of people in real time.

Robotization of management and production is a challenge for humanity, in which we have to answer the questions "why do people work?", "why do the business work?" Many actual tasks will be moved inside the box (below the API); then people will have a chance to set more complex tasks that cannot be imagined today.

"Civilian Strategy" projects

The project "Civilian Strategy" (Kosner, 2015) was held by public associations during election campaigns: to intensify the political activity, to gather information about the mood, problems, aspirations of the citizens. "Civilian Strategy" was a

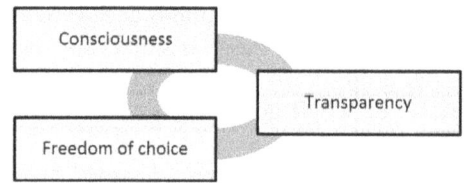

Figure 65.1 The relationships between participants

platform for dialogue between society, government and business. The project provides the ground for the simulation of control systems and self-management of society. It raised the questions – "how to include the people in management as subjects?" (not professionals, but ordinary people); "how to start the process of goal setting from below?" The path from the problems and aspirations of each of the participants has been proposed and implemented – to the regional development strategy.

During the sessions participants primarily needed to realize the strategies of their own lives and feel themselves as citizens and residents of the city (Ototsky & Manenkov, 2011).

Theoretically the project is based on Dudchenko's (1996) works where the development of the business is connected to the personal development of the participants, which in turns means the development of the relationships between participants (see Figure 65.1). The development of the organization should be connected to a personal realization of each participant as the activities inside of organization are a major part of their lives. Three factors are the key: consciousness of what you are doing, transparency in communication as a best way of defence, understanding that everyone make a choice itself.

Note

1 Systeco Ltd.

Bibliography

Ackoff, R.L., & Emery, F.E. (1972). *On Purposeful Systems: An Interdisciplinary Analysis of Individual and Social Behavior as a System of Purposeful Events.* Chicago: Aldine-Atherton.

Beer, S (1981). *Brain of the Firm*, 2nd edn. Wiley, Chichester.

Dudchenko, V.S. (1996). *Bases of Innovative Methodology.* Moscow: Institute of Sociology of RAS.

Espejo, R., & Reyes, A. (2011). *Organisational Systems: Managing Complexity with the Viable System Model.* Heidelberg, etc.: Springer.

Espejo, R., Schuhmann, W., Schwaninger, M., & Bilello, U. (1996). *Organizational Transformation and Learning: A Cybernetic Approach to Management.* Chichester: Wiley.

European Commission. (2007). *Digital Business Ecosystem*. www.digital-ecosystems. org/

Jacobs, J. (1961). *The Death and Life of Great American Cities*. Random House.

Kosner, A.W. (2015). Google Cabs and Uber Bots Will Challenge Jobs "below the API". *Forbes*. Retrieved 22.9.2018 from https://books.google.si/books

Losev, A. (1989). *History of Ancient Philosophy*. Thought

Luis von Ahn (2011). *Massive-Scale Online Collaboration*. TEDxCMU.

Mulej, M. (2013). *Dialectical Systems Thinking and the Law of Requisite Holism Concerning Innovation*. ISCE.

Ototsky, P., & Manenkov, S. (2011). Cognitive Centres: Technology for Designing the Future: Methodology and Implementation Experience. *Kybernetes* 40(3/4), 528–535.

Uexküll, J. von (1982). The Theory of Meaning. *Semiotica* 42(1), 25–82.

Punished by punishment

How model-based policies counteract delayed feedback in prison overcrowding

Camilo Olaya,[1] Juliana Gomez-Quintero[2] and Andrea Navarrete

Keywords: *system dynamics; prison overcrowding; crime; justice; simulation; public policy*

1 Prison overcrowding: a persistent social system

The WOSC 2017 Congress invites individuals "to contribute to the debate of the dynamics underpinning contemporary societal problems". One example of a pressing problem is prison overcrowding, whose numbers are rising almost everywhere, and more than ever in history, in spite of well-intended policies (Giertz & Nardulli, 1985; Kennedy, 1985; MacDonald et al., 2012; Simon, 2016; Snacken & Beyens, 1994; Woods, 2016). Pitts et al. (2014) summarize the history of policy making for improving prison overcrowding as recurrent "short-term fixes to a perpetual problem". Popular approaches reflect simplistic assumptions that typically assume a simple linear cause-and-effect thinking in which apparently the "discovery" of key factors or determinants would allow society to improve a problem. Indeed the question "what are the causes of overcrowding" ends up *framing* the way in which the problem of prison overcrowding is tackled, as several works show (e.g. Cameron, 1989; Guetzkow & Schoon, 2015; Kuhn, 1994; Lösel, 2007; Lynch, 2011; Suhling, 2003). Even previous systemic approaches seem to find difficulties (Kaufman, 1985).

In this work, we highlight "the bigger picture", the bigger systemic picture, that is. Jay Forrester (1971) underlined what he called the "counterintuitive behaviour of social systems" since such systems show a conflict between the short and long term:

> a policy which produces improvement in the short run . . . is usually one which degrades the system in the long run. . . . Likewise, those policies and programs which produce long run improvement may initially depress the behaviour of the system. This is especially treacherous. The short run is more visible and more compelling. It speaks loudly for immediate attention.
>
> (p. 122)

Such conflict ends up producing results very different from the ones intended in the first place. In short, social systems show "policy resistance": "people seeking to solve a problem often make it worse" (Sterman, 2000, p. 5). Prison overcrowding is a good example.

Why do systems surprise us? This is one of the driving questions of the conference. In this work we present a way to understand the counterintuitive dynamics of the increase of prison overcrowding in Colombia as the result of the long-term dynamics of a social system whose boundary includes the reaction of society to crime through harsher punishment. We use system dynamics both to organize knowledge and as tool for building a simulation model for identifying key systemic aspects in order to propose concrete actions for improving such situations.

2   Systemic modelling

Prison overcrowding is a permanent problem in Colombia. We start by recognizing a possible social system driving such problem. Consequently we identify actors that drive the dynamics of such a penitentiary system through their goals and actions, as a first step for identifying the operations that drive the dynamics of prison overcrowding (Olaya, 2015; Olaya & Gomez-Quintero, 2016). Key actors are: Infrastructure Directorate of the Ministry of the Interior (DIN), INPEC (National Penitentiary and Prison Institute), Congress of the Republic, Attorney General's Office, Public Ministry and Civil Society and prisoners.

Prison overcrowding in Colombia has followed an increasing oscillatory pattern over the last twenty years. Such behaviour has been boosted by a wide variety of laws that strengthen the sanctions to criminal conduct since the reaction of society and policy makers against crime tends to be anchored in static and linear thinking that leads to just "more punishment" – harsher legislation and longer prison terms (e.g. increase of the minimum and maximum terms for major crimes, obligatory preventive detention for more types of crimes, restrictions on benefits, creation of new criminal offences). The belief that more punishment leads to less crime is a perfect example of linear thinking which, in this case leads to more people in jail. In turn, prison overcrowding is confronted by policy makers with more prison infrastructure and more quota construction – a further linear thinking example. These control mechanisms constitute a negative feedback loop that involves also various material and information delays (serving of prison sentences, prison quotas construction delays, overcrowding perception by the government). However, negative feedback loops that contain material or information delays promote oscillatory patterns (Sterman, 2000); this dynamic archetype matches major feedback loops in our modelled system and therefore constitutes our working dynamic hypothesis for explaining the increasing oscillations of prison overcrowding in Colombia (Figure 66.1). In combination with those negative feedback loops, there are dominant reinforcing loops that end up boosting the oscillatory pattern; in the long-term they can act either as virtuous or vicious circles depending upon the effectiveness of inmate resocialization (Figure 66.1).

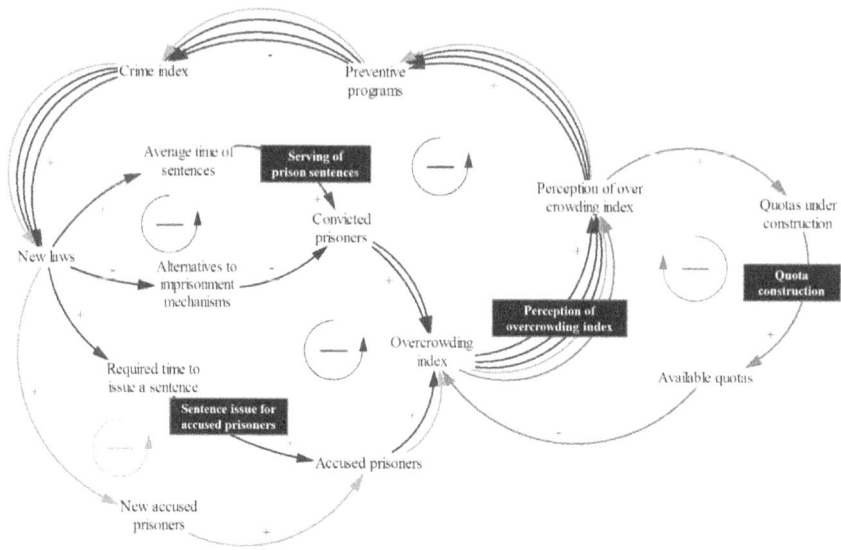

Figure 66.1 Key delayed control feedbacks
Loops are highlighted in different shades of gray.

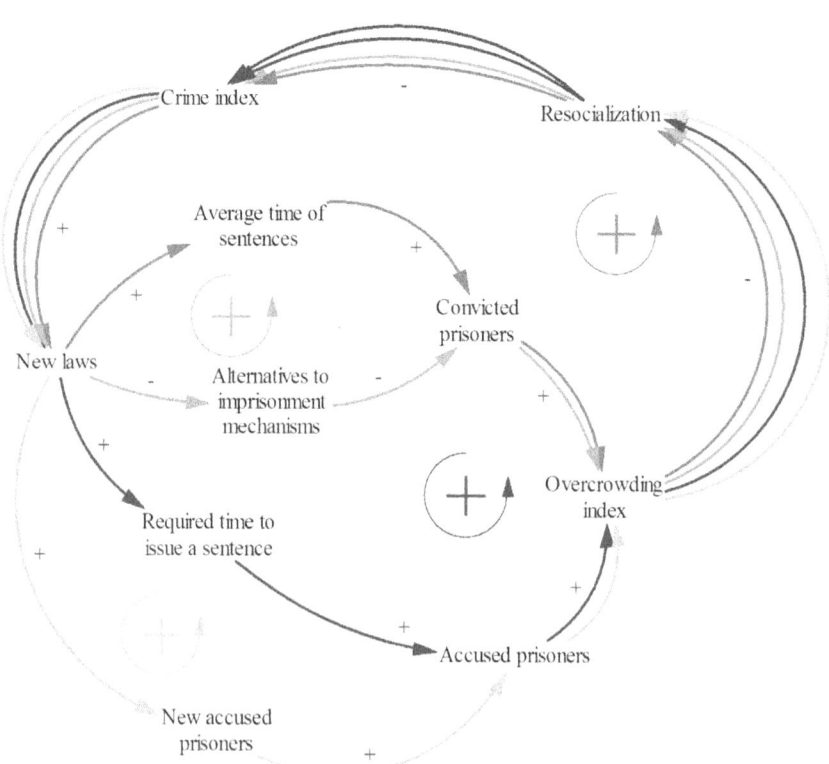

Figure 66.2 Key reinforcing feedbacks

Our work favours model-based governance. We built a system dynamics simulation model that helps to explain the dynamic hypothesis and serves as a tool for exploring and designing policies intended to tackle prison overcrowding. We tested two types of policies: (i) Punitive policies, which include limits on preventive detention – and therefore fewer accused prisoners (i.e. fewer suspects in jail that have not been sentenced) and wider access to prison benefits (non-jail alternatives to incarceration such as home detention, fines, electronic monitoring, restorative programs, etc.); both lessen reinforcing feedback and boost negative feedback of key material delays; (ii) preventive policies, such as new or better preventive programs that strengthen crime control feedback.

Simulations show better results whenever punitive and preventive policies are combined along with the reduction of delay times: faster quota building and earlier reports on crime and overcrowding, in conjunction with stronger negative feedback, reduce oscillations (and hence less overreaction to initial problematic increments) which helps to tackle prison overcrowding and therefore better resocialization and consequently less crime. In the long run preventive policies warrant the promotion of reinforcing feedback as virtuous cycles of effective resocialization and less crime as time goes by. Table 66.1 summarizes the implications for policy design whose language reflects a dynamic perspective of prison overcrowding as the outcome of a social system.

Table 66.1 Policy design implications based on policies to tackle prison overcrowding

Type of policy	*Strategy*	*Policy design implications*
Punitive policies	Limits on preventive detention	Recognition of reinforcing feedbacks that overshoot behaviour faster than delayed control loops.
	Increase of prison benefits	Reduction of delay time to improve response time to changes in behaviour.
Preventive policies	New or better preventive programs	Recognition of delays and focus of policy design on avoiding delayed decision reactions (breaking the feedback loop that responds to delayed, obsolete, information).
Combination of punitive and preventive policies	Limits on preventive detention Increase of prison benefits New or better preventive programs	Combination of stronger and faster control feedback along with long-term virtuous cycle based on effective resocialization. Preventive policies are required to be sustained through time even if immediate results are not visible. Perhaps the most challenging issue is to avoid overreaction to crime and to overcrowding (e.g. through communication campaigns, political leverage, education of the public, etc.)
Additional alternatives	Investment in reduction of delay times	Faster and more precise estimation of overcrowding index, faster prison building, etc. . . . *as much as possible.*

3   Outlook

Modelling and simulation tools boost systemic and operational thinking and allow a better recognition of social systems as drivers of societal problems. The main challenge is the improvement of the mental models of citizens and policy makers that seem to be anchored in simple cause-and-effect thinking (more punishment for combating crime; construction of prison capacity as the main answer for dealing with overcrowding) that do not involve dynamics and the long-term shocks produced by such short-term "solutions". The case of prison overcrowding in Colombia shows the relevance of systems thinking and the understanding of dynamic complexity, which should be incorporated in the cognitive portfolio of citizens and policy makers if they aspire to deal successfully with complex systems.

Notes

1 Departamento de Ingeniería Industrial, Universidad de los Andes, Colombia.
2 Programa de Ingeniería Industrial, Escuela de Ciencias Exactas e Ingeniería, Universidad Sergio Arboleda, Colombia.

Bibliography

Cameron, S. (1989). Determinants of the Prison Population: An Empirical Analysis. *International Journal of Social Economics*, *16*(8), 17–25. doi: 10.1108/EUM0000000007300

Forrester, J. W. (1971). Counterintuitive Behavior of Social Systems. *Theory and Decision*, *2*(2), 109–140. doi: 10.1007/bf00148991

Giertz, J. F., & Nardulli, P. F. (1985). Prison Overcrowding. *Public Choice*, *46*(1), 71–78. doi: 10.1007/bf00150965

Guetzkow, J., & Schoon, E. (2015). If You Build It, They Will Fill It: The Consequences of Prison Overcrowding Litigation. *Law & Society Review*, *49*(2), 401–432. doi: 10.1111/lasr.12140

Kaufman, G. (1985). The National Prison Overcrowding Project: Policy Analysis and Politics: A New Approach. *The Annals of the American Academy of Political and Social Science*, *478*, 161–172.

Kennedy, E. M. (1985). Prison Overcrowding: The Law's Dilemma. *The Annals of the American Academy of Political and Social Science*, *478*, 113–122.

Kuhn, A. (1994). What Can We Do about Prison Overcrowding? *European Journal on Criminal Policy and Research*, *2*(4), 101–106. doi: 10.1007/bf02249443

Lösel, F. (2007). Counterblast: The Prison Overcrowding Crisis and Some Constructive Perspectives for Crime Policy. *The Howard Journal of Criminal Justice*, *46*(5), 512–519. doi: 10.1111/j.1468-2311.2007.00497.x

Lynch, M. (2011). Mass Incarceration, Legal Change, and Locale. *Criminology & Public Policy*, *10*(3), 673–698. doi: 10.1111/j.1745-9133.2011.00733.x

MacDonald, M., Greifinger, R., & Kane, D. (2012). The Impact of Overcrowding. *International Journal of Prisoner Health*, *8*(1), null. doi: 10.1108/ijph.2012.62108aaa.001

Olaya, C. (2015). Cows, Agency, and the Significance of Operational Thinking. *System Dynamics Review*, *31*(4), 183–219. doi: 10.1002/sdr.1547

Olaya, C., & Gomez-Quintero, J. (2016). *Conceptualization of Social Systems: Actors First*. Proceedings of the 34th International Conference of the System Dynamics Society. Delft University of Technology, The Netherlands.

Pitts, J. M. A., Griffin, O. H., & Johnson, W. W. (2014). Contemporary Prison Overcrowding: Short-Term Fixes to a Perpetual Problem. *Contemporary Justice Review*, *17*(1), 124–139. doi: 10.1080/10282580.2014.883844

Simon, J. (2016). The New Overcrowding. *Connecticut Law Review*, *48*(4), 1191–1216.

Snacken, S., & Beyens, K. (1994). Sentencing and Prison Overcrowding. *European Journal on Criminal Policy and Research*, *2*(1), 84–99. doi: 10.1007/bf02249251

Sterman, J. (2000). *Business Dynamics: Systems Thinking and Modeling for a Complex World*. Boston, MA: McGraw-Hill.

Suhling, S. (2003). Factors Contributing to Rising Imprisonment Figures in Germany. *The Howard Journal of Criminal Justice*, *42*(1), 55–68. doi: 10.1111/1468-2311.00265

Woods, C. S. (2016). Addressing Prison Overcrowding in Latin America: A Comparative Analysis of the Necessary Precursors to Reform. *ILSA Journal of International & Comparative Law*, *22*(3), 533–561.

Unleashing local democratic potential

Democratic knowledge in community self-organising processes

Zoraida Mendiwelso-Bendek[1]

Keywords: *self-organisation, community-based research, Participatory Action Research, Community and University Partnership*

1 Introduction

To co-produce knowledge with the power to address societal challenges and inequities is a key factor to support local democratic knowledge (Hall et al. 2015). Citizens construct identity in the process of extending the boundaries of their power and their citizenship emerges from their stable interactions (Mendiwelso-Bendek, 2015). This construction is the outcome of communication processes in which citizens relate to one another in their moment-to-moment interactions (Espejo and Mendiwelso-Bendek 2011). In this construction citizens experience a gap between what are considered their universal rights and their capacity to exercise them effectively (Bauman 2011). The topic of this paper is the construction of democratic values through self-organisation, supported by community-based research. The challenge for citizens, researchers and practitioners is co-producing knowledge about local communities, and their empowerment emerges from self-organisation as they unfold the local democratic potential in processes that build the values of acting locally but thinking and connecting globally.

Advocacy movements interact across those spaces creating new – more multidimensional – identities and understanding of citizenship (Gaventa and Mayo 2010). At the same time global alliances are exploring how university and community partnerships are contributing to communities' empowerment (Hall et al. 2015) and especially to making visible the importance of a wider reflective process about participatory approaches and new processes of co-creation of knowledge in social transformations. A key point is the role of community university research in supporting community self-organisation processes to increase promoting democratic values like social justice, intergenerational solidarity, diversity and equality (Mayo, Mendiwelso-Bendek and Packham 2013).

Communities are constituted by people in interaction. Shared interests trigger their interactions with each other and with agents from other communities and organisations (e.g. public and private agents). The evolution of interactions is by and large the outcome of self-organisation. Interactions may be enabled and supported but not planned. Their own resources and creativity, as well as external agents, such as universities, third-sector organisations, government agencies and others, may enable this self-organisation. Self-organisation is inherent to social processes; the challenge is to make interactions effective. How can citizens improve the quality of their interactions with each other and others?

Participatory action researchers' aim, in its different expressions, is to increase self-awareness of the context which involves day-to-day living dialectical encounters directing common people to have sufficient control over the generation of their own new knowledge (Fals-Borda 1990). If we perceive the reality as the dialectical relationship between subjects and objects then people should take part in the research themselves (Freire 1982), so participatory research seeks to break down the limit between researchers and the researched knowledge production by people's producing knowledge themselves (Gaventa 1990).

2   Self-organisation

Accepting that self-organisation is inherent to the complexity of social processes, the challenge is to work out how citizens of a community can improve the quality of their own interactions, dialogues within a dialectic reality. The self-organising process often highlights imbalances in power relations (Mayo, Mendiwelso-Bendek and Packham 2013). The communications among citizens, stakeholders require enhancing their competences to enable their organisations and overcome isolation and powerlessness (Espejo and Mendiwelso-Bendek 2011).

In this paper I'm asking how community-based research can contribute to support community self-organisation to reduce inequalities. How can community-based research contribute to increasing the collective impact in policy development and how can a stronger civil and civic citizenship enroot political participation to improve knowledge, skills, structures and processes? How can citizens co-create desirable values through their interactions with other agents? For instance, policies in a community may be directed to benefiting those that are better prepared to articulate their needs. The better education and competencies of these groups make them more visible and influential to governmental agents, which require organised citizens in order to direct their resources and achieve better policy performance. Not surprisingly resources will be directed towards the citizens with more interaction capabilities, at the expense of those with fewer competencies. In these circumstances local self-organisation may be precisely the detriment of those in more need. Well-intentioned policies may end up increasing operational imbalances within the community to the detriment of justice and fairness. In practice this requires that in addition to allocating resources to education, health, housing and so forth, it is necessary to direct resources to enable balanced self-organisation in the community, aiming at community members with similar participatory capabilities regardless of history, race or gender. It is necessary to redress inadequate processes of self-organisation. This strategy improves not only self-organisation within the community but also the quality of this community's relations with those organisations creating, regulating and producing policies relevant to the community. In summary it is necessary to improve self-organisation processes within community groups at the same time of improving their influence in value co-creation with external agents redressing imbalances in power relations.

Note

1 Lincoln Business School.

Bibliography

Bauman, Z. (2011). *Collateral Damage: Social Inequalities in a Global Age.* Cambridge: Polity Press.

Espejo, R. and Mendiwelso-Bendek, Z (2011). An argument for active citizenship and organisational transparency, *Kybernetes*, Vol. 40, Issue 3/4, pp. 477–493.

Fals-Borda, O. (1990). Remaking Knowledge, in O. Fals-Borda and Md. Rahman (eds.), *Anisur Action and Knowledge: Breaking the Monopoly with Participatory Action-Research.* New York: The Apex Press, pp. 146–167.

Freire, P. (1982). Creating Alternative Research Methods: Learning to Do It by Doing It, in H. Budd, A. Gillette, and R. Tandon (eds.), *Creating Knowledge: A Monopoly? Participatory Research in Development.* Participatory Research Network Series 1. New Delhi: Society for Participatory Research in Asia, pp. 29–37.

Gaventa, J. (1990) Toward a Knowledge Democracy: Viewpoints on Participatory Research in North America, in O. Fals-Borda and Md.A. Rahman (eds.), *Action and Knowledge: Breaking the Monopoly with Participatory Action-Research.* New York: The Apex Press.

Gaventa, J. and Mayo, M. (2010). Spanning Citizenship Spaces through Transnational Coalitions: The Global Campaign for Education, in J. Gaventa and R. Tandon (eds.), *Globalizing Citizens.* Zed.

Hall, B., Tandon, R., & Tremblay, C. (2015). *Strengthening Community University Research Partnerships: Global Perspectives.* University of Victoria and PRIA.

Mayo, M., Mendiwelso-Bendek, Z., and Packham, C. (eds.) (2013). *Community Research as Community Development.* London: Palgrave.

Bendek, Z. M. (2002). Citizens of the future: Beyond normative conditions through the emergence of desirable collective properties. *Journal of Business Ethics*, 39(1-2), 189-195.

Mendiwelso-Bendek, Z. (2015). Community-Based Research: Enabling Civil Society's Self-Organisation, *Kybernetes*, Vol. 44, Issue 6/7, pp. 903–912.

Mendiwelso-Bendek, Z. and Espejo, R. (2015). *Community Self-Organisation: How to Make It More Effective?* Paper 001, ISSS Conference Berlin 2015.

Interactions revolution

Alberto Marino and Manlio Del Giudice

Systems thinking and organisational change in the NHS

From heroic to system leadership

David Michael Cooper

Keywords: *NHS; management cybernetics; strategy; dialogue*

1 Introduction

The phrase "systems thinking" has been increasingly evoked by politicians, policy makers and practitioners aiming to promote and implement new ways of delivering public services in England. They claim that they are being holistic, not looking at individual components in isolation, but looking at the wider system in which they are embedded. Moving from the local and sub-optimal to the "whole system" approach.

The purpose of this paper is to look beyond this narrow interpretation of systems thinking and to explore how a management cybernetics approach can help in developing a cohesive response to the significant design challenges currently facing the NHS.

We begin by summarising the key social and political drivers of change in the NHS, which will be broadly similar in other Western economies, and the response of the major political parties, particularly since the instruction of the Health and Social Care Act in England in 2010.

Second we look at the way the emergence and self-organisation of the local health economy needed to provide integrated community health care but has been constrained and distorted by the dominant social and political discourses that have become established in the UK and elsewhere. We look in particular at the impact on the acute sector and illustrate the value of a management cybernetics perspective in appreciating the underlying issues and promoting the search for sustainable solutions, which are owned by all the key players.

Finally, we explore some recent system-theoretic contributions to the current debate on the future of the NHS and alternative organisational outcomes or trajectories that may emerge from the conflicts inherent in the current systems.

2 Setting the agenda

Since its formation in 1946, the main political parties have been continually reforming and reorganising the NHS. These changes have been driven by a number of financial, organisational and ideological pressures including:

- Increasing costs driven by demographic and technological changes
- Variability in the provision of care
- The need to improve access to services and cut down waiting times
- Inequalities in health outcomes
- Professional autonomy

Following deregulation and the introduction of privatisation in the 1980s, the public framing of these changes has been dominated by three main discourses:

- A neo-liberal narrative of marketisation and competition.
- Government and political pressures to adopt a performance-driven, command and control, managerial culture, where heroic-leadership is expected.
- Media focus on high-profile failures of care

With private health care organisations and powerful lobbies promising to deliver cheaper or better services, both Conservative and Labour Governments have promoted the wider use of the private sector to deliver care – justified either as a way of getting extra capacity or as a competitive stir to existing organisations. This approach is known in the UK as the New Public Management (NPM). During the Labour administration over the period 1997–2010, the concept of quasi-markets provided the framework for a wide range of reforms, including Payment by Results, and Foundation Hospital Trusts regulated by an external organisation, Monitor, which became part of NHS Improvement (NHSI) in April, 2016.

3   A management cybernetics perspective

Using a Viable Systems Model (VSM) model of the organisational, local and national levels of the NHS (Beer, 1979; Beer, 1990; Hoverstadt, 2008), we show how these multiple discourses limit the collaborative space available for the design of effective coordination and control processes that are needed for an evolving health economy. These often conflicting discourses also generate continuing uncertainties and ambiguities in strategy and policy making, raise questions about governance and accountability and promote an increasingly anxious, stressed and defensive public sector work climate. Such a climate is not conducive to disinterested, collaborative systemic thinking, and is likely to encourage a more inward-looking, self-protective stance.

We demonstrate how the VSM provides a useful framework for the essential dialogue between the actors in emerging structures, where there are multiple levels of accountability and coordination and often ambiguous objectives.

We then examine the wider challenges to systemic practice that spring from specific, historically determined, features of provider organisations. We look at acute hospitals, which can be characterised as loosely connected, multi-professional, service organisations. In some cases, key staff have dual loyalties to public and private sector providers.

Nigel Crisp (Crisp, 2011), surveying the changes introduced by the Blair Government, argued that in, the later part of their administration, the Labour Government recognised the need to move away from centralised, bureaucratic control and move towards more local decision-making. This view was part of a wider approach to managing public services, set out by Michael Barber (Barber, 2007), who saw command and control as only relevant to failing institutions. This notion was taken further by the Coalition Government in 2010, which abolished a whole layer of government, the Strategic Health Authorities.

In the early days of this legislation, with strategy having no clear organisational base, it was believed that each local economy would work out its own solution to dealing with strategic questions. This was part of the new "localism", which its advocates saw as liberating local decision-makers from central government interference.

It soon became apparent that this solution was unworkable and a variety of models and organisational forms have been proposed in recent years to aid multi-organisational strategy development.

Despite this, getting staff engagement in strategy and the activities needed for effective coordination and control can be particularly problematic, calling for exceptional leadership skills; an understanding of strategy as a collaborative journey; a systemic mind-set both at the local and regional level, and a high level of maturity in institutional and inter-organisational governance – a fairly rare combination of factors.

Differing emphases on competition versus collaboration between different levels of the NHS system, resulting from the Health and Social Care Act (2012), have added a further layer of complexity, ambiguity and risk to this task.

We give examples of how these challenging and problematic issues can be effectively addressed if there is a shift to a systems perspective and a change in the mind-set of key actors. We then look at some recent promising contributions to the public debate on the future of the NHS, where a more far-reaching systems view – including changes in culture, and the role of professional associations – has been adopted (Iles, 2011; Seddon, 2008).

These contributions provide a useful framework for considering alternative emergent scenarios. These range from a temporary disruption, and sustained recovery – through systemic leadership and good governance – to a continuing period of instability, leading to declining confidence, the loss of key skills and experience and a downward spiral in performance, resulting in the need for external intervention.

This leads us to ask how the systemic wisdom needed to sustain effective organisations can be facilitated and we look at a number of strategies that can be adopted to promote systemic leadership, drawing on the experience of a variety of practitioners in the public sector (Mulgan, 2006; Oldham, 2014).

Bibliography

Barber, M. (2007). *Instruction to Deliver*. London: Methuen(Politico's).
Beer, S. (1979). *The Heart of Enterprise*. Chichester: Wiley.

Beer, S. (1990). *Diagnosing the System for Organisations*. Chichester: Wiley.

Crisp, N.(2011). *24 Hours to Save the NHS*. Oxford: Oxford University Press.

Hoverstadt, P. (2008). *The Fractal Organisation*. Chichester: Wiley.

Iles, V. (2011). *Why Reforming the NHS Doesn't Work*. Retrieved from www.really learning.com

Mulgan, G. (2006). *Good and Bad Power*. Allen Lane: Penguin Books.

Oldham, Sir J. (2014). *Report on the Independent Commission on Wholeperson Care*. Report published by UK Labour Party. Retrieved 22.9.2018 from https://www.birmingham.ac.uk/Documents/college-social-sciences/public-service-academy/portal/oldham-report.pdf

Seddon, J. (2008). *Systems Thinking in the Public Sector*. Axminster: Triarchy Press.

Innovative approach to social media as communication channels

Zdenka Zenko[1] *and Vanesa Stefanovski*[2]

Keywords: *innovation; diffusion; communication channels; social media; systemic approach*

1 Social media's inclusion in communication channels

Our research anticipated to find and select relevant theory to increase our understanding of how well social media are understood by managers and governors and included into communication channels. The purpose of our systematic research is to contribute to the enhancement of requisite variety of communications in organizations and society. Communication channels (Rogers 2003) are used in social systems also to introduce inventions. They have significantly changed in the recent decade due to technological advancement (Oslo Manual 2005) and other social changes (PEW 2015; Boyd and Ellison 2013). Theoretical frameworks found in the relevant studied literature could not include all the complexity of this movement (Rogers 2003; Dearing and Meyer 2006; Hall 2006; Štefanovski 2016; WeAreSocial 2015; Ženko and Mulej 2011; Štrukelj and Šuligoj 2014).

In our theoretical research we have included the invention-to-innovation diffusion process, idea-creating processes, social responsibility and sustainability in systemic approach. The last part of the research contributed to understanding of social media phenomena focusing only on Instagram. We aimed to discover reasons for its thriving acceptance by the users in only a few years (Quora 2013; BBC 2012; ReadWrite 2012). We were searching the areas, industries, services and products which are more active on Instagram (Forbes 2016b).

Our study of diffusion processes using social media revealed that societies, companies and organizations lack the understanding of social media; do not know the characteristics of participants in social media; do not recognize the leaders of social groups as change agents and/or opinion leaders. As a result, social media are not well included in the mix of communication channels in innovative processes.

2 Research design and limitations

In the first part, we applied conceptual generalizations to build a theoretical framework, presenting this complex phenomenon using system theory and innovation management. The second part is focused on a literature review of social media to test our hypothesis on social media as communication channels in innovative processes.

We limited our theoretical approach to complexity, dialectical system theory, innovation management, diffusion of innovations theory. We have found a lack of literature on the above selected viewpoints of social media used as one of the communication channels in innovative processes. Instagram is a global and quite recent phenomenon and thus its social changes are very complex.

3 Some discussion and concluding

Our literature research has not revealed any systematic approach to studying social media focusing on building our understanding of them as communication channels. Use of social media is rapidly growing among certain social groups. Our selected theoretical framework provided new insights on diffusion processes of social media. We further focused our search on Facebook, Pinterest, Instagram and Twitter (ReadWrite 2012; Chaey 2013). Instagram (introduced in October 2010) gained until 2012 150 million users and was in 2012 purchased by Facebook. Now it has more than 400 million users globally. We applied the selected theoretical approaches to study Instagram. A comparative study of its users and selected theoretical framework revealed that they have some characteristics similar to and others different from change agents and opinion leaders of social changes.

Well planned and managed combinations of communication channels that include social media present a great challenge for managers and governors. Many of the leading companies on Forbes' list for 2015 are using Instagram for advertising. Findings of our research can support strategic planning in different areas like corporate, scientific, educational, healthcare and others.

Notes

1 University of Maribor.
2 University of Maribor.

Bibliography

BBC. (2012). Facebook Buys Instagram App for $1bn. Retrieved from www.bbc. com/news/business-17667099

Belak, J., Duh, M., Mulej, M. & Štrukelj, T. (2010). Requisitely Holistic Ethics Planning as Pre-Condition for Enterprise Ethical Behaviour. *Kybernetes*, ISSN 0368-492X, *39*(1), 19–36.

Boyd, D. & Ellison, N. B. (2013). Sociality through Social Network Sites. *The Oxford handbook of Internet studies*, 151–172. Retrieved 22.9.2018 from http://www. oxfordhandbooks.com

Chaey, C. (2013). *Facebook Introduces 15-Second Video on Instagram with 13 New Filters and Cinema Effects.* Retrieved from www.fastcompany.com/3013325/ facebook-introduces-15-second-video-on-instagram-with-13-new-filters-and-cinema-effects

Dearing, J. W. & Meyer, G. (2006). Revisiting Diffusion Theory. *Communication of Innovations: A Journey with Ev Rogers*. Ed. Arvind Singhal & James W. Dearing, 29–60. New Delhi, Thousand Oaks, London: SAGE Publications.

Drucker, F. P. (1998). *The Discipline of Innovation*. Boston: Harward Business Review. Retrieved from http://ogsp.typepad.com/focus_or_die_ogsp/files/drucker_1985_the_discipline_of_innovation.pdf

Forbes. (2016a). The Influencer Marketing Gold Rush Is Coming: Are You Prepared? Retrieved from www.forbes.com/sites/johnhall/2016/04/17/the-influencer-marketing-gold-rush-is-coming-are-you-prepared/#7da0a1042964

Forbes. (2016b). The World's Most Valuable Brands. Retrieved from www.forbes.com/powerful-brands/list/#tab:rank

Hall, B. H. (2006). Innovation and Diffusion. *The Oxford Handbook of Innovation*, Ed. J. Fagerberg, D. C. Mowery & R. R. Nelson, 459–484. Oxford: Oxford University Press.

Oslo Manual. (2005). *The Measurement of Scientific and Technological Activities: Proposal Guidelines for Collecting and Interpreting Technological Innovation Data*. European Commission and Eurostat. Retrieved from http://ec.europa.eu/eurostat/documents/3859598/5889925/OSLO-EN.PDF

PEW. (2015). Demographics of Key Social Networking Platforms. Retrieved from www.pewinternet.org/2015/01/09/demographics-of-key-social-networking-platforms-2/

Quora. (2013). How Did Instagram Build Up It's Community in Its Early Days? Retrieved from www.quora.com/How-did-Instagram-build-up-its-community-in-its-early-days

ReadWrite. (2012). Instagram Growth Far Outpaces Facebook or Twitter. Retrieved from http://readwrite.com/2012/07/26/instagram-growth-far-outpaces-facebook-or-twitter/

Rogers, E. M. (2003). *Diffusion of Innovation*, 5th ed. New York: The Free Press.

Štefanovski, V. (2016). *Teorija difuzije inovacij in družbeno omrežje Instragam = Diffusion of Innovation Theory and Social Network Instragam*. Master thesis. University of Maribor: EPF Maribor.

Štrukelj, T. & Šuligoj, M. (2014). Holism and Social Responsibility for Tourism Enterprise Governance. *Kybernetes*, *43*(3/4), 394–412.

WeAreSocial. (2015). Smaller Networks Boon as Facebook Dips. Retrieved from http://wearesocial.com/uk/blog/2015/01/smaller-networks-boom-facebook-dips-2

Ženko, Z. & Mulej, M. (2011). Diffusion of Innovative Behaviour with Social Responsibility. *Kybernetes*, *40*(9/10), 1258–1272.

Observing engagement in systems

Roberto Bruni,[1] *Luca Carrubbo*[2]
and Debora Sarno[3]

Keywords: *system engagement; system viability; consonance; resonance*

1   Introduction

Ongoing studies on systems and eco-systems usually focus on the role and relevance of interactions among involved actors operating within. Win-win logic and modern value generation process foster the participation of any entity in making new propositions and solutions for markets (Araujo et al. 2010; Kjellberg and Helgesson 2007; Storbacka and Nenonen 2011; Storbacka et al. 2008; Vargo et al. 2008; Vargo and Lusch 2011), deepening new future lines for competitiveness and allowing the vision of "co", dealing deeply with co-design, co-production, co-marketing, co-creation (Vargo and Lusch 2004; Prahalad and Ramaswamy 2004; Ballantyne and Varey 2006; Grönroos 2008).

According to systems thinking, actors (as providers, users and consumers as well) interact between each other appearing more and more involved in the resources integration. Some interactions are characterized by engagement (Appelbaum 2001; Harvey 2005; Haven 2007).

Worldwide scholars and outlooks state that engaged actors are able to better contribute to the value perceived effectively; in particular, Storbacka et al. (2016) provide general perspectives from literature on actors' engagement and contributions to actor engagement as microfundation of value co-creation. Following the service systems' approach and its focus on systems' levels, a number of authors argue that new researches and models around the theme of engagement in systems are needed (Brodie et al. 2011).

The aim of this work is to identify the key drivers for engagement in systems. The topic is approached by a theoretical analysis and conceptual development of the integrative framework related to engagement in management. Some dimensions of the engagement in relation to the systems characteristics are highlighted. Then, engagement footprints are noticed in system actors that interact not only for their individualistic goals but for the system's survival, identifying the system's survival as a focal object of their engagement. The research is based on literature in service systems and builds on the Fundamental Propositions about the Customer Engagement of Brodie et al. (2011) by filtering them with the Viable System

Approach (VSA). The result of this analysis is the generalization of the key drivers of the customer engagement attitude to system engagement. Specifically, the role of consonance (structural property of relations) and resonance (active and mutual involvement of entities with a common aim) are taken into account in explaining the engagement attitude of systems (consonance) and the effective actions and tension-to-system survival (resonance) of systems within the finality of viability.

2   Engagement in service systems and system viability

Engagement is a multidimensional concept subject to a context- and/or stakeholder-specific expression related to the cognitive sphere, behavioural intensity and emotional quality of a person's active involvement during a task which has an interpersonal component, since the interaction among the actors involved is an important part of the engagement experience (Connell and Wellborn 1991). It has been a topic of interest of many researchers in different environments, from teaching to working, to playing or receiving care.

A *service system* is defined as "configuration of people, information, organizations, and technologies that operate together for mutual benefit" (Maglio et al. 2009, p. 1). In the operations management sense, indeed, a service system can be modelled as a network of agents, resources and engagements (Badinelli 2011). Then, a *service system engagement* is the process of integrating resources from the service recipient and one or more other resource providers in a network in order to *co-create value*. Such integration is the result of a designing and planning process of the network agent (which can be the recipient of the service in case of a self-service system or the provider), finalized to identify the most appropriate resource, access it and use it.

Advancement in this field has been reported by Storbacka et al. (2016) who explored actor engagement as a microfundation for value co-creation. Indeed, *microfundation* research constitutes a bridge for empirical investigation, explaining the origin of more abstract macro concepts in strategy and organizational theory.

Brodie et al. (2011) identified from literature five fundamental propositions able to develop a general definition of customer engagement conceptual relationship and find out that, unlike the traditional relationship concept of involvement and participation, customer engagement is based on an interactive and co-creative experience with an *engagement object*. In the case of Brodie et al. such object is the brand, while in the Storbacka et al. (2016) analysis it can be seen as the resource integration process.

In the systems perspective, some affirmed theories such as the Viable System Model – from Beer (1979) with a special focus on business management scenarios – and Viable Systems Approach (VSA) – easily applicable to any social entity as individuals or organizations (Barile and Polese 2010) – considered the "*viability*" as the ultimate goal of every systemic entity in competitive contexts. As reported by Barile and Gatti (2007), VSA appears coherent with a complex value-creation process in which viability and competitive behaviour are linked, among others, with the ability to identify and manage functions and relationships. According to VSA, system viability is determined by the ability of showing consonant and resonant behaviour relationships, where *consonance* is the potential (structural) compatibility between elements, and *resonance* is the eventual harmonic interaction between elements (activation of the consonant relations).

In this viewpoint, engagement is the way to make the system viable by pursuing a common purpose over time. The engagement emerges from the resonance but the conditions are generated by the consonance. Single actors are involved in organizations/systems by means of a series of agreements, behaviours, shared values, intentionality and motivations, activities and relationships, in the shared purpose to stimulate the system survival over time.

3   Systems engagement: discussions and final remarks

Summarizing, literature on engagement is focused on the characteristics of the single subjects engaged toward value proposition and their specific causes; relevant research on the engagement topic is based on the customer engagement. Going beyond the participation and the involvement, the engagement is related to the *focal interactive experience* (Brodie et al. 2011).

While it is not possible to directly transfer the psychological state of the single person to an organization (or system), it is possible to assume that the concepts of relationship, resource integration and value exchange are part of the engagement emersion in actors and eventually in organizations/systems. Moreover, the VSA shows that the system's behaviour is the result of a decision-making activity (which depends on humans' perceptions and psychological states), and contributes to the interpretation of the focal interactive experience of Brodie et al. (2011) identifying it in the searching for survival of the system. The searching for system's survival is a continuous, progressive and iterative process of resource integration and value exchange adapting the system to the context variety and variability.

The interactive experience emerges from the resource integration and value exchange between parts of the system showing a structural compatibility (consonance), and it is represented by the shared purpose of the system elements (the searching for the system survival). Each part in the system needs to be interested in the system survival because the system has much more value than the value of the single parts.

The engagement of a system within other systems depends on the contextualization of the parts, knowledge, perceptions, behaviour and capabilities of them. The resonance (state of engagement of the system with other systems) can be stimulated if the decision maker is able to manage the complexity of the context, as it happens for an orchestra and its director (Ruggiero et al. 2016).

By introducing a modelling approach to explain the attitude to be engaged of actors and systems, this research will contribute to deepen the knowledge of the discipline, representing the first steps into the modelling of the engagement in systems and giving new managerial tools to practitioners.

Notes

1 University of Cassino and Southern Lazio.
2 University of Salerno.
3 University of Foggia.

Bibliography

Appelbaum, A. (2001). *The Constant Customer*, Available at http://gmj.gallup.com/content/745/constant-customer.aspx

Araujo, L., Finch, J. H. & Kjellberg, H. (eds.) (2010). *Reconnecting Marketing to Markets*. Oxford, UK: Oxford University Press.

Badinelli, R. D. (2011). *Fuzzy-Control Models of Service-System Engagements*. The Naples Forum on Service.

Ballantyne, D. & Varey, R. J. (2006). Creating value-in-use through marketing interaction: The exchange logic of relating, communicating and knowing. *Marketing Theory*, 3, 335–348.

Barile, S. & Gatti, M. (2007). Corporate governance e creazione di valore nella prospettiva sistemico-vitale. *Sinergie*, (73–74), 151–168.

Barile, S. & Polese, F. (2010). Smart service systems and viable service systems. *Service Science*, 2(1), 21–40.

Beer, S. (1979). The heart of enterprise. Chichester: Wiley.

Brockmyer, J. H., Fox, C. M., Curtiss, K. A., McBroom, E., Burkhart, K. M., Pidruzny, J. N. (2009). The development of the game engagement questionnaire: A measure of engagement in video game-playing. *Journal of Experimental Social Psychology*, 45, 624–634.

Brodie, R. J., Hollebeek, L. D., Juric, B., & Ilic, A. (2011). Customer engagement: Conceptual domain, fundamental propositions and implications for research. *Journal of Service Research*, 14(3), 252–271.

Connell, J. P. & Wellborn, J. G. (1991). Competence, autonomy, and relatedness: A motivational analysis of self-esteem processes. In M. R. Gunnar & L. A. Sroufe (Eds.), *Self Processes in Development: Minnesota Symposium on Child Psychology* (23, 167–216). Hillsdale, NJ: Erlbaum.

Grönroos, C. (2008). *Adopting a Service Business Logic in Relational Business-to-Business Marketing: Value Creation, Interaction and Joint Value Co-Creation*, in Proceedings of the Otago Forum 2, 269–287.

Harvey, B. (2005). *What Is Engagement?*, Available at www.nextcenturymedia.com/2005/12/what-is-engagement.html

Haven, B. (2007). *Marketing's New Key Metric: Engagement*, Available at www.forrester.com/Research/Document/Excerpt/0,7211,42124,00.html

Kjellberg, H. & Helgesson, C. F. (2007). On the nature of markets and their practices. *Marketing Theory*, 7(2), 137–162.

Maglio, P. P., Vargo, S. L., Caswell, N. & Spohrer J. (2009). The service system is the basic abstraction of service science. *Information Systems and E-Business Management*, 7, 395–406.

Prahalad, K. C. & Ramaswamy, V. (2004). *The Future of Competition: Co-Creating Unique Value with Customers.* Cambridge: Harvard University Press.

Ruggiero, A., De Simone, M. C., Russo, D. & Guida, D. (2016). Sound pressure measurement of orchestral instruments in the concert hall of a public school. *International Journal of Circuits, Systems and Signal Processing, 10,* 75–812.

Storbacka, K., Brodie, R. J., Böhmann, T., Maglio, P. P. & Nenonen, S. (2016). Actor engagement as a microfoundation for value co-creation. *Journal of Business Research, 69*(8), 3008–3017.

Storbacka, K. & Nenonen, S. (2011). Markets as configurations. *European Journal of Marketing, 45*(1/2), 241–258.

Storbacka, K., Nenonen, S. & Korkman, O. (2008). *Markets as Configurations: A Research Agenda for Co-Created Markets.* Proceedings of Forum of Markets and Marketing: Extending Service-Dominant Logic, Sydney, December 4–6.

Vargo, S. L. & Lusch, R. F. (2004). Evolving to a new dominant logic for marketing. *Journal of Marketing, 68*(January), 1–17.

Vargo, S. L. & Lusch, R. F. (2011). It's all B2B . . . and beyond: Toward a systems perspective of the market. *Industrial Marketing Management, 40*(2), 181–187.

Vargo, S. L., Maglio, P. P. & Akaka, M. A. (2008). On value and value co-creation: A service systems and service logic perspective. *European Management Journal, 26*(3), 145–152.

Artificial intelligence and constructed-language emergence

Overtaking synthetic conlangs with Self-Organized Linguistic Systems

Diego Gonzalez-Rodriguez[1] and Jose-Rodolfo Hernandez-Carrion[2]

Keywords: *artificial intelligence; self-organization; constructed languages; agent-based modelling; Conlangs*

This work is oriented to explore the potential of bottom-up generative processes in the context of conlang production, aiming to describe the basis of a new field of research: Self-Organized Linguistic Systems or SOLS.

The Self-Organized Linguistic Systems approach provides a framework for the creation of self-generated artificial languages, emergent and dynamic alternatives to contemporary synthetic conlangs such as Esperanto, Toki Pona or Lojban, which have been precisely engineered with specific prefixed goals.

SOLS may serve as a starting point for the development of context-dependent or domain-specific languages which may be used for several purposes such as (i) second language acquisition, (ii) communications between robots within unpredictable situations, (iii) knowledge representation, (iv) linguistic emergence in social agents, (v) visual language production and generative art.

Rather than relying on traditional artificial intelligence approaches to knowledge representation, this new field of research is specified under the perspective of both self-organized systems and constructed languages and considers that the development of conlangs can happen in artificial societies of simple agents, that is, as the output of social interactions in computational simulations under the agent-based modelling paradigm. Emergent conlangs share some of the elements of natural languages and can provide a live and dynamic communication system for artificial agents in situations in which objects and actions are not preexistent and in which a lexicon should emerge from the interactions with the environment.

This work starts analyzing some of the already existing constructed languages, under the label of synthetic conlangs, and the value provided by each one of them. It explains historical and technical reasons for the design of synthetic conlangs and exposes some of the advantages and disadvantages of them.

The paper also discusses how traditional ways of knowledge representation can be overtaken under the design of unconventional systems, and describes how language can be revisited with new systems of representation that go beyond the visual and audible options and become multidimensional.

Finally, the paper introduces the notion of emergent languages and exposes the foundations of Self-Organized Linguistic Systems. SOLS rely on the theoretical framework of Complex Adaptive Systems in conjunction with the development of computational tools and simulations such as agent-based models.

SOLS can be helpful to understand better the nature of language and sense-making as emergent processes. This paper exposes possible implementations and further developments of the initially presented concepts. Rather than developing a linguistic theory or using pre-existing frameworks such as Paul Hopper's functional work on emergent grammar, this work exposes some of the first steps in the development of straightforward algorithmic approaches for the generation of both descriptive and imperative lexicons in the context of agent-based models.

In the proposed initial SOLS model, automatic generation of a lexicon takes place in the context of a digital environment with objects, actions and agents with embodied cognition through interactions between agents and objects.

Specifically, this paper exposes how SOLS can be developed with bi-dimensional games and simulations. Non-interactive agent-based SOLS can allow artificial agents to independently evolve emergent languages as part of their self-organizing or adaptation processes. However, game-style SOLS can also provide an open environment in which artificial agents can also interact with human agents, leading to human-machine communications and peer learning processes.

In conclusion, SOLS are the intersection of artificial agent-based models with constructed languages, and may provide new ways for knowledge representation, artificial systems development and visual languages design (2017).

Notes

1 University of Valencia.
2 University of Valencia.

Bibliography

Gobbo, F., & Durnová, H. (2014). *From Universal to Programming Languages.* Informal Proceedings of Computability in Europe.

Goertzel, B. (2013). Lojban++: An Interlingua for Communication between Humans and AGIs. *Artificial General Intelligence*, *7999*. https://doi.org/10.1007/978-3-642-39521-5

Goertzel, B., Pennachin, C., Araujo, S., Silva, F., Queiroz, M., Lian, R., & Senna, A. (2010). *A General Intelligence Oriented Architecture for Embodied Natural Language Processing.* Proceedings of the 3d Conference on Artificial General Intelligence (AGI-10), 1–6. https://doi.org/10.2991/agi.2010.16

Gonzalez-Rodriguez, D., & Hernandez-Carrion, J. R. (2014). A Bacterial-Based Algorithm to Simulate Complex Adaptive Systems. *Lecture Notes in Artificial Intelligence (LNAI)*, *8575*, 250–259. https://doi.org/10.1007/978-3-319-08864-8_24

Gonzalez-Rodriguez, D., & Kostakis, V. (2015). Information Literacy and Peer-to-Peer Infrastructures: An Autopoietic Perspective. *Telematics and Informatics*, *32*, 586–593. https://doi.org/10.1016/j.tele.2015.01.001

Harrison Kuhn, T. (2016). The Controlled Natural Language of Randall Munroe's Thing Explainer. In B. Davis, G. J. Pace, & A. Wyner (Eds.), *Controlled Natural Language* (pp. 102–110). Springer International Publishing. https://doi.org/10.1007/978-3-319-41498-0_10

Isenberg, R. (2000). Universitetet i Bergen. *Artificial Languages.* Retrieved September 2016, from http://folk.uib.no/hnohf/artlang.htm

Kuhn, T. (2014). A Survey and Classification of Controlled Natural Languages. *Computational Linguistics, 40*(1), 121–170. https://doi.org/10.1162/COLI_a_00168

SciArt Lab. (2016). 2D Simulator: Hacking Science, Art and Technology. Retrieved October 2016, from www.sciartlab.com/?page_id=517

Speer, R., & Havasi, C. (2004). *Meeting the Computer Halfway: Language Processing in the Artificial Language Lojban.* In Proceedings of MIT Student Oxygen Conference.

Tily, H., Frank, M. C., & Jaeger, T. F. (2011). *The Learnability of Constructed Languages Reflects Typological Patterns.* In Proceedings of the 33rd annual conference of the cognitive science society (pp. 1364–1369). Citeseer.

The interaction type approach to relationships management

Giancarlo Nota[1] and Rossella Aiello[2]

Keywords: *viable system approach, interaction type, relationships management*

1 Introduction

The problem of relationship management has received great attention from researchers and practitioners in several sectors of the economy (Allen et al., 2009): public administration (Roberts, 2015), (Schellong, 2009), (Stromback and Kiousis, 2011); social networks (Kilduff and Tsai, 2006; Scott, 2013) and many others (Waters and Bortree, 2012). In general, the motivations behind this growing interest are dependent on the context in which the relationship management is pursued. Seeking long-lasting relationships with customers to achieve a competitive advantage in a given market (Lambert, 2014), or involving citizens in the decision process on how public resources are used (Aiello et al., 2016), provides examples of motivations that encourage the implementation of systems aiming at the relationship management.

Given such a variety of settings and applications, it is certainly not surprising that a universally accepted definition of relationships on which to build a general theory of relationship management does not exist. Indeed, within each application domain, specialized models, languages and theories are developed so that they work in a satisfactory way with respect to the chosen domain but cannot be used when the application domain changes. Nevertheless, common concepts and characteristics relying on relationship management could be used to build a meta-model that exploits the new concept of interaction type introduced in this paper.

Starting from the viable system approach (VSA) we adopt the structure-system point of views in order to get a deeper insight on the structural and behavioural properties of a relationship and to can shift from the concept of relationship to that of interaction and vice versa using interaction types.

2 Modelling the interactions

After a review of the scientific literature from which two orientations of the definition of relationship emerge that we call structural and behavioural, the paper will introduce the concept of interaction type as a bridge linking the relationship (structural concept) to that of interaction (systemic concept). The interaction type is first introduced in its simplest form using an UML class diagram that points out its role when the observer view changes from structure to system and from system to structure. The hierarchical nature of interaction types is then discussed in order to show how complex relationships, that involve more parties at several interaction levels, can be better represented and exploited. Finally, a model of transactional environment is proposed in which a transaction type can be interpreted as a particular case of interaction type and a transaction as an ordered set of interactions.

3   Case studies

Two case studies are presented; the first concerns an example of fund transfer and points out how the interaction type hierarchy can be applied. The second case study is from the port community system of Salerno and Brindisi – South Italy – and shows how a clearance to enter or leave national waters can be described as "departure from" and "arrival to" interaction types.

4   The implications

The purpose of this work is twofold. From one side, we wish to contribute to the traditional VSA view that considers relation and interaction respectively as structural and systemic concepts; to this purpose, we introduce a new structural concept, the interaction type, acting as a bridge between a relationship and an interaction. On the other side, we posted an operational goal. The definitions proposed are expressed in terms of metadata so that they could be used as starting point to the specification and coding relationship management software systems.

Notes

1 University of Salerno.
2 University of Salerno.

Bibliography

Aiello, R., Bisogno, M., and Nota, G. (2016). *Accountability, Transparency and Open-Data Philosophy in the Public Sector: A Knowledge Network Model*. EGPA Spring workshop, Modena, Italy.

Allen, S., Deragon, J., Orem, M., and Smith, C. (2009). *The Emergence of the Relationship Economy: The New Order of Things to Come*. Silicon Walley, Link to your world.

Barile, S. (2008). *L'impresa come sistema – Contributi sull'approccio Sistemico Vitale (ASV)*. Giappichelli.

Bruegge, B. and Dutoit, A. (2010). *Object-Oriented Software Engineering Using UML, Patterns, and Java*. 3rd Edition. Pearson.

Buttle, F. and Maklan, S. (2015). *Customer Relationship Management: Concepts and Technologies*. Abingdon, UK: Routledge.

Chen, I. J. and Popovich, K. (2003). Understanding customer relationship management (CRM): People, process and technology. *Business Process Management Journal*, 9(5), 672–688.

Damkuviené, M. and Virvilaité, R. (2007). The concept of relationship in marketing theory: Definitions and theoretical approach. *Economics and Management*, 12, 318–325.

Dwyer, F., Schurr, P., & Oh, S. (1987). Developing buyer-seller relationships. *Journal of Marketing*, 51(2), 11–27.

Elmasri, R. and Navathe, S. (2010). *Fundamentals of Database Systems*. 6th Edition. Addison Wesley.

Foo, M.-H., Douglas, G., and Jack, M. A. (2008). Incentive schemes in the financial services sector: Moderating effects of relationship norms on customer brand relationship. *International Journal of Bank Marketing*, 26(2), 99–118.

Ford, D., Gaddeb, L.-E., Håkansson, H., Shehota, I., and Waluszewski, A. (2010). Analysing business interaction. *The IMP Journal*, 4(1), 82–103.

Gadde, L. G. and Mattsson, L. G. (1987). Stability and change in network relationships. *International Journal of Research in Marketing*, 4(1), 29–41.

Gamma, E., Helm, R., Johnson, R., and Vlissides, J. (1994). *Design Patterns: Elements of Reusable Object-Oriented Software*. Addison Wesley.

Golinelli, G. (2010). *Viable Systems Approach (VSA)*. CEDAM.

Golinelli, G., Barile, S., Saviano, M., and Polese, F. (2012). Perspective shifts in marketing: Toward a paradigm change? *Service Science*, 4(2), 121–134.

Golinelli, G. and Gatti, M. (2001). The firm as a viable system: SYMPHONYA. *Emerging Issue in Management*, 2.

Grönroos, C. (1994). From marketing mix to relationship marketing: Towards a paradigm shift in marketing. *Management Decision*, 32(2), 4–20.

Grönroos, C. (1999). Relationship marketing: Challenges for the organization. *Journal of Business Research*, 46, 327–335.

Gummesson, E. (2012). *Total Relationship Marketing*. Elsevier Ltd.

Håkansson, H. (1982). *International Marketing and Purchasing of Industrial Goods*. Hoboken, NJ: John Wiley & Sons, Inc.

Håkansson, H. and Shenota, I. (1995). *Developing Relationships in Business Networks*. Abingdon, UK: Routledge.

Holmlund, M. (2004). Analyzing business relationships and distinguishing different interaction levels. *Industrial Marketing Management*, 33(4), 279–287.

Holmlund, M. and Törnroos, J. (1997). What are relationships in business networks? *Management Decision*, 5(4), 304–309.

IMO (2011). Guidelines for Setting Up a Single Window System in Maritime Transport: fal.5/circ.36. www.imo.org/en/OurWork/Facilitation/docs/FAL%20related%20nonmandatory%20instruments/FAL.5-%20Circ.36.pdf.

IMO (2016a). Fal Forms and Certificates. www.imo.org/en/ourwork/facilitation/formscertificates/pages/default.aspxh.

IMO (2016b). Guidelines for the Use of Electronic Certificates: fal circ.39, rev.2. www.imo.org/en/ourwork/facilitation/electronic%20business/documents/fal.5-circ.39-rev.2%20-%20guidelines%20for%20the%20use%20of%20electronic%20certificates%20(secretariat).pdf

Kilduff, M. and Tsai, W. (2006). *Social Networks and Organizations*. Thousand Oaks, CA: SAGE Publications Ltd.

Kleinaltenkamp, M., Plinke, W., and Söllner, A. (2015). Theoretical Perspectives of Business Relationships: Explanation and Configuration. In *Business Relationship Management and Marketing*, pp. 27–54. Berlin: Springer.

Lambert, D. (2014). *Supply Chain Management: Processes, Partnerships, Performance.* 4th Edition. Fung Global Institute and World Trade Organization.

Palmatier, R. W. (2008). *Relationship Marketing.* Marketing Science Institute.

Pedersen, J. T. (2012). One Common Framework for Information and Communication Systems in Transport and Logistics: Facilitating Interoperability. In *Sustainable Transport: New Trends and Business Practices.* Berlin and Heidelberg: Springer, pp. 165–196.

Roberts, N. C. (2015). *The Age of Direct Citizen Participation.* Abingdon, UK: Routledge.

Schellong, A. (2009). *Citizen Relationship Management: A Study of CRM in Government.* European University Studies. Series XXXI, Peter Lang.

Scott, J. (2013). *Social Network Analysis.* Thousand Oaks, CA: SAGE Publications Ltd.

Stromback, J. and Kiousis, S. (2011). *Political Public Relations: Principles and Applications.* Abingdon, UK: Routledge Communication Series.

Waters, R. D. and Bortree, D. S. (2012). Advancing Relationship Management Theory: Mapping the Continuum of Relationship Types. *Public Relations Review*, 38(1), 123–127.

Zablah, A. R., Bellenger, D. N., and Johnston, W. J. (2004). An Evaluation of Divergent Perspectives on Customer Relationship Management: Towards a Common Understanding of an Emerging Phenomenon. *Industrial Marketing Management*, 33(6), 475–489.

Zolkiewski, J. (2004). Relationships Are Not Ubiquitous in Marketing. *European Journal of Marketing*, 38(1/2), 24–29.

The relationship between hotels and online travel agencies

Opportunities and challenges for long-term strategies

Antonio Iazzi,[1] *Pierfelice Rosato*[2]
and Oronzo Trio[3]

Keywords: *marketing relationship; viable system; online travel agency; tourism*

1 Introduction

The aim of this paper is to analyze the characteristics of the relationships between hotels and online travel agencies. Traditional travel agents have played an important role as proxies enabling travelers to make connections with hotels. Information and Communication Technologies have deeply transformed business strategies and practices in the tourism and hospitality industries (Ip et al., 2011). Specifically, the Internet has served as one of the most effective marketing and communication tools for hoteliers in facilitating information sharing and online transactions (Tse, 2013), thus leading businesses into a new era. In response to the new challenges (Buhalis and Law, 2008) and in order to increase their competitiveness, hotels have established various online presences. In this field Online Travel Agencies (OTAs) have a strategic role in inducing travelers to make reservations mainly through discounts on travel packages (Toh et al., 2011), coupons or cash-back (Guo et al., 2014). The online travel represents about 20% of the Italian market, for a value higher than EUR 10 billion buying through OLTA grew by 14% in the Italian travel market (Politecnico di Milano – Observatory of Digital Innovation 2016). Hotels pay them a commission fee to ensure such a desirable achievement, and face a trade-off between reservations and a high commission fee.

2 Literature reviews

According to transaction cost theory hotels should benefit from a collaborative relationship since they could focus on providing their guest experience while travel agents manage the distribution functions (Donaldson, 1990). But this theory cannot completely explain the relationships between hotels and OTAs cause informative asymmetry and agents opportunistic behavior.

Instead of a traditional agent–principal relationship, online travel agents (OTAs) act just as business partners or as vendors. Consequently, they have a powerful role in that process and a strong bargaining power that affects the hotels' vertical competition (Middleton et al., 2009; Lee et al., 2013).

Starting from this assumption, we aim to identify and analyze: (i) the sources of a different bargaining power in the Hotel-OTAs relationship; (ii) the power asymmetry effects in setting B2B tourism contractual elements; (iii) the strategies that enable hotels to fill the information gap in order to balance the bargaining power.

Some studies have examined the relation between hotels and OTAs, but have overlooked principal-agent theory, failed to specify the factors that cause the conflicts (Lee et al., 2013). Information asymmetry may mean that transaction cost theory fails to model the agent–principal relationship in this instance. Especially when agents have more information than principals, principals are likely to have difficulties in governing their agents, thereby raising the prospect that the agent is not serving the principal's best interest (Priego and Palacios, 2008).

We adopt a Viable Systems Approach (Golinelli, 2010) to analyze the determinants of power asymmetry and the reasons of adversarial relationships. According to the Viable Systems Approach an actor has greater bargaining power than its partner when it is able to manage a critical resource and to "influence" or define contractual conditions. "Influence" and "criticality" can be considered the main dimensions in defining relational perspectives. A different "bargaining power" affects the degree of trust and the development of cooperative relationships in business-to-business markets that are generally founded on the alignment of interests.

3   Methodology

A descriptive research has been carried out through a semi-structured questionnaire to three-, four- and five-star Italian hotels. The structure of the questionnaire has been defined by a previous explorative research. We asked hotel managers to score from 1 to 7, where "1" means "this element is not at all important", "4" "This item is neither important nor unimportant", "7" "This item is essential".

SPSS software (version 20.0) was used to perform a correlation analysis on the data, in order to estimate the direction and strength of the linear association between the two variables, that could be positive (i.e., higher levels of one variable are associated with higher levels of the other) or negative (i.e., higher levels of one variable are associated with lower levels of the other). In the following step, a descriptive analysis was carried out in order to assess the importance of the variables linked to intention in the choice of hotels.

4   First result

Firstly, the questionnaire was completed by 200 Italian hotel managers. The sample was mainly composed by three-star hotels (45,5%) and four-stars (32,1%). 81,8% of the sample argue that they mediate through OTAs. 44,4% use more than three OTAs; 48% of hotels reserve up to 20 rooms for OTAs and 53% up to 50 rooms.

In the hotel–OTAs relationship, informational resources are strongly managed by OTAs, as they have a deep knowledge of consumption dynamics and behaviors. This is basically due to the role of web that enables OTAs to increase their competitiveness and the bargaining power towards hotels. The hotels' decisions to cooperate with OTAs is prevalently motivated by their reputation and by the opportunity to get qualified and systematic reviews.

The correlation analysis about the reasons why OTAs are used is a stronger linear association between the "revenue growth variables" and "profit growth" and between "service improvement" and "efficiency". As regards the negative effects of the relationship, it shows a stronger linear correlation between "decrease visibility" and "decrease reputation" and between "decrease of decision-making power" and "decrease tourist loyalty".

Reading the relationship between hotels and OTAs with the Viable Systems Approach, the current status appears to be conflictual, because of the criticality of the resource owned by OTAs, as well as the ability of the same to influence the relationship. This generates bargaining power for the same OTAs.

A different mode of governance of the relationships requires hotels to reduce information asymmetry and improve the reputation in order to increase their competitiveness.

To reach this goal hotels could constitute a territorial aggregation for their commercial and marketing strategies to improve the uniqueness of the product for tourists. Territorial aggregation should share information, increase the knowledge of the tourists and contribute to develop a spatial planning.

It is essential to use web and social media strategies, involving the whole regional system of receptivity. Despite 80% of hotels are having a website, 25% of the reservations made through the site; 50% of the bookings and 20% of the sites use web analytics tools. There are few hotels that use social media strategies.

Notes

1 University of Salento.
2 University of Salento.
3 University of Salento.

Bibliography

Buhalis, D. and Law, R. (2008). "Progress in information technology and tourism management: 20 years on and 10 years after the Internet: The state of eTourism research", *Tourism Management*, 29 (4), 609–623.
Donaldson, L. (1990). "The ethereal hand: Organizational economics and management theory", *Academy of Management Review*, 15 (3), 369–381.
Dubois, A. and Gadde, L.E. (2013). "Systematic combining: A decade later", *Journal of Business Research*, 67 (6), http://dx.doi.org/10.1016/j.jbusres.2013.03.036.
Eisenhardt, K.M. and Graebner, M.E. (2007). "Theory building from cases: Opportunities and Challenges", *Academy of Management Review*, 50 (1), 25–32.
Ganesan, S. (1994). "Determinants of long-term orientation in buyer-seller relationships', *Journal of Marketing*, 63 (2), 1–19.

Golinelli, G.M. (2010). *Viable Systems Approach (VSA): Governing Business Dynamics*, Kluwer, Cedam, Padova.

Gummesson, E. (2000). *Qualitative Methods in Management Research*, Sage Publications, Thousand Oaks, CA.

Guo, X., Zheng, X., Ling, L. and Yang, C. (2014). "Online marketing cooperation and competition between hotels and online travel agencies: From the perspective of cash back after stay", Working paper. University of Science and Technology of China, http://dx.doi.org/10.2139/ssrn.2296447.

Harrison, P.J. and Stevens, C.R. (1971). "A Bayesian approach to short term forecasting", *Operational Research Quarterly*, December.

Holtham, C., Lampel, J., Brady, C., and Rich, M. (2003). "How far can business war-rooms pro-vide an effective environment for management learning?", *Educational Innovation in Economics and Business (EDINEB 2003), Salzburg*, June.

Ip, C., Law, R. and Lee, H. (2011). "A review of website evaluation studies in the tourism and hospitality fields from 1996 to 2009", *International Journal of Tourism Research*, 13 (3), 234–265.

Lee, H.A., Guillet, B.D. and Law, R. (2013). "An examination of the relationship between online travel agents and hotels: A case study of choice hotels international and expedia.com", *Cornell Hospitality Quarterly*, 54 (1), 95–107.

Maturana, H. and Varela, F. (1992). *The Tree of Knowledge*, Shambhala, Boston & London.

Middleton, V.T.C., Fyall, A., Morgan, M. and Ranchhod, A. (2009). *Relationships in the Distribution Channel of Tourism*. 4th ed., Butterworth-Heinemann, Oxford, UK.

Priego, M.J.B. and Palacios, C.A. (2008). "Analysis of environmental statements issued by EMAS-certified Spanish hotels", *Cornell Hospitality Quarterly*, 49 (3), 381–394.

Saumure, K. (2001). *Focus Group: An Overview*, University of Alberta.

Stake, E.R. (2013). *Multiple Case Study Analysis*, The Guilford Press, New York.

Toh, R.S., Raven, P. and DeKay, F. (2011). "Selling rooms: Hotels vs. third-party websites", *Cornell Hospitality Quarterly*, 52 (2), 181–189.

Tse, T.S.M. (2013). "Chinese outbound tourism as a form of diplomacy", *Tourism Planning & Development, Special Issue: China Outbound Tourism*, 10 (2), 149–158.

Yin, R.K. (1984). *Case Study Research: Design and Methods*, Sage Publications, Thousand Oaks, CA.

Yin, R.K. (2012). *Applications of Case Study Research*, Sage Publications, Thousand Oaks, CA.

The role of banks in the sustainable development of territory

An ecosystems view

Vincenzo Formisano,[1] *Marialuisa Saviano,*[2]
Francesco Caputo[3] *and Maria Fedele*[4]

Keywords: *bank; sustainability; interaction; complex adaptive systems; ecosystem*

1 Introduction

In the last years, several issues, ranging from globalization to sustainable development, have increased the level of complexity in managing social and economic dynamics by underlying the need for a systems approach (Espejo 1994; Barile 2009; Golineli 2010; Espejo & Reyes 2011; Olsson & Bosch 2014; Saviano & Caputo 2012). The progressive widening of the interconnectedness between phenomena makes apparent the complex nature of socio-economics by highlighting their multi-dimensional dynamics (Akaka et al. 2012; Barile et al. 2015, 2016; Caputo 2016a).

In this context, the traditional view of territory as an objective area in which actors live has been surpassed by recognizing its irreducible systemic functioning (Barile et al. 2013b). This perspective has been progressively developed by researchers from different disciplinary domains leading to recognizing the need for an ecosystems view in the study of social and economic dynamics (Akaka et al. 2013; Vargo et al. 2015).

A new vision of territory development in which actors are engaged in a collaborative relationship to foster a sustainable development of territories is emerging (Cooke & Lazzeretti 2008). By embracing the ecosystems view, it becomes apparent that the territory development is an emerging outcome that requires the engagement of all the interested actors in a shared value co-creation logic (Barile & Saviano 2013, 2014; Barile et al. 2013a; Lusch et al. 2016). Economic, social, and environmental dynamics are intertwined in a complex adaptive systems functioning (Holland 1992). Therefore, each actor needs to harmonize his/her behavior with social, economic, and environmental dynamics to contribute to triggering virtuous circles of development (Akaka & Vargo 2015; Di Nauta et al. 2015; Saviano 2016).

Among the key actors for territory development, banks are expected to play a relevant role (Formisano et al. 2016). Over time, banks have progressively developed competitive logics based on a view of belonging to an autonomous economic sector in which to develop and maintain a competitive advantage (Hempel & Simonson 1999; Di Fatta et al. 2016). Although the banks' willingness to pursue their survival is legitimate and correct, in many cases, the ways in which they have pursued this aim have weakened their traditional mission of serving and sustaining the development of territory (Rich 2013) by contributing to triggering a vicious circle of stagnation (Palley 2012). Conversely, as the ecosystems view highlights, banks are one of the numerous actors (from policy makers to the citizens themselves) that must be engaged to share the mission of territory development. Development must be economically, socially, and environmentally sustainable (Girard & Nijkamp 2009; Hu & Scholtens 2014; Ioppolo et al. 2016). These requirements are strongly interrelated and imply all the territorial actors be involved to ensure the sustainable development of territory (Kloot & Martin 2000; Barile et al. 2014; Saviano et al. 2016).

By analyzing banks' communication strategies, it is possible to note that they ever more explicitly express their attention to the issues of territory, highlighting their role in promoting its sustainable development (Scholtens 2009; Sansone & Formisano 2016; Caputo et al. 2016; Evangelista et al. 2016). However, this approach appears generally inspired more by the banks' willingness to build profitable relationships with the market in the light of the dominant competition logic than by a true co-creation view (Bouma et al. 2001). In other words, this orientation does not seem to substantially change the view of their role in territory development.

With the aim of reinterpreting the traditional banks' 'service' mission of sustaining the economic development of territories, in the light of a new ecosystems view, this work proposes an exploratory study directed to investigate the role of banks in promoting the sustainable development of territory also from a social and an environmental perspective (Hu & Scholtens 2014). More specifically, the paper analyzes the relevance of social development and environmental protection for a sample of Italian banks and its impact on territory development. Moreover, the paper investigates the propensity of banks to adopt a collaborative approach in favoring the emergence of a shared ecosystems view of territory development (Válek & Jašíková 2013).

2 Hypotheses

On the basis of a literature review on the ecosystems view of sustainable development of territory and on the management of banks from a managerial and sustainability perspective, the paper states the following hypotheses:

H1: There is a positive and direct relationship between the relevance of environmental protection for banks and the number of green investments in the territory.

H2: There is a positive and direct relationship between the relevance of environmental protection for banks and the number of social investments in the territory.

    H3: There is a positive and direct relationship between the relevance of social development for banks and the numbers of green investments in the territory.

    H4: There is a positive and direct relationship between the relevance of social development for banks and the number of social investments in the territory.

3 Methodology

The paper builds on a questionnaire survey directly investigating the perceptions of banks' top management and employees of a sample of Italian banks. A questionnaire based on the literature review is used to measure the following dependent variables: (i) the relevance of environmental protection for banks and (ii) the relevance of social development for banks. Each dependent variable is measured using three questions scaled through a 7-point Likert Scale (Matell & Jacoby 1971).

The proposed hypotheses are tested using the Structural Equation Modeling (SEM) in order to verify if there is a positive relationship between the 'sensibility' of banks' employees about some drivers of sustainability and the local investment and attention about the topics of sustainability and the ecosystem.

The reliability of the model is verified using the Cronbach's alpha (Bland & Altman 1997) while the Convergent Validity (CV) of model is verified using the Composite Reliability (CR) and the Average Variance Extracted (AVE) (Cunningham et al. 2001). Finally, the fitness of model is verified using some fitness indexes such as the Adjusted Goodness of Fit Index (AGFI) (Tanaka & Huba 1989), the Comparative Fit Index (CFI) (Bentler 1990), and the Normed Fit Index (NFI) (Bollen 1989).

The conceptual model is reported in Figure 75.1.

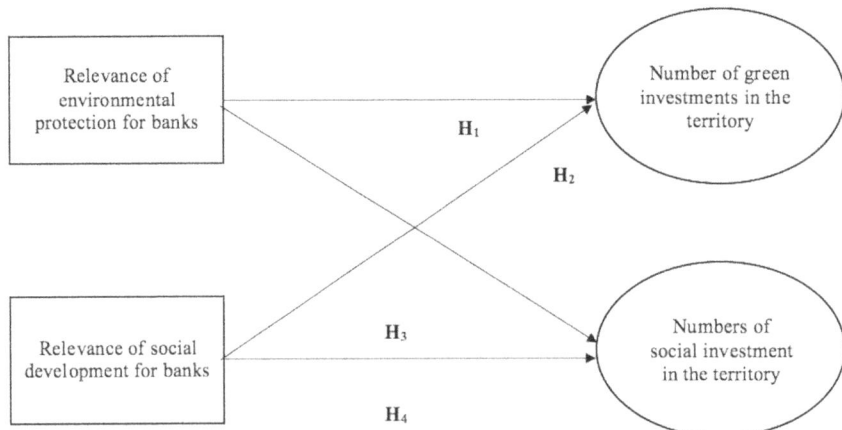

Figure 75.1 The conceptual model

4 Expected results

The paper offers exploratory advancements in knowledge about the role of banks in the sustainable development of territorial ecosystems.

5 Theoretical and practical implications

The paper highlights the role of banks in promoting a sustainable development of territory and the opportunity of a transition toward a shared ecosystem view of territory development. Key variables are identified that can support decision makers in this evolution.

6 Conclusions and future lines for research

With the aim of investigating the role of banks in supporting the emergence of a shared approach to territory development inspired by an ecosystems view, this work highlights the opportunity for banks to widen their mission including the promotion of social development and environmental protection as key factors of territory development.

Notes

1 University of Cassino and Southern Lazio.
2 University of Salerno.
3 Masaryk University.
4 University of Cassino and Southern Lazio.

Bibliography

Akaka, M. A., Vargo, S. L. (2015). Extending the context of service: From encounters to ecosystems. *Journal of Services Marketing*, 29(6/7), 453–462.

Akaka, M. A., Vargo, S. L., Lusch, R. F. (2012). An exploration of networks in value cocreation: A service-ecosystems view. *Review of Marketing Research*, 9(Special Issue), 13–50.

Akaka, M. A., Vargo, S. L., Lusch, R. F. (2013). The complexity of context: A service ecosystems approach for international marketing. *Journal of Marketing Research*, 21(4), 1–20.

Barile, S. (2009). *Management sistemico vitale*. Turin: Giappichelli.

Barile, S., Carrubbo, L., Iandolo, F., Caputo, F. (2013a). From 'EGO' to 'ECO' in B2B relationships. *Journal of Business Market Management*, 6(4), 228–253.

Barile, S., Saviano, M., Polese, F., Di Nauta, P. (2013b). Il rapporto impresa-territorio tra efficienza locale, efficacia di contesto e sostenibilità ambientale. *Sinergie rivista di studi e ricerche*, 90(1), 25–49.

Barile, S., Lusch, R., Reynoso, J., Saviano, M., Spohrer, J. (2016). Systems, networks, and ecosystems in service research. *Journal of Service Management*, 27(4), 652–674.

Barile, S., Saviano, M. (2013). An introduction to a value co-creation model: Viability, syntropy and resonance in dyadic interaction. *Syntropy*, 2, 69–89.

Barile, S., Saviano, M. (2014). Resource integration and value co-creation in cultural heritage management. In Aiello, L. (Ed.), *Handbook of Research on Management of Cultural Products: E-Relationship Marketing and Accessibility Perspectives* (pp. 58–82). Hershey: IGI Global.

Barile, S., Saviano, M., Caputo F. (2015). How Are Markets Changing? The Emergence of Consumers Market Systems. In Dominici, G., Evangelista, F. (Eds.), *Third International Symposium Advances in Business Management: 'Towards Systemic Approach'* (pp. 203–207). Avellino: Business Systems Laboratory, E-book Series.

Barile, S., Saviano, M., Iandolo, F., Calabrese, M. (2014). The viable systems approach and its contribution to the analysis of sustainable business behaviors. *Systems Research and Behavioral Science*, 31(6), 683–695.

Basile, G., Caputo, F. (2017). Theories and challenges for systems thinking in practice. *Journal of Organisational Transformation & Social Change*, 14(1), 1–3.

Bentler, P. M. (1990). Comparative fit indexes in structural models. *Psychological Bulletin*, 107(2), 238–246.

Bland, J. M., Altman, D. G. (1997). Statistics notes: Cronbach's alpha. *Bmj*, 314(7080), 572–614.

Bollen, K. A. (1989). A new incremental fit index for general structural equation models. *Sociological Methods & Research*, 17(3), 303–316.

Bouma, J. J., Jeucken, M., Klinkers, L. (Eds.). (2001). *Sustainable Banking: The Greening of Finance*. London: Greenleaf Pub.

Calabrese, M., Iandolo, F., Caputo, F., Sarno, D. (2017). From mechanical to cognitive view: The changes of decision making in business environment. In Barile, S., Pellicano, M., Polese, F. (Eds.), *Social Dynamics in a System Perspective* (pp. 223–240). New York: Springer.

Caputo, F. (2016a). A focus on company-stakeholder relationships in the light of the Stakeholder Engagement framework. In Vrontis, D., Weber, Y., Tsoukatos, E. (Eds.), *Innovation, Entrepreneurship and Digital Ecosystems* (pp. 455–470). Cyprus: EuroMed Press.

Caputo, F. (Ed.). (2016b). *Governing Business Systems: Theories and Challenges for Systems Thinking in Practice 4th Business Systems Laboratory International Symposium* (pp. 124–129). Avellino: Business Systems Laboratory.

Caputo, F. (2017). Reflecting upon knowledge management studies: Insights from systems thinking. *International Journal of Knowledge Management Studies*, 8(3/4), 177–190.

Caputo, F., Del Giudice, M., Evangelista, F., Russo, G. (2016). Corporate disclosure and intellectual capital: The light side of information asymmetry. *International Journal of Managerial and Financial Accounting*, 8(1), 75–96.

Caputo, F., Evangelista, F. (2017). Information sharing and cognitive involvement for sustainable workplaces. In Leon, R. D. (Ed.), *Managerial Strategies for Business Sustainability during Turbulent Times* (pp. 122–139). New York: IGI Global.

Caputo, F., Evangelista, F., Perko, I., Russo, G. (2017). The role of big data in value co-creation for the knowledge economy. In Vrontis, S., Weber, T., Tsoukatos, E. (Eds.), *Global and National Business Theories and Practice: Bridging the Past with the Future* (pp. 269–280). Cyprus: EuroMed Press.

Caputo, F., Evangelista, F., Russo, G., Buhnova, B. (2017). A systems view of companies' communication in online social environment. *Journal of Organizational Transformation & Social Change*. http://dx.doi.org/10.1080/14779633.2017.1291144.

Caputo, F., Perano, M., Mamuti, A. (2017). A macro-level view of tourism sector: Between smartness and sustainability. *Enlightening Tourism: A Pathmaking Journal*, 7(1), 36–61.

Caputo, F., Walletzky, L. (2017). Investigating the users' approach to ICT platforms in the city management. *Systems*, 5(1). doi:10.3390/systems5010001.

Carayannis, E. G., Caputo, F., Del Giudice, M. (2017). Technology transfer as driver of smart growth: A Quadruple/Quintuple innovation framework approach. In Vrontis, S., Weber, T., Tsoukatos, E. (Eds.), *Global and National Business Theories and Practice: Bridging the Past with the Future* (pp. 295–315). Cyprus: EuroMed Press.

Cooke, P. N., Lazzeretti, L. (Eds.). (2008). *Creative Cities, Cultural Clusters and Local Economic Development*. New York: Edward Elgar Publishing.

Cunningham, W. A., Preacher, K. J., Banaji, M. R. (2001). Implicit attitude measures: Consistency, stability, and convergent validity. *Psychological Science*, 12(2), 163–170.

Del Giudice, M., Ahmad, A., Scuotto, V., Caputo, F. (2017). Influences of cognitive dimensions on the collaborative entry mode choice of small and medium-sized enterprises. *International Marketing Review*, 34(5), 652–673.

Del Giudice, M., Scuotto, V., Khan, Z., Caputo, F., Carayannis, E. (2017). Micro level actions by owners-managers in improving sustainability practices of culture and creative Small and Medium Enterprises: UK-Italy Comparison. *Journal of Organizational Behaviour*, 1–19. doi:10.1002/job.2237.

Di Fatta, D., Caputo, F., Evangelista, F., Dominici, G. (2016). Small world theory and the World Wide Web: Linking small world properties and website centrality. *International Journal of Markets and Business Systems*, 2(2), 126–140.

Di Nauta, P., Merola, B., Caputo, F., Evangelista, F. (2015). Reflections on the role of university to face the challenges of knowledge society for the local economic development. *Journal of the Knowledge Economy*, 1–19. doi:10.1007/s13132-015-0333-9.

Dominici, G., Yolles, M., Caputo, F. (2017). Decoding the dynamics of value cocreation in consumer tribes: An agency theory approach. *Cybernetics and Systems: An International Journal*, 48(2), 84–101.

Espejo, R. (1994). What is systemic thinking? *System Dynamics Review*, 10(2–3), 199–212.

Espejo, R., Reyes, A. (2011). *Organizational Systems: Managing Complexity with the Viable System Model*. New York: Springer Science & Business Media.

Evangelista, F., Caputo, F., Russo, G., Buhnova, B. (2016). Voluntary corporate disclosure in the Era of Social Media. In Caputo, F. (Ed.), *The 4rd International Symposium Advances in Business Management: 'Towards Systemic Approach'* (pp. 124–128). Avellino: Business Systems, E-book series.

Formisano, V., Fedele, M., Antonucci, E. (2016). Innovation in financial services: A challenge for start-ups growth. *International Journal of Business and Management*, 11(3), 149–162.

Girard, L. F., Nijkamp, P. (Eds.). (2009). *Cultural Tourism and Sustainable Local Development*. London: Ashgate Publishing, Ltd.

Golineli, G.M. (2010). *Viable Systems Approach (VSA): Governing Business Dynamics*. Padova: Kluwer (Cedam).

Hempel, G. H., Simonson, D. G. (1999). *Bank Management: Text and Cases*. New York: Wiley.

Holland, J. H. (1992). Complex adaptive systems. *Daedalus*, 121(1), 17–30.

Hu, V. I., Scholtens, B. (2014). Corporate social responsibility policies of commercial banks in developing countries. *Sustainable Development*, 22(4), 276–288.

Ioppolo, G., Cucurachi, S., Salomone, R., Saija, G., Shi, L. (2016). Sustainable local development and environmental governance: A strategic planning experience. *Sustainability*, 8(2), 180–203.

Kloot, L., Martin, J. (2000). Strategic performance management: A balanced approach to performance management issues in local government. *Management Accounting Research*, 11(2), 231–251.

Lusch, R. F., Vargo, S. L., Gustafsson, A. (2016). Fostering a trans-disciplinary perspectives of service ecosystems. *Journal of Business Research*, 69(8), 2957–2963.

Matell, M. S., Jacoby, J. (1971). Is there an optimal number of alternatives for Likert scale items? Study. *Educational and Psychological Measurement*, 31(3), 657–674.

Olsson, H. H., Bosch, J. (2014, June). Ecosystem-driven software development: A case study on the emerging challenges in inter-organizational RD. In *International Conference of Software Business* (pp. 16–26). New York: Springer International Publishing.

Palley, T. I. (2012). *From Financial Crisis to Stagnation: The Destruction of Shared Prosperity and the Role of Economics*. Cambridge: Cambridge University Press.

Rich, B. (2013). *Mortgaging the Earth: The World Bank, Environmental Impoverishment, and the Crisis of Development*. Washington: Island Press.

Sansone, M., Formisano, V. (2016). Marketing innovation and key performance indicator in banking. *International Journal of Marketing Studies*, 8(1), 44–56.

Saviano, M. (2016). Il valore culturale del patrimonio naturale nella promozione dello sviluppo sostenibile. *Sinergie Italian Journal of Management*, 34(99), 163–190.

Saviano, M., Barile, S., Caputo, F. (2017). Re-affirming the need for systems thinking in social sciences: A viable systems view of smart city. In Vrontis, S., Weber, T., Tsoukatos, E. (Eds.), *Global and National Business Theories and Practice: Bridging the Past with the Future* (pp. 1552–1567). Cyprus: EuroMed Press.

Saviano, M., Barile, S., Spohrer, J., Caputo, F. (2017). A service research contribution to the global challenge of sustainability. *Journal of Service Theory and Practice*, 27(5), 951–976.

Saviano, M., Caputo, F. (2012). Le scelte manageriali tra sistemi, conoscenza e vitalità. Management senza confini. Gli studi di management: tradizione e paradigmi emergenti, XXXV Convegno annuale AIDEA, University of Salerno, 4–5 October, pp. 1–21.

Saviano, M., Caputo, F., Formisano, V., Walletzký, L. (2016). From theory to practice: Applying systems thinking to Smart Cities. In Caputo, F. (Ed.), *The 4rd International Symposium Advances in Business Management: 'Towards Systemic Approach'* (pp. 35–40). Avellino: Business Systems, E-book series.

Saviano, M., Nenci, L., Caputo, F. (2017). The financial gap for women in the MENA region: A systemic perspective. *Gender in Management: An International Journal*, 32(4), 203–217.

Scholtens, B. (2009). Corporate social responsibility in the international banking industry. *Journal of Business Ethics*, 86(2), 159–175.

Tanaka, J. S., Huba, G. J. (1989). A general coefficient of determination for covariance structure models under arbitrary GLS estimation. *British Journal of Mathematical and Statistical Psychology*, 42(2), 233–239.

Tronvoll, B., Barile, S., Caputo, F. (2017). A systems approach to understanding the philosophical foundation of marketing studies. In Barile, S., Pellicano, M., Polese, F. (Eds.), *Social Dynamics in a System Perspective* (pp. 1–18). Springer.

Válek, L., Jašíková, V. (2013). Time bank and sustainability: The permaculture approach. *Procedia-Social and Behavioral Sciences*, 92, 986–991.

Vargo, S. L., Wieland, H., Akaka, M. A. (2015). Innovation through institutionalization: A service ecosystems perspective. *Industrial Marketing Management*, 44, 63–72.

Credit risk and commercial banks' performance in Albania

A state model approach

Fatmira Kola,[1] Arsena Gjipali[2] and Erjon Sula[3]

Keywords: *credit risk; profitability; non-performing loans; loan loss provisions; moral hazard*

The banking sector has a key role in the economic development of each country. The banking sector is a complex system composed of a large number of stakeholders that interact in a non simple way continuously. The banking sector's classification as complex results, in other words, from varieties heritage. The economies of developing countries like Albania are characterized by high demand for credit due to increasing investment. The revenues are even higher when the risk is greater. The high risk-associated credits leads to high returns. Credit risk is one of the most important kinds of risk in the banking sector that affects bank performance, as it exhibits the loss probability because of the failure of the debtor to fulfill its obligations to the bank. In June 2016 the level of Non Perfoming Loans (NPL) in Albania appeared in 24.4% of total loans, representing a major obstacle to the development and performance of the banking sector in Albania.

The non-performing loan problem is considered to be one of the biggest causes behind the Albanian banking difficulties. Commercial banks in Albania appear very profitable, significantly higher than bank returns in other parts of the world. The relationship between the business cycle and banks' loan losses was one of the hot debates in recent economic literature especially in relation to financial stability analysis. The quality of loans can be one of the factors that limit the banks' loan supply and effect on investment spending. A recent decline in revenue was observed due to higher provisioning expenses, which reduce banks' profits. The control on bank risks is one of the most important factors the profitability of the bank depends on. Profitability is one of the main reasons for the existence of business enterprises, and business enterprises continue their operation by making profits. It is one of the most important indicators for the investors.

A recent decline in revenue was observed due to higher provisioning expenses, which reduce banks' profits. To identify factors affecting bank profitability, we have got to study the: bank-specific (internal) factors, industry-specific and macroeconomic (external) factors. The internal factors that influence profitability are expressed in terms of efficiency, productivity, competition, concentration, soundness, safety and profitability. An industry-specific factor is the market concentration, while macroeconomic factors are Gross Domestic Product (GDP), Inflation Rate and Real effective exchange rate (REER). In this paper, we will test whether lending decisions of all banks operating in Albania exhibit moral hazard. The study carried out an empirical investigation into the quantitative effect of credit risk on the performance of 16 commercial banks in Albania organized quarterly for 14 years (2002–2015).

The objective of the present study is to investigate the factors affecting bank performance in Albania. We use secondary data of the micro- and macroeconomic level. The macroeconomic data are drawn from the databases of the National Institute of Statistics (INSTAT) and the Bank of Albania. The bank-level data are extracted from each of the 16 actual banks in the market Report, Albanian Association of Banks (AAB) and Bank of Albania.

The purpose of the estimable model outlined in this section is to capture the effects of macroeconomic, bank-specific factors and the Herfindahl-Hirschmann index (HHI) in the industry of bank performance. We also include a range of bank-specific variables that have been used in previous empirical studies that examine the drivers of bank performance.

We found strong impact on the bank performance of the various macroeconomic, bank-specific factors and HHI. According to the results, a slight increase of risks to the financial stability in the banking sector, expressed in the model with the increase of non-performing loans, negatively affects bank performance. The non-performing loans are associated with a slight reduction of their coverage with reserve and capital funds. Increase of foreign currency loans in the banking sector contributed negatively in the reduction of risks evaluation. Credit to enterprises, foreign currency credit and medium-term credit give the main contribution to this increase. For individual banks, the increase in provisioning expenses for credit affected the establishment of the financial loss.

Notes

1 Tor Vergata University.
2 University of Tirana.
3 University Metropolitan Tirana.

Bibliography

Angbazo, L. (1997). Commercial bank net interest margins, default risk, interest rate risk and off balance sheet banking, *Journal of Banking and Finance*, 21 (1), 55–87.

Barile, S. (2009). Verso la qualificazione del concetto di complessità sistemica, *Sinergie*, 79.

Bourke, P. (1989). Concentration and other determinants of bank profitability in Europe, North America and Australia, *Journal of Banking and Finance*, 13, 65–79.

Fiordelisi, F., Marques-Ibanez, D., & Molyneux, P. (2010). Efficiency and risk in European banking, *European Central Bank*, (1211), June.
Molyneux, P. & Thornton, J. (1992). Determinants of European bank profitability, *Journal of Banking and Finance*, 16 (6), 1173–1178.

Websites

www.bankofalbania.org/web/Raporti_i_Stabilitetit_Financiar_per_gjashtemujorin_e_pare_te_vitit_2016_7596_1.php.

Towards a smart systems view of museum networks

Francesco Caputo,[1] *Marta Maria Montella,*[2]
Leonard Walletzky[3] *and Biagio Merola*[4]

Keywords: *Cultural Heritage Management; museum networks; systems thinking; smart technologies; management of variety*

1 An introductive overview

The domain of cultural heritage management includes the complex of activities planned and implemented with the aim to define more efficient, effective, and sustainable approaches in the management of products, services, and traditions that have a high value for a specific culture (Stovel 1998; Cameron & Kenderdine 2007; Montella 2010a, 2011; Cerquetti 2011).

Over time, different research streams have focused the attention on the opportunity to define new pathways and perspectives in the management of cultural heritage as a way to increase its positive effects on the social and economic development of territories (Bessière 1998; Scott 2004; Hampton 2005; Cerquetti 2010).

In such a line, many contributions have been offered with reference to the implementation of more efficient communication strategies (Kalay et al. 2007), to the development of more appealing experiences (Otnes & Maclaran 2007; Pietroni et al., 2012), and to the adoption of technology-based innovations to increase the attractiveness of cultural heritage (Meyer et al. 2007; Ott & Pozzi 2011).

Despite the relevance of all these contributions, it is possible to underline the existence of a dominant object-based approach in which the cultural heritage is viewed as a sort of 'good to sell' by acting on the traditional marketing levers (Barile and Saviano 2012; Barile & Saviano 2014). Conversely, little attention is paid to the opportunities related to a radical change in perspective in the way in which the cultural heritage is perceived (Inglehart 1990; Kreps 2003; Barile 2012).

In order to bridge this gap, the paper builds upon the conceptual and interpretative framework offered by the Systems Thinking (Beer 1979; Espejo 1990, 1994; Espejo & Reyes 2011; Barile & Saviano 2010, 2011; Golinelli 2010; Barile et al. 2012b, 2015, 2016; Basile & Caputo 2017) and the Self-Organization Theory (Witt 1997; Foster & Metcalfe 2003; Ulrich & Probst 2012) in order to identify possible approaches that can support the widening of perspective in the management of cultural heritage.

More specifically, by shifting the focus from the management of cultural heritage items to the management of cultural heritage systems, the paper focuses the attention on the topic of museum networks (Lorenzoni 1987; Crooke 2006; Pencarelli & Splendian 2011). In such a line, the work investigates the conditions required for the emergence and survival of museum networks in order to highlight the possible contribution of Information and Communication Technologies (ICTs) in defining more performant managerial models by acting on collaboration, information sharing, and communication (Ing 1999; Lemelin & Bencze 2004; Marty & Jones 2008; Caputo & Walletzký 2017).

To this aim, the work adopts the interpretative lens of the Viable Systems Approach (VSA) (Barile 2009a; Golinelli 2010; Barile et al. 2014) in order to 'observe' museum networks as viable systems (Barile 2009b) in terms of systems able to survive in their contexts by establishing effective relationships with their relevant suprasystems.

2   Heading

By adopting the interpretative contributions offered by Systems Thinking, the paper proposes a literature review on the topic of cultural heritage management in order to: (i) define criteria and guidelines for an effective systems management of museum networks (Van Huy 2006; Cerquetti & Montela 2015); (ii) develop a framework of reference useful to investigate the museum network as a viable system (Barile 2013; Barile et al. 2014; Saviano & Caputo 2012); and (iii) underline the advantages related to the network configuration in the cultural heritage management (Thorelli 1986; Lorenzoni 1992; Latin 1991; Polese 2004; Di Fatta et al. 2016; Dominici et al. 2017).

The theoretical reflections herein are verified using a qualitative method approach (Gubrium & Holstein 1997) based on the analysis of a single case study (Flyvbjerg 2006): the Sistema Museale Regionale dell'Umbria. The empirical observation is to directly investigate what are the elements able to affect the emergence of a viable system from the management of a museum network. Moreover, a technology-based view is adopted in order to define possible contributions of ICT in improving network museums' ability to dynamically adapt themselves to the changes of 'markets' through a more efficient, effective, and sustainable approach to the management of variety (Dickover 1994; Montella 2010b; de Oliveira & da Silva 2011; Espejo & Reyes 2011).

3   Theoretical and practical implications

The interpretative framework herein supports a better understanding of the variety that affects the interactions among the various actors that are involved in the articulated scenario of cultural heritage (Barile & Saviano 2012; Golinelli 2012) and of the opportunities offered by the ICTs in ensuring a smart approach in the management of cultural heritage (Frattasi et al. 2006; Duff et al. 2010).

In the same direction, the paper offers useful indications to decision makers interested in adopting an innovative approach to the management of museum networks able to overcome the limitations of traditional transactional views (Gouthier & Schmid 2003; Van de Werfhorst & Hofstede 2007).

4 Conclusions and future directions for research

In a context like Italy in which the social and economic development of a territory can significantly rely on cultural heritage, the development of adequate managerial approaches for museums is acquiring an increasing relevance. In this direction, the paper highlights the contribution of systems thinking to the management of museum networks by underlining the relevant opportunities offered by a smart approach to ICT in improving collaboration, information sharing, and communication between different cultural heritage units in the building of shared pathways to address the challenges of increasing market variety (Caputo et al. 2016a, 2016b).

Notes

1 Masaryk University.
2 Sapienza University of Rome
3 Masaryk University.
4 University of Foggia.

Bibliography

Barile, S. (2009a). *Management Sistemico Vitale*. Torino: Giappichelli.
Barile, S. (2009b). The dynamic of informative varieties in the processes of decision making. In The 3rd International Conference on Knowledge Generation, Communication and Management, Orlando, FL, July.
Barile, S. (2012). Verso una novata ipotesi di rappresentazione del concetto di bene culturale. In Golinelli, G.M. (Eds.), *Patrimonio culturale e creazione di valore. Verso nuovi percorsi* (pp. 71–96). Padova: Cedam.
Barile, S. (Ed.) (2013). *Contributions to Theoretical and Practical Advances in Management: A Viable Systems Approach (VSA)*, Vol. 2. Roma: Aracne.
Barile, S., Lusch, R., Reynoso, J., Saviano, M., Spohrer, J. (2016). Systems, networks, and ecosystems in service research. *Journal of Service Management*, 27(4), 652–674.
Barile, S., Montella, M., Saviano, M. (2012a). A service-based systems view of cultural heritage. *Journal of Business Market Management*, 5(2), 106–136.
Barile, S., Pels, J., Polese, F., Saviano, M. (2012b). An introduction to the viable systems approach and its contribution to marketing. *Journal of Business Market Management*, 5(2), 54–78.
Barile, S., Saviano, M. (2010). A new perspective of systems complexity in service science. *Impresa, Ambiente, Management*, 4(3), 375–414.
Barile, S., Saviano, M. (2011). Foundations of systems thinking: The structure-system paradigm. In Vv. Aa, *Contributions to Theoretical and Practical Advances in Management: A Viable Systems Approach (VSA)* (pp. 1–24). Avellino: International Printing.

Barile, S., Saviano, M. (2012). Dalla Gestione del Patrimonio di Beni Culturali al Governo del Sistema dei Beni Culturali. In Golinelli, G.M. (Ed.), *Patrimonio culturale e creazione di valore, Verso nuovi percorsi* (pp. 97–148). Padova: Cedam.

Barile, S., Saviano, M. (2014). Resource integration and value co-creation in cultural heritage management. In Aiello, L. (Ed.), *Handbook of Research on Management of Cultural Products: E-Relationship Marketing and Accessibility Perspectives* (pp. 58–82). Hershey, PA: Business Science Reference.

Barile, S., Saviano, M., Caputo, F. (2014). A systems view of customer satisfaction. In National Conference "Excellence in quality, statistical quality control and customer satisfaction", University Campus "Luigi Einaudi", University of Turin, September 18–19.

Barile, S., Saviano, M., Caputo, F. (2015). How are markets changing? The emergence of consumers market systems. In Dominici, G. (Ed.), *The 3rd International Symposium Advances in Business Management: "Towards Systemic Approach"* (pp. 203–207). Avellino: Busyness Systems, E-book Series.

Basile, G., Caputo, F. (2017). Theories and challenges for systems Thinking in practice. *Journal of Organisational Transformation & Social Change*, 14(1), 1–3.

Beer, S. (1979). *The Heart of Enterprise*. New York: John Wiley & Sons.

Bessière, J. (1998). Local development and heritage: Traditional food and cuisine as tourist attractions in rural areas. *Sociologia ruralis*, 38(1), 21–34.

Cameron, F., Kenderdine, S. (2007). *Theorizing Digital Cultural Heritage: A Critical Discourse*. Cambridge, MA: MIT Press.

Caputo, F., Evangelista, F., Russo, G. (2016b). Information sharing and communication strategies: A stakeholder engagement view. In Vrontis, D., Weber, Y., Tsoukatos, E. (Eds.), *Innovation, Entrepreneurship and Digital Ecosystems* (pp. 436–442). Cyprus: EuroMed Press.

Caputo, F., Giudice, M.D., Evangelista, F., Russo, G. (2016a). Corporate disclosure and intellectual capital: The light side of information asymmetry. *International Journal of Managerial and Financial Accounting*, 8(1), 75–96.

Caputo, F., Walletzký, L. (2017). Investigating the users' approach to ICT platforms in the city management. *Systems*, 5(1), 1–15.

Cerquetti, M. (2010). Dall'economia della cultura al management per il patrimonio culturale: presupposti di lavoro e ricerca. Il capitale culturale. *Studies on the Value of Cultural Heritage*, 1, 23–46.

Cerquetti, M. (2011). L'innovazione del prodotto culturale. *Economia, cultura, territorio*, 55–69.

Cerquetti, M., Montela, M.M. (2015). Museum networks and sustainable tourism management: The case study of Marche region's museums (Italy). enlightening tourism. *A Pathmaking Journal*, 5(1), 100–125.

Crooke, E. (2006). Museums and community. *A Companion to Museum Studies*, 171–185.

de Oliveira, J.A., da Silva, A.J. (2011). Arts, culture and science and their relationships. *Systemic Practice and Action Research*, 24(6), 565–574.

Dickover, N. (1994), Reflection-in-action: Modelling a specific organization through the Viable Systems Model. *Systems Practice*, 7(1), 43–62.

Di Fatta, D., Caputo, F., Evangelista, F., Dominici, G. (2016). Small world theory and the World Wide Web: Linking small world properties and website centrality. *International Journal of Markets and Business Systems*, 2(2), 126–140.

Dominici, G., Yolles, M., Caputo, F. (2017). Decoding the dynamics of value cocreation in consumer tribes: An agency theory approach. *Cybernetics and Systems*, 48(2), 84–101.

Duff, W.M., Carter, J., Howarth, L., Ross, S., Dallas, C. (2010). The museum environment in transition: The impact of technology on museum work. *Cultural Heritage on Line*, 1000–1005.

Espejo, R. (1990). The viable system model. *Systemic Practice and Action Research*, 3(3), 219–221.

Espejo, R. (1994). What is systemic thinking? *System Dynamics Review*, 10(2–3), 199–212.

Espejo, R., Reyes, A. (2011). *Organizational Systems: Managing Complexity with the Viable System Model*. New York: Springer Science & Business Media, Heidelberg, &Springer.

Flyvbjerg, B. (2006). Five misunderstandings about case-study research. *Qualitative Inquiry*, 12(2), 219–245.

Foster, J., Metcalfe, J.S. (Eds.) (2003). *Frontiers of Evolutionary Economics: Competition, Self-Organization, and Innovation Policy*. London: Edward Elgar Publishing.

Frattasi, S., Fathi, H., Fitzek, F.H., Prasad, R., Katz, M.D. (2006). Defining 4G technology from the users perspective. *IEEE Network*, 20(1), 35–41.

Golinelli, G.M. (2010). *Viable Systems Approach: Governing Business Dynamics*. Padova: Cedam.

Golinelli, G.M. (Ed.) (2012). *Patrimonio culturale e creazione di valore, Verso nuovi percorsi*. Padova: Cedam.

Gouthier, M., Schmid, S. (2003). Customers and customer relationships in service firms: The perspective of the resource-based view. *Marketing Theory*, 3(1), 119–143.

Gubrium, J.F., Holstein, J.A. (1997). *The New Language of Qualitative Method*. Cambridge: Oxford University Press on Demand.

Hampton, M.P. (2005). Heritage, local communities and economic development. *Annals of Tourism Research*, 32(3), 735–759.

Ing, D.S. (1999). Innovations in a technology museum. *IEEE Micro*, 19(6), 44–52.

Inglehart, R. (1990). *Culture Shift in Advanced Industrial Society*. New York: Princeton University Press.

Kalay, Y., Kvan, T., Affleck, J. (Eds.) (2007). *New Heritage: New Media and Cultural Heritage*. London: Routledge.

Kreps, C.F. (2003). *Liberating Culture: Cross-Cultural Perspectives on Museums, Curation, and Heritage Preservation*. London: Psychology Press.

Latin, R.V. (1991). Cybernetics and network management (viable system modeling). *Systems Practice*, 4(4), 339–360.

Lemelin, N., Bencze, L. (2004). Reflection-on-action at a science and technology museum: Findings from a university-museum partnership. *Canadian Journal of Math, Science & Technology Education*, 4(4), 467–481.

Lorenzoni, G. (1987). Costellazione di imprese e processi di sviluppo. *Sviluppo e Organizzazione*, 102, 59–72.

Lorenzoni, G. (Ed.) (1992). *Accordi, reti e vantaggio competitivo. Le innovazioni nell'economia d'impresa e negli assetti organizzativi*. Milano: Etas Libri.

Marty, P.F., Jones, K.B. (2008). *Museum Informatics: People, Information, and Technology in Museums*. London: Taylor & Francis.

Meyer, É., Grussenmeyer, P., Perrin, J.P., Durand, A., Drap, P. (2007). A web information system for the management and the dissemination of Cultural Heritage data. *Journal of Cultural Heritage*, 8(4), 396–411.

Montella, M. (2010a). Le scienze aziendali per la valorizzazione del capitale culturale storico. Il capitale culturale. *Studies on the Value of Cultural Heritage*, 1(1), 11–22.

Montella, M. (2010b). Arte, comunicazione, valore: una conversazione. Il capitale culturale. *Studies on the Value of Cultural Heritage*, 1(1), 149–161.

Montella, M. (2011). Conoscenza e informazione del cultural heritage come spazio d'impresa. *Sinergie rivista di studi e ricerche*, 76, 91–111.

Otnes, C.C., Maclaran, P. (2007). The consumption of cultural heritage among a British Royal Family brand tribe. *Consumer Tribes*, 51–66.

Ott, M., Pozzi, F. (2011). Towards a new era for Cultural Heritage Education: Discussing the role of ICT. *Computers in Human Behavior*, 27(4), 1365–1371.

Pencarelli, T., Splendiani, S. (2011). Le reti museali come "sistemi" capaci di generare valore: verso un approccio manageriale e di marketing. Il capitale culturale. *Studies on the Value of Cultural Heritage*, (2), 227–252.

Pietroni, E., Ray, C., Rufa, C., Pletinckx, D., Van Kampen, I. (2012). Natural interaction in VR environments for Cultural Heritage and its impact inside museums: The Etruscanning project. In Virtual Systems and Multimedia (VSMM), 2012 18th International Conference on, IEEE, September, pp. 339–346.

Polese, F. (2004). *L'integrazione sistemica degli aggregati reticolari di impresa*. Padova: Cedam.

Ronald, C. (1937). The nature of the firm. *Economica*, 4(16), 386–405.

Saviano, M., Caputo, F. (2012). Le scelte manageriali tra sistemi, conoscenza e vitalità. Management senza confini. Gli studi di management: tradizione e paradigmi emergenti, XXXV Convegno annuale AIDEA, University of Salerno, 4–5 October, pp. 1–21.

Scott, A.J. (2004). Cultural-products industries and urban economic development prospects for growth and market contestation in global context. *Urban Affairs Review*, 39(4), 461–490.

Stovel, H. (1998). *Risk Preparedness: A Management Manual for World Cultural Heritage*. London: ICCROM.

Thorelli, H.B. (1986). Networks: Between Market and Hierarchies. *Strategic Management Journal*, 7(1), 37–51.

Ulrich, H., Probst, G. (Eds.) (2012). *Self-Organization and Management of Social Systems: Insights, Promises, Doubts, and Questions*. New York: Springer Science & Business Media.

Van de Werfhorst, H.G., Hofstede, S. (2007). Cultural capital or relative risk aversion? Two mechanisms for educational inequality compared. *The British Journal of Sociology*, 58(3), 391–415.

Van Huy, N. (2006). The role of museums in the preservation of living heritage: Experiences of the Vietnam Museum of Ethnology. *International Journal of Intangible Heritage*, 1, 35–41.

Witt, U. (1997). Self-organization and economics: What is new? *Structural Change and Economic Dynamics*, 8(4), 489–507.

Managing variety in healthcare through personalized medication

The contribution 3D-printing technologies

Rita Patrizia Aquino,[1] *Sergio Barile,*[2]
Antonio Grasso[3] *and Marialuisa Saviano*[4]

Keywords: *healthcare; variety management; smartness; sustainability; personalized medication; 3D-printing*

Among the numerous humankind problems, healthcare represents a key issue to manage. Various managing systems are implemented reflecting different views, approaches and the socio-economic conditions of the country (Glouberman and Mintzberg, 1996; Plsek & Wilson, 2001; Wendt, 2009; Swayne et al., 2012). However, a fundamental problem affects almost every healthcare system: managing the trade-off between the need to deliver effective healthcare services and the need to control expenses and to ensure the overall sustainability of the system (France et al., 2005). This problem is an expression of the complexity of healthcare systems.

Healthcare systems, in fact, have been studied as complex adaptive systems whose management is required to adopt an adaptive approach based on incentives and inhibitions instead of command and control (Rouse, 2008). Despite the massive organizational efforts, healthcare processes are hard to be effectively designed, controlled and, most of all, optimized. Decision makers have to manage high levels of variety and variability (Espejo, 1994, 2015a, 2015b; Blecker & Abdelkafi, 2006; Gershenson, 2015), taking into account not only the irreducible information asymmetry in the relationship with patients/users (Bloom et al., 2008; Barile et al., 2014a, Barile et al. 2014b) but also the diverging interests and needs that characterize actors involved in the healthcare system dynamics, starting from the private and public ones (Savas & Savas, 2000; Barlow et al., 2013; Saviano et al., 2014).

In the Italian healthcare system, like in many others, the need to control expenses and to ensure the sustainability of the system has led to the introduction of the economic logic of management (France et al., 2005; Anessi-Pessina & Cantù, 2006; Borgonovi et al., 2008). This logic, however, has been mainly interpreted in terms of cutting expenses instead of improving efficiency, reducing waste, etc., so ending up reducing the level of service offered (e.g. number hospital beds, days of hospitalization for surgery, number of expensive new drugs provided by NHS, etc.). To improve efficiency, instead, the managerial logic has oriented towards the standardization of processes by introducing the use of protocols, standards, etc. in the service delivery. The definition of protocols and standards is expected to find the 'best way' to implement processes (Kongstvedt, 2001). Although conceived to widely ensure a standard level of quality of service, the standardization of healthcare does not imply improvements of the effectiveness of service.

In terms of management of variety, the standardization of healthcare services implies a reduction of the variety to manage, especially the variety that emerges from the patient side. Patients would appreciate a personalized service; they, instead, are classified as 'groups of diseases' on the basis of protocols systems, e.g. the DRGs (Diagnosis Related Groups) that define standardized procedures of treatment to determine the payment due to healthcare providers. Although the DRG-based payment systems have been adopted with the aims of improving transparency, efficiency and quality in hospitals, the pressure for efficiency they introduce, often leads hospitals "to skimp on quality as a way of saving costs by manipulating the services/care provided to patients [and] technology adoption rates may decelerate if new technologies do not induce cost-savings" (Busse et al., 2011: 156). Hence, we wonder:

What are the implications of the standardization of healthcare from the perspective of the effectiveness of the service?

Although many efforts are in place to reconcile effectiveness, efficiency and sustainability of healthcare (Saviano et al., 2010, 2015), the trade-offs remain making complex the management of healthcare. As mentioned, these trade-offs are substantially expressions of the multiple perspectives and interests that are in play: the healthcare organizations, on the one hand, and the users/patients, on the other hand, represent the two sides of a relationship that is becoming ever more problematic (Barile et al., 2014a). Relevant variety to manage (and reconcile) appears in between the healthcare service provider-client relationship. This variety is the outcome of multiple overlapping dynamics that recursively emerge from within the two sides of the relationship as an expression of suprasystems influences that affect needs, orientations and choices. Clearly, in this context, the pressure for cutting expenses is making dominant the 'efficiency' perspective of the healthcare provider that, through standardization, tends to reduce the variety potentially emerging from the patient side, who, instead, takes an 'effectiveness' perspective (Saviano et al., 2010).

How can the effectiveness, efficiency and sustainability of healthcare be harmonized?

A well-established research stream looks at healthcare systems as 'service systems', i.e. types of complex adaptive systems that are value co-creation configurations of people, technology, internal and external service systems connected by value propositions, and shared information (such as language, laws, measures, models, etc.) (IfM & IBM, 2008; Iandolo et al., 2013). Accordingly, people and technology represent the two basic components of healthcare systems, like in any organization. Subsequently, given that people are the main source of variety, both on the side of the provider and on the side of the client, what is the role of technology in healthcare from a variety management perspective?

In the design of service systems, technology (especially Information and Communication Technology) is expected, on the one hand, to support interaction through effective information management (Chaudhry et al., 2006; Sabatino et al., 2014; Corrente et al., 2015), but on the other hand, to allow cost reduction through making organizations more efficient (Skinner, 2003). The question, then, becomes:

What can be the contribution of technology in harmonizing the effectiveness, efficiency and sustainability of healthcare?

Massive investments in smart health technologies and an increasing use of Health Technology Assessment methods (Philips et al., 2004) characterize the interest of healthcare for technology, whose positive impact on quality, Efficiency and costs is widely recognized. The contribution of technology to the effectiveness of healthcare has received less attention, although increasing in particular in terms of service personalization.

One of the healthcare services in which the need of customization would be great, is 'personalized medicine', a model for healthcare where the patient's individual profile leads decision making (Rodriquez et al., 2015). This model has been proposed in various areas such as tissue and organ fabrication; creation of customized prosthetics, implants and anatomical models; surgical instrumentation, reconstructive surgery, rapid prototype service etc. (FDA October 8–9, 2014 Additive Manufacturing of Medical Devices: An Interactive Discussion on the Technical Considerations of 3D Printing) whereas the potential of application in 3D-printed medical in pharmaceutics (devices and drugs) is recently emerging (FDA, May 4th 2016, Technical Considerations for Additive Manufactured Devices; Draft Guidance for Industry and Food and Drug Administration Staff; Availability). In effect, the path to personalized custom 3d-printed dosage forms and drug delivery devices has been already traced with the aim to deliver "the right drug at the right dose at the right time" (Hamburg & Collins, 2010: 301). Personalized medicine, in fact, would be an excellent answer to the need of appropriateness of healthcare. However, many obstacles are on the way towards the shared application of personalized drugs and devices both from scientific, regulatory and social viewpoints. The key challenge is "how are the requisite 'unique' medicines for each patient to be manufactured on a routine basis?" (Khaled et al., 2014: 105). For example, oral tablets are the most popular drug dosage form because of ease of manufacture, pain avoidance, accurate dosing and good patient compliance. However, no viable method is available that could routinely be used to make personalized tablets. In effect,

> tablets are almost universally manufactured at large centralized plants via . . . processes using tablet presses essentially unchanged in concept for well over a century. This route to manufacture is clearly unsuited to personalized medication and in addition provides stringent restrictions on the complexity achievable in the dosage form.
>
> (Khaled et al., 2014: 105)

Moreover, it has been observed that "Health care services need to be customized to fit not only a patient's medical condition but also the patient's age, mental condition, personal traits, preferences, family circumstances, and financial capacity" (Berry & Bendapudi, 2007: 115).

It clearly appears a relevant variety to manage in the personalization of healthcare. Managing this variety, however, is expected to imply an increase in costs (Puschmann & Alt, 2001; Holmes et al., 2009; Swan, 2009). Yet, using technology to manage this variety may help to avoid this increase in costs, i.e. to harmonize the efficiency and effectiveness of healthcare. In fact, health technology assessment can be approached on the basis of cost-effectiveness evaluations: focus is both on increasing effectiveness and reducing costs (Ash et al., 2004; Chaudhry et al., 2006).

With the purpose of exploring possibilities offered by technology for reconciling standardization and personalization (i.e. efficiency and effectiveness) of healthcare, under a general view of sustainability, this work aims to investigate the criticalities and potentialities of the use of 3D-printing technology in personalized medication (Khaled et al., 2014; Choonara et al., 2016).

Our exploratory study focuses attention on the context in which basic needs and opportunities of personalization in dosage forms mainly emerge. Advantages of 3D printing include precise control of droplet size and dose, high reproducibility, complex standardized drug manufacturing processes and the ability to produce dosage forms, that are likely to challenge conventional drug fabrication, characterized by innovative and unique dosage morphology/forms, personalized drug dosing, complex drug-release profiles. 3D-manufacturing of personalized medication is the final stage of the drug prescription and delivery processes that involve doctors, patients and pharmacists. As to issues concerning 3D drug printing, they comprise: safety and security issues over the long-term effects, which will clearly need to be monitored; regulatory barriers both in the case of a large availability which requires randomized controlled trials, time and funding and in the case of pharmacists dispensing, which must also be legally better defined as manufacturing or compounding equipment; finally, patents and copyrights may be an issue encountered in 3D printing for personal use, non-profit or NHS distribution. A deep analysis of the criticalities and potentialities of personalization in medicine as well as possible advantages in terms of both effectiveness and cost of healthcare needs to be performed by taking the perspective of key stakeholders in order to identify obstacles, risks and opportunities of personalized medicine as well as possibilities offered by the introduction of 3D-printing technologies in pharmacies' drug delivery process.

Envisioning a smart and sustainable healthcare is our aim.

Notes

1 Pharma_nomics Interdepartmental Center, University of Salerno.
2 Sapienza, University of Rome.
3 Pharma_nomics Interdepartmental Center, University of Salerno.
4 Pharma_nomics Interdepartmental Center, University of Salerno.

Bibliography

Anessi-Pessina, E., & Cantù, E. (2006). Whither managerialism in the Italian national health service? *The International Journal of Health Planning and Management*, 21(4), 327–355.

Ash, J. S., Berg, M., & Coiera, E. (2004). Some unintended consequences of information technology in health care: The nature of patient care information system-related errors. *Journal of the American Medical Informatics Association*, 11(2), 104–112.

Barile, S. (2013). *Contributions to Theoretical and Practical Advances in Management: A Viable Systems Approach (VSA)*. Aracne, Roma.

Barile, S., Saviano, M., & Caputo, F. (2014a). How are markets changing? The emergence of consumers market systems. In Gandolfo, D. (Ed.), *3rd International Symposium: Advances in Business Management: Towards Systemic Approach* (pp. 203–207). Business Systems Laboratory, E-book Series, Avellino.

Barile, S., Saviano, M., & Polese, F. (2014b). Information asymmetry and co-creation in health care services. *Australasian Marketing Journal (AMJ)*, 22(3), 205–217.

Barlow, J., Roehrich, J., & Wright, S. (2013). Europe sees mixed results from public-private partnerships for building and managing health care facilities and services. *Health Affairs*, 32(1), 146–154.

Berry, L. L., & Bendapudi, N. (2007). Health care a fertile field for service research. *Journal of Service Research*, 10(2), 111–122.

Blecker, T., & Abdelkafi, N. (2006). Complexity and variety in mass customization systems: Analysis and recommendations. *Management Decision*, 44(7), 908–929.

Bloom, G., Standing, H., & Lloyd, R. (2008). Markets, information asymmetry and health care: Towards new social contracts. *Social Science & Medicine*, 66(10), 2076–2087.

Borgonovi, E., Fattore, G., & Longo, F. (2008). *Management delle istituzioni pubbliche*. Egea, Milano.

Busse, R., Geissler, A., & Quentin, W. (2011). *Diagnosis-Related Groups in Europe: Moving towards Transparency, Efficiency and Quality in Hospitals*. McGraw-Hill Education, London.

Chaudhry, B., Wang, J., Wu, S., Maglione, M., Mojica, W., Roth, E., & Shekelle, P. G. (2006). Systematic review: Impact of health information technology on quality, efficiency, and costs of medical care. *Annals of Internal Medicine*, 144(10), 742–752.

Choonara, Y. E., du Toit, L. C., Kumar, P., Kondiah, P. P., & Pillay, V. (2016). 3D-printing and the effect on medical costs: A new era? *Expert Review of Pharmacoeconomics & Outcomes Research*, 16(1), 23–32.

Corrente, M. I., Grasso, A., Villecco, F., d'Amore, M., Aquino, R. P. (2015). Phytovigilance of soy isoflavones products: Use of the Business Intelligence for designing efficient Monitoring Forms. *International Journal of Mechanical Engineering and Industrial Design*, 3(3), 1–8.

Espejo, R. (1994). What is systemic thinking? *System Dynamics Review*, 10(2–3), 199–212.

Espejo, R. (2015a). An enterprise complexity model: Variety engineering and dynamic capabilities. *International Journal of Systems and Society*, 2(1),1–22.

Espejo, R. (2015b). Performance for viability: Complexity and variety management. *Kybernetes*, 44(6/7), 1020–1029.

France, G., Taroni, F., & Donatini, A. (2005). The Italian health-care system. *Health Economics*, 14(S1), S187–S202.

Gershenson, C. (2015). Requisite variety, autopoiesis, and self-organization. *Kybernetes*, 44(6/7), 866–873.

Glouberman, S., & Mintzberg, H. (1996). *Managing the Care of Health and the Cure of Disease, Part I: Differentiation*. INSEAD, Fontainebleau, France.

Golinelli, G. M. (2010). *Viable Systems Approach (VSA): Governing Business Dynamics*. Cedam, Padova.

Hamburg, M. A., & Collins, F. S. (2010). The path to personalized medicine. *New England Journal of Medicine*, 363(4), 301–304.

Holmes, M. V., Shah, T., Vickery, C., Smeeth, L., Hingorani, A. D., & Casas, J. P. (2009). Fulfilling the promise of personalized medicine? Systematic review and field synopsis of pharmacogenetic studies. *PLoS One*, 4(12), e7960.

Iandolo, F., Calabrese, M., Antonucci, E., & Caputo, F. (2013). Towards a value co-creation based healthcare system. In Gummesson, E., Mele, C., & Polese, F. (eds.), *The 2013 Naples Forum on Service: Service Dominant Logic, Networks & Systems Theory and Service Science: Integrating three Perspective for a New Service Agenda*, Giannini, Napoli.

IfM & IBM (2008). *Succeeding through Service Innovation: A Service Perspective for Education, Research, Business and Government*. University of Cambridge, Cambridge. ISBN: 978-1-902546-65-0.

Khaled, S. A., Burley, J. C., Alexander, M. R., & Roberts, C. J. (2014). Desktop 3D printing of controlled release pharmaceutical bilayer tablets. *International Journal of Pharmaceutics*, 461(1), 105–111.

Kongstvedt, P. R. (2001). *The Managed Health Care Handbook*. Jones & Bartlett Learning, New York.

Michalski, M. H., & Ross, J. S. (2014). The shape of things to come: 3D printing in medicine. *JAMA*, 312(21), 2213–2214.

Philips, Z., Ginnelly, L., Sculpher. M., Claxton, K., Golder, S., Riemsma, R., Woolacott, N., & Glanville, J. (2004). Review of guidelines for good practice in decision-analytic modelling in health technology assessment. *Health Technology Assessment*, 8(36), iii–iv, ix–xi, 1–158.

Plsek, P. E., & Wilson, T. (2001). Complexity, leadership, and management in healthcare organisations. *British Medical Journal*, 323(7315), 746–768.

Puschmann, T., & Alt, R. (2001, January). Customer relationship management in the pharmaceutical industry. In System Sciences, 2001. Proceedings of the 34th Annual Hawaii International Conference on (pp. 9–17). IEEE.

Rodriquez, M., Aquino, R. P., & D'Ursi, A. M. (2015). Is it time to integrate sex and gender into drug design and development? *Future Medicinal Chemistry*, 7(5), 557–559.

Rouse, W. B. (2008). Health care as a complex adaptive system: Implications for design and management. *Bridge-Washington-National Academy of Engineering*, 38(1), 17.

Sabatino, A., Grasso, A., Micera, D., & Villecco, F. (2014). A decision support system for primary care. *International Journal of Mechanical Engineering and Industrial Design*, 2(3), 5–24.

Savas, E. S., & Savas, E. S. (2000). *Privatization and Public-Private Partnerships*. Chatham House, New York.

Saviano, M. (2012). *Condizioni di efficacia relazionale e di performance nelle aziende sanitarie*. Giappichelli, Torino.

Saviano, M., Bassano, C., & Calabrese, M. (2010). A VSA-SS approach to healthcare service systems the triple target of efficiency, effectiveness and sustainability. *Service Science*, 2(1–2), 41–61.

Saviano, M., Caputo, F., & Napoli, B. (2015). Addressing the social and economic challenges of orphan drugs: A managerial perspective. *International Journal of Pharmaceutical Sciences & Business Management*, 3(12), 1–26.

Saviano, M., Parida, R., Caputo, F., & Kumar Datta, S. (2014). Health care as a world-wide concern: Insights on the Italian and Indian health care systems and PPPs from a VSA perspective. *EuroMed Journal of Business*, 9(2), 198–220.

Skinner, R. I. (2003). The value of information technology in healthcare/reply. *Frontiers of Health Services Management*, 19(3), 3.

Swan, M. (2009). Emerging patient-driven health care models: an examination of health social networks, consumer personalized medicine and quantified self-tracking. *International Journal of Environmental Research and Public Health*, 6(2), 492–525.

Swayne, L. E., Duncan, W. J., & Ginter, P. M. (2012). *Strategic Management of Health Care Organizations*. John Wiley & Sons, Hoboken, NJ.

Symes, M. D., Kitson, P. J., Yan, J., Richmond, C. J., Cooper, G. J., Bowman, R. W., & Cronin, L. (2012). Integrated 3D-printed reactionware for chemical synthesis and analysis. *Nature Chemistry*, 4(5), 349–354.

Wendt, C. (2009). Mapping European healthcare systems: A comparative analysis of financing, service provision and access to healthcare. *Journal of European Social Policy*, 19(5), 432–445.

Theme VI

Knowledge and organisation

*Andrée Piecq, Claude Lambert
and Sergio Boria*

The emergence and the interpretative process in the systemic approach of the organization

Claude Lambert[1]

Keywords: *semiotics; pragmaticism; process; entanglement; abduction*

1 Context

Emergence is studied here as an interpretant which is logically involved in the interpretive process initiated in the systemic approach of a human organization. This process is an essential step for the professional as for the researcher that addresses an organization. This is even more relevant when the organization faces uncertainties or when the shared conception of it by the members is a source of difficulties and irritations caused by doubt on the current operating mode. The choice of the word "approach" is deliberate because it is not about applying a linear tool. The proposal is to cause a process of reinterpretation of the organization that must lead, amaze and marvele, and thus form a new conception of the observed organization. This new approach should allow the implementation of a change. In other words, verification and explanation will not be discussed here. The purpose is to bring the observer to a new design. To do this, I propose to apply the interpretive process developed by C.S. Peirce in the study of signs (semiotics) to the systemic approach of the organization.

2 How does emergence meet the requirements of the systems approach?

I retain as a definition: "The emergence is defined as the qualities or properties of a system that present a novelty compared to the properties of the components taken separately or arranged differently in another type of system".

This definition highlights the complexity of the concept: In order to relate the concept with an effective organization, it is necessary to be able to recognize the qualities of "Whole" but also its "Parts", as if they were isolated or part of another system or context. Then we must be able to relate the qualities of the "Whole" and those of the "Parts". Finally, the observer-designer should be able to infer a new character and/or sudden qualities of the Whole. Qualities of the Whole, the Parts, relations between its qualities and novelty: here is the challenge of emerging as a concept.

It is understood that the philosophical point of view of the concept of emergence is equivocal. It is the subject of divergences between various trends (holism versus atomism). In the social sciences, the totality is at the heart of the debate between holism and methodological individualism. Following the trends, precedence is given to one of the two poles. We also find these tensions at the heart of organizations with the tension between individual and collective interest. These tensions are valid only in a linear perspective, binary and linear causality. Therefore, I chose to deviate myself to take the road of the ternary, the circularity, of the dialogic in a pragmatic perspective.

3   Role of the emergence in the pedagogy of the systems approach

A pragmatic process

Before coming to the emergence, it is helpful to place the intervention in an organization from a pragmatic point of view (not utilitarian) in a philosophical sense. Thus I refer to the maxim C.S. Peirce related to its pragmatic design: "Consider what are the practical effects that we expect to be produced by the object of our conception. The design of these effects is the complete design of the object". I note that it should not be understood as definitely complete. In fact, for Peirce, the semiotic process is infinite. It stops only temporarily by intersubjective agreement of researcher-actors who believe they have reached a representation that allows them to establish new habits enabling them to act.

The semiotic approach

Continuing the peircean pragmatism, the pedagogical approach proposed here is based on the triadic semiotics of the same author. Peirce developed a semiotic theory that is simultaneously general, triadic and pragmatic. It is a theory that envisages together the emotional life, practical and intellectual and considers all components of semiotics. His approach generalizes the concept of sign. This is a triadic theory based on three philosophical categories: firstness, secondness and thirdness. It connects three terms of different types: the sign or representamen, object and interpreting. Peirce's semiotics takes into consideration the context of production and reception of signs and defines the sign by its action on the interpretant.

The interpretive process as a guide

In his first semiotics, Peirce proposed the study of relationships between signs. In a second semiotics, Peirce proposes to study the interpretive process as it occurs inside the sign. In this second semiotics, the sign of the definition is always "relative to . . . ". The sign is thus entered as a logical position to the process.

If we want to represent the interpretive process as proposed in this second semiotics, it must be at once circular, level and tangled. Jean Fisette proposes the following representation in Figure 79.1:

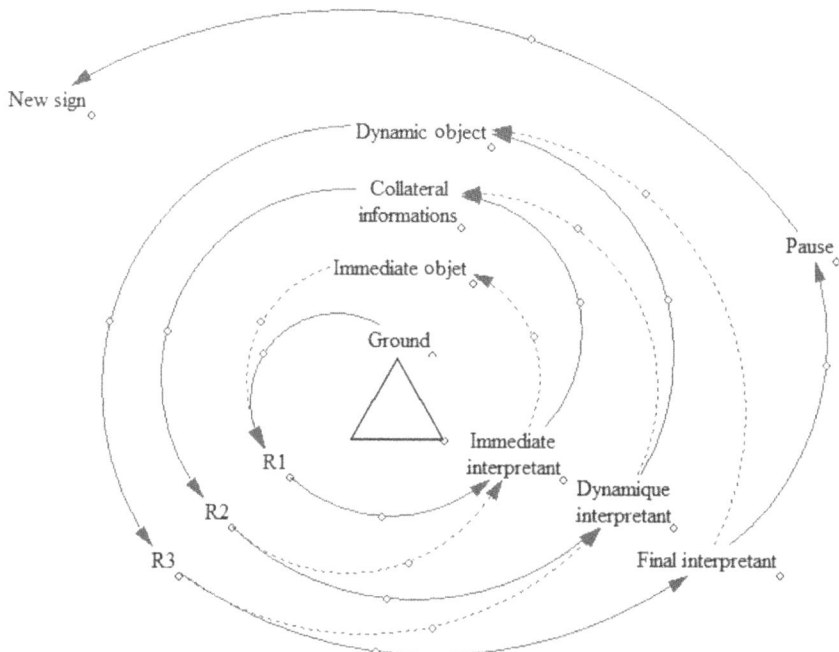

Figure 79.1 The interpretive process
Source: Jean Fisette, 1989

Starting from the ground, the organization is recognized in its uniqueness, singularity. From this first interpretation, the organization is considered in its context, its components (Immediate interpretant) which in our case its interpreted by observing the behaviors that allow us to infer the rules, relationships, . . . (the guidelines, cf. A. Piecq "The giroscope").

The dynamic interpretant is building relationships between components and representimg the structure of the organization as well as the system properties. This dynamic interpretant meets the definition of emergence involving a process of identification of the parts, the properties of the whole and relationship between whole and parts.

The emergence is the final interpretant and seen as a cultural symbol, a complex sign that presents to a conscious mind. From this interpretant in his relation to the constructed object, the observer-designer made a "pause" in having created a new sign allowing it to consider the effects of this renewed conception of the organization.

I note that the process through dynamic interpretant, the dynamic object and the ground can continue indefinitely until the irritation caused by the initial doubt has passed. It is understood that the process can resume based on the changes observed.

Assuming that the final interpretation has an emerging nature, this implies that the transition to the new sign implies an abductive inference, i.e.: making assumptions.

Thus, it can be argued that systemic organizational approach is not directed by deduction and the application of an existing model. On the other hand, the approach is not either within an exhaustive analysis by an inductive approach. We follow Peirce which situates the relationship between these forms of inference: The abduction must have the character of the unexpected, the induction helps validate the assumptions produced by the abduction. Finally, the deduction applies as the facts do not contradict habits permitted by the rules of deduction.

4   Role of the emergence in the pedagogy of the systems approach

This proposal makes from the approach of systemic consultant a circular tangled process that must achieve assumptions for a renewal of the understanding of the system action. This experience is not always easy to transmit in a rigorous way by not giving in to linear forms and causalities. The second semiotics C.S. Peirce is enlightening opens up educational opportunities. In addition, this combination allows to refer to other elements of Peirce's theory that are shining for systemician: uncertainty, doubt, belief, context. . . . This alliance opens a wide field of research and applications in a perspective of a renewal of managerial practice.

Note

1 IT Consultant, President of S&O (Belgium), Trainer at G.I.R.O.S. (Belgium), initiator of "Complexitude Workgroup" (France).

Bibliography

Deledalle, Gérard, and Joëlle Réthoré (1979), *Théorie et pratique du signe. Introduction à la sémiotique de Charles S. Peirce*. Paris: Payot.

de Waal, Cornelis (2013), *Peirce: A Guide for the Perplexed*. 1st ed. Bloomsbury Academic.

Fisette, Jean (2005a), *Pour une pragmatique de la signification*. Montréal, Québec: Xyz.

Fisette, Jean (2005b), *Introduction a la Semiotique de C S Pierce*. Montréal: Xyz.

Martine Arino (2007), *La subjectivité du chercheur en sciences humaines*. Paris: L'Harmattan.

Nicole Everaert-Desmedt (1995), *Le processus interprétatif*. Liège: Editions Mardaga.

Peirce, Charles S., and Gérard Deledalle (1978), *Écrits sur le signe*. Paris: Seuil.

Piecq, Andrée (2011), *De la pensée systémique à la pratique de l'organisation: le giro-scope*. Paris: l'Harmattan.

Tiercelin, Claudine (2013a), *C. S. Peirce et le pragmatisme*. Collège de France.

Tiercelin, Claudine (2013b), *La pensée-signe: Études sur C. S. Peirce*. Collège de France.

Vincent Descombes (1996), *Les Institutions Du Sens*. Collection "Critique". Paris: Editions de minuit.

Artifact-dependent dialectic processes as enablers of self-organization in knowledge-driven teams

Sven-Volker Rehm[1]

Keywords: *coordination; dialectics; information technology; knowledge manage-ment; new product development; pattern; self-organization*

1 Introduction

There is a limited understanding of the dynamics of inter-organizational new product development (NPD) teams whose cooperation is essentially based on and directed by their members' expertise. In particular, there is no common approach that can guide management interventions. We address this need through a study of three NPD teams, each composed of members from a network of small and medium-sized enterprises (SMEs) during the knowledge-intensive development of a radical innovation. We draw on data from a four-year qualitative, interpretive case study. By closely following and observing the team members' interactions and uses of information technology (IT), we explicate phenomena of complexity, emergence and self-organization related to the use of IT and other artifacts. From these observations we formulate a conceptual model intended to describe the dynamics of inter-organizational, knowledge-based teams. This model informs theory with respect to development of a communicative theory of the firm; and we contribute to management practice by better understanding the dynamics occurring at interactions across firm boundaries.

Firms often rely on cooperative innovation for competitive advantage (Bierly et al., 2009; Chesbrough, 2003a; Powell, 1990; Teng, 2007). One option for firms is to engage in inter-organizational new product development (NPD) projects (Chesbrough, 2003b; Davenport et al., 2006). In such projects, generating and systematically developing innovative ideas largely require transfer, sharing and integration of knowledge (Easterby-Smith et al., 2008; Grant, 1996; Nonaka, 1994). The development of new products and services through cooperative work involves the need for exploring and exploiting knowledge from diverse fields of expertise that is difficult for a single firm to achieve, but can be accomplished through inter-organizational NPD teams (Baum et al., 2010; Lorenzoni and Lipparini, 1999). Such teams are usually project-specific groups of experts from concerned – often multiple – domains; and generally, no single team member or firm takes a lead in managing the team.

2 Dialogical team interactions

Throughout the team's cooperation, decisions about formation, distribution and assignment of tasks, as well as their subject matters and sequence have to be taken by the NPD team (Christensen and Raynor, 2003; Staples and Webster, 2008). These decisions are interwoven with the progressing creation of new knowledge of the innovation process (Hardy et al., 2003; Oshri et al., 2008). Literature on knowledge creation assumes that this is accomplished through dynamic interactions of individuals in a dialogical process (Nonaka and Toyama, 2003). However, the driving forces behind knowledge creation in an expertise-driven NPD are the resulting, altered awareness of team members towards the innovation process and the team, and the processes for arranging tasks within the NPD team (Kotlarsky et al., 2014; Okhuysen and Bechky, 2009). This creates a basis for effects of self-organization within the teams. However, popular project management approaches intend to implement rather strict coordination structures, in order to enable process measurability and support through information systems (IS). In management practice, this is often accomplished as a management task that relies on plans, protocols, rules and procedures, which are regularly implemented in a top-down fashion (Faraj and Xiao, 2006). However, such "managed" coordination does not apply in the context of expertise-driven NPD teams that exist for the duration of the project, and that do not have a single member or firm taking a lead in coordination. While prior literature has hinted at the elements of task arrangement and communication practices as important in expertise-driven teams (Chou and He, 2011; Faraj and Xiao, 2006), an understanding of emerging effects of self-organization has not been developed. Moreover, while the significant role of IS in facilitating coordination is known, how IS take effect on dialogical team interactions in expertise-driven NPD teams, and how IS can be selected in order to meet managerial requirements for managing inter-organizational NPD projects, is still an open issue.

3 Self-organization phenomena

Our research addresses this issue through a study of three inter-organizational NPD teams, each composed of members from a network of SMEs, during the knowledge-intensive development of a radical innovation. Using a four-year, intensive qualitative case study, we find evidence on three separate but mutually reinforcing, emergent effects on team organization: From the use and modification of IT artifacts in the teams, dialectic processes emerge that serve as carriers of self-organization phenomena. These materialize as (i) dialectic restructuration of communication patterns and coordination processes, (ii) dialectic restructuration of work arrangements, and (iii) emergent patterns of IT artifact uses.

As contribution to research, we integrate these effects by proposing a conceptual model describing dialectic patterns of work arrangement in the context of expertise-driven NPD teams. For management practice, our interpretation of the case study identifies critical indications for how information systems facilitate coordination in cross-company NPD teams that are expertise-driven.

Note

1 WHU – Otto Beisheim School of Management.

Bibliography

Baum, J. A. C., Cowan, R., & Jonard, N. (2010). Network-Independent Partner Selection and the Evolution of Innovation Networks. *Management Science*, 56(11), 2094–2110.

Bierly, P. E., Damanpour, F., & Santoro, M. D. (2009). The Application of External Knowledge: Organizational Conditions for Exploration and Exploitation. *Journal of Management Studies*, 46(3), 481–509. doi:10.1111/j.1467-6486.2009.00829.x

Chesbrough, H. W. (2003a). *Open innovation: The new imperative for creating and profiting from technology.* Boston, MA: Harvard Business School Press.

Chesbrough, H. W. (2003b). The Era of Open Innovation. *MIT Sloan Management Review*, 44(3), 35–41.

Chou, S.-W., & He, M.-Y. (2011). The Factors That Affect the Performance of Open Source Software Development: The Perspective of Social Capital and Expertise Integration. *Information Systems Journal*, 21(2), 195–219. doi:10.1111/j.1365-2575.2009.00347.x

Christensen, C. M., & Raynor, M. (2003). *The innovators solution: Creating and sustaining successful growth.* Boston, MA, London: Harvard Business School; McGraw-Hill.

Davenport, T. H., Leibold, M., & Voelpel, S. (2006). *Strategic management in the innovation economy: Strategy approaches and tools for dynamic innovation capabilities.* Erlangen: Publicis.

Easterby-Smith, M., Lyles, M. A., & Tsang, E. W. K. (2008). Inter-Organizational Knowledge Transfer: Current Themes and Future Prospects. *Journal of Management Studies*, 45(4), 677–690.

Faraj, S., & Xiao, Y. (2006). Coordination in Fast-Response Organizations. *Management Science*, 52(8), 1155–1169. doi:10.1287/mnsc.1060.0526

Grant, R. M. (1996). Toward a knowledge-based theory of the firm. *Strategic Management Journal*, 17, 109–122.

Hardy, C., Phillips, N., & Lawrence, T. B. (2003). Resources, Knowledge and Influence: The Organizational Effects of Interorganizational Collaboration. *Journal of Management Studies*, 40(2), 321–347.

Kotlarsky, J., Scarbrough, H., & Oshri, I. (2014). Coordinating Expertise across Knowledge Boundaries in Offshore-Outsourcing Projects: The Role of Codification. *MIS Quarterly*, 38(2), 607.

Lorenzoni, G., & Lipparini, A. (1999). The Leveraging of Interfirm Relationships as a Distinctive Organizational Capability: A Longitudinal Study. *Strategic Management Journal*, 20(4), 317–338.

Nonaka, I. (1994). A Dynamic Theory of Organizational Knowledge Creation. *Organization Science*, 5(1), 14–37.

Nonaka, I., & Toyama, R. (2003). The Knowledge-Creating Theory Revisited: Knowledge Creation as a Synthesizing Process. *Knowledge Management Research & Practice*, 1(1), 2–10.

Okhuysen, G. A., & Bechky, B. A. (2009). Coordination in Organizations: An Integrative Perspective. *The Academy of Management Annals*, 3(1), 463–502. doi:10.1080/19416520903047533

Oshri, I., van Fenema, P., & Kotlarsky, J. (2008). Knowledge Transfer in Globally Distributed Teams: The Role of Transactive Memory. *Information Systems Journal*, 18(6), 593–616. doi:10.1111/j.1365-2575.2007.00243.x

Powell, W. W. (1990). Neither Market Nor Hierarchy: Network Forms of Organization. *Research in Organizational Behavior*, 12, 295–336.

Staples, D. S., & Webster, J. (2008). Exploring the Effects of Trust, Task Interdependence and Virtualness on Knowledge Sharing in Teams. *Information Systems Journal*, 18(6), 617–640. doi:10.1111/j.1365-2575.2007.00244.x

Teng, B. (2007). Corporate Entrepreneurship Activities through Strategic Alliances: A Resource-Based Approach toward Competitive Advantage. *Journal of Management Studies*, 44, 119.

Complexity, autopoiesis and governing issues of company people organization

Luciano Martinoli[1]

Keywords: *company organization; Luhmann System Theory; autopoiesis; mainstream management; Quantum Field Theory*

This paper aims to show the complex nature of a specific social system: the Company People Organization. This will be achieved identifying the Theoretical Incompleteness (Minati 2016) of such system as a concise and all-inclusive feature of complexity. Successively, thanks to the Social System Theory from Niklas Luhmann (1986, 1995, 1997), which is grounded on autopoietic system theory (Maturana and Varela 1980, 1992), it will be showed that the autopoiesis nature of this peculiar social system matches the Theoretical Incompleteness. That is to say that autopoiesis, in case of Company Organization, and generally for any social system, is the feature that fully, and deeply, describes its complexity. Thanks to this result it will be easier and more handy to discuss the related governance issue. Unfortunately due to the "self-development" root of autopoietic systems, any approach based on causality governance is bound to fail.

Despite this, currently the mainstream managerial thinking, which is the current steering paradigm for such systems, proposes a "planning and control" philosophy of governance and many related managerial tools based on it: training, human resources management, change management and so on. Comparing the management practice with the insightful viewpoint provided by Luhmann, namely the inner autopoietic (complex) nature of company organization, the paper will provide evidences about how mainstream managerial thinking is far from being in whatsoever way effective and meaningful.

In order to address the issue of handling this kind of system, a redefinition of the concept and practice of "governing": the "Quantum Governance of Human Systems" (Zanotti 2014) will be mentioned. Key features of this, in the case of the Company Organization, will be shown.

Note

1 CSE Crescendo, Expo della Conoscenza, via Aurispa 7 20122 Milano, luciano. martinoli@expoconoscenza.org, +390245479800.

Bibliography

Bonometti, P., (2012). Improving Safety, Quality and Efficiency through the Management of Emerging Processes: The Tenaris Dalmine Experience. *The Learning Organization*, Vol. 19(4), pp. 299–310.

Luhmann, N., (1986). *The Autopoiesis of Social Systems, Sociocybernetic Paradoxes: Observation*. Control and Evolution of Self-Steering Systems, 171–192. London, Sage.

Luhmann, N., (1995). Why Does Society Describe Itself as Postmodern? *Cultural Critique*, pp. 171–186.

Luhmann, N., (1997). Limits of Steering. *Theory, Culture and Society*, Vol. 14, pp. 41–57.

Maturana, H. R. and Varela, F. J., (1980). *Autopoiesis and Cognition*. Dordrecht, Holland, D. Reidel.

Maturana, H. R. and Varela, F. J., (1992). *The Tree of Knowledge*. Boston & London, Shambhala.

Minati, G., (2016). Knowledge to Manage the Knowledge Society: The Concept of Theoretical Incompleteness. *Systems*, Vol. 4, p. 26. www.mdpi.com/2079-8954/4/3/26

Minati, G., Penna, M. P., and Pessa, E., (1998). Thermodynamic and Logical Openness in General Systems. *Systems Research and Behavioral Science, John Wiley and Sons Ltd.*, Vol. 15(3), pp. 131–145.

Minati, G., Zanotti, F., and Martinoli, L., (2016). Approccio Direttivo *Direttività e non direttività nell'approccio ai sistemi sociali, Riflessioni Sistemiche n.14*. AIEMS. Retrieved 23.9.2018 from http://www.aiems.eu/archivio/files/riflessioni_sistemiche_n_14.pdf .

Moeller, H.G., (2011). *The Radical Luhmann*, Columbia University Press.

Moeller, H.G. and Zanotti, F., (2016). *Per Comprendere Luhmann, una necessità per le classi dirigenti*. Milan, IPOC.

Taylor, F., (1911). *The Principle of Scientific Management*. New York & London, Harper & Brothers.

Zanotti, F., (2014). "Quantum Governance of Development": Prolegomena for a General Theory and the Case of an Enterprise. *International Journal of Public and Private Management*, July–December.

Theoretical notes regarding the practical application of Stafford Beer's Viable System Model

Markus Orengo[1]

Keywords: *viable system model; VSM; organization development; algedonic channel; cyclic recursion*

1   Background

The viable system model (VSM) of Stafford Beer has been deployed in various contexts during the previous decades. Equally, a variety of related contributions in literature has been published. However, compared to the classical organization theory, the concept has still a very limited diffusion among managers and organization developers. The present contribution is an attempt to capture current difficulties with the use of the VSM in practical applications. On this basis, a set of suggestions towards a more effective application of the model is made.

2   Findings

Based on an observational study, it is claimed that the VSM is currently stuck in the typical chasm of a bell-shaped diffusion curve. In order to diffuse from early adopters to an early majority, one has to consider that the two target groups have different needs and have therefore to be approached differently. Accordingly, a set of suggestions is made. First, instead of (over)-simplifying the VSM or (over)-stretching its scope, the abstract nature of complexity balances of the model should be (re)-emphasized and better linked to the established tools and methods of classical organization theory. A second suggestion is to build a standardized case library. A third suggestion is to formally compare the VSM from a practitioner's perspective to other systems oriented approaches, such as Teal-Organizations, Holacracy, Sociocracy and agile methods. A fourth suggestion is to close some theoretical gaps linked to the subject of algedonic channels and cyclic recursion. Eventually, the paper suggests that, in order to boost the diffusion of the VSM, a community of practice is needed that focuses on the needs of the early majority.

Note

1 Social systems engineering GmbH.

Bibliography

Bachmann, M. and Michel, D. (2001). *Das Pentagramm der Komplexitätsbewälti-gung*. Basel, CH: Verlag Paraplegie.

Bartlog, H. and Lambertz, M. (2016). Das SCRUM VSM Spiel. Retrieved form www.youtube.com/watch?v=vETg2q1irs8

Beer, S. (1972). *Brain of the Firm*. Chichester et al.: John Wiley & Sons.

Beer, S. (1975). *Platform for Change*. Chichester et al.: John Wiley & Sons.

Beer, S. (1979). *The Heart of Enterprise*. Chichester et al.: John Wiley & Sons.

Beer, S. (1984). The Viable System Model: Its Provenance, Development, Methodology and Pathology. In Espejo, R. & Harnden, R. (Eds.), *The Viable System Model* (pp. 11–37). Chichester et al.: John Wiley & Sons.

Bleicher, K. (1991). *Organisation: Strategien Strukturen Kulturen*. Frankfurt: Gabler Verlag.

Bockelbrink, B. (2015). Comparing Different Models of Management. Retrieved from http://evolvingcollaboration.com/comparing-different-models-of-management

Brocklesby, J. et al. (1995). Demystifying the Viable System Model. *In Asia Pacific Journal of Operational Research* (January).

Caplow, T. (1968). *Two against One, Coalitions in Triad*. London et al.: Prentice-Hall.

Cockton, G. and Jones, S. (without date). *Making the Viable System Model (VSM) More Accessible to Senior Managers via Multimedia*. School of Computing and Technology, University of Sunderland.

Cohen, M., March, J. and Olsen, J. (1972). A Garbage Can Model of Organizational Choice. *Administrative Science Quarterly*, Vol. 17, No. 1 (March), pp. 1–25.

Durkheim, E. (1898). Représentations individuelles et représentations collectives. *Revue de Métaphysique et de Morale*, tome 6, numéro de mai.

Endenburg, G. (1998). *Sociocracy*. Delft: Eburon.

Espejo, R. and Reyes, A. (2011). *Organizational Systems*. Heidelberg et al.: Springer-Verlag.

Foss, R. (1989). The Organization of a Fortress Factory. In *The Viable System Model* (pp. 121–143). Chichester et al.: John Wiley & Sons.

Gamson, W. (1961). A Theory of Coalition Formation. *American Sociological Review*, Vol. 26, No. 3 (Junuary), pp. 373–382.

Gartner (2015). Understanding Gartner's Hype Cycles. Retrieved from www.gartner.com

Hildbrand, S. and Bodhanya, S. (2015). Guidance on Applying the Viable System Model. *Kybernetes*, Vol. 44, No. 2, pp. 168–201.

Hoverstadt, P. (2008). *The Fractal Organization*. Chichester et al.: John Wiley & Sons.

Korotayev, A. and Tsirel, S. (2010). A Spectral Analysis of World GDP Dynamics. *Structure and Dynamics*, Vol. 4, No. 1.

Laloux, F. (2014). *Reinventing Organizations*. Brussels: Nelson Parker.

Lambertz, M. (2016). SCRUM & das Viable System Model. Retrieved from www.youtube.com/watch?v=vETg2q1irs8&noredirect=1

Leonard, A. (1994). The Very Model of a Modern System-General. In Beer, S. (Ed.), *Beyond Dispute* (pp. 346–356). Chichester et al.: John Wiley & Sons.

Leonard, A. (2007). Symbiosis and the Viable System Model. *Kybernetes*, Vol. 36, No. 5/6, pp. 571–582.

Leonard, A. (without date). To Change Ourselves: A Personal VSM Application. Retrieved from https://web.archive.org/web/20150205152907/http://allennaleonard.com/PersVSM.html

Nefiodow, L. and Nefiodow, S. (1996). *Der sechste Kondratieff.* Sankt Augustin, Deutschland: Rhein-Sieg Verlag.

March, J. and Simon, H. (1958). *Organizations.* Cambridge, MA: Blackwell Publishers.

Marchetti, C. (1981). Society as a Learning System. *Technological Forecasting and Social Change*, Vol. 18, No. 4, pp. 267–282.

McCandless, D. (2012). *Information is Beautiful.* UK: Harper Collins Publ.

Moore, G. (1991). *Crossing the Chasm.* New York et al.: HarperCollins Publishers.

Orengo, M. (2011). (Re)-Organisation von Unternehmen. *SEM|Radar*, No. 1, pp. 127–155.

Parsons, T. (1951). *The Social System.* New York: Free Press.

Pérez Ríos, J. (2012). *Design and Diagnosis for Sustainable Organizations.* Berlin: Springer-Verlag.

Pfiffner, M. (2010). Five Experiences with the Viable System Model. *Kybernetes*, Vol. 39., No. 9/10, pp. 1615–1626.

Ramage, M. and Shipp, K. (2010). *Systems Thinkers.* London: Springer.

Reynolds, M. and Holwell, S. (2010). *Systems Approaches to Managing Change: A Practical Guide.* London: Springer.

Robertson, J. (2015). *Holacracy.* New York: Henry Holt and Company.

Rosa, N. (2016). *Barriers to the Diffusion of the VSM.* London: UCL Center for Systems Engineering.

Schumpeter, J. (1911). *Theorie der wirtschaftlichen Entwicklung.* Leipzig: Verlag von Dunker & Humblot.

Schwaninger, M. (2004). Methodologies in Conflict: Achieving Synergies between System Dynamics and Organizational Cybernetics. *Systems Research and Behavioral Science*, 23(2006), pp. 337–347.

Schwaninger, M. (2006). Theories of Viability: A Comparison. *Systems Research and Behavioral Science*, Vol. 21(2004), pp. 411–431.

Schwaninger, M. (2009). *Intelligent Organization.* Berlin et al.: Springer Verlag.

Schwaninger, M. and Scheef, C. (2016). *A Test of the Viable System Model: Theoretical Claim vs. Empirical Evidence.* Berlin et al.: Springer Verlag.

Scott, R. (2014). *Institutions and Organizations.* Los Angeles et al: Sage Publications.

Skyttner, L. (2005). *General Systems Theory.* Singapore: World Scientific Publishing.

Walker, J. (2006). The VSM Guide. Retrieved from www.esrad.org.uk/resources/vsmg_3

Ware, C. (2012). *Information Visualization.* Oxford: Elsevier Ltd.

Westphal, R. (2016). Offene Fragen zu Scrum und dem Viable System Model. Retrieved from http://ralfw.de/2016/08/offene-fragen-zu-scrum-und-dem-viable-system-model/

A path towards sustainability through social innovation

Evidences from Italy

Rossella Canestrino,[1] *Primiano Di Nauta*[2]
and Pierpaolo Magliocca[3]

Keywords: *social innovation; learning systems; sustainability; SfSI*

1 Theoretical background

The concept of Social Innovation (SI) has been one of the most discussed in the field of innovation in the last years (Ashta et al., 2014) and is even getting stronger in the debate about social development (Rüede and Lurtz, 2012). This is particularly due, on one side, to the relevance that knowledge and innovation have among the most important strategically significant resources for both firms and local systems competitiveness (Barile and Di Nauta, 2011; Barile et al., 2013; Calza et al., 2015; Canestrino, 2008; Di Nauta et al., 2015) and, on the other side, to the challenges that are affecting the worldwide social development and sustainability (Canestrino et al., 2015; Canestrino et al., 2016). Despite these trends, the emerging process of SI (Goldenberg et al., 2009; Murray et al., 2010; Westley and Antadze, 2010; Rüede and Lurtz, 2012) seems to be understudied, particularly referring to the conditions upon which it may be sustained and reinforced.

Understanding SI is not easy, not only because of the existing overlapping between the theoretical backgrounds – namely innovation and social responsibility – usually used to explain the concept, but also because of the high number of actors – social entrepreneurs, investors, incubators, intermediary organizations and transnational networks – generally involved in the process (Avelino and Wittmayer, 2015).

According to Mulgan (2006), SI refers to innovation activities and services that are motivated by the aim of meeting a social need, and that are predominantly diffused through organizations primarily focused on social purposes. In this direction, managing SI requires the evaluation, in an integrated manner, of both organizations' innovation activities and social responsibility.

With reference to innovation, several models have been proposed in literature (Nonaka and Takeuchi, 1995; Tsai and Ghoshal, 1998). The authors usually consider knowledge – mainly new knowledge – as the outcome of a learning process that is implicitly equated with innovation. In a broad sense, local availability of know-how is the background for setting up the learning process, which can lead to a social innovation. In fact, every innovation arises and includes knowledge; at the same time, every innovation is responsible for new knowledge diffusion (Canestrino et al., 2015; Grant, 1991). Since very few firms are able to develop internally a wide range of knowledge, the interaction among actors is required to foster knowledge creation and, consequently, innovation: thanks to proximity, individuals, as well as organizations, are able to get in contact one to each other, sharing resources, know-how and capabilities. Then, innovation generally arises thanks to a system of actors who relate one to each other producing new patterns of learning (Canestrino and Magliocca, 2016). In doing this, both individuals and organizations are strongly influenced and shaped by institutions; that means they are "embedded" in an institutional environment, or set of rules, which include the system of laws, norms and standards (Lundvall, 1992; Edquist, 1997). Not surprising, a well-established body of literature recognizes the important linkage among knowledge, networks and innovation. Within this field, for example, the concept of National Systems of Innovation (Freeman, 1987; Nelson, 1992, Lundvall, 1992) explains the process of innovation, by considering it as the outcome of the interaction among firms, organizations and institutions.

In the path of the mentioned perspective, the locus of innovation lies no longer within the boundaries of a single firm, but within the nexus of the (potential) relationships among several actors which are able to favour fruitful learning interactions in a systems' perspective (Canestrino et al., 2016; Del Giudice et al., 2011). Actually, analyzing the way the actors that belong to a given system – being it an organizational or social one – create, transfer and share knowledge for innovation, it seems not new within the field of knowledge management: what is new, in the authors' opinion, is the attempt to explore the way they achieve sustainability by the means of SI. It means, therefore, to investigate the way some kinds of innovation networks shift their own aim from "improving the productivity" to "solving a problem for a better quality of life for the community", thus creating novel social patterns.

2   Purpose

Following the above reflections, this paper aims at exploring the way SI arises within the locus of collaborations, which means the way some networks turn into the so-called Systems for SI – SfSIs – finally enabling the involved actors to reach both economic and social sustainability. In doing this, a case study analysis has been discussed.

3   Design/methodology/approach

A two step-based approach was designed to reach the aimed research's goals. Firstly, a literature review has been carried out in order to shape a wider understanding of SI, as well as to picture the characteristics and the dynamics of the networks for SI (called Systems for SI – SfSIs). Secondly, a qualitative method has been adopted by the means of face-to-face interviews, visits and meetings to the selected key-actors belonging to the SfSI "La Paranza" – located in Naples, in the South of Italy. The key-actors were interviewed, following "a conversation with purpose" (Burgess et al., 1991).

The adopted method allowed: (i) to learn about the network's activity and how it has been started up; (ii) to explore it in detail, with reference to both the actors involved and the role they play; (iii) to understand the way it has developed over time, turning into a SfSI.

4 Findings

Examining the experience of "La Paranza" enabled the authors to highlight the way a small network, based on personal and informal relationships among a few number of people, got the chance to turn into a SfSI able to reach both economic and social sustainability.

The presence of a *shared vision* and *joining common experiences* were finally identified as key drivers in the emergence of a SfSI: the *shared vision* particularly acted as glue, supporting the emergence of a "core" network, but it was only *joining common experiences* that the selected network turned into a SfSI, allowing the members to acquire the identity of a "community". It means therefore that a SfSI arose when a Community of Practice (CoP) established among the networks' members, inspiring people to create value for society. In such circumstances, collaborations, exchange of ideas and learning processes were reinforced, transforming the set of relationships and interactions among the actors into the locus for knowledge creation and innovation.

5 Originality/value

This contribution has both theoretical and practical implications.

From a theoretical perspective, the proposed contributes to the literature improvement about SI by developing a more comprehensive knowledge-based and systems-oriented framework of the concept, increasing the scholars' ability to recognize and understand SI, as well as the SfSIs and their dynamics. Besides, it allows also to show the way private players may act, as well as the way they should act to grant sustainable development, widening the actual debate about sustainability and urban re-vitalization.

From a practical perspective, it expands the ability of practitioners to manage a SfSI, supporting the effectiveness of firms' innovation practices with reference to both organizations' goals and societies' aims.

This contribution is the first output of an ongoing research about SfSIs, based on the selection and the investigation of empirical evidences located in the city of Naples, in the South of Italy. The research is still ongoing, mainly because of the long time required to analyze the networks' dynamics, as well as to explore the way they change time by time. Accordingly, the proposed findings may be reasonably considered as a useful starting point for future deepening about the topic.

6   Research limits

The contribution presents the same limits that every single case study analysis has, the most common of which concern the inter-related issues of methodological rigour, researcher's subjectivity and external validity (results generalization). Despite this, dealing with a single case study allowed the authors to collect much more information otherwise not available, as well as to fully understand the dynamics of the selected SfSI.

Notes

1   Department of Administrative and Quantitative Studies; Parthenope University of Naples; Via Generale Parisi 13, 80132 Naples (Italy); E-mail: rossella.canestrino@uniparthenope.it.
2   Department of Economics, University of Foggia; Via Caggese 1, 71121 Foggia (Italy); E-mail: primiano.dinauta@unifg.it.
3   Department of Economics, University of Foggia; Via Caggese 1, 71121 Foggia (Italy); E-mail: pierpaolo.magliocca@unifg.it.

Bibliography

Ashta, A., Couchoro, M. and Musa, A. S. M. (2014). Dialectic Evolution through the Social Innovation Process: From Microcredit to Microfinance. *Journal of Innovation and Entrepreneurship*, 3(1). 1–23.
Avelino, F. and Wittmayer, J. M. (2015). Shifting Power Relations in Sustainability Transitions: A Multi-Actor Perspective. *Journal of Environmental Policy & Planning*, 18(5). 1–22.
Barile, S. and Di Nauta, P. (2011). Viable Systems Approach for territory development. In Barile, S., Bassano, C., Calabrese, M., Confetto, M. G., Di Nauta, P., Piciocchi, P., Polese, F., Saviano, M., Siano, A., Siglioccolo, M., Vollero, A. (Eds.), *Contributions to Theoretical and Practical Advances in Management: A Viable Systems Approach (vSa)*, ASVSA – Association for Research on Viable Systems, International Printing, 199–243.
Barile, S., Saviano, M., Polese, F. and Di Nauta, P. (2013). Il rapporto impresa-territorio tra efficienza locale, efficacia di contesto e sostenibilità ambientale. *Sinergie Italian Journal of Management*, 90(Gennaio-Aprile). 25–49.
Burgess, N., Shapiro, J. L. and Moore, M. A. (1991). Neural Network Models of List Learning. *Network: Computation in Neural Systems*, 2(4). 399–422.
Calza, F., Canestrino, R. and Cannavale, C. (2015). A Cultural Insight for Knowledge Transfer: An Interpretative Model of Innovation Spreading at Local Level. *Journal of Global Economics, Management and Business Research*, 2(3). 129–142.
Canestrino, R. (2008). *Il Trasferimento della Conoscenza nelle reti di Imprese*, Giappichelli Editore, Torino.
Canestrino, R., Bonfanti, A., Magliocca, P. and Oliaee, L. (2016). Networks for Social Innovation: Devoting "Learning Spaces" to Social Aims, Conference Book of Proceedings of 11th International Forum on Knowledge Asset Dynamics "Towards a New Architecture of Knowledge: Big Data, Culture and Creativity", Dresden, Germany, June 15–17. Proceedings edited by Spender J.C.; Schiuma G.; Noenning J.R.; pp. 953–965.

Canestrino, R., Bonfanti, A. and Oliaee, L. (2015). Cultural Insights of CSI: How Do Italian and Iranian Firms Differ? *Journal of Innovation and Entrepreneurship*, 4(12). 1–9.

Canestrino, R. and Magliocca, P. (2016). Transferring Knowledge through Cross-Border Communities of Practice. In Buckley, S., Majewski, G., Giannakopoulos, A. (Eds.), *Organizational Knowledge Facilitation through Communities of Practice in Emerging Markets*, Vol. 1, IGI Global, Hershey, PA, 1–30.

Carroll, A. B. (1991). The Pyramid of Corporate Social Responsibility: Toward the Moral Management of Organizational Stakeholders. *Business Horizons*, 34(4). 39–48.

Del Giudice, M., Carayannis, E. G. and Della Peruta, M. R. (2011). *Cross-Cultural Knowledge Management: Fostering Innovation and Collaboration Inside the Multi-cultural Enterprise* (Vol. 11). Springer Science & Business Media.

Di Nauta, P., Merola, B., Caputo, F. and Evangelista, F. (2015). Reflections on the Role of University to Face the Challenges of Knowledge Society for the Local Economic Development. *Journal of the Knowledge Economy*. 1–19.

Edquist, C. (1997). *Systems of Innovation: Technologies, Institutions, and Organizations*. Psychology Press.

Freeman, C. (1987). Technical Innovation, Diffusion, and Long Cycles of Economic Development. In *The Long-Wave Debate*, Springer, Berlin, Heidelberg, 295–309.

Goldenberg, M., Kamoji, W., Orton, L. and Williamson, M. (2009). *Social Innovation in Canada: An Update, CPRN Research Report*, Canadian Policy Research Networks.

Grant, R. M. (1991). The Resource-Based Theory of Competitive Aadvantage: Implications for Strategy Formulation. *Knowledge and Strategy*, 33(3). 3–23.

Lundvall, B. A. (1992). *National Systems of Innovation: Towards a Theory of Innovation and Interactive Learning*, Pinter Publishers, London.

Mulgan, G. (2006). The Process of Social Innovation, Technology, Governance, Globalization. *MIT Press*, 1(2). 145–162.

Murray, R., Caulier-Grice, J. and Mulgan, G. (2010). *The Open Book of Social Innovation*, The Young Foundation.

Nelson, R. R. (1992). National Innovation Systems: A Retrospective on a Study. *Industrial and Corporate Change*, 1(2), 347–374.

Nonaka, I. and Takeuchi, H. (1995). *The Knowledge-Creating Company*, Oxford University Press, Oxford.

Rüede, D. and Lurtz, K. (2012). Mapping the Various Meanings of Social Innovation: Towards a Differentiated Understanding of an Emerging Concept. EBS Business School Research Paper Series 12–03, pp. 1–51.

Tsai, W. and Ghoshal, S. (1998). Social Capital and Value Creation: The Role of Intra-Firm Networks. *Academy of Management Journal*, 41(4). 464–476.

Westley, F. and Antadze, N. (2010). Making a Difference: Strategies for Scaling Social Innovation for Greater Impact. *The Innovation Journal: The Public Sector Innovation Journal*, 15(2). 3–20.

Using a systems-thinking approach to identify needs of an organization in order to enable its transformation

Alexander Kaiser,[1] *Florian Kragulj,*[2]
Florian Fahrenbach,[3] *Thomas Grisold*[4]
and Roman Walser[5]

Keywords: *systems-thinking; stakeholder-approach; need-knowledge; decision support system*

1   On prioritizing needs in social systems

From a management perspective, knowing what members of a system need can lead to non-directive and very effective management decisions as one need (e.g. mobility) can be satisfied in many ways. Therefore, if the shared needs in a social system are explicitly known and prioritized, an organization can focus at satisfying needs and thus, dramatically increase the range of potential satisfiers. However, knowing the needs of a system is only one side of the coin. How do we know the order of their satisfaction? In other words, where should we direct limited time and resources in need of satisfying strategies? Many streams of research like Maslow's hierarchy of need (Maslow, 1943) proposed approaches to prioritize needs, but they lack applicability for organizational learning purposes and strategic decision making. Consequently, the following research question can be derived:

> *How to prioritize needs in a social system in order to guide effective need-satisfaction strategies?*

A reasonable number of approaches assess needs which are hidden or we are not aware of in order to make them explicit (Goffin, Lemke, & Koners, 2010; Goffin & Lemke, 2004; Kaiser, Fordinal, & Kragulj, 2014; Kaiser & Kragulj, 2016). We developed a method called Bewextra that embraces needs in bottom-up, democratic and non-directive organizational learning processes.

2   Introducing a method to generate a catalogue of validated needs

Our method consists of three steps:

- In Bewextra-Collect, satisfiers are acquired in a future-learning-approach (Scharmer, 2009) by asking system members to report their dreams and wishes in a workshop setting (Kragulj, 2014).

- In Bewextra-Analytic, hypotheses about needs are generated using an abductive reasoning and a haptic clustering approach based on grounded theory (Charmaz, 2006; Reichertz, 2007).
- In Bewextra-Validate, system members are asked to validate the need hypotheses by the means of an online questionnaire.

Bewextra results in a catalogue of explicated and validated catalogue of needs that are shared between members of an organization. We use a systems-thinking approach to develop a better understanding what motivates and drives the actions of members within a system (Mele, Pels, & Polese, 2010). We reason that once we know the specific needs of a system, we have to prioritize them (Thomson, 2005) in order to facilitate the transformation of a system with respect to the satisfaction of the needs.

3 Introducing a decision support system to prioritize needs from three perspectives

In this work, we propose a decision support system, the Bewextra Need Priority Index (BNPI) to prioritize the outcome of Bewextra in a methodologically coherent and replicable way. To take into account all relevant stakeholder groups (Achterkamp & Vos, 2007; Freeman, 1984), and consequently account for a systems science perspective (Mele, Pels, & Polese, 2010), it is designed in a way that reflects the following three perspectives:

- The internal view of a system assesses the importance and relevance of a need from the viewpoint of the members of an organization (role of the client: those who are affected by organizational decisions).
- The systemic view refers to the inherent relation of needs in a system and assesses the potentiality and leverage effects of one need on another. It is assessed by decision-makers who are also members of the organization but have the power to influence organizational decisions (role of the decision maker).
- The external view of a system considers the expertise of outsiders who are acquainted with the organization and who gained significant knowledge about the system but are themselves not members like experts, customers or facilitators (role of the planner).

All views are assessed on separate scales. In order to combine, i.e. add up them, they have to be normalized and re-scaled. All views are connected to weights which can be flexibly adjusted and allow the decision maker to reflect his/her preferences and strategy. The BNPI results in a ranking that reflects the importance of needs from the perspective of all previously identified relevant stakeholder groups.

Furthermore, we present results from a case study with the scientific board of the Institute for Applied Business Research which is part of Austrian Federal Economic Chamber, in which the BNPI was successfully applied. The aim was to explicate and prioritize knowledge about substantial needs to guide a bottom-up vision development process.

Notes

1 Vienna University of Economics and Business; Welthandelsplatz 1, 1020, Vienna, Austria.
2 Vienna University of Economics and Business; Welthandelsplatz 1, 1020, Vienna, Austria.
3 Vienna University of Economics and Business; Welthandelsplatz 1, 1020, Vienna, Austria.
4 Vienna University of Economics and Business; Welthandelsplatz 1, 1020, Vienna, Austria.
5 Vienna University of Economics and Business; Welthandelsplatz 1, 1020, Vienna, Austria.

Bibliography

Achterkamp, M. C., & Vos, J. F. J. (2007). Critically identifying stakeholders evaluating boundary critique as a vehicle for stakeholder identification. *Systems Research and Behavioral Science*, 24(1), 3–14.

Charmaz, K. (2006). *Constructing grounded theory: A practical guide through qualitative analysis*. London: Sage.

Freeman, R. E. (1984). *Strategic management: A stakeholder approach*. Marshfield: Pitman Publishing Inc.

Goffin, K., & Lemke, F. (2004). Uncovering your customer's hidden needs. *European Business Forum*, (18), 45–47.

Goffin, K., Lemke, F., & Koners, U. (2010). *Identifying hidden needs: Creating breakthrough products*. London: Palgrave Macmillan.

Kaiser, A., Fordinal, B., & Kragulj, F. (2014). Creation of need knowledge in organizations: An abductive framework. In R. H. Sprague (Ed.), *47th Hawaii International Conference on System Science* (pp. 3499–3508). Los Alamitos: IEEE Computer Society Press.

Kaiser, A., & Kragulj, F. (2016). Bewextra: Creating and inferring explicit knowledge of needs. *Journal of Futures Studies*, 20(4), 79–98.

Kragulj, F. (2014). Interacting with the envisioned future as a constructivist approach to learning. *Constructivist Foundations*, 9(3), 439–440.

Maslow, A. (1943). A theory of human motivation. *Psychological Review*, 50(4), 370–396.

Mele, C., Pels, J., & Polese, F. (2010). A brief review of systems theories and their managerial applications. *Service Science*, 2(12), 126–135.

Reichertz, J. (2007). Abduction: The logic of discovery of grounded theory. In A. Bryant & K. Charmaz (Eds.), *The Sage handbook of grounded theory* (pp. 214–229). London: Sage.

Scharmer, O. (2009). *Theory U: Learning from the future as it emerges*. San Francisco: Berrett-Koehler Publishers.

Thomson, G. (2005). Fundamental needs. In S. Reader (Ed.), *The philosophy of need*. Cambridge: Royal Institute of Philosophy Supplement.

Co-constructing a learning organization as an opportunity to participate in an horizontal, collaborative system

Umberta Telfener[1]

Keywords: *learning organization; hierarchical and heterarchical organizations; second order practices; honor complexity; generative change*

1 Hierarchical and heterarchical organizations

Life allows hierarchical and heterarchical commands. If we consider the Midway battle in the Pacific Ocean, during the Second World War, we are told that in June 1942 the US Navy had her command ship bombarded. At that point ships started responding not to a central command but to what was happening in real time: the control got mobile and passed to the ship which was nearer to the battle field. The direction passed from a hierarchical and top-down modality, organized by forehand communication, to a horizontal one, based on constant feedbacks and on the moves of the enemies. This allowed the battle to be won.

The push towards a more shared way of managing organizational systems is more and more felt as important, and I personally consider the Learning Organization (L.O.) the metaphor that most underlines the co-construction of a cybernetic brain, the chance to build a second order non directive way of leading. It is a horizontal organization which allows to enhance the dialogue among participants and to learn from each other. As managers or called-in consultants in an Organization/Institution we help a L.O. emerge each time we: (i) organize and explicitly connect the beliefs and the presuppositions of all the people who are dancing together; (ii) are able to organize a communal praxis; (iii) tune in on relationships and processes; (iv) enhance the levels of awareness and of reflection on the common construction of a shared processual and evolutive "reality"; so that (v) everything each person learns and understands will become a shared contribution to the wellness of all.

To decide if an organization works horizontally or vertically is not a decision an outside observer can take arbitrarily. The observer will be only able to describe what she sees, aware that she will not be able to access one only truth. Systems theory teaches us that every organization is a non-trivial system, dependent on its history and not determinable analytically. The type of structure an organization has is decided by leaders of that specific system: if he defines it as part or apart from the organization. The epistemological choice defines the type of manager a person will be and the type of organizational world that will emerge from his arbitrary choices and from the mandate a manager has received.

It is different if one thinks to be apart from the team, in a power position, in a world organized by a priori rules which one needs to implement and pass to the workers, who are considered under one's management or instead one feels and considers oneself as part of a team, with whom it is possible to share decisions and negotiate world views. We are lately asked as psychologists to enter Organizations and consult, in order to perturb vertical structures towards non directive strategies of organization and culture. We therefore set up consultations to try and perturb the usual mode of working, hoping to open up to more horizontal and participative relationships. Senge (1999) suggests the term "metanoia" (from Greek μετανοεῖν, metanoein, changing one's thoughts), as the process of intellectual/moral/spiritual conversion through shared intuitions and choices that go towards evolution. To become able to do things together that one did not even think possible.

In order to perturb an Institution and build a L.O., in order to try and work in an horizontal mode, I believe that some assumptions must be respected: think systemically and honor complexity; take into consideration the premises that guide the actions of all participants; participate dynamically to the process, co-constructing shared visions and plans despite different positioning; activate motivation, and allow emotions to stem. All this is done by accessing to a reflexive thinking, which means allowing second order practices to emerge: creating inquiries in a recursive loop, taking the design turn and accepting our deep ignorance (von Foerster 1990).

2   An example

I was called as a consultant in a public hospital in Sardinia, Italy, to work specifically with three wards in which the chief physicians felt they couldn't delegate to anybody and had all the responsibilities and decisions on their shoulders. The quest was to alleviate this distress and to think together on the organizational procedures. I did a long analysis of the quest (Telfener 2011) with the three chief doctors and made my proposal: (i) to pass from a vertical to a horizontal organization, (ii) to try and make a L.O. emerge from the common encounters and from the conversational work we were going to do together over time. I therefore organized a systemic inquiry (Ison 2010) as a context of ethically related processes which emerged from common actions.

In order not to shut down the wards, the people who worked there were divided into two groups and would come to one of the two groups, one in the morning and one in the afternoon (four hours each). The participants could choose according to the convenience of the ward, since we had a clear common objective and a very clear mandate, which we all shared.

We needed to consider such meetings as a challenge and an opportunity, I therefore explained to the whole group my premises, intentions and aims and got a shared plan to build reciprocal respect and presence, as the first steps in our work. I organized the common work to activate the dialogue and the reciprocal trust and acquaintance. We scheduled organizational meetings and other reunions on specific relational shared issues; we read together the rules and regulations from the State Department of Health and discussed their consequences. We spoke about the culture of the ward and the one of patients and other stakeholders. We explicated expectations, chores and mandates and organized common exercises and common challenges. We worked together in a three day program repeated three times apart from one another, with Skype contacts in between, homework and specific chores for all the workers, despite their role in the ward.

3   How does a Learning Organization work

How did I try and produce a change in this organization? (i) Reflecting on the premises which were present in the morphic field that we all shared (Sheldrake 1995), (ii) making them explicit to all, through active role-playing and common conversations, (iii) enhancing in this way the coherence of the system and their gains from the work done till this moment; (iv) never undermining the individual and collective resources in the field and (v) organizing a common future planning, through clear procedures and objectives. To do this we took some time, certainly not all the time needed: the participants had three encounters with me on a monthly basis and met among themselves once a week for two hours to discuss what was going on and what they expected and wished for. The Skype meetings with me allowed them to tell me what were their difficulties and their moments of distress.

In order to enhance this process each participant had to take responsibility of its own knowing and being and we all felt the need to create a non judgmental setting, where people were asked to explore and define themselves, the work procedures and their common relationships.

As the implementer, I realized how much I needed to put myself in the picture and be transparent about the process that was emerging from our common dance. We worked a lot on future dynamics and possible problems, not because we knew already what would happen but because we were eager to explore our common point of view and share possible strategies to overcome difficulties. We imagined the evolution of this ward toward a more participative, dialogical practice through many devices: theater, role-playing, dramatization, active exercises and much else. We were ready to have clear objectives that indicated a possible evolution. This did not mean to know what the future would be but rather enhance a commitment, an intention, a project.

4   Conclusions

Every organization is a human community created together that becomes organized by itself, by laws and processes that are no longer under the domain of single individuals. Wholes and parts are interrelated and when trust organizes the flux of actions and practices the organization will adapt to a generative process which will be coherent with the individuals that live within it.

There are two possible learning strategies, one we could call reactive and the other generative. If all learning integrates thinking and doing, the first learning is organized by a priori models and retraces established and shared habits. Generative learning introduces novelty, detaches from the past and builds a growing awareness of the processes to which one participates by learning. This type of knowledge is very interesting since it introduces novelty and offers the possibility to go beyond what has been already said and done. It is a process which needs learning and is possible in an organization where participants share conversations and have reciprocal trust. These objectives are more easily reached in a system organized as a Learning Organization, where people interact and exchange capacities and emotions, narratives and actions. Because a L.O. offers the opportunity to transcend the dichotomy between individual and collective.

Note

1 U.T., clinical psychologist, teacher of the Systemic Milan school of family therapy, teaches at the post-graduate school in Health Psychology of the University of Rome La Sapienza. utelfener@gmail.com.

Bibliography

Ison, R. (2010). *Systems Practice: How to Act in a Climate-Change World*. England, London: Springer.

Senge, P. (1999). *The Dance of Change*. New York: Doubleday/Currency.

Sheldrake, R. (1995). *Seven Experiments That Could Change the World*. Rochester, VT: Park Street Press.

Telfener, U. (2011). *Apprendere i contesti*. Milano: Cortina Editore.

Von Foerster, H. (1990). Ethics and second order cyberneticfs. International Conference Systems & Family Therapy, Paris, 4–6 October.

Uncertainty and democratic governances

Interactions between "Vision of Democracy" and its contexts

Andrée Piecq[1]

Keywords: *emergence; uncertainty; interactions; contexts; democracy; complex system*

Introduction

Democracy, "the government of the people, by the people for the people" is, according to Winston Churchill, "the worst form of government – except for all those other forms, which have been tried from time to time".

The assumption of this article is "Uncertainty related to Democratic Governances emerges from interactions between the 'Vision of Democracy' and its contexts".

Observation of various democratic instances shows that they seem to be in troubles since the end of the 20th century until today.

From these observations, assumptions about "Vision of Democracy" are made regarding its structure and interactions with its "guiding principles"[2] (Piecq, 2011).

The methodology is explained in "De la pensée systémique à la pratique de l'organisation – Le 'giroscope'" (Piecq, 2011) and in two articles (Piecq and Lambert, 2013, 2014a).

The modeling tool is constituted by 12 main concepts used by the systemic community: the "guiding principles". They are accessible by observing behaviors. Their interactions are studied.

The contexts of the organization are also taken into account.

The structure of the organization is what emerges from the interactions between the "guiding principles" and their contexts.

To test the hypothesis of this article, Democracy is studied as a system formed from two complementary subsystems in interaction. It is analyzed from observations made on documents that explore its history throughout centuries.

1 "Vision of Democracy" as a rules system;
2 "Democratic Governances" as a human system.

Brief historical reminders

Athens V century BC

Constitutional reforms emerge from social explosions begun in the sixth century (political, social and financial inequalities).

Middle Ages in Europe

Inhabitants in certain villages and certain abbeys administer themselves.

England

1689 "Bill of Rights" – Begins the Constitutional Monarchy of today. Only the election of Members of Parliament by universal suffrage was added.

France

1830 Revolution – Nation becomes again sovereign: Parliamentary system.

United States

1787 Constitution (27 amendments still current) – Power of sovereign people, separations of legislative, executive and judicial powers with the President the head of the government. Presidential Democracy. The first modern Democracy.

Democracy in Europe in the 21st century

"Democratic" regimes "with a representative system of parliamentary type" are the most numerous in Europe and in the world.

The vision "Democracy"

Some benchmarks over the concepts of vision and myth.

The term vision is used as something which emerges in the mind and which over time is transformed in a myth.

In the study of organizations, the vision is a prerequisite for the elaboration of their finality (Piecq, 2011).

The myth:

- Is an explanatory system, systematized;
- Is shared by all members of an organization;
- Regulates both collective and inter-personal relations;
- Ensures social cohesion;
- Generates mythical rules accessible only by the phenomenological rules.

(Piecq, 2011)

The study of the myth "vision of Democracy" in a historical context:

- Shows cyclical developments. Citizens' attempts to establish a Democracy are followed by rejections, which in turn are followed by joining and so on through the centuries and in all countries;
- Emerges from wars, social conflicts and the rejection of tyranny.

In the 21st century, in Europe, the context of the myth "vision of Democracy" is different, because of the global context:

- Demographic expansion is galloping;
- Climate context is changing;
- Digital age appears as well as the universalization of communication through social networks;
- Commercial, economic and financial interests are at the forefront;
- Globalization (economic, cultural, political . . .) is becoming the law;
- Creation of the European Community changes the political balance;
- The peace between the European States is established;
- They are involved in external conflicts;
- They have to fight terrorism to defend the "Vision of Democracy" even in their own countries.

The "Vision of Democracy" cannot emerge unscathed from this context.

To have an operational consistency, the "Vision of Democracy" as all systems must be expressed in a finality and it must be defined in goals, objectives and actions. From its origins to today, the finality of "Vision of Democracy" is "sovereignty for all citizens", a mythical rule (Piecq, 2011) directly in line with the myth of "Vision of Democracy" To achieve this finality, five goals are required and are also mythical rules:

- Citizen's equality of rights;
- Individual freedom;
- Freedoms and rights protections;
- Tolerant pluralism;
- Free elections.

Three objectives are necessary to achieve goals and to implement necessary actions to achieve them.

These objectives are expressed in mythical rules:

- Establishment of the sovereignty of the people;
- Separation of powers in a representative democracy;
- Establishment of fundamental freedom.

The actions (human behaviors) can be directly observed today:

1   Establishment of people's sovereignty to:

- Create a constitution, govern and organize relations between citizens and protect them.

   Observations show people's sovereignty is threatened and usurpations of power remain possible;

- Create a representative parliamentary system. Observations show abuses of power or deviations can occur when the right to dissolve the chambers is discretionary;
- Guarantee general interest and protection of the individual interest. Observations show these interests may be contradictory and even paradoxical (interest of the elect does not always correspond to those of electors). Mandates are cumulate by elected representatives; financial lobbies become too powerful;
- Establish free universal suffrage with the obligation to vote after the presentation of political programs;

   Observations show it is not enough. "A tyranny" of the majority may appear; in certain so-called "democratic" countries elections are held in high scrutiny and ballots are neither secretive nor free.

2   Separation of powers in a representative democracy:

- Separate executive judicial and legislative powers.

   Observations show there are interdependencies and reciprocal actions; in the European Union, three institutions are working together to draw up regulations and directives: European Commission, European Union Council and the European Parliament;

- Create new international institutions above national institutions being accepted by all. Observations show this is possible only if a European and even a world identity exists, which is not the case;

3   Fundamental freedoms:

- Creating the Charter of Fundamental Rights of the European Union. Observations show the difficulty of executing such a Charter because each country wants to retain its rights;
- Changing wealth sharing to enable development and growth for all. Observations show it may have a communicating base effect that would reverse the procession of wealth without bringing equilibrium.

These observations show actions desired and carried out to achieve the objectives are contradictory human behaviors.

   The observations between goals, objectives and actions showed two different fields in interaction: the field of ideas (rules) and the field of humans' behavior.

From this non-exhaustive reading, hypotheses emerge from observations of the interactions between rules with other "guiding principles". This choice is only illustrative and not exhaustive:

1   It is as if a minority with power acts independently from voters.

   This hypothesis emerges from interactions between:

   - Rules (implicit phenomenological rules) concerning the role and function of subsystems;
   - Boundaries (permeable or impermeable) between subsystems "minority with power" and "voters";
   - Members (their roles and functions);
   - Members (communication – direct versus indirect, and the pitfalls that may entail);
   - Feedback (positive-oriented change versus negative-oriented homeostasis).

2   It is as if the diversity of voters were not represented.

   This hypothesis affects the representativeness and emerges from interactions between:

   - Rules concerning the status of subsystems and members;
   - Subsystems: minorities with power and voters;
   - Boundaries (permeable or impermeable) between subsystems "minority with power" and "voters";
   - Members (the "minority with power" differs from members of "voters" in their roles and functions);
   - Members (the principle of direct versus indirect communication, and the pitfalls that may entail);
   - Feedback (positive-oriented change versus negative-oriented homeostasis).

3   It is as if risks of corruption could appear.

   This hypothesis emerges from interactions between:

   - Phenomenological rules that affect both function, role and actions of elected officials and which are contradictory to the mythical rule;
   - Members (their roles and functions);
   - Members (the principle of direct versus indirect communication, and the pitfalls that may entail).

4   It is as if procedures put boundaries between elected representatives and voters.

   These boundaries would be emerging from interactions between boundaries (permeable or impermeable) between subsystems "minority with power" and "voters";

   - Subsystem differences between "minority with power" and "voters";
   - Members (their roles and functions);

- Members (the principle of direct versus indirect communication, and the pitfalls that may entail);
- The "guiding principle" of feedback (positive-oriented change versus negative-oriented homeostasis).

Hypotheses

It is as if the uncertainty emerges from interactions between "guiding principles" and historical contexts and as if it emerges from the incoherence of the actions done to achieve objectives (desired actions and actions really done).

The Democratic Governances

Democratic Governances (human systems) are managed by individuals. Every human system has a vision of what it is or should be. This vision is transformed into a myth, which obeys mythical rules that generate phenomenological rules observed through the actions elaborated to achieve the desired objectives.

The myth of "Democratic Governances" is the "Vision of Democracy" and its mythical rules are:

- Individual freedom;
- Protection of freedoms and rights;
- Tolerant pluralism and free elections.

Desired actions and real actions are contradictory. They lead to the emergence of uncertainty, described as a factor of destabilization that emerges from facing unknown difficulties.

After its emergence, the "Democratic Governance" passes through multiple changes and its practice is altered.

The practice of democracy consists of "behaviors" made by the members of the governances (human systems). "Practice" is not "Democracy", which is an idea, a concept, a vision that has emerged in particular contexts, as explained above.

Talking about "Democracy" with the same words as "Democratic Governance" is a semantic confusion, a target error, a logical-level confusion. It leads to criticisms that should be addressed to members of democratic governance and not to the democracy.

The Democratic Governances throughout the world do not necessarily have the same finality. Even if they have the same finality, objectives and actions used to achieve them are different. Sometimes there are contradictory and even paradoxical. What can give the illusion of a shared direction in democracies is having the same opinions. Nevertheless, these opinions are not operative: common goals and actions do not follow them.

The emergence of uncertainty in democratic governance

Each system structure is unique, original, without a double. They all are the product of a meta emergence: an emergence of a higher logical level produced by the collision between the "guiding principles" and contextual elements from the system. The structure of the visible system "Democratic Governances" emerged from shock produced by the meeting and the densification of forces (interactions) that bind the two sources: "guiding principles" and historical context.

Living in 2017 and being governed by a "Vision of Democracy" is difficult, whether at the national or international level. It is as if this value is flouted as if its application in current context could only cause suffering; it is as if individual liberty, of persons and, in some cases, of government, had become preponderant. Everywhere crises appear: separation desires (the Brexit), difficulties in governments finding their mark in federalism (Flanders, Wallonia and Brussels); arbitrary appointments or dismissals, financial crises, problems of religious extremism to name a few.

The certainty of living in a society without war, that accepts new political and religious balances, living in a prosperous society by being able to use, without counting, the natural reserves, no longer exists. This loss of certainties and benchmarks creates feelings of doubt and fear about tomorrow. It provokes turmoil among "Democratic Governances" and some defection. Some media and politicians question "Democratic Governances" and even the "Vision of Democracy". Some call for a strong regime without individual or collective freedom.

In a democratic society, the sovereignty of a nation, equality of citizens' rights, freedom of the individual, protection of freedoms and rights, pluralism, universal suffrage, freedom of expression, resort to the rule of the majority, carry in them the uncertainty of the result.

In other words, it is as if certainty only exists in tyrannical regimes such as absolute monarchies, totalitarian parliaments, bloody republics, despotic presidencies or any other strong political system. "In this world, there can be no certainty about anything, except death and taxes", wrote Benjamin Franklin.

Will the analysis of complex systems prove that is right?

The analysis of the systems (Democratic Governances of the world) is confused with the analysis of complex systems.

1   These governances (systems) are embedded in multiple contexts, interacting with each other and with the "guiding principles" (the subsystems, the mythical rules, the phenomenological rules . . .) which make up the system;
2   All these interactions form their structure, which sometimes evolves towards change or homeostasis;
3   Their analysis is at Meta level.

Any structure evolves in response to changing contexts and with their interactions with "guiding principles"

The observations of the structure of a complex system shows:

*   At time t1 its basic state;
*   At time t2, it has undergone modifications. Indeed if its contexts change, all the structure evolves in response to this change.

The new structure t2 is not predictable (Piecq and Lambert, 2014a).

In democratic governance, uncertainty is the emergence of interactions between the "Vision of Democracy" and the historical contexts in which it has evolved over centuries.

Finally, how to live in a democratic state with uncertainty?

Uncertainty is a discrepancy between an anticipation and events that actually occur. It is a "defect" in relation to what is expected in relation to the unknown. An unknown which is present in all the elements and events of life. This anticipation can be linked, as in Democracy, to a myth and one or more beliefs.

The question is what organizational way to propose for managing these discrepancies, and how accept them.

To conclude

Living in a world of certitudes is like living in a perfect homeostatic world where everything is foreseen and where change will not exist, as in absolute governances or in tyrannical governments, where uncertainty has no place.

As the history of democratic government shows, when citizens resume their rights and enter a revolutionary process, the uncertainty of making the revolution disappears, but the uncertainty of the outcome appears. The victory is a desire, an expectation, an uncertainty.

There will always be uncertainty!

The states in which governance is democratic are complex systems that are not frozen: they live. Their contexts change; they are part of globalization. They change with the elections and the parties put in place. Their structure changes and they live with the unpredictability of the emergence of this new form. At this level, it is a freedom rather than a difficulty.

In 2017, in our everyday life, full of fears, anxieties and anger, it is as if this freedom, which is carried by uncertainty, could be a hope for the future.

To tame uncertainty is to allow the emergence of a paradigm that integrates indeterminacy and renunciation of certainty and which justifies action. Today, citizens resume their rights in the face of attacks, catastrophes of all kinds. It is part of the community action.

The myth regains strength, resumes vigor and revives. It is ephemeral, but is it not the index of changes in citizen's behavior that could breathe life back into democratic governance?

Notes

1 G.I.R.O.S. (Groupe d'Intervention et de Recherche en Organisation des Systèmes).
2 The 12 "guiding principles": system and subsystem, members, finality, totality, circularity, borders, rules, feedback, homeostasis, equifinality, information sending, information delivery.

Bibliography

Boutinet, J.-P. (1990), *Anthropologie du Projet*, PUF, Paris, pp. 49–81.

DeWiel, B.(2005), *La démocratie : histoire des idées*, coll. Zêtêsis, Les Presses de l'Université Laval, Québec.

Finley, I.M. (2003), *Démocratie antique et démocratie moderne*, Poche.

Lemoigne, J.-L. (2006), *Théorie du système général*, mcxapc, France.

Menant, F. (2005), *L'Italie des communes, 1100–1350*, Belin, DL 2005, Paris.

Minati, G. and Licata, I. (2013), Emergence as Mesoscopic Coherence, *Systems*, 1(4), 50–65.

Piecq, A. (2011), *De la pensée systémique à la pratique de l'organisation, Le giroscope*, L'Harmattan, Paris.

Piecq, A. and Lambert, C.L. (2013), An essay of a systemic reading that can support a paradigm shift, *Proceedings of the 1st European Systemic Seminar (ESS2013) of the European Systemic Union (UES-EUS), Acta Europeana Systemica n°3*, Charleroi, Belgium.

Piecq, A. and Lambert, C.L. (2014a), How to develop an "open" future: Is it possible to take advantage of coherence and incoherence?, *Proceedings of the Congress EMCSR 2014 Civilization at the Crossroads-Response and Responsibility of the Systems*, Vienna.

Piecq, A. and Lambert, C.L. (2014b), The project: How to deal with coherence and incoherence, *Proceeding of the 9th Congress of the (UES–EUS) Globalization and Crisis: Systems Complexity and Governance Acta Europeana Systemica n°4*, Valencia, Spain.

Prigogine, I. (2001), *L'homme devant l'incertain*, Jacob O., Paris.

Tocqueville, A. D. (1981), *De la démocratie en Amérique*, GF Flammarion, Paris.

Watzatzlawick, P. and Weakland, J.H.(1977), *Sur l'interaction*, Le Seuil, Paris.

Watzlawick, P., Helmick Beavin, P., and Jackson, D.J. (1972), *Une logique de la communication*, Le Seuil, Paris.

Watzlawick, P., Weakland, J., and Fisch, R. (1975), *Changements paradoxes et psychothérapies*, Le Seuil, Paris.

Winock, M. (2008), *Histoire de la France politique: Tome 3, L'invention de la démocratie (1789–1914)*, Poche.

Theme VII

Systems thinking and system dynamics

Marie Noelle Sarget, Jose-Rodolfo Hernandez-Carrion and Stefano Armenia

Technological unemployment as frictional unemployment

From Luddite to Routine-Biased Technological Change

Federico Fiorelli[1]

Keywords: *technological unemployment; luddite fallacy; ICT revolution; routine-biased technological change; labour market*

The current digital revolution is driving more and more social researchers to question whether the introduction of ICT may lead to a reduction in employment levels. The relationship between technological innovation and unemployment is an extremely controversial subject in the socio-economic literature (Rifkin 1996). Historically, during the first industrial revolution, fear of machines drove many workers to destroy them. The Luddite movements at the turn of the eighteenth and nineteenth centuries were concerned that their work would be superfluous in the new industrial factories. Fortunately, the growth of productivity has allowed for relative growth of production and employment in the last two centuries. The phenomenon of technological unemployment has almost entirely disappeared from both the fears of workers and the writings of economists.

However, the digital revolution has raised new doubts in relation to this type of unemployment (Aronowitz and Di Fazio 1994). The fact that unemployment rates are considerably higher than the middle of the last century suggests a certain relationship between the new economic structure based on digital automation and the economic system's ability to absorb the work force.

New economic theories mostly agree that ICT technologies are impairing the professional structure of advanced economies. According to the theory of Routine-Biased Technological Change developed by researchers at MIT, the labour market is undergoing a process of polarisation (Acemoglu and Autor 2011). High skilled workers carry out non-routine jobs on the one hand, while low skilled workers carry out routine work on the other. The formers are benefiting from growth in productivity caused by digital technologies, while the latter are experiencing longer periods of unemployment or wage stagnation.

Now medium-high educational qualifications and the chance to participate in lifelong learning courses allow non-routine workers to be complementary to the machines. Conversely, a low school education and an inability to participate in such training courses mean the routine workers are replaced by machines (Brynjolfsson and McAfee 2011).

In this way, technological unemployment is transformed into frictional unemployment. The balance between the jobs created and destroyed by digital technologies is positive. However, the new jobs require human capital with much more training than in the past. The slowness with which governments change the education and training paths, paired with the acceleration of technological change, increases the mismatching between the skills demanded of the labour forces and the educational credentials offered. A future *robot-society* or a *second machine age* require human capital capable of running *with* the machines and not *against* them (Brynjolfsson and McAfee 2014).

On the sociological level, apart from the pause in the 1990s, the percentage of researchers who might think that a worker-robot interaction is possible has increased (Johnson et al. 2009). In fact, the new digital technologies intended to increase the productivity of the economic system require continuous organisational changes and more highly educated human capital (David 1990). The level of automation is not yet sufficient for machines to carry out a job independently. Their main function remains that of assisting the employee. New technologies represent a possible connection between the worker and the object of labour both in working practices and in their predictive simulation (Sheridan 2016).

These phenomena do not tend towards a possible end of labour (Kern and Schumann 1984) but rather towards a new division of work in which human intelligence is supported by its artificial equivalent. The workers can therefore be divided between those who are required for programming and controlling machines and those who can benefit from digital assistance in many stages of their work. Therefore, it remains paramount that education is provided to ensure well-trained human capital in terms of technical and operational expertise and also in terms of creativity and willingness to change.

The adoption of digital technologies and the modification of work processes represent both a challenge and an opportunity. They are a challenge because workers are required to continue the training courses throughout their professional life. They are an opportunity in that digital technologies enrich and distinguish business activities by reducing the amount of time allocated to routine work (Huws 2006).

The current spreading of technologically dense workplaces, a phenomenon due both to widespread digital culture and a reduction in the cost of new technologies with increasing computing power (Moore 1965), is profoundly changing the ways in which companies relate to clients, how they determine their decision-making procedures, hire staff and communicate with the various business sectors. The presence of working environments that are technologically and socially more complex brings about a technological socialisation and the technologisation of social organisation. In the former case, the mediation between the various members of the corporate structure promotes technological changes (co-working strategies), while in the latter case the need to digitise the various stages of production results in sudden changes to the labour ecosystem (Bernstein et al. 2007).

The main objective of new digital technologies, in particular those which favour the management, storage and analysis of information, is not to replace the skills of workers with those of the machines, but, on the contrary, to improve the quality of human labour through the use of technologies that allow the employee to engage in the most creative and pleasant work activities. Thus so-called tacit knowledge, especially that of a socio-relational nature, continues to account for the added value of human labour in future automated production (Baumard 1999).

In conclusion, technological unemployment is only a possible threat to employment in the short term. Technologies, and digital technologies in particular, represent the primary tool for increasing productivity possessed by contemporary economies. Automation offers the benefit of safer working environments and, at the same time, frees workers from the more tedious and repetitive tasks. Even in a context of strong technological expansion, adequate training of workers and the unemployed can secure future growth in terms of the economy and employment.

Note

1 University La Sapienza (Rome).

Bibliography

Acemoglu, D. and Autor, D. (2011). Skill, Tasks and Technologies: Implications for Employment and Earnings. *Handbook of Labor Economics*, 4B, 1043–1171.

Aronowitz, S. and Di Fazio, W. (1994). *The Jobless Future*. Minneapolis: University of Minnesota Press.

Baumard, P. (1999). *Tacit Knowledge in Organizations*. London: Sage Publications.

Bernstein, D., Crowley, K. and Nourbakhsh, I. (2007). Working with a Robot: Exploring Relationship Potential in Human-Robot Systems. *Interaction Studies*, 8(3), 465–482.

Brynjolfsson, E. and McAfee, A. (2011). *Race against the Machine*. Boston: Digital Frontier Press.

Brynjolfsson, E. and McAfee, A. (2014). *The Second Machine Age*. New York: W.W. Norton & Co. Inc.

David, P.A. (1990). The Dynamo and the Computer: An Historical Perspective on the Modern Productivity Paradox. *American Economic Review*, 80, 355–361.

Frey, C.B. and Osborne, M. (2013). *The Future of Employment: How Susceptible Are Jobs to Computeristaion?* Oxford: Oxford University Press.

Huws, U. (2006). *The Transformation of Work in a Global Knowledge Economy: Towards a Conceptual Framework*. Leuven: WORKS.

Johnson, R., Saboe, K., Prewett, M., Coovert, M. and Elliott, L. (2009). Autonomy and Automation Reliability in Human-Robot Interaction: A Qualitative Review. *Proceedings of the Human Factors and Ergonomics Society Annual Meeting*, 53(18), 1398–1402.

Kern, H. and Schumann, M. (1984). *Das Ende der Arbeitsteilung? Rationalisierung in der industriellen Produktion: Bestandsaufnahme, Trendbestimmung*. Munich: Verlag C.H. Beck.

Moore, G. (1965). Cramming More Components onto Integrated Circuits. *Electronics Magazine*, 38(8), 114–117.

Rifkin, J. (1996). New technology and the end of jobs. *The case against the global economy*, 108–121. Retrieved 23.9.2018 from http://www.exponentialgoverning.com/uploads/7/4/8/2/74826303/new_technology_and_the_end_of_jobs.pdf

Sheridan, T.B. (2016). Human-Robot Interaction: Status and Challenges. *Human Factors*, 58(4), 525–532.

Evaluation of measures to combat poverty on the basis of multi-country global hybrid econometric model

Abdykappar Ashimov,[1] *Yuriy Borovskiy*[2]
and Dauren Aidarkhanov[3]

Keywords: *multi-country global hybrid econometric model; macroeconomic analysis; parametric control*

One of the global problems of humanity is the poverty of some regions.

Measures to combat poverty at the country level, regional economic unions and all over the world are still ineffective.

Assessment of the poverty problem and the possibility of developing measures to combat poverty can be done on the basis of a cyber-systematic approach simulation.

The paper presents the results of macroeconomic analysis of poverty problems at the level of world countries, within the Eurasian Economic Union (EAEU) and the world at large. The approach is proposed for evaluation measures against poverty at the level of countries, EAEU and the world on the basis of parametric control theory and a multi-country global hybrid econometric model (hereinafter the Model) of the world economy. The results of comparing the effectiveness of measures to combat poverty are provided: at the level of regional unions and at the level of countries which are part of regional unions; at the world level and at the level of countries which are forming the world economy.

The proposed approach to develop measures to combat poverty is implemented on the basis of GDP per capita at purchasing power parity (Deaton 2010), adopted as a measure of poverty. The developed model is a hybrid of the advanced multi-country econometric Fair model (Fair model) and multi-country econometric model of the macroeconomic balance of the IMF (IMF model). The mentioned Model is constructed on the basis of:

1   Development of Fair model structure (Fair 2004, 2013) subject to conditions of: government budget formation and countries' government debt; integration of countries to the EAEU; forecasting performance of Fair model and conditions of elasticities equality of a trade balance indicators at a nominal exchange rate in Fair model and in the IMF model.

2   Taking to the account in the structure of the IMF model (Lee et al. 2008), the conditions for the elasticities equality of the trade balance indicators at the nominal exchange rate in the Fair model and in the IMF model.

3   Development of the two interfaces of interaction of Fair model and the IMF model.

The first interface is designed to convert the endogenous variable of annual values of equilibrium exchange rate of the IMF model for input into the Fair model in the form of quarterly exchange rates values. The design of the first interface is on the basis of Foroni et al. (2015) taking into account the effects of monetary and exchange rate policies indicators, as well as the cyclical and temporary factors.

The second interface is designed based on the approach of Baxter and King (1995), for the conversion of endogenous quarterly values of Fair model indices for input to the IMF model in the form of annual values of fundamental factors.

Parameters estimation of advanced Fair model, the IMF model and interfaces is implemented on the basis of the statistical data of the period 1960–2014, mainly with the methods which are used in basic versions of Fair model and the IMF model. The approach (Beidas-Strom and Cashin 2011) was used in the IMF model for more accurate estimation of panel regressions parameters of the balance of savings and investment. It should be also noted that for the Fair model and the IMF model values of elasticities indicators of exports and imports on the nominal exchange rate were estimated based on the approach which is suggested in Tokarick (2014).

Using the parametric control theory (Ashimov et al. 2013) approach the possibility of transferring the computational experiments results based on the Model to the studied subject area was evaluated.

Based on the estimated Model the evaluation of poverty indicators was implemented at the level of countries of the world, within the EAEU and the world as a whole for the past and medium-term forecast period, which are consistent with the known data, and also investigated was the impact of economic policies taking into account the adopted exchange rate policy in each country on poverty indicators at the country level, the EAEU and the world at large.

Assessments of undervaluation or overvaluation of the current exchange rates, compared with the equilibrium exchange rates for each individual country were also obtained.

A number of extreme problems to combat poverty at the country level, within the EAEU and the world were formulated and solved based on the estimated Model and parametric control theory, where the optimization criterion is GDP per capita at purchasing power parity; solutions are sought on economic policy instruments in the areas of fiscal and monetary policy in the given areas of value of the instruments, taking into account the adopted exchange rate policy in each country.

Based on the results of extreme problems solving the comparative analysis of the measures, effectiveness was implemented at the country level within the regional unions and concerted measures at the level of regional unions, and also the comparative analysis of the measures effectiveness was implemented at the country level, which are forming the world economy within the Model under consideration and concerted measures at the world level in general.

Extreme problems of convergence on the adopted poverty indicator within the EAEU were also considered.

It is shown the effectiveness of extreme problems solutions at the level of regional unions, than at the level of individual countries, belonging to the relevant regional union; the effectiveness of the extreme problems solutions at the world level than at the level of regional unions and countries.

Using the approach of parametric control theory the possibility of practical implementation of obtained optimal solutions corresponding to the extreme problems was assessed.

Notes

1 Kazakh National Technical University.
2 Kazakh National Technical University.
3 Kazakh National Technical University.

Bibliography

Ashimov, A.A., Sultanov, B.T., Adilov, Z.M., Borovskiy, Y.V., Novikov, D.A., Alshanov, R.A., and Ashimov, A.A. (2013). *Macroeconomic Analysis and Parametric Control of National Economy*. New-York, NY: Springer.

Baxter, M., and King, R. G. (1999). Measuring business cycles: approximate band-pass filters for economic time series. *Review of Economics and Statistics*, 81(4), 575–593.

Beidas-Strom, S., and Cashin, P., (2011). Are Middle Eastern Current Account Imbalances Excessive? (Working paper No. 11–195). Retrieved from IMF website: www.imf.org/external/pubs/ft/wp/2011/wp11195.pdf

Deaton, A. (2010). Price indexes, inequality, and the measurement of world poverty. *American Economic Review*, 100(1), 5–34.

Fair, R.C. (2004). *Estimating How Macroeconomy Works*. Cambridge, MA: Harvard University Press.

Fair, R.C. (2013). Macroeconometric Modeling. Retrieved from Yale University website: https://fairmodel.econ.yale.edu/mmm2/mm.pdf

Foroni, C., Guerin, P., and Marcellino, M. (2015). Using Low Frequency Information for Predicting High Frequency Variables (Working paper No. 13). Retrieved from http://static.norges-bank.no/pages/103966/Working_Paper_13_15.pdf?v=12/10/2015105048AM&ft=.pdf

Lee, J., Milesi-Ferretti, G.M., Ostry, J., Prati, A., and Ricci, L.A. (2008). *Exchange Rate Assessments: CGER Methodologies* (Occasional Paper No. 261). Washington, DC: International Monetary Fund.

Tokarick, S. (2014). A method for calculating export supply and import demand elasticities. *The Journal of International Trade & Economic Development*, 23(7), 1059–1087. doi: 10.1080/09638199.2014.920403

Global problems and parametric control theory

Abdykappar Ashimov[1] and Yuriy Borovskiy[2]

Keywords: *global problem; the theory of parametric control of macroeconomic systems; computable general equilibrium model*

The formed world economy (after independence of former colonies, collapse of the Soviet Union, emergence of new states and economic unions) for many years has kept the imbalances in the economic development of individual countries and regions of the planet. These imbalances formulate global and regional problems (including individual countries' problems); without their solution further humanity advancement along the path of economic progress is impossible. The most urgent problems are the "North-South" problem – the gap in economic and social development between rich and poor countries and the associated food problem, which lies in the inability of poor countries to support themselves completely even with vital food. It is generally recognized that for the solution of these global problems it is required to have all humanity coordinate efforts and, in particular, efforts of all concerned states. This interest in coordinated efforts can be improved if: (i) reasonable optimal steps will be offered (steps which will be defined by economic policy instruments) and which should be taken in order to solve these global challenges; (ii) the justified assumptions that coordinated actions of all states (or at least states of the countries within the regional economic union) will give a greater effect for each country than uncoordinated action.

The objective of this paper is to show the capabilities of parametric control theory (Ashimov et al. 2013) on the basis of the developed macroeconomic model for the optimal solution of these global problems and to prove the benefits of joint action to address them. It should be noted that the main difference of parametric control theory from other well-known macroeconomic theories is the use of only those mathematical models that have been tested for the possibility of their practical application through a series of original and classical methods.

The results of this study were obtained on the basis of a developed multi-country and multi-sector computable general equilibrium dynamic model (the Model) which is the development of the Globe model (GLOBE 1 2016). The Model describes the functioning of interacting economies of nine regions (countries): Kazakhstan, Russia, Belarus, Armenia, Kyrgyzstan, the European Union (as one country), the USA, China; the rest of the world (as one country) from 2004 to 2021. Here the first five countries form the Eurasian Economic Union (EAEU). The economy of each Model's region includes 16 Branches (producers). Each Branch produces one type of produce corresponding to its name. In addition to the producers in each region there are consumers agents: Households and the State. During the output of products Branches use two production factors: labor and capital, which are owned by Households in their region. There is also another special agent-region Globe, which receives income from markups of trade and transport during the goods export and import between other regions. The Model describes the accepted practice of providing financial services and floating exchange rates in all regions.

The Model verification (testing the possibility of its practical application) was successfully implemented with the help of the following papers: evaluating stability of smooth maps defined by the model (Ashimov et al. 2014; Golubitsky and Gueillemin 1973); assessment of mappings' sustainability indicators defined by the model (Ashimov et al. 2014; Orlov 2002); a series of counterfactual and forecast scenarios.

The estimation of optimal measures to reduce the gap in economic development between rich and poor regions (countries), as well as the measures to increase agricultural output has been performed with the help of solving of a number of parametric control problems in finding public policy instruments values based on the Model, which are maximizing the relevant criteria in the appropriate restrictions.

The main (global) P_w problem is the problem of the optimal coordinated economic policy across all regions of the Model in 2015–2021. The criterion of this problem includes: growth rates of per capita GDP in all regions, the rates of growth of agricultural output per capita in all regions, as well as indicators characterize the economic convergence of the regions (with a minus sign). Here we have the following economic policy instruments: effective tax rates differentiated by sector and the share of government consumption in public spending in all regions. Similar parametric control problems were formulated and solved at the level of all countries of the Eurasian Economic Union (P_{EAEU} problem), and also at the level of each region r (P_r problem).

The results analysis of solving P_w, P_{EAEU}, P_r problems based on the Model shows that global approach of P_w problem can increase the GDP of each region, as well as the agricultural output in the amount of 1, 5–3, 4% compared to the baseline calculation. At the same time for all regions the corresponding increase in these indicators, which was obtained by solving the P_{EAEU} problem is less than for P_w problem and the effects from the solution of P_r problem is less than for P_{EAEU} problem.

The originality of the approach of parametric control theory to the solution of global nature problems is justified by verification of the reviewed model which is not only for the basic calculation, but it is also for optimal scenarios obtained as a result of solution of specified parametric control problems.

Notes

1 Kazakh National Research Technical University.
2 Kazakh National Research Technical University.

Bibliography

Ashimov, A.A., Adilov, Z., Alshanov, R., Borovskiy, Yu. and Sultanov, B. (2014). The Theory of Parametric Control of Macroeconomic Systems and Its Applications (I). *Advances in Systems Science and Application*, 14(1), 1–21.

Ashimov, A.A., Sultanov, B.T., Adilov, Z.M., Borovskiy, Y.V., Novikov, D.A., Alshanov, R.A. and Ashimov, A.A. (2013). *Macroeconomic analysis and parametrical control of a national economy*. New York, NY: Springer.

GLOBE 1. (2016, September 21). Retrieved from www.cgemod.org.uk/globe1.html

Golubitsky, M. and Gueillemin, V. (1973). *Stable mappings and their singularities*. New York, Heidelberg, Berlin: Springer-Verlag.

Orlov, A.I. (2002). *Econometrics: A textbook*. Moscow: Ekzamen (in Russian).

Towards forward looking policy activities

A meta-model and a case study

Simone Landini[1] *and Sylvie Occelli*[2]

Keywords: *road safety policy activity; meta-model; learning; anticipation; reflexive systems*

1 Background

As interaction possibilities increase and the reaction times of many socioeconomic phenomena accelerate, modern societies are facing escalating levels of complexity. Their acknowledgment is a main source of uncertainties that challenge the conventional observation approaches. They form what has been recently called by some scholars post-normal times which urge a sense-based and value-attentive re-appraisal in the unfolding of the thinking-acting-outcome circular relationships, i.e. the what, why, how and for whom actions should be undertaken (Aaltonen, 2005; Umpleby, 2016).

In the policy domain, the investigation of the topic has been mainly revolved on how to fine-tune action design to problem handling, a main focus of attention in several study fields such as cybernetics, system analysis and organizational studies (Ashby, 1956; Beer, 1973; Espejo and Gill, 1997).

Notwithstanding its longstanding relevance, the "how to adequately design and implement, in situated context, the appropriate action mix in order to meet people's needs" is still high on the research agenda. Besides acknowledging the wickedness of the problems (Conklin, 2006), it calls for a fresh look at information requirements and data-extraction. Furthermore, it urges us to pay attention at the alignment of stakeholder networks, and, even more importantly, at the assessment of agents' interaction process and system overall outcomes (Miceli and Castelfranchi, 2000). Failing to address the latter, in fact, is likely to negatively affect the co-evolution between policy design and process, thus hampering the transformative process necessary for innovation to occur in public administration (Osborne and Brown, 2013; Reynolds, 2011).

Information and Communication Technologies (ICT) have a main responsibility in modulating and/or enabling each of these aspects (Coleman, 2008; Gil-Garcia, 2012). Their impact is even more dramatic as they may change the very sources of that process and continuously reshape the ways it unfolds over time in the different contexts. A few noticeable changes relate, for example, to the relationships between observer and problems, the action commitment by decision-makers and the structuring of those intangible functional and communicative parts of the policy environment that are increasingly important in the reproduction of today social systems (Poli, 2010; Umpleby, 2007).

2   Motivations and conceptual underpinnings

Often neglected is the fact that the generally larger and multi-actor policy partnerships created by ICTs are likely to adopt different stances in defining public value, depending on the role (position) of the engaged actors (Talbot, 2008). Consequently, appraisals of what the outcome of transformative process is, in terms of success and failure, may take many forms (Smith and Stirling, 2007). In addition it varies across recipients and stakeholders, depending on the dimensional levels underpinning policy activity, such as process, programs and politics (McConnell, 2010).

As it happens with other current social phenomena such as societal risks and climate change adaptation (Anderson, 2010; McGrail, 2012), ICT-based government transformative process is therefore confronted with kinds of questions that reveal an intrinsic uncertainty (Walker, 2001).

To cope with the quandaries of uncertainty, the notion of adaptive/anticipatory governance has been suggested (Fuerth and Faber, 2012), which emphasizes the role and limitations of knowledge on which policy choices are based, encourages monitoring of outcomes and broad participation in policy-making processes.

Learning plays an utterly central role in all the notions of adaptive/anticipatory governance, whereby the latter implies notions of future, modelling (Rosen, 1985) and enabling conditions. It is commanded to happen regularly and self-consciously, being both a task of any government organization and an integrant part of policy implementation.

Depending on the reflexive undertaking in linking human cognitive and operational functions, different forms of learning are recognized. For example, Cooney and Lang (2007) distinguish between simple and complex learning. The former is an activity mainly involving data gathering, and improvement of skills, and competencies. It is typically aimed to update the profile of policy instruments, or the techniques used to achieve certain goals. The latter is more encompassing and responds to the fundamental limitations of human cognition. It entails redefining the problem to be addressed, and critically revisiting the determinants of what constitutes relevant knowledge by the problem's owners.

A salient feature of a complex learning processes is that the yielded knowledge is not solely technically and expert-driven (Tsoukas, 2005). As it acknowledges the incompleteness of any single knowledge view it necessarily entails a pluralist approach, which engages stakeholders and policy recipients (Orlikowski, 2002; van Dijk and Winters-van Beek, 2009). The recent plea for innovation in the public sector just makes a claim that a whole government approach is necessary, capable of both steering the directions of change and creating the conditions for actions (OECD, 2011).

3 Study contents

The arguments addressed in this paper are a contribution in this endeavour. They build upon three main tenets.

The first is that for a policy activity system, under certain circumstances, factors of uncertainty are a source of variety; they can make available new alternatives for action and enable conditions for transformative processes to unfold (Walker, 2001). This challenges traditional notions of control and shifts the attention to softer but no lesser important conceptions of learning and engagement (Landini and Occelli, 2010).

The second makes it explicit that a policy activity system, as any other human organization, is endowed with a reflexive ability which gives sense to the directionality of the evolution paths as the system transforms itself (Espejo, 2015; Leydesdorff, 2005; Soros, 2013).

The third tenet acknowledges the basic function of ICT's pervasive presence in providing a system infrastructure to avoid discrepancies between design and implementation (Occelli, 2006, 2013). Insofar as the system reflexive ability is leveraged, furthermore, ICT based learning is an essential support to the steering of the transformative process. When considered at the interface between a decision-making activity and what is done in current practices, ICT usages can yield new system capabilities for action. Whenever action alternatives do exist, furthermore, such capabilities do not solely enable their realization, but in situated social practices they also pave the ways to change anticipatory processes.

The study strategy develops in two steps.

First a conceptualization exercise is carried out. It builds upon the system analysis approach proposed by Le Moigne (2006) which helps recognize the maturity levels in the reflexive ability of living systems. A conceptual framework is outlined that describes an ideal learning process a policy activity is likely to undertake over time. A set of meta-models is focused in identifying policy problems with owners progressively might devise as changes occur in issues, affecting engaged stakeholders and context (Michiotis and Cronin, 2011). We argue that the analytic exercise offers insights into a system's viability conditions – an aspect of anticipatory behaviour – thus sharpening the understanding of the co-evolution between policy design and process.

Then, to provide ground to the conceptual discussion the framework is used in a case study and applied to analyze the role of the Piedmont monitoring centre in the regional road safety policy. As IRES has been assigned the task to set up and manage the centre, it has a unique vantage point for observing the centre's activities, from both the inside and outside (Occelli, 2012). The application of the framework made available a clearer description of how the activities developed since the centre was established. More importantly, it allowed us to grasp a few difficulties that are slowing down the progress. Some of them can reasonably be ascribed to the organizational and financial worries currently distressing many planning authorities. Others relate to problems in knowledge sharing across stakeholders (Banathy, 2000) and to bottlenecks in channelling the centre's information feedbacks to road safety design and management. The latter in fact can be considered as an issue of the current epistemological debate in second-order cybernetics and reflexivity (Kauffman, 2016).

Notes

1 IRES – Istituto di Ricerche Economico Sociali del Piemonte.
2 IRES – Istituto di Ricerche Economico Sociali del Piemonte.

Bibliography

Aaltonen, M. (2005). How Do We Make Sense of the Future? An Analysis of Futures Research Methodology-V2.0. *Journal of Futures Studies*, 9(4), 45–60.
Anderson, B. (2010). Preemption, Precaution, Preparedness: Anticipatory Action and Future Geographies. *Progress in Human Geography*, 34(6), 777–798.
Ashby, W.R. (1956). *An Introduction to Cybernetics*. London, UK: Chapman & Hall.
Banathy, B.A. (2000). Navigating Bounded and Unbounded Spaces. *Systems Research and Behavioural Science*, 17, 481–484.
Beer, S. (1973). Designing Freedom. Retrieved from http://ada.evergreen.edu/~arunc/texts/cybernetics/beer/book.pdf.
Coleman, S. (2008). Governance and/as Technology. In H. Chen, L. Brandt, V. Gregg, R. Traunmüller, S. Dawes, E. Hovy, A. Macintosh, & C.A. Larson (Eds.), *Digital Government e-Government Research, Case Studies, and Implementation* (Chapter 1, pp. 3–20). New York, NY: Springer.
Conklin, J. (2006). Wicked Problems and Social Complexity. In J. Conklin (Ed.), *Dialogue Mapping: Building Shared Understanding of Wicked Problems*. New York: John Wiley & Sons.
Cooney, R., & Lang, A.T.F. (2007). Taking Uncertainty Seriously: Adaptive Governance and International Trade. *The European Journal of International Law*, 18(3), 523–551.
Espejo, R. (2015). Good Social Cybernetics Is a Must in Policy Processes. *Kybernetes*, 44(6/7), 874–890. doi.org/10.1108/K-02-2015-0050.
Espejo, R., & Gill, A. (1997). The Viable System Model as a Framework for Understanding Organizations. Retrieved from www.moderntimesworkplace.com/good_reading/GRRespSelf/TheViableSystemModel.pdf.
Fuerth, L.S., & Faber, E.M.H. (2012). Anticipatory Governance. *Practical Upgrades*. Retrieved from www.gwu.edu/~igis/assets/docs/working_papers/.
Gil-Garcia, J.R. (2012). *Enacting Electronic Government Success*. New York, NY: Springer.
Kauffman, L.H. (2016). Cybernetics, Reflexivity and Second-Order Science. *Constructivist Foundations*, 11(3), 489–497. Retrieved from http://constructivist.info/11/3/489.kauffman.
Landini, S., & Occelli, S. (2010). Control as a Learning Process in Human Complex Adaptive Systems: Some Remarks for Policy Making. Paper Presented at ECCS 2010, 11–17 Septembre, Lisbon.

Le Moigne, J.L. (2006). La théorie du système général: théorie de la modélisation, Collection Les Classiques du Réseau Intelligence de la Complexité. Retrieved from www.mcxapc.org/inserts/ouvrages/0609tsgtm.pdf.

Leydesdorff, L. (2005). Anticipatory Systems and the Processing of Meaning: A Simulation Study Inspired by Luhmann's Theory of Social Systems. *Journal of Artificial Societies and Social Simulation*, 8(2). Retrieved from http://jasss.soc.surrey.ac.uk/8/2/7.html.

McConnell, A. (2010). Policy Success, Policy Failures and Gray Areas in-between. *Journal of Public Policy*, 30(3), 345–362.

McGrail, S. (2012). 'Cracks in the System': Problematisation of the Future and the Growth of Anticipatory and Interventionist Practices. *Journal of Futures Studies*, 16(3), 21–46.

Miceli, M., & Castelfranchi, C. (2000). The Role of Evaluation in Cognition and Social Interaction. In K. Dautenhahn (Ed.), *Human Cognition and Social Agent, Technology* (pp. 225–300). Amsterdam: John Benjamin Publishing Company.

Michiotis, S., & Cronin, B. (2011). Assessing Capacity and Maturity for Change in Organizations: A Pattern-Based Tool Derived from Complexity and Archetypes. In A. Tait & K.A. Richardson (Eds.), *Moving Forward with Complexity: Proceedings of the 1st International Workshop on Complex Systems Thinking and Real World Applications* (pp. 113–130). Litchfield Park: Emergent Publications.

Occelli, S. (2006). A Framework for a reflexive information wired environment. Cognitica, Actes du Colloque de l'Association pour la Recherche Cognitive, 6–8 December, Bordeaux, 111–122.

Occelli, S. (2012). Monitoring Road Safety in an Information Wired Environment. *NETCOM*, 16(3–4), 201–220.

Occelli, S. (2013). Knowledge Evidence for Progressive Policy Making. In I. Zalisova, I.S. Walterova, & R. Bejdak (Eds.), *Digital Governance: From Local Data to European Policies* (pp. 23–36). Praha: Agama.

OECD. (2011). Government at a Glance 2011. *OECD Publishing*. Retrieved from http//dx.doi.org/10.1.1787/gov_glance-2011-en.

Orlikowski, W.J. (2002). Knowing in Practice: Enacting a Collective Capability in Distributed Organizing. *Organization Science*, 13(3), 249–273.

Osborne, S.P., & Brown, L. (Eds.) (2013). *Handbook of Innovation in Public Services*. Cheltenham, UK: Elgar.

Poli, R. (2010). The Complexity of Self-Reference: A Critical Evaluation of Luhmann's Theory of Social Systems. *Journal of Sociocybernetics*, 8(1–2), 1–23.

Reynolds, M. (2011). Bells That Still Can Ring: Systems Thinking in Practice. In A. Tait & K.A. Richardson (Eds.), *Moving Forward with Complexity: Proceedings of the 1st International Workshop on Complex Systems Thinking and Real World Applications* (pp. 328–352). Litchfield Park: Emergent Publications.

Rosen, R. (1985). *Anticipatory Systems*. New York, NY: Pergamon.

Smith, A., & Stirling, A. (2007). Moving Outside or Inside? Objectification and Reflexivity in the Governance of Socio-Technical Systems. *Journal of Environmental Policy & Planning*, 9(3–4), 351–373.

Soros, G. (2013). Fallibility, Reflexivity, and the Human Uncertainty Principle. *Journal of Economic Methodology*, 20(4), 309–329.

Talbot, C. (2008). *Measuring Public Value: A Competing Values Approach*. London: The Work Foundation. Retrieved from www.theworkfoundation.com/download publication/report/202_measuring_pv_final2.pdf.

Tsoukas, H. (2005). *Complex Knowledge*. New York, NY: Oxford University Press.

Umpleby, S.A. (2007). Unifying Epistemologies by Combining World, Description and Observer. Retrieved from www.gwu.edu/~umpleby/recent_papers/2007_ASC_Unifying_Epistemologies.pdf.

Umpleby, S.A. (2016). Second-Order Cybernetics as a Fundamental Revolution in Science. *Constructivist Foundations*, 11(3), 455–465. Retrieved from http://constructivist.info/11/3/455.umpleby.

van Dijk, J., & Winters-van Beek, A. (2009). The Perspective of Network Government: The Struggle between Hierarchies, Markets and Networks as Modes of Governance in Contemporary Government. In A. Meier, K. Boersma, & P. Wagenaar (Eds.), *ICTs, Citizens & Governance: After the Hype!* (pp. 235–255). Amsterdam: IOS Press.

Walker, W.E. (2001). Uncertainty: The Challenge for Policy Analysis in the 21st Century. Retrieved from www.rand.org/content/dam/rand/pubs/papers/2009/P8051.pdf.

Nobody deserves this fate
The vicious cycle of low human development in Guinea-Bissau

Bassiro Só,[1] Eduardo Ferreira Franco,[2] Hamilton Carvalho[3] and Joaquim Rocha dos Santos[4]

Keywords: *Guinea-Bissau; human development; political instability; system dynamics*

1   Introduction

Nobody chooses her birthplace. A child born to a family in Guinea-Bissau has a life expectancy of only 55 years (World Bank, 2014). Moreover, she can expect a life course marked by several stressors, suffering, and absence of social conditions to achieve full human potential. Guinea-Bissau perhaps epitomizes a group of countries that exposes the sheer incapacity of humanity to bring about progress and minimum living conditions to all human beings on this planet.

The country is relatively young, becoming independent from Portuguese colonial occupation in 1973. From that date until 1994, the country lived under a dictatorial one-party regime. Because of a confluence of several factors, Guinea-Bissau remains one of the world's poorest countries. In 2014, the country's Human Development Index (HDI) was 0.402, below the average of the group of countries with the lowest HDI, which was 0.505. This result puts the country at position 178th in the United Nations' world ranking, out of 188 countries and recognized territories (UNPD, 2015).

The HDI is a summary measure for assessing long-term progress in three basic dimensions: a long and healthy life; access to knowledge; and a decent standard of living. The Guinea-Bissau's Gross National Income (GNI) per capita is U$ 1,362 (below the average for sub-Saharan countries, which is U$ 1,637) and 80.4% of the population are considered multidimensionality poor. Life expectancy at birth is more than 16 years lower than the world average (World Bank, 2014). Educational levels are very low: people aged 25 years old have on average only 2.8 years of formal education. In practical terms, while several African countries have increased their standards of living in recent years, Guinea-Bissau failed to present any improvement in human development from 2005 to 2014.

Its current economic and social situation cannot be dissociated from the political-military context that the country has experienced. Since its independence, no president completed his term. From 1974 to 2014, there were five elected presidents, five interim presidents, a transitional president, four successful coups and several other unsuccessful or frustrated attempts. This scenario puts the political system of the country as one of the weakest in the world (IMF, 2015).

In 2014, a new government was democratically elected, bringing hope of change. Taking advantage of its popularity, it formulated a new strategic political program to promote development, entitled "Strategic and Operational Plan, Guinea-Bissau 2025, Sun In Iardi" (Vaz, 2015), which was approved by international donors in March 2015. The goal was to overcome for good the seemingly perpetual vicious cycle of low human development in the country. However, its future is still uncertain, considering not only the historical pattern of political instability in Guinea-Bissau but, especially, as we will discuss below, the lack of consensus among renowned specialists about the proper ways to break a vicious macrocycle of low human development in poor countries.

2   Objective and contributions

Within this broad context, this paper aims, first, at developing a model that captures the interplay between macro factors in determining the prevalence of low human development in Guinea-Bissau. The model is developed following the sound tradition of the system dynamics methodology. Second, the paper presents the simulation of two scenarios based on the proposed model: one considering "business as usual" and the other testing the comprehensive intervention in the country, named "Terra Ranca."

3   Material and methods

System Dynamics is the method employed to model Guinea-Bissau's poverty trap. The method was developed in the middle of last century by Jay Forrester (1961) to study complex business problems and was later expanded to investigate problems associated with sustainability of population growth, global warming, and many others. The approach consists of an iterative process to define a dynamic hypothesis, develop a formal model, test it, and to formulate and evaluate different intervention policies (Sterman, 2000). Complex social problems that produce vicious cycles, such as poverty in Sub-Saharan Africa, benefit from the system dynamics modeling approach, which allows the integration of several theoretical frameworks to explain the dynamics of such cycles.

The approach requires the formalization of the model's structure, the identification of causal relationships between the elements, the understanding of decision rules and delays, and the analysis of the dynamic behaviors underlying poverty in the country.

4   Results and discussions

Traditional approaches to fighting poverty traps seem to be incomplete, failing to account for vital parts of the system and to deal the dynamics of change

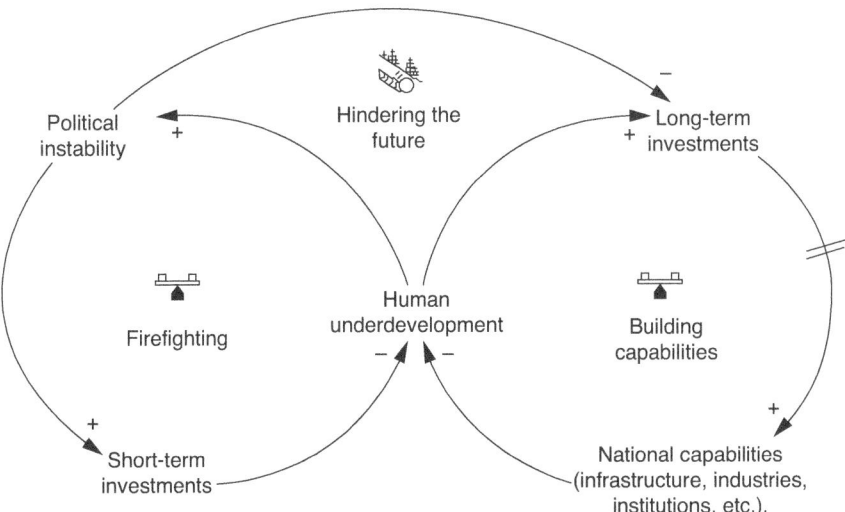

Figure 90.1 Shifting the burden in the Guinea-Bissau macro system

over time. A formalized system dynamics model, on the other hand, allows the explicit representation of mental models of the system causing the problematic behavior. It also opens the possibility of exploring different scenarios in response to interventions aiming at changing the system.

Therefore, we model the Guinean problem through two stages. The first one presents a causal loop diagram (CLD) that shows how the poverty trap in Guinea-Bissau fits a common systems archetype, known as "Shifting the Burden" (Senge, 2006). According to Figure 90.1, political instability leads to a short-term focus, which prevents the proper development of longer-term solutions.

The CLD also allows the specification of the dynamic hypothesis for the Guinean problem discussed in this paper: Underdevelopment and poverty in Guinea-Bissau lead to political instability and short-term focus, preventing the country from developing its national capabilities, which require a long-term focus and strenuous multidimensional efforts. Under the short-termism that characterizes unstable regimes, the more resources are allocated to firefighting, the less are allocated to "building capabilities," the only viable way out of the poverty trap. Even international pressure has been insufficient to reverse this trend, which is accentuated when resources from donors are used to pay ongoing state expenses, such as salaries of civil servants.

The situation in Guinea-Bissau is far from trivial. The country has been trapped in a situation of very low human development since the beginning of its existence. Poverty or, broadly, low human development, is a complex social problem that has been challenging governments and international agencies. The paper developed a model to explain the permanence of alarming social conditions in one of the poorest countries in the world. The model suggests that interventions that ignore the systemic nature of the problem, the interrelations among the several dimensions underlying it and the policy resistance that characterizes complex social systems are bound to fail. That is the case of the usual recommendations from international aid agencies, such as the World Bank and IMF, which tend to present a "laundry list" of reforms. Such lists do not account for thresholds – such as the minimum scale of international aid necessary to produce significant effects – and the complex interplay between social, political, geopolitical, economic, and cultural dimensions that has characterized true development processes in other countries. The paper thus argues against false hope policies, without denying the importance of humanitarian aid to alleviate a situation of catastrophic human suffering.

Notes

1 Faculdade de Economia, Administração e Contabilidade of University of São Paulo.
2 Escola Politécnica of University of São Paulo.
3 Escola Politécnica of University of São Paulo.
4 Faculdade de Economia, Administração e Contabilidade of University of São Paulo.

Bibliography

Forrester, J. (1961). *Industrial Dynamics*. Productivity Press.
IMF, I. M. F. (2015). *Guinea-Bissau: Selected Issues*. Retrieved from www.imf.org/external/pubs/ft/scr/2015/cr15195.pdf
Senge, P. M. (2006). *The Fifth Discipline: The Art & Practice of the Learning Organization* (Revised &). Doubleday.
Sterman, J. (2000). *Business Dynamics: Systems Thinking and Modeling for a Complex World*. McGraw-Hill/Irwin.
UNPD. (2015). *Human Development Report 2015: Work for Human Development: Guinea Bissau*. Retrieved from http://hdr.undp.org/sites/all/themes/hdr_theme/country-notes/GNB.pdf
Vaz, C. (2015). *Guiné-Bissau Submete Bruxelas Sua Visão Estratégica de Desenvolvimento*. Bissau. Retrieved from http://goo.gl/DFQR76
World Bank. (2014). *Life Expectancy at Birth, Total (Years)*. Retrieved August 29, 2016, from http://data.worldbank.org/indicator/SP.DYN.LE00.IN

System dynamics and agents-based simulation as tools for characterising intangible process assets in organisations

Maria-Isabel Sanchez-Segura,[1] *German-Lenin Dugarte-Peña*[2] *and Fuensanta Medina-Dominguez*[3]

Keywords: *strategic management; modelling and simulation of intangible assets; intangible assets valuation; knowledge management; intangible assets engineering*

1 Introduction

Small and medium enterprises have followed organisational models over time intending to guarantee sustainability and profit of organisations concerning their organisational strategic goals. Strategic goals are the essence of the organisation and define the direction that all activities and policies shall point to. Despite organisational business goals seeming clear, strategies and actions for conducting the organisation to achieve them have become insufficient, and the complexity concerning the organisation (as system), mostly ignored, has increased the difficulties in the endeavour of understanding in an effective way the organisation's functioning.

One of the main factors that have influenced the misunderstanding of strategic management in organisations is the fact that intangible process assets (knowledge, organisational models, policies, practices, repositories, etc.) have a direct effect in an organisation performance, economics and functioning (Marr 2008; Stewart and Ruckdeschel 1998). In addition to tangible, intangible process assets contribute to the organisational behaviour indicators' values and they must be considered if a real and effective strategy is desired to be implemented. An organisation with better intangible process assets has better prospects of long-term success (Andrews and Serres 2012; Axtle 2006; Greco et al. 2013; Khan 2014; Lerro et al. 2012; Li et al. 2010). In the information technology industry, intangible assets have also been recognised as strategic components (Castro et al. 2013; Saunders and Brynjolfsson 2016). By mostly considering only tangible assets of the organisation, common studies have provided valuable but improvable knowledge of the organisation and have served for the design of valuable but improvable strategies as well. Innovative studies considering also intangible process assets have produced better results.

This work aims to present a systemic approach for simulating through System Dynamics and Multi-agent Simulation the interaction of intangible process assets in their assignment of leveraging strategic and business goals of an organisation for guaranteeing its sustainability, survivability and continuous improvement, thus promoting the improvement of the organisation environment's wellness. Several fields of knowledge (Swarm intelligence, System dynamics, Multi-agent simulation, strategic management, etc.) have influenced this work, giving a multi-disciplinary and complexity-oriented focus for understanding the relations between process assets and strategic goals.

Swarm intelligence contributes to this work by providing ideas and methods of cooperative work that have worked for thousands of years in nature, guaranteeing survivability, autonomy, cooperativeness and evolutionary learning to communities of honey-bees, ants, termites, etc. (Sanchez-Segura et al. 2016a; Zhang et al. 2013). By observation of the behaviour of these communities, a new rearrangement of the intangible process assets collection of an organisation intends to take the swarm intelligence and bring it into a new intangible assets ideal model that levers up the achievement of strategic and business goals (Benyus 2005; Bonabeau et al. 1999; Sanchez-Segura et al. 2016b; Zhang et al. 2013). The Strategic management field addresses the conception of intangible assets as important factors of the constitution of an organisation's intellectual capital (Marr 2008), thus as levers for economic and wealth indicators of the society that the organisation is related to. Proven as contributors to the gross domestic product of nations (Ståhle and Ståhle 2012), the improvement of intangible process assets' health will redound in benefits for the organisation itself and for its environmental communities as well (Axtle 2006; Clements and Bass 2010; Hill et al. 2013).

In this work, System Dynamics (SD) (Ford and Sterman 1998; J. a Y. W. Forrester 1995; J. W. Forrester and Forrester 1994; Sterman 2000) and Multi-agent (Gilbert 2008; Gilbert and Terna 2000; Teran-Villegas 2001) models are used for representing the dynamics of the organisation, in which relations between process assets and strategic goals become visible, enabling the assessment of process assets and the exploration of scenarios and the system configuration that will allow the improvement of the organisation. These simulation models are capable to give decision makers valuable information about effects and relations that cannot easily be explored or known through the real system due to its complexity and the economic consequences and risks for the organisation that such exploration would imply. The characterisation of process assets allows decision makers and other stakeholders to know how good or bad a process asset is based on its quality and impact and under determined conditions and parameters of the organisation, and the effect that these assets have in the achievement of strategic goals, allowing, for example, the identification of which intangible process assets must be improved for attending special interests or strategies.

2   Basic operation of the models

For these models, each process asset (PA) has an indicator of quality (Q) and impact (I) with respect to each of the strategic goals (SG) defined for the organisation. Being a number between 0 and 1, all relational coefficients sum up a total of 1 (pondering the importance of each IA for the achievement of strategic goals). Dynamic bio-inspired relations between PA and SO consider these coefficients and compute a performance for strategic goals and an indicator of the wellness of each intangible asset. The simulation model allows the modeller (and any other stakeholder) to dynamically modify coefficients and weights in order to envisage "in real-time" how the intangibles assets wellness and the strategic goals performance is affected as a consequence of varying parameters of the models (hence, the impact and quality indicators of process assets).The characterisation of process assets is given as a function of their indicators and the wellness computed for each of them, allowing the dynamic classification of each of them as "Warning", "Replaceable", "Evolving" or "Stable" process asset, providing an input for the process of decision-making regarding strategic valuation of the organisation.

3 Future and related work

This work represents an innovative systemic proposal for organisational process assets characterisation since it is based in a multi-causal approach for representing a system that has traditionally been considered as linear and (unconsciously) unsystemic. Two models, one based on System Dynamics and other based on an Agents-based approach, have been developed and tested in a small enterprise, deriving in contributions to the consolidation of the SIPAC methodology (Strategic Intangible Process Assets Characterisation), which is part of a new wider field called Systemic Process Assets Engineering (SPAE), being developed at the University Carlos III of Madrid (http://spaengineering.sel.inf.uc3m.es/) (Sanchez-Segura et al. 2016a; Sanchez-Segura et al. 2016b).

Notes

1 Carlos III University of Madrid.
2 Carlos III University of Madrid.
3 Carlos III University of Madrid.

Bibliography

Andrews, D., & Serres, A. (2012). Intangible Assets, Resource Allocation and Growth: A Framework for Analysis. *OECD Economics Department*, 989.

Axtle, M. (2006). Intellectual Capital (Intangible Assets) Valuation Considering the Context. *Journal of Business*, 4(9), 35–42.

Benyus, J. (2005). Biomimicry's Surprising Lessons from Nature's Engineers. *TED: Technology, Entertainment and Design*. Retrieved from www.ted.com/talks/janine_benyus_shares_nature_s_designs

Bonabeau, E., Dorigo, M., & Theraulaz, G. (1999). *Swarm Intelligence: From Natural to Artificial Systems*. New York, NY, USA: Oxford University Press.

Castro, G. M., Delgado-Verde, M., Amores-Salvadó, J., & Navas-López, J. E. (2013). Linking Human, Technological, and Relational Assets to Technological Innovation: Exploring a New Approach. *Knowledge Management Research & Practice*, 11(January), 123–132. http://doi.org/10.1057/kmrp.2013.8

Clements, P., & Bass, L. (2010). The Business Goals Viewpoint. *IEEE Software*, *59*(Suppl 1), S19–S23. http://doi.org/10.1016/S0003-3928(10)70004-4

Ford, D. N., & Sterman, J. D. (1998). Dynamic Modeling of Product Development Processes. *System Dynamics Review*, *14*(1), 31–68. http://doi.org/10.1002/(SICI)1099-1727(199821)14:1 < 31::AID-SDR141 > 3.0.CO;2-5.

Forrester, J. a Y. W. (1995). Counterintuitive Behavior of Social Systems, *The System Dynamics Road Maps*,1–29. Retrieved 23.9.2018 from https://ocw.mit.edu/courses/sloan-school-of-management/15-988-system-dynamics-self-study-fall-1998-spring-1999/readings/behavior.pdf

Forrester, J. W., & Forrester, J. W. (1994). Learning through System Dynamics as Preparation for the 21st Century by Jay W. Forrester. *Systems Thinking and Dynamic Modeling Conference for K-12 Education*, 1–22.

Gilbert, N. (2008). Agent-Based Models. *Sage Publications*, *153*, 98. http://doi.org/10.4135/9781412983259

Gilbert, N., & Terna, P. (2000). How to Build and Use Agent-Based Models in Social Science. *Mind & Society*, *1*, 57–72. http://doi.org/10.1007/bf02512229

Greco, M., Cricelli, L., & Grimaldi, M. (2013). A Strategic Management Framework of Tangible and Intangible Assets. *European Management Journal*, *31*(1), 55–66. http://doi.org/10.1016/j.emj.2012.10.005

Hill, C., Jones, G., & Schilling, M. (2013). *Strategic Management: An Integrated Approach*. Boston: Houghton Mifflin. Retrieved from www.jstor.org/stable/41782301

Khan, M. W. J. (2014). Identifying the Components and Importance of Intellectual Capital in Knowledge-Intensive Organizations. *Business and Economic Research*, *4*(2), 297.

Lerro, A., Iacobone, F. A., & Schiuma, G. (2012). Knowledge Assets Assessment Strategies: Organizational Value, Processes, Approaches and Evaluation Architectures. *Journal of Knowledge Management*, *16*(4), 563–575.

Li, S.-T., Tsai, M.-H., & Lin, C. (2010). Building a Taxonomy of a Firm's Knowledge Assets: A Perspective of Durability and Profitability. *Journal of Information Science*, *36*(1), 36–56. http://doi.org/10.1177/0165551509347955

Marr, B. (2008). Impacting Future Value: How to Manage your Intellectual Capital. *The Society of Management Accountants of Canada, the American Institute of Certified Public Accountants and the Chartered Institute of Management Accountants.* Retrieved from www.journalofaccountancy.com/content/dam/jofa/archive/issues/2008/09/mag-intcapital-eng.pdf

Sanchez-Segura, M.-I., Dugarte-Peña, G.-L., Medina-Dominguez, F., & Ruiz-Robles, A. (2016a). A model of Biomimetic Process Assets to Simulate Their Impact on Strategic Goals. *Information Systems Frontiers*, *19*(5), 1067–1084.

Sanchez-Segura, M.-I., Medina-Dominguez, F., & Ruiz-Robles, A. (2016b). Uncovering Hidden Process Assets: A Case Study. *Information Systems Frontiers*, *18*(6), 1041–1049.

Saunders, A., & Brynjolfsson, E. (2016). Valuing Information Technology Related Intangible Assets. *Management Information Systems Quarterly*, *40*(1), 83–110. Retrieved from http://misq.org/valuing-information-technology-related-intangible-assets.html\nhttp://aisel.aisnet.org/misq/vol40/iss1/6

Ståhle, S., & Ståhle, P. (2012). Towards Measures of National Intellectual Capital: An Analysis of the CHS Model. *Journal of Intellectual Capital*, *13*(2), 164–177. http://doi.org/10.1108/14691931211225012

Sterman, J. (2000). *Business Dynamics: Systems Thinking and Modeling for a Complex World*. USA: McGraw-Hill.

Stewart, T., & Ruckdeschel, C. (1998). *Intellectual Capital: The New Wealth of Organizations*. Performance Improvement.

Teran-Villegas, O.-R. (2001). Emergent Tendencies in Multi-Agent-Based Simulations Using Constraint-Based Methods to Effect Practical Proofs over Finite Subsets of Simulation Outcomes Table of Contents. *Facilities, PhD*, 296. Retrieved from http://cfpm.org/cpmrep86.html

Zhang, Y., Agarwal, P., Bhatnagar, V., Balochian, S., & Yan, J. (2013b). Swarm Intelligence and Its Applications. *Hindawi Publishing Corporation: The ScientificWorld Journal, 2013*, 3. Retrieved from http://dx.doi.org/10.1155/2013/528069

The architecture for innovation and change

A viable system analysis

Paul James Jackson and Diana Limburg[1]

Keywords: *Viable System Model; PPPM; business architecture; innovation and change*

1 Overview

This paper explores the contribution of cybernetics to what we call the "architecture for innovation and change" (or A4IC). In so doing, it draws together ideas from Beer's Viable System Model (VSM) (Beer, 1979, 1985) with literature on both portfolio, program and project management (PPPM) and business architecture (for example, Whelan and Meaden, 2012). The paper contends that successful projects and programs depend crucially on the architecture firms used to support and coordinate innovation and change in the round. It posits that PPPM operates through a mediating architecture that brings together R&D, organizational change, innovation and knowledge management, business strategy and strategic marketing. These must be well-designed and integrated as a system, which will vary across organizations depending on: the content of change, the context (such as industry), and the business lifecycle stage. These combinations, we argue, create a range of A4IC configurations, or archetypes, which affect how PPPM should be organized and managed. The notion of archetypes draws in part from architecture (Alexander et al., 1977) and the idea that practitioners often face recurrent patterns of problems and solutions, depending on the context and content at hand.

Using the above theory, the paper assembles a framework of concepts with which to explore A4IC issues. The aim is not to "describe" the architectural configurations as such, but rather, first, to enrich and evolve a generic approach to exploring A4IC (underpinned by ideas from VSM), and, secondly, identify possible archetypes that illustrate how the business architecture might be configured in different situations.

The literature drawn upon (and contributed to) in the paper is necessarily cross-disciplinary, exploring theories and frameworks in adjacent bodies of knowledge. This includes the world of PPPM (Project Management Institute [PMI], 2013) and related ideas of P3O: "portfolio, programme and project offices" (Axelos, 2013). It also draws on ideas from the growing area of "business architecture" (for example, Whelan and Meaden, 2012), as well as innovation and change management (see, Bessant and Tidd, 2007). However, the paper is also trans-disciplinary, in that it uses the Viable System Model (VSM) as a meta theory for designing effective organizational structures (Beer, 1985; Hoverstadt, 2008; Espejo and Reyes, 2011).

The paper is rooted in ideas about systemic thinking (for example, Checkland, 1981). Here, problem situations, whether with projects, organizations or business ecosystems, are analyzed in terms of "wholes", within which subsystems (people, business units or entire organizations) exist in a dynamic relation to other parts of the system. In so doing, the parts of a system help to produce "emergent" – or synergistic – properties and capabilities that characterize the system at a higher, collective level of analysis.

2 Program and project management

Within this broad domain, the field of management cybernetics offers significant potential to problem-solving in the world of PPPM (for example, see Saynisch, 2010). In particular here, VSM provides a powerful way of looking at the effective design of organizations, including identifying their necessary component parts (Beer, 1985; Hoverstadt, 2008; Espejo and Reyes, 2011). For a system to be viable in these terms it must have all the requisite capabilities that enable it to survive and grow as an independent entity. VSM conceives of organizations (and the viable subsystems within them) as existing within a wider environment and having five component systems: (1) operations, (2) coordination, (3) delivery management, (4) development and (5) policy.

According to this account, PPPM is systemically related to other development functions. While it clearly interfaces with Systems 3 and 1 (in order to manage and introduce project outputs and outcomes – and thus support the implementation of change and innovation), it is "cut from the cloth" of a broader function involved in anticipating future needs and opportunities, and preparing the organization for them. Cohesion at this level is therefore critical, but – in our view – often ignored in PPPM (and other) literature.

Because Systems 4 exist at each level of recursion, organizations face a choice as to what elements of the development function (including PPPM) should sit at each level. Put differently, organizations face the ongoing question of how centralized should these functions be. Clearly, given what we have said about how projects and programmes can vary because of the innovation and change involved, the answer is "it depends". A range of (contextual) factors will affect what is appropriate, too.

Much management literature on organizational structures and change, we would argue – such as Mintzberg (1992) and Greiner (1998) – emphasizes the need for organizational redesign at critical stages (or "crises") in their growth cycle, or in response to changing environmental circumstances. Such thinking, though, focuses on governance and management structures (Systems 5 and 3, in VSM terms) and operational units (System 1) to the neglect of the functions needed to help the organization detect the need for change and respond accordingly (System 4). This function exists in different guises across organizational layers, and provides the mechanisms for innovation and change. Getting that architecture right, and the PPPM elements within it, is crucial to corporate success.

3   Summary

The paper concludes by arguing that how the elements of A4IC are integrated – for instance how the different roles, structures, business systems and communications channels interface and relate to one another – is central to successful corporate evolution and competitive success. Such components also need to be "plumbed into" the broader system, including strategic planning and approval, and operational delivery. As we show, VSM provides a detailed and robust means of understanding how this could be done.

Note

1 Oxford Brookes University.

Bibliography

Alexander, C., Ishikawa, S. & Silverstein, M. (1977). *A Pattern Language: Towns, Buildings, Construction*. New York: Oxford University Press.
Axelos (2013). *Portfolio, Programme and Project Offices*. Norwich: TSO.
Bessant, J. & Tidd, J. (2007). *Innovation and Entrepreneurship*. Chichester: Wiley.
Beer, S. (1979). *The Heart of Enterprise*. Chichester: Wiley.
Beer, S. (1981). *Brain of the Firm*. 2nd edition. Chichester: Wiley.
Beer, S. (1985). *Diagnosing the System for Organizations*. Chichester: John Wiley.
Bicheno, J. & Holweg, M. (2008). *The Lean Toolbox*. 4th edition. Buckingham: Picsie Books.
Checkland, P.B. (1981). *Systems Thinking, Systems Practice*. Chichester: Wiley.
Espejo, R. & Reyes, A. (2011). *Organizational Systems*. Heidelberg: Springer.
Freeman, C. (1982). *The Economics of Industrial Innovation*. London: Frances Pinter.
Greiner, L.E. (1998). Evolution and Revolution as Organizations Grow. *Harvard Business Review*, May–June.
Hoverstadt, P. (2008). *The Fractal Organization*. Chichester: Wiley.
Jenner, S. (2012). *Managing Benefits*. London: TSO.
Johnson, G., Whittington, R., Angwin, D. & Scholes, K. (2013). *Exploring Strategy*. London: Pearson.
Kotter, J.P. (2008). *A Sense of Urgency*. Boston, MA: Harvard Business School Publishing.
Kotter, J.P. (2012). *Leading Change*. Boston, MA: Harvard Business Review Press.
Lincoln, Y.S. & Guba, E.G. (2000). The Only Generalization Is: There Is No Generalization. In: Gomm, R., Hammersley, M. and Foster, P. (Eds). *Case Study Method: Key Issues, Key Texts*. London: Sage. Pp. 27–44.
Mintzberg, H. (1992). *Structure in Fives: Designing Effective Organizations*. London: Prentice Hall.

Morris, B. (2013). *Business Architecture Made Easy: A Journey from Complexity to Simplicity.* UK: Book Baby.

OGC (2003). *Managing Successful Programmes.* Norwich: TSO.

OGC (2005). *Managing Successful Projects with PRINCE2.* Norwich: TSO.

Osterwalder, A. & Pigneur, Y. (2010). *Business Model Generation: A Handbook for Visionaries, Game-Changers and Challengers.* Hoboken, NJ: Wiley.

Pichler, R. (2010). *Agile Project Management with Scrum: Creating Products that Customers Love.* Upper Saddle River, NJ: Addison Wesley.

PMI (2013). *A Guide to the Project Management Body of Knowledge (PMBOK Guide).* PA: PMI.

Ries, E. (2011). *The Lean Startup: How Constant Innovation Creates Radically Successful Businesses.* London: Portfolio Penguin.

Rose, S., Spinks, N. & Canhoto, A.I. (2015). *Management Research: Applying the Principles.* Abingdon: Routledge.

Saynisch, M. (2010). Beyond Frontiers of Traditional Project Management: An Approach to Evolutionary, Self-Organizational Principles and the Complexity Theory: Results of the Research Program. *Project Management Journal*, 41(2), 21–37.

Schwaber, K. & Sutherland, J. (2012). *Software in 30 Days: How Agile Managers Beat the Odds, Delight Their Customers, and Leave Competitors in the Dust.* Hoboken, NJ: Wiley.

Sull, D., Homkes, R. & Sull, C. (2015). Why Strategy Execution Unravels. *Harvard Business Review*, March.

Whelan, J. & Meaden, G. (2012). *Business Architecture: A Practical Guide.* Farnham: Gower.

Womack, J., Jones, D. & Roos, D. (2007). *The Machine That Changed the World: The Story of Lean Production.* New York: Free Press.

Yin, R.L. (2013). *Case Study Research: Design and Methods.* 5th edition. London: Sage.

Current trends in international trade

Challenges and opportunities for development

Nino Papachashvili[1]

Keywords: *trade policy; non-tariff barriers; development; global values chain*

1   Introduction

After the global financial-economic crisis of 2008, new "landscape" has been formed in the world trade system, which creates new tasks for policymakers.

The wide network of the global value chain (GVC) has been established during the recent decades, entailing new reality for economic development. As well as importance of the intermediate product trade has significantly increased in between the countries. Despite the fact that along with reduction of the total trade value, the total value of the global value chain has also been reduced; seeking development opportunities in the new reality remains relevant. The GVC researcher is right to indicate that:

> The importance of global value chains will continue to increase in our increasingly interdependent economic world, and the need to have a better understanding of all its implications, including in particular for trade policy, is a critical task for policy makers.
>
> (Stephenson 2015)

We shall take the circumstance into account that the non-tariff barriers are at the global scale increased in trade and the greater part of them brings damage to the commercial interest. The question, how the trade policy shall respond to new challenges and where to find the development opportunities, will at some extent clearly reveal if we consider the international trade in the global value chain perspective on the one hand and if we report the non-tariff barriers hindering the international trade flows on the other hand.

2   The modern trends of international trade in the global value chain perspective

Despite the fact that the value of merchandise trade and trade in commercial services in 2015 is nearly twice as high as in 2005, the value declined in 2015, following modest growth in 2012 and 2014. This condition is maintained for the next year, which was estimated as "Global Trade Plateaus" by the leading expert in international trade, S.J. Evenett (Evenett 2016).

As far as international trade is considered to be the means of creating jobs, increasing wages and improving working conditions, the issue deserves a lot of attention of the scientists and policymakers.

A consensus has been reached that new technologies changed the shape of international trade and GVCs have a major impact on the location of goods and services. Trade in intermediate products represents about 40% of world merchandise trade. International trade in intermediate goods grew from about US $1 trillion in 1993 to roughly US $6 trillion in 2008, before falling during the crisis of 2009 (Nicita et al. 2013).

Pascal Lamy (former director of WTO) indicates:

> In actual fact, the opening up of trade and the development of value chains have had such a major impact on the location of goods and services that it is no longer as easy as it used to be to gauge the added value of a product (whether we are talking about goods or services) whose component parts come from all around the world. A worldwide average import content of exports has shot up from 20% twenty years ago to 40% today and it looks set to hit the 60% mark in twenty years' time.
>
> (Lamy 2014)

Most scientists and experts shared the view that new technological developments will benefit overall trade volumes: Industrial evolution – digital innovation and the drive to sustainability, reverse innovation and mass customization, the rise of micro-multinationals; trade reclassified – the falling cost and rising speed of trade (Persio 2015). ICC Business World Trade Agenda also highlights that global value chains have become a dominant feature of today's global economy, and as a result there is growing evidence and recognition that the nature of trade is changing.

The main peculiarities of nowadays is the context where Global Supply Chains are increasingly fragmented across the many countries, each involved in the assembly process at a different stage, thus resulting in parts and components crossing multiple borders before being incorporated into the final product.

The analysis of the available statistic data base reveals that: At the global level, the average foreign value added in exports constitutes approximately 28%. The developed countries amongst them are characterized with the high share (their share constitutes 5 of total world 31) and the share of the developing countries is relatively low (25%). Consideration of the value added formation in the sector perspective will clearly reveal the competitive advantages of the countries and the regions, which is important for development of the relevant policy. The domestic value share in export depends on the number of factors: the size of the country, complexity of export, the position in the global value chain perspective, etc.

The policymakers shall necessarily take the fact into account that inclusion into the global value chain is the significant source for economic growth and job opportunities. Sometimes, we might doubt who benefits most from the global value chain. Approximately 80% of the world trade (export) comes to the transnational corporations in the international industrial network; this circumstance often serves the significant source of establishment of new technologies and inflow of knowledge in the developing countries. Governments need to think about how to encourage a range of GVCs in order to improve prospects for growth and development.

Despite dealing with the existing problems that appear difficult, we share the opinion that a more rapid way for development is inclusion into the chain. The scientist Richard Baldwin expresses his opinion that:

> Global supply chains have transformed the world. They revolutionized development options facing poor nations; now they can join supply chains rather than having to invest decades in building their own. The offshoring of labor-intensive manufacturing stages and the attendant international mobility of technology launched era-defining growth in emerging markets, a change that fosters and is fostered by domestic policy reform (Elms and Low 2013).

International trade and industrial links are not a novelty for sure. The technological progress allowed further fragmentation thereof amongst the countries and accelerated inter-relation process.

Deriving from the hereof, it is evident that inclusion into the global value chain is important. New challenges of the international trade shall be properly assessed and the new trade policy shall be adjusted. For instance, some researchers point to the outdated trade order:

> trade policies must be re-evaluated and updated to reflect the new structure of world trade and the operation of GVCs. Current trade rules were designed for the 20th century, where goods were made and exported either fully or primarily by one country. They may thus be out of synch for disciplining and monitoring current patterns of international trade. GVCs have created a dichotomy between the reality of trade and the existing normative framework that governs it at the WTO level, which needs to be addressed. Likewise, international cooperation in trade policy issues must be rethought in the light of GVCs (Stephenson 2015).

It is natural to ask the question about the latest challenges of the foreign trade policy. What are the new tasks encountering the policymakers.

3   Non-tariff barriers

The benefit predicted by the free trade supportive models often remains within the theoretical scales. The exporters and importers encounter a multitude of expenditures in practice consideration of which within the model scales cannot be duly achieved. At that, regardless of the expected benefit of the free trade, most of the world's countries upon application of the foreign trade policy often refer to the protectionist remedies. It became particularly evident in the post-2008 financial-economic crisis period, when the drastic growth of non-tariff barriers took place.

According to the WTO, a total of 1441 trade-restrictive measures have been introduced by the G20 since the 2008 standstill pledge. Of these, 354 had been removed by mid-October 2015. In other words, the remaining stockpile of new trade-restrictive measures was 1087 in their latest report, a figure that represents more than 75 percent of the measures introduced since 2008 (WTO 2015).

The analysts of the Economic Policy Research Center do not agree with the WTO data. They indicate to the fact that the WTO conducts incomplete description of trade-restrictive events. According to the database generated thereby, the non-tariff barriers applied for protectionist purposes exceeded 4000 and the measures made in view of trade liberalization stay far behind. According to the Global Trade Alert

> Resort to protectionism in 2015 is 50% up on that seen in 2014. Policy initiatives harming foreign commercial interests in 2015 outnumbered trade liberalization three-to-one. Since 2010 between 50 and 100 protectionist measures were implemented in the first four months of each year. In 2016 the total had exceeded 150. G20 members were responsible for 81% of protectionist measures implemented in 2015.
>
> (Evenett and Fritz 2016)

It seems that new forms of protectionism will remained as a challenge for development in the near future.

A number of researches point to the restrictive nature of the non-tariff barriers, triggered by the intensive application practice thereof. For instance, the OECD/WTO survey conducted for the Fifth Global Review of Aid for Trade. According to the findings of this survey, which include ten regional economic communities or transport corridors and 62 developing countries, considering trade facilitation in the narrow sense (i.e. border procedures) is the most important source of trade costs for goods exports (83.3% of respondents), together with transport infrastructure (80.6% of respondents) and other non-tariff measures (79.2% of respondents). Other types of trade costs such as tariffs or access to trade finance are reported as less important (Aid for Trade 2015).

According to the OECD recommendation, trade simplification and regional cooperation can serve as the effective strategy for inclusion into the value chain. It allows mitigation of the number of risks and relevant expenditures at some extent.

4   Summary

Consideration of the current state of international trade assures us in the greatest importance of the global value chain. Dealing with the increased non-tariff barriers remains a challenge for the benefit of the international trade.

Against the background of growth on non-tariff barriers, the governments should better maintain open markets;

Maintenance of open markets is a necessary but insufficient pre-condition for economic growth and development. Coordinate activity between the economic policymakers on the local and global level is of crucial importance for elimination of the impediments;

Consideration of the global value chain issue on the junction of non-tariff barriers assures us that international coordination is far more important to deal with the current situation;

Increased numbers of the hidden and visible non-tariff barriers indicate that enhancement of cooperation within economic integration scopes is easier to overcome and is far more perspective, conditioned by the uniform standards and harmonized rules.

Note

1 Iv. Javakhishvili Tbilisi State University.

Bibliography

Aid for Trade (2015). Aid for Trade at a Glance 2015: Reducing Trade Costs for Inclusive, Sustainable Growth. OECD, WTO, p. 172.

Evenett, S.J. (2016). Running Out of Tools: The G20 and the Global Trade Plateau: G20 Hangzhou Summit (2016). Proposals for Trade, Investment, and Sustainable Development Outcomes, ICTSD, July 2016.

Evenett, S.J., Fritz, J. (2016). *Global Trade Plateaus: The 19th GTA Report.* CEPR Press.

Elms, D. K., & Low, P. (Eds.). (2013). *Global Value Chains in a Changing World.* Geneva: World Trade Organization.

Lamy, P. (2014). *The World Trade Organization: New Issues, New Challenges* (September 4).

Nicita, A., Ognivtsev, V., Shirotori, M. (2013). *Global Supply Chains: Trade and Economic Policies for Developing Countries.* New York and Geneva, UCTAD.

Persio, S.L. (2015). *Five Trends That Will Shape the Future of Trade.* Retrieved from www.gtreview.com/news/global/five-trends-that-will-shape-the-future-of-trade/.

Stephenson, S. (2015). *Global Value Chains: The New Reality of International Trade, in the Book "Global Value Chains: Development Challenges and Policy Options, Proposals and Analysis": E15 Initiative* (p. 7). Geneva, International Centre for Trade and Sustainable Development (ICTSD) and World Economic Forum, 2013. Retrieved from www.e15initiative.org/.

WTO (2015). *The World Trade Report 2015.* Geneva, WTO. Retrieved from www.wto.org.

Systems/systemic approach to scientific investigation

A case study of competitive intelligence

Stanislava Mildeova[1]

Keywords: *science; systems approach; systemic thinking, competitive intelligence; system dynamics modelling*

1   Introduction

The necessity of modern scientific work is to understand and analyse problems holistically, i.e. to see them as a network of interconnected parts. The aim of this paper is to discuss the necessity of systems/systemic approach to scientific investigation. In this sense, the paper examines a case study of Competitive Intelligence as the current progress in the field of ICT. In the context of system archetypes search, own model building, and conducting experiments simulating competitive struggle, the author of the paper shows systems thinking and system dynamics tools that are able to support systems/systemic approach to scientific investigation.

2   Systems/systemic approach to scientific problems

The common deficiencies in solving contemporary problems are the lack of systems approach, limited focus on the context, and also the inability to properly understand other systems. We are not able to differentiate the significant parts of systems from their insignificant counterparts; therefore, all that we see is considered important. This reality should be reflected in the demands for adequate methodological and methodical support of scientific work. According to Mildeová (2013), when solving scientific problems we come across situations where conventional approaches do not produce the desired results. The scientist formulates the purpose of the research and then explains why it is important to solve the problem. And of course, they formulate research questions and their relevant working hypotheses. The research problem (question) requires several main methodological steps. We should apply the appropriate methods and procedures that help us clearly define the problem, identify its boundaries, and at the same allow enough time to get to the root of the matter.

3   Case study: competitive dynamics

Olivier and Howard (2000) state that the literature provides a range of different tools to analyse rivalry and competitive dynamics. In the paper the term Competitive advantage is understood, according to the Institute-for-Competitive-Intelligence (2014), as an advantage gained by exploiting the unique blend of activities, assets, attributes, market conditions, and relationships that differentiate an enterprise from its competitors, and may include: access to natural resources, specific locations, skilled workforce, lower costs, better quality products, unique technologies, or exceptional customer service. So far, the paper has the objective to focus on the system dynamics model as an ICT tool that may be helpful in the competitiveness of enterprises.

Let us imagine a situation where our production company together with another firm is practically in a monopolistic position in the market. Therefore, only the single company is our competitor. Although a given production also takes place in other firms, their products occupy only a bare minimum of the total market share. The competitor will suddenly introduce a new product into the market. This becomes a threat to our business. The typical management reaction is to reduce the price and accelerate the development of our own product. After the launch of this new product, the other business is once again in danger of competition which will reduce their price and accelerate development.

The description of this situation may be based on the Escalation systems archetype by Senge (1990). Escalation systems archetype could be seen as a non-cooperative game where both players (competitors) assume that only one of them can win. Thus, they respond to actions of the other player (competitor) in order to "defend themselves". Subsequently, the aggression of both players (competitors) grows and can result in self-destructive behaviour in competition. The logic of the Escalation systems archetype and its feedback structure is reflected in the system dynamics model created by the author. The model is based on simulation modelling using a continuous simulation (Sterman 2000). The model was calibrated with real data; part of the data was obtained by the heuristics methods. Random elements represent external influences on the model. This "forecast" approach provides predictions of the enterprise's competitiveness and the impact of various management decisions.

A good researcher must not only possess an investigative and creative style of thinking, they must also think in the context of the problem they are solving. When we say that a scientist must think in context, we mean that they have to think systematically. The article is largely devoted to the problem of systemic thinking in scientific research. The author has shown that it is not just about systems thinking, but also about the use of modelling and simulation experiments for the search for answers to research questions to be successful. Science is mainly quantitative. That means it needs to know the facts and answers to such questions as "What?", "When?", "Where?", "How much?", etc. At the same time, science must be predictive, meaning, giving the same answer to the same question in the future. In this context, within systems approaches a system dynamic was introduced, which combines theories, methods, and philosophy for systems behaviour analysis and hence represents an integrated, multidisciplinary approach for systems modelling.

Note

1 University of Finance and Administration.

Bibliography

Institute-for-Competitive-Intelligence. (2014). [on-line], Retrieved from www.institute-for-competitive-intelligence.com/downloads/categories/0_4e800718d67165f73 c2eff7dfb4d33b3.html [15 Aug. 2013].

Mildeová, S. (2013). Research Problem Description and Definition: From Mental Map to Connection Circle. *Journal on Efficiency and Responsibility in Education and Science*, 6(4), 328–335. ISSN 1803–1617. [on-line], Retrieved from www.eriesjournal.com/_papers/article_230.pdf [31 Dec. 2013]. doi: 10.7160/eriesj.2013.060409

Olivier, F., Howard, T. (2000). The Rivalry Matrix: Understanding Rivalry and Competitive Dynamics. *European Management Journal*, 18(6), 619–637.

Senge, P. (1990). *The Fifth Discipline*. New York: Currency Doubleday.

Sterman, J.D. (2000). *Business Dynamics: Systems Thinking and Modeling for a Complex World*. USA: McGraw-Hill Higher Education.

Financial market in emerging economies

Drawing lessons for Albania

Arsena Gjipali[1] and Valbona Karapici[2]

Keywords: *stock exchange; financial market; banking system; former transition economies*

1 Introduction

It is generally agreed that the main role of the financial system is to spur economic growth through the efficient allocation of the financial sources. For this purpose, it is important that an economy has a sound and stabilized financial system. The first two decades of transition have seen remarkable progress in financial sector reform for the former-socialist countries of Central and South Eastern Europe (CSEE) and the former Soviet Union. The policy towards capital account liberalization of these countries has been cautious, despite some differences across the regions as regards the use of individual controls. Most of the CSEE economies abolished restrictions on foreign direct investment (FDI) inflows at the beginning of the transition early in the 1990s. Since early in the transition process, most countries have also guaranteed the free repatriation of both profits (current account convertibility) and FDI capital. Individuals are allowed to hold and operate foreign exchange accounts at local banks and treatment of trade credits has also been liberal in most countries. However, as will be shown in this paper, the progress has been uneven across regions, countries and market segments.

Given the transformation in ownership that needed to take place from the state to the private sector many observers considered well-functioning stock markets essential to the process of transition from central planning. Some even agree that stock exchange positively affects main macroeconomic variables, especially regarding economic development. This essay highlights the role the stock market has played in the CSEE economies. The rationale is that, if stock markets have been successful and have positively affected economic development in other countries of the region suffering similar economics transformation processes, the lack of such a market in Albania perhaps limits possibilities for further economic performance. Hence, the very purpose of this paper is to empirically investigate how stock market indexes in the economies that introduced stock markets only recently affect industrial indexes as a proxy to the economic growth.

2 Context

Stock markets after the 1990s were not new in the economies that transformed from the central planning during the last decade of the last century. But during the transition from plan to market economy, stock exchanges have re-emerged or were created in almost all of the countries of the Central and South East Europe. However, for at least the first decade of transition, many of these markets were still undeveloped or dormant (Claessens et al. 2001). Exchanges had mainly been used for the mandatory listing of shares of mass-privatized companies and for voluntary initial public offerings (IPOs). Initially liquidity of newly established stock markets was relatively low and trading was thin with the result that in the early days at least, markets tended to be open for only a few hours a day and only one or two days a week. Consequently stock prices were volatile compared with developed stock markets and it seems likely that this inhibited the growth of trade because of the increased risk (Harrison and Moore 2009).

In Albania, never were any initial public offers (IPOs) made for state-owned companies to be privatized. Most important state companies were privatized through the formula of selling to "strategic foreign investors" 60%–100% of the shares, sometimes with very cheap prices. In some other state companies the presence of such international financial institutions like IFC, EBRD, etc. is high (20%–40%). For SMEs, Albania used management and employee buyout privatization methods during the 1990s.

Currently the CSEE countries have all created functioning stock markets with rules similar to those in developed economies. Furthermore, as transition has progressed, the economies of CSEE have gradually integrated into the world economy increasing their trade especially with the rest of Europe and reducing their trade with the former planned economies of the Soviet Union (Harrison and Moore 2009). In the context of EU accession, the Czech Republic, Hungary and Poland have made early progress in liberalizing capital movements. Estonia and Latvia liberalized capital transactions quickly in the early 1990s. Capital flows into Central and Eastern Europe started to become sizeable only in 1993. FDI was initially much more important than portfolio flows (Gelos and Sahay 2001). Yet, in Albania, a central stock exchange is totally missing.

Recent research on the role of stock market has related its performance with macroeconomic variables. Theoretically, stock markets may affect economic activity through the creation of liquidity (Levine 1991). Many profitable investments require a long-term commitment of capital, but investors are often reluctant to relinquish control of their savings for long periods. There is the rationale that liquid equity markets make investment less risky – and more attractive – because they allow savers to acquire an asset, equity and to sell it quickly and cheaply if they need access to their savings or want to alter their portfolios. At the same time, companies enjoy permanent access to capital raised through equity issues. By facilitating longer-term, more profitable investments, liquid markets improve the allocation of capital and enhance prospects for long-term economic growth. Further, by making investment less risky and more profitable, stock market liquidity can also lead to more investment.

According to Baier et al. (2004), the more efficient allocation of resources rather than more capital accumulation is the primary channel through which a stock exchange affects output growth. That meaning that the link of stock market development on the economy is based on the premise that the presence of stock markets would mitigate the principal agent problem. Given that the stock price at any time is a mirror of firm performance, weakening corporate governance would be reflected as a fall in share price. Petros (2012) analyzed the relationship between stock market development and economic growth both in the short run and in the long run in Zimbabwe, suggesting that there exists a significant positive relationship between stock market development and economic growth.

3   Methodology and estimation

The choice of methodology is limited by data availability. The variables taken into consideration are stock market index (STEX), inflation rate (INFL), exchange rate (EXR), balance of trade (BOT), index of industrial production (INDP). The latter measures the output of businesses integrated in the industrial sector of the economy such as manufacturing mining, and utilities. Theoretically, financial markets contribute to economic development through enhancing physical capital accumulation. While banks finance only well-established, safe borrowers, stock markets can finance risky, productive and innovative investment projects (Caporale et al. 2004). This in turn stimulates investment and lowers the cost of capital, contributing in the long-term to economic growth. The argument here is that if economic growth is a function of stock market development (for example), then it is at least a plausible hypothesis that stock markets may cause economic growth.

As we are interested in analyzing the impact of variables that vary over time on the index of industrial production, fixed-effects (FE) estimation is used on secondary data collected from the website of trading economics in the form of unbalanced panel data. The time series are monthly data from 2000 to December 2012. The beginning time period varies according to country. The countries taken into consideration are: Bulgaria, Estonia, Lithuania and Serbia. Besides restrictions regarding data availability, such a choice of countries can be argumented given their heterogeneity regarding geographical, political and social characteristics as well as the different periods they developed their stock markets. For example, Estonia and Lithuania are the countries that were admitted in the European Union in 2004, Bulgaria only later in 2007. On the other side, these countries have similar economic system backgrounds and experienced similar transformation processes to the market economy, and stock exchanges opened only after 1990 (1992–1997). The data for Lithuania are from January 2000, for Estonia from January 2001, for Bulgaria from October 2002 and for Serbia only from January 2007. Monthly data allow for a relatively large span of observations for each of the countries. Unit root tests indicate for the existence of non-stationarity and the presence of unit roots in the variables of the inflation rate, stock exchange and exchange rate.

Coefficient results show that the stock exchange index is the most significant variable in explaining the variation of industrial production as a share of GDP. That means that the good performance of the former would lead in an increase in the industrial production with two months' delay. This is very encouraging also for the other economic aggregates related, such as employment, investment and overall performance of the economy. Exchange rate variables positively affect industrial production only at a 10% level of significance, indicating that a depreciation of domestic currency is positive for the part of the economy that involves production, making the latter cheaper to the foreigners. Inflation rate negatively affects performance of industrial share, perhaps through increasing the costs of inputs.

Given the above, the study calls for new policies to be implemented in Albania to encourage the creation and the well-functioning of a stock exchange and other similar agents. In particular, policies should be geared towards careful regulation of the market. Furthermore, incentives could be provided by the government to reduce the informal economy as well as help business to be familiar with a new way of financing, different from the banking sector. Taking to account successful experiences of East Europe countries, government could enforce legal incentives to oblige companies which exceed a certain level of capital to be listed in TSE.

Notes

1 University of Tirana.
2 University of Tirana.

Bibliography

Baier, S. L., Dwyer, G. P., & Tamura, R. (2004). Does opening a stock exchange increase economic growth? *Journal of International Money and Finance*, 23(3), 311–331.

Caporale, G. M., Howells, P. G., & Soliman, A. M. (2004). Stock market development and economic growth: The causal linkage. *Journal of Economic Development*, 29(1), 33–50.

Claessens, S., Djankov, S., & Klingebiel, D. (2001). Stock markets in transition economies. *Financial Transition in Europe and Central Asia: Challenges of the new Decade*, 109–137.

Gelos, R. G., & Sahay, R. (2001). Financial market spillovers in transition economies. *Economics of Transition*, 9(1), 53–86.

Harrison, B., & Moore, W. (2009). Stock market comovement in the European Union and transition countries. *Financial Studies*, 13(3), 124–151.

Levine, R. (1991). Stock markets, growth, and tax policy. *The Journal of Finance*, 46(4), 1445–1465.

Petros, J. (2012). The effect of the stock exchange on economic growth: a case of the Zimbabwe stock exchange. *Research in Business and Economics Journal*, 6, 1.

A system dynamics approach to the analysis of food insecurity

Stefano Armenia[1] *and Antonella Passarelli*[2]

Keywords: *food security; poverty and malnutrition; food-sheds analysis; Haiti; systems thinking; system dynamics*

Food insecurity, together with poverty, is now one of the most serious problems facing the world's population and a real cancer in developing countries. Many organizations have analyzed this problem and, from 1992 to date, tried to achieve improvements to such a phenomenon by reducing the number of undernourished people, through policies mainly based on the feedback obtained by the use of the so-called "Food Security indicators" (FAO, 2014).

Literature and various data-statistics sites of important organizations such as FAO (which has been trying, for several years, so far, to analyze and mitigate the problem), also provide additional data such as the "GNI per capita" (i.e. the gross national income per person), which shows that countries that manage to get the best results in their fight against undernutrition are those who display higher levels of economic development. With respect to this, the main problem would not be exactly the availability of food, rather it is its uneven distribution to the population, in turn mainly due to the inequality of income. Therefore even if the current food production is enough to feed all the population, but unequally distributed, social, political and economic interventions are of primary importance. Equally important is the Key Perfomance Indicator (KPI) named "Food Balance Sheets" (see "Food Balance Sheets – A Handbook" – FAO, 2001), which represents the amount of food available, overall quantifiable, as calories available per capita, and that allows analyzing the qualitative aspect of food security, linked to nutrition security and, therefore, to the problem of malnutrition in developing countries.

FAO has analyzed the problem by supporting a study for the creation of a qualitative model (Understanding the dynamics of Food Supply and Distribution Systems [FSDS] – Armendariz, Armenia and Atzori, 2015b) based on a causal-relationship diagram, that is capable of highlighting the endogenous and exogenous variables characterizing the Food Supply and Distribution Systems (FSDS) in developing countries (Aragrande and Argenti, 1999). In line with that approach and in order to constitute a quantitative extension of that research, in this paper our aim is to show the effectiveness of the System Dynamics (SD) modelling and simulation methodology for the analysis of the Food Insecurity problem. We argue that through the

SD approach it is in fact possible to provide relevant organizations, such as FAO, with a valid instrument capable of supporting the development and testing of policies aimed at mitigating the problem (Armendariz, Armenia, Atzori and Romano, 2015). In fact, SD allows describing the structure of the system, assessing its behaviour on a long-term and test (in virtual environments) possible policies.

In other words, in this paper we have expanded the previously proposed Casual-relationship Diagram by adding some additional relevant aspects as: (i) the importance of local food-sheds (intended as alternative food systems with the capacity to produce food locally and reduce the negative social and environmental impacts of farming), and (ii) the importance of nutrition security (as a qualitative aspect of food security).

We also have proposed mitigation policies of the problem, e.g. time for agricultural technology development, demand of agricultural land and waste disposal, as well as effort of Hazard Analysis Critical Control Points (HACCP) controls, improvement of basic services (health and education), activation of workforce migration and food imports, based on the statement of Müller O. and Krawinkel M. (2005) that malnutrition is one of the major public health problems in developing countries (in fact it is responsible for disease and deaths, especially for young children). It stems from a diet deficient in nutrients and its main cause is poverty, but it depends on many other factors, such as the political and economic situation, education and sanitation, climate conditions and cultural food customs.

In order to support our dynamical hypothesis, we have tested our model by simulating a real case regarding the dramatic situation in Haiti, an island in Central America (about 10 million inhabitants, the least developed country in the Western Hemisphere and one of the poorest in the world), and whose results show how the activation of some policies can mitigate the problem over time. For example, the implementation of policies, related to the development of agricultural technology and demand for land for waste disposal, shows an improvement on food quality, one of the main variables analyzed in our model, while after the implementation of policies, related to increased productivity and workforce migration, the value of the food security gap falls.

However, in this paper, only the qualitative model will be presented, both for brevity's sake and in order to provide readers with the possibility to understand in detail the systemic relationships that exist in such complex and dynamic contexts.

Notes

1 System Dynamics Italian Chapter of the System Dynamics Society (SYDIC); Sapienza University of Rome.
2 Sapienza University of Rome.

Bibliography

Aragrande, M., Argenti, O. (1999). *Studying Food Supply and Distribution Systems to Cities in Developing Countries. Methodological and Operational Guide. Food into Cities Collection*, Rome: FAO.

Armendariz, V., Armenia, S., Atzori, A. S. (2015a). System dynamics updates to FAO's methodological guide to understand food supply and distribution systems (FSDS) in developing and transition countries. http://systemdynamics.org/conferences/2015/proceed/papers/P1153.pdf

Armendariz, V., Armenia, S., Atzori, A. S. (2015b). Understanding the dynamics of Food Supply and Distribution Systems (FSDS). www.fao.org/fileadmin/templates/ags/docs/MUFN/CALL_FILES_EXPERT_2015/CFP3-18_Full_Paper.pdf

Armendariz, V., Armenia, S., Atzori, A. S., Romano, A. (2015). Analyzing food supply and distribution systems using complex systems methodologies. http://centmapress.ilb.uni-bonn.de/ojs/index.php/proceedings/article/view/448

Aronson, D. (1996). Overview of systems thinking. www.thinking.net/Systems_Thinking/OverviewSTarticle.pdf

Carletto, C., Zezza, A., Banerjee, R. (2013). Towards better measurement of household food security: Harmonizing indicators and the role of household surveys. *Global Food Security* 2(1), pp. 30–40. www.elsevier.com/locate/gfs

Christian, J. P., Bills, N. L., Wilkins, J. L., and Fick, G. W. (2008). Food-shed analysis and its relevance to sustainability. *Renewable Agriculture and Food Systems*, pp. 1–7. https://doi.org/10.1017/S1742170508002433

FAO (from 1961 to 2011). Food balance indicators. http://fao.org

FAO (2001). Food balance sheets: A handbook: Fao corporate document repository. www.fao.org/3/a-x9892e.pdf

FAO (2014). Strengthening the enabling environment for food security and nutrition. *The State of Food Insecurity in the World*, pp. 1–57. www.fao.org/publications

FAO (2015). Meeting the 2015 international hunger targets: Taking stock of uneven progress. *The State of Food Insecurity in the World*, pp. 1–62. www.fao.org/publications

Gardner, G., Halweil, B., Peterson, J. A. (2000). *Underfed and Overfed: The Global Epidemic of Malnutrition*. WorldWatch Institute. Retrieved 23.9.2018 from http://www.worldwatch.org/system/files/EWP150.pdf

Ihan, A. N., Rohana, A. J., Wan Manan, W. M. (2015). Concept and measurements of household food insecurity and its impact on malnutrition: A review. *International Medical Journal* 22(6), pp. 509–516.

Kennedy, G., Nantel, G., Shetty, P. (2003). The scourge of hidden hunger: Global dimensions of micronutrient deficiencies. *Food Nutrition and Agriculture* 32, pp. 8–16.

Nordin, S.M., Boyle, M., Kemmer, T.M. (2013). Position of the academy of nutrition and dietetics: Nutrition security in developing nations: Sustainable food, water, and health. *Journal of the Academy of Nutrition and Dietetics* 113(4), pp. 581–595.

Müller, O., Krawinkel, M. (2005). Malnutrition and health in developing countries. *Canadian Medical Association Journal* 173(3), pp. 279–286.

Peterson, Worldwatch Paper 150. www.worldwatch.org/system/files/EWP150.pdf

Reutlinger, S., Selowsky, M. (1976). Malnutrition and poverty: Magnitude and policy options. World Bank Staff Occasional Papers No 23 Johns Hopkins, Baltimore and London.

Sterman, J. (2000). *Business Dynamics: Systems Thinking and Modeling for a Complex World*. New York, NY: McGraw-Hill Higher Education.

WFP (2007). *World Hunger Series 2007: Hunger and Health*. UK: Earthscan.

World Health Organization (WHO) (2003). *Diet, nutrition and the prevention of chronic diseases: Report of a Joint WHO/FAO Expert Consultation* (Vol. 916). World Health Organization.

Zupi, M. (2014). La sicurezza alimentare in 13 paesi asiatici in via di sviluppo dell'ASEM. *CeSPI* (101). www.cespi.it/ITALIA.html

Websites

http://cespi.it
http://cia.gov
http://fao.org
http://ifad.org
http://ifpri.org
http://indexmundi.com
http://powersim.it
http://wfp.org
http://who.int
http://worldbank.org
http://worldwatch.org

Systemic characteristics of a Human and Organizational Factors (HOF) approach of safety management

Jean-François Vautier,[1] Nicolas Dechy,[2] Thierry Coye de Brunélis,[3] Guillaume Hernandez[4] and Richard Launay[4]

Keywords: *human and organizational factors; systemic approach; risk management; safety; industrial accident; emergence; variety; coherence; model*

1   Introduction

A Human and Organizational Factors (HOF) approach of safety management is dedicated to study, with human and social sciences lenses, the unsafe acts of workers and the factors of working situations, departments, companies . . . which influence directly or indirectly the human performance (Tosello et al. 2003).

This paper presents qualitative research based on the authors' experience feedback. Its purpose is to investigate and enhance the use of some systemic concepts which are currently implemented by HOF specialists and to illustrate them from day-to-day human activities to activities of workers during severe industrial accidents. The authors are HOF specialists who belong to different companies. The aim is to gather different implementations of systems thinking in HOF approaches.

Five systemic concepts were identified in this research: definition, limit, emergence, variety and models of the systems. This study presents, for each concept, its meaning, one or two illustrations in the HOF field and the interest of using the concept more explicitly in this latest field.

2   The definition of a system

The set of four causes proposed by Aristotle is here examined as a way to define and describe a system (Vautier 2015). In this way, several items have to be considered:

- The material cause: the elements of the system,
- The formal cause: the structure of the set of elements (the arrangement of the elements),

- The final cause: the reason why this set of elements, assembled as a system, exists,
- The moving cause: the process of designing, making and maintaining the system.

As an illustration, in a training session a question may be asked: "On what does the nuclear safety of a facility depend?" The usual responses are:

- Technical elements like barriers, machines . . . ,
- Quality of competences, tools, procedures . . . of the working situations to avoid human mistakes . . . ,
- Good safety culture, priority to safety . . .,
- Adequate monitoring of the interactions between the departments of the facility,
- Efficient ways of interaction with the external systems like the subcontractors,
- . . .

These items may be related to the four causes of Aristotle.

As an interest of using more explicitly this concept of definition of a system, it could help to think about a lot of parameters and hope not to forget some important ones . . .

3   The limit of a system

Here, it is considered the system of causes which is taken into account during an unwanted event analysis.

In this context, a similar question to the previous one may be asked: "On what does an unwanted event depend?" Different kinds of responses come often from different kinds of actors:

- A technical actor focuses usually on a technical aspect e.g. the technical barriers,
- A HOF specialist may enlarge the system of causes in focusing on the role of humans and organizations concerning for example the maintenance of these barriers,
- An economist may continue to enlarge the system in focusing on the costs of the technical and human elements and on the budget of the facility . . .

It shows that the limit of a system depends on the point of view of the actor . . . Then, it means that the limit of a system of causes depends on the variety of actors which contribute to the analysis.

As an interest of using more explicitly this concept of limit, it could help to increase the variety of causes by recruiting different kinds of actors for an analysis.

4   The emergence

Different definitions of this concept exist . . . Here, the emergence is defined as a performance of a system which changes if the arrangement of its elements changes as well.

This arrangement of the elements may be spatial, temporal or functional (in this latest case, it concerns how the elements fit one to each other in order to make something efficient together).

As an illustration:

- About spatial aspect: in a training room, a U shape row of tables makes it possible to maximize the communication between the students by contrast with several straight rows of tables positioned one behind the other,
- About functional aspect: the analysis of the Three Mile Island nuclear accident in 1979 (Llory 1999) shows the effect of the lack of matching between the man-machine interface of the control room and the abilities of workers to build an adequate mental representation. Indeed, the workers thought that when a specific light was activated in the control board it meant that an automaton was operating to close a valve . . . But, it was not true . . . The light only indicated that the order of closing the valve was sent by the automaton . . . And with this sequence . . . an inadequate cognitive automatism was raised in the brain of workers . . . The problem was that during the accident the valve was in reality blocked in a middle open position even if the order of closing had been sent . . .

Hence, the importance of acting on the matching between the elements of a system (here humans and machines) is recalled . . .

As an interest of using more explicitly this concept of emergence, it could help to emphasize the necessity for the designers to use a large scale of HOF standards in addition to the feedback experience (Tosello et al. 2012).

5   The variety of a system

Here the focus is on the concept of "requisite variety" (Ashby 1956) which postulates that any system X to control any system Y must have a variety greater than or equal to it. "Variety" can be understood as the amount of different behaviors and states of a system. This set of behaviors and states is similar to the number of degrees of freedom of the system.

As an illustration, the Tenerife accident, on March 27, 1977, was a collision between two airplanes (collision of two Boeing 747s) during take-off which caused the death of 583 people. One of the causes of the accident was the inadequate mental representation of the pilot-in-command of one of the Boeing 747s due in particular to the lack of information he took from his flight engineer . . . Then afterwards, a set of technics of communication between people was introduced (named the crew resources management [CRM]) to prevent communication flaws. For example, the pilot-in-command has now to follow a procedure of cross control before taking off. Then, all the members of the cockpit may give sufficient information to the others.

In other words, all the members of the cockpit have more degrees of freedom to express the knowledge they have about the situation. As an interest of using more explicitly this concept of variety, it could help to improve the communication inside a team.

The application of this concept of "requisite variety" shows also the importance of avoiding "the out-of-the-loop syndrome". It means that a driver (for example a pilot . . .) has not enough degrees of freedom to pilot correctly a technical system since he does not know its current state and then cannot anticipate . . . In this way, as an interest of this concept of variety, it could induce to apply a human-centered design (ISO 9241–210) to achieve the design variety required by involving different profiles of designers and disciplines (such as HOF for example) and involving the end-users in order to benefit from their feedback.

6   The models of a system (descriptive, explicative and normative ones)

A difference is shown between:

- A descriptive model which gathers often boxes connected together by arrows that represent relationships of matter (for example a document is sent by post mail) or energy (an email is sent)

and

- An explicative model which gathers often boxes connected together by arrows that represent causal relationships (the boxes represent often characteristics of a system). In this kind of model, the arrows may also represent adaptability relationships between boxes. Then, the overall representation shows in this latest case what is considered as an effective functioning of the system. It is why this model may also be considered as a normative one. Echoing to systemic triangulation (Le Moigne 1977) which proposes to describe systems in three ways (functional, structural, historical), some multi-axis representations are dedicated to explain or understand an event from an organizational perspective.

Hence, the three dimensions of "organizational analysis" of an event proposed by Dien et al. (2012).

Different kinds of models with adaptability relationships between the components of a working situation (tasks, humans, tools, work environment) may be used in the HOF field . . .

As an illustration, consider again a group of students in a training room . . . A U shape row of tables makes it possible to maximize the communication between

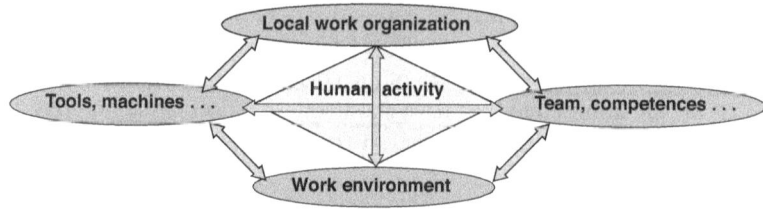

<table>
<tr><td>⬭</td><td>: Component of the model of a working situation</td></tr>
<tr><td>⟺</td><td>: Interaction between two components of the model of a working situation</td></tr>
<tr><td>◇</td><td>: Emergence as a result of the interactions between the components of a working situation</td></tr>
</table>

Figure 97.1 Model of a working situation
Source: Authors' elaboration

people (students and trainer) . . . But, other shapes exist, dedicated to other finalities. . . . For example, if the students have to pass an examination, it is better to position the students the most distant from one another and locate them in several straight rows of tables positioned one behind the other. It is why the HOF specialists look for the coherence between the components of this kind of model.

As an interest of using more explicitly the concept of model, it could show the importance of defining the meaning of the arrows in a model and it makes it possible to propose a practical definition of the concept of interaction.

7   Conclusion

For each of the five systemic concepts, the study shows the interest of using them more explicitly in a HOF approach. Then, it indicates a way of introducing the systems thinking in the HOF field, i.e. in the field of industrial risk management focused on the study on human performance.

In other words, the study contributes to show how to improve the HOF approaches by highlighting the interest of systems thinking. It illustrates also some ways to implement the recent recommendations of the International Atomic Energy Agency (IAEA) about the systemic approaches. Indeed, IAEA (2016) proposed to define a systemic approach as an approach relating to the system as a whole in which the interactions between technical, human and organizational factors are duly considered.

Only a few systemic concepts are studied here. Other systemic concepts may be developed like for example the system dynamics or the dialogical pairs (Vautier et al. 2016).

Notes

1 CEA and AFSCET.
2 IRSN.

3 THALES.
4 CEA.

Bibliography

Ashby, W.R. (1956). *An Introduction to Cybernetics*. Chapman & Hall.

Dien, Y., Dechy, N. and Guillaume, E. (2012). Accident investigation: From searching direct causes to finding in-depth causes. *Problem of Analysis or/and of Analyst? Safety Science*, 50(6), 1398–1407.

IAEA (2016). *Leadership and Management for Safety. GSR Part 2*. IAEA publications.

ISO (2011) iso9241–210, *Ergonomics of human-system interaction, Part 210: Human-centred design for interactive systems*, ISO

Le Moigne, J.-L. (1977). *La théorie du système général, théorie de la modélisation* (2nd édition, 1994). jeanlouis le moigne-ae mcx.

Llory, M. (1999). *L'accident de la centrale nucléaire de Three Mile Island*. Éditions L'Harmattan, Paris.

Tosello, M., Lévêque, F., Dutillieu, S., Hernandez, G. and Vautier, J.-F. (2012). Conditions for the successful integration of Human and Organizational Factors (HOF) in the nuclear safety analysis. *Work*, 41, 2656–2660.

Tosello, M., Vautier, J.-F. and Sevestre, B. (2003). A new study of human factors in the nuclear safety field. XVth Congress of the International Ergonomics Association (IEA), Seoul, Korea, SAFETY V, August 24–29.

Vautier, J.-F. (2015). Making a causal contextualization with the four causes of Aristotle. *Advances in Systems Science and Application*, 15(2), 176–187.

Vautier, J-F., Tosello, M., Hernandez, G., Dutillieu, S., Quiblier, S., Sylvestre, C., Lévêque, F., Barnabé, I., Baussart, N., Paulus, V., Lipart, C., Barrière, V. and Dupont, M. (2016). A synchro-diachro approach to question the development of a Human and Organizational Factors (HOF) network. International Conference on Human and Organizational Aspects of Assuring Nuclear Safety: Exploring 30 Years of Safety Culture, IAEA, Vienna, Austria, 22–26 February.

How to overcome systemic obstacles to problem solving of the inclusion of disabled people

Patrick Farfal[1]

Keywords: *disability; citizenship; inclusion; system loops; spin-offs*

1   Need for a systemic approach of disability

The approach to disability must resolutely be systemic. At least because the matter of disability obviously and immediately addresses the question of citizenship and social link, which is reciprocal by definition.

Also because disability as a fact is far from being marginal: one European out of ten is concerned by disability; nearly 10 million disabled persons (in a broad sense) can be counted in France. Only 15% of disabled persons contract disability at birth, so, any valid person may contract a disability any day. Differences, also diversity, factors of complexity, demand a systemic approach.

> Disability is complex, dynamic, multidimensional, and contested.
> (World Health Organization and the World Bank, 2011)

Lastly, disability needs compensation (sensory or motor aid . . . , desk fitting out . . .): it is the environment which adapts itself to the disabled person!

2   System interactions and system loops

The "disability system" can be defined in a more or less extended way (well-known question of the "system of interest" in systems engineering), either reduced to disabled people, facing their environment, or extended to the whole "disability system" (disabled people, hospitals, law, insertion bodies, . . . , in interaction with one another), facing the rest of society and environment ("valid" people, city spaces, world of work . . .).

Also, in all those interactions, circular causality can be found: vicious (blocking) loops: social exclusion, shame ⇨ deprivation of opportunities ⇨ lack of rights ⇨ reduced participation ⇨ social exclusion, etc., and virtuous loops too (compensation is one: compensation for accessibility to buildings, transports, life spaces, desk adaptations . . .).

3   Facts

In practice, and, generally speaking, in the society, individualism takes the lead over "living together". Stereotypes of disabled persons (deemed less performative, generating extra costs . . .) become widespread among people both in everyday life and at work. Answers provided by some elected members or administrators are not sufficient because they are fragmentary (for example limited to training to solve the problem of high unemployment rate), while a set of consistent and complementary answers are needed.

4   Components of a comprehensive (holistic) view

The whole of those answers must include time factor (Durand, Nunez, 2002); the point of view on disabled persons must be educated from childhood, from primary education.

"Changing attention" needs a switch from an "individual and medical model" (where the disabled person, merely considered as to be cared, is the problem) to a "social model" of disability, where the environment (to be adapted) is the problem.

So, a systemic treatment of disability implies coordinated actions in the following fields: children (welcome, education . . .), companies and employment (competences acknowledgement, recruitment . . .), administration (welcome and support, recognition of disabled worker status . . .), training (of disabled people, nursing staff, but also recruiting people and employers . . .), accessibility (to housing, buildings, transports, cultural and associative life, and of course cure and care), right to compensation (of sensory or motor disability . . .).

Even the component cure and care is of systemic nature: the person must be treated in her whole (therapeutic education, medicine acting at each step of the care path, care directed towards the transition to social and occupational rehabilitation, the disabled person acting throughout her path . . .).

Associations dedicated to disability, who treat, educate, train, insert, support, and those who, in their sports, cultural, or artistic activities, include a disability part, obviously play a major role in that approach.

5   Unexpected spin-offs of a systemic approach

Unexpected spin-offs of the compensation of disability can be seen: the adaptability of some space (building, transport) to the needs and constraints of a person with a loss of autonomy is not a simple respect of law as regards accessibility, but is broadened to the quality of use of "life spaces" (CRIDEV, 2008) for everybody, taking into account the needs and constraints of the whole of people: the disabled person often appears to enlighten the needs of the whole (example: access platforms to busses). Considering system engineering vocabulary, that amounts to speaking of taking into account the needs and constraints of all the stakeholders, which is an essential condition of the secure outcome of a project (INCOSE, 2012).

The adaptation of the environment to the disabled person, in the very scope of the February 11th 2005 French law, as well as the claim of her full citizenship (schooling, employment . . .), with its consequences onto the whole of people is not the least surprise arisen from thinking about disability, which once again, as a virtuous loop, emphasizes the systemic nature of the disability question.

Considering the systemic features of the question of disability would make it possible for some elected or administration people not to immediately focus on solutions, often fragmentary, without any care of other relations between the actors of the field and their environment, but on the contrary tackle the question as a whole, and think about the benefits induced on "valid" people, a major part of the population.

Note

1 PATSYS, 25 rue Jean Leclaire, 75017 Paris, pfarfal.patsys@sfr.fr, +33(0)672148240.

Bibliography

Bricage, Pierre, The Metamorphoses of the Living Systems: The Associations for the Reciprocal and Mutual Sharing of Advantages and of Disadvantages, *6e congrès européen de science des systèmes*, 19–22 septembre, 2005.

Durand, Daniel and Nunez, Emmanuel, *An Operative Pedagogy of the Systemscience Approach*, 2002. Retrieved 23.9.2018 from http://afscet.asso.fr/resSystemica/Crete02/DurandNunez.pdf

INCOSE, *Systems Engineering Handbook: A Guide for System Life Cycle Processes and Activities*, 2012. Retrieved 23.9.2018 from http://disi.unal.edu.co/dacursci/sistemasycomputacion/docs/SystemsEng/SEHandbookv3_2006.pdf

Loi n° 2005–102 du 11 février 2005 *pour l'egalite des droits et des chances, la participation et la citoyennete des personnes handicapees.* Legifrance. Retrieved 23.9.2018 from https://www.legifrance.gouv.fr/affichTexte.do?cidTexte=JORFTEXT000000 0809647&categorieLien=id

Martin, James, *Information Engineering: A Trilogy.* Introduction-Prentice Hall, 1989.

World Health Organisation, *International Classification of Functioning, Disability and Health.* ICF. Geneva: World Health Organization, 2001.

World Health Organisation and the World Bank, *World Report on Disability*, 2011. Retrieved 23.9.2018 from http://www.who.int/disabilities/world_report/2011/report.pdf

Exploration of application stores' business model deploying value network and system dynamics approach
Case of Apple App Store

Nastaran Hajiheydari and Reza Alibakhshi[1]

Keywords: *application stores; business model; e3Value network; mobile application; mobile application ecosystem; system dynamics*

1 Introduction

Application stores and their importance in mobile industry, alongside their emergence from an electronic commerce point of view are amongst the noteworthy trends of digital economy. Accordingly, this paper is dedicated to extract the most suitable explanation representing mobile application stores' business model considering mobile and smart phones' value network. In this research we have reviewed and extracted the main role players of mobile industry value networks and their elements, and accordingly a dynamic model of value exchange for this network has been created applying a System Dynamics approach. Due to the fact that mobile industry value network enjoys complex dynamics which elaborates the intricacy of relationships in value exchange, we have used a System Dynamics approach to create a thorough understanding of value streams. Based on the proposed qualitative model, the business model of mobile application stores has been extracted with concentration on the case of the Apple App Store and this model hs been evaluated using a thematic analysis approach with second-hand data. According to the findings of this study, deploying a value-oriented perspective in designing the business model of mobile application stores helps us to build a model with comprehensive understanding of the value proposition of this electronic business as well as its main partners and customers. Additionally, regarding the dynamic nature of business environment, this paper indicates that using a System Dynamics approach in defining business models can be of great value.

2 Smart devices

With the rapid growth of portable smart devices, the dependent development of their required applications has been one of the digital booms of the past few years. Since the introduction of the Apple App Store in 2008, numerous application stores from different companies with expertise in device manufacturing, telecommunication, operating systems, and so forth have been developed which endorses the attractiveness on these electronic markets (Müller, 2011). Only during the six years between 2007 and 2013 the number of applications for smart devices has reached to around 900.000 applications (Perez, 2013). According to an infographic report provided by University of Alabama at Birmingham's Online Masters in Management Information Systems, by 2017 over 268 billion application downloads will generate $77 billion worth of revenue (Birmingham, n.d.) Additionally, the business potential of app stores is important to every role player in the customer value network, from device producers to developers and e-tailers. The emergence of mobile applications has also made the entire structure of the value network of mobile industry undergo change as well which indicates the importance of recognizing the market of mobile applications (Holzer, 2009). Considering the still-growing number of smart devices and mobile platforms behooves us to study their businesses and their value network through a rigorous academic point-of-view.

3   Dynamic business models

In realizing the way businesses functionalize themselves, as Chesbrough and Teece believe, whether or not recognized by the company, every single company has a framework and a model for their businesses (Chesbrough, 2013; Teece, 2010). Despite the vast study of business models, the very concept contains opaque factors and elements, which have ended in multiple definitions and classifications (Al-Debei, 2010; Morris, 2005; Shafer, 2005). Regardless of all these definitions and classifications, mutual understanding of business models is the logic of creating and capturing value from customers and partners (Fielt, 2014). According to the importance of business models in electronic markets, identifying application stores' business models is one of the main approaches to identify the functionality and procedures of application stores, especially in terms of value proposition by different actors and roles. As Teece defines, in short, a business model defines how the enterprise creates and delivers value to customers, and then converts payments received to profits (Teece, 2010). Therefore, identifying the value network of a mobile application enables us to extract the main actors in their value network and their relationship in the ecosystem of mobile application. Furthermore, since the business models are dynamic due to their interaction with their environment, every business model needs to be considered based on their dynamic nature in organizations (Sosna, 2010; Chesbrough, 2013; Doz, 2010). This idiosyncrasy of business models accentuates the need to have a dynamic point-of-view in evaluating and designing business models, not only in strategic decision making of business models, but also in their generation and design (Hajiheydari and Zarei, 2013).

4   Apple App Store case

Accordingly, this paper tries to identify the main actors and their contributions in the business model of application stores in general and the Apple App Store as a case. Considering the elements of a business model, as Osterwalder and

Pigneur describe, this paper aims to represent a proposed business model for the App Store with identified sections of customer segments, customer relationships, communication and distribution channels, value propositions, key resources, key activities, key partnerships, revenue streams, and cost structures (Osterwalder and Pigneur, 2010). Consequently, this research would be started with identifying key role players and their main contribution in the mobile application value network. Later, we will concentrate on mobile app stores as one of the main actors in this value network to investigate meaningful business models based on literature and with the contribution of domain experts. The descriptive business model will then be modeled with a System Dynamics approach as an appropriate approach of modeling complex systems. Therefore, in addition to identifying a suitable business model for the App Store, the effectiveness of System Dynamics in modeling business models will be considered as well.

Note

1 University of Tehran.

Bibliography

Adrian, B. (2002). *Overview of the mobile payments market 2002 through 2007.* Gartner.

Al-Debei, M. M. (2010). Developing a unified framework of the business model concept. *European Journal of Information Systems, 19*(3), 359–376.

Allan, A. (2003). *Internet business models and strategies.* New York: McGraw-Hill.

Anurag Tewari, P. S. (2014). Platform business models and mobile ecosystem. *Thesis,* 1–24.

Apple. (2015). *TestFlight.* Retrieved from https://developer.apple.com/testflight/

Ballon, P. N. (2008). The reconfiguration of mobile service provision: Towards platform business models. *Available at SSRN 1331549.*

Barnes, S. J. (2002). The mobile commerce value chain: Analysis and future developments. *International Journal of Information Management, 22*(2), 91–108.

Basole, R. C. (2012). Value transformation in the mobile service ecosystem: A study of app store emergence and growth. *Service Science, 4*(1), 24–41.

Bergvall-Kåreborn, B. (2011). Mobile applications development on Apple and Google platforms. *Communications of the Association for Information Systems, 29*(1), 565–580.

Bic, L. (1988). *The logical design of operating systems.* Vol. 2. Englewood Cliffs, NJ: Prentice Hall.

Bieger, T. D.-A. (2011). *Innovative Geschäftsmodelle – Konzeptionelle Grundlagen.* Berlin: Heidelberg ua.

Birmingham, U. O. (n.d.). *The future of mobile application.* Retrieved from http://businessdegrees.uab.edu/resources/infographics/the-future-of-mobile-application/

Buellingen, F. (2004). Development perspectives, firm strategies and applications in mobile commerce. *Journal of Business Research, 57*(12), 1402–1408.

Chen, P.-T. (2010). Unlocking the promise of mobile value-added services by applying new collaborative business models. *Technological Forecasting and Social Change, 77*(4), 678–693.

Chesbrough, H. (2013). *Open business models: How to thrive in the new innovation landscape.* Brighton: Harvard Business Press.

Chung, S. H. (2010). A dynamic forecasting model for nursing manpower require-ments in the medical service industry. *Service Business, 4*(3), 225–236.

Cuadrado, F. (2012). Mobile application stores: Success factors, existing approaches, and future developments. *IEEE Communications Magazine, 50*(11), 160–167.

Dedrick, J. K. (2011). The distribution of value in the mobile phone supply chain. *Telecommunications Policy, 35*(6), 505–521.

De Reuver, M. (2009). Designing viable business models for context-aware mobile services. *Telematics and Informatics, 26*(3), 240–248.

Doz, Y. L. (2010). Embedding strategic agility: A leadership agenda for accelerating business model renewal. *Long Range Planning, 43*(2), 370–382.

Dubosson-Torbay, M. A. (2002). E-business model design, classification, and mea-surements. *Thunderbird International Business Review, 44*(1), 5–23.

Erman, B. A. (2011). Mobile applications discovery: A subscriber-centric approach. *Bell Labs Technical Journal, 15*(4), 135–148.

Feijóo, C. I.-L.-B. (2009). Exploring a heterogeneous and fragmented digital ecosys-tem: Mobile content. *Telematics and Informatics, 26*(3), 282–292.

Fielt, E. (2014). Conceptualising business models: Definitions, frameworks and clas-sifications. *Journal of Business Models, 1*(1), 85–105.

Forrester, J. W. (2007). System dynamics: A personal view of the first fifty years. *System Dynamics Review, 23*(2), 345–358.

Funk, J. L. (2009). The emerging value network in the mobile phone industry: The case of Japan and its implications for the rest of the world. *Telecommunications Policy, 33*(1), 4–18.

Ghezzi, A. F. (2009). Value networks: Scenarios on the Mobile Content market configurations. *In 2009 Eighth International Conference on Mobile Business*, 35–40.

Gonçalves, V. N. (2010). "How about an app store?" Enablers and constraints in platform strategies for mobile network operators. *Mobile Business and 2010 Ninth Global Mobility Roundtable (ICMB-GMR), 2010 Ninth International Conference*, 66–73.

Guest, G., MacQueen, K. M., and Namey, E. E. (2011). *Applied thematic analysis*. Thousand Oaks, CA: Sage.

Hajiheydari, N., and Zarei, B. (2013). Developing and manipulating business models applying system dynamics approach. *Journal of Modelling in Management, 8*(2), 155–170.

Heitkoetter, H. K. (2012). *Mobile platforms as two-sided markets*. Retrieved 23.9.2018 from https://aisel.aisnet.org/amcis2012/proceedings/AdoptionDiffusionIT/11/

Holzer, A. (2009). Mobile application market: A mobile network operators' perspec-tive. *In Workshop on E-Business*, 186–191.

Holzer, A. (2011). Mobile application market: A developer's perspective. *Telematics and Informatics, 28*(1), 22–31.

Hong, S.-J. J.-Y.-Y. (2008). Understanding the behavior of mobile data services con-sumers. *Information Systems Frontiers, 10*(4), 431–445.

Joe Peppard, A. R. (2006). From value chain to value network: Insights for mobile operators. *European Management Journal, 24*(2–3), 128–141.

Kenney, M. (2011). Structuring the smartphone industry: Is the mobile internet OS platform the key? *Journal of Industry, Competition and Trade, 11*(3), 239–261.

Kim, C. (2012). A database: Centred approach to the development of new mobile service concepts. *International Journal of Mobile Communications, 10*(3), 248–264.

Kim, J. K. (2011). Value network of mobile content providers. *2011 International Conference on Social Science and Humanity*, 5.

Kim, J. Y. (2014). Mobile application service networks: Apple's App Store. *Service Business*, *8*(1), 1–27.

Kimbler, K. (2010). App store strategies for service providers. *In Intelligence in Next Generation Networks (ICIN), 2010 14th International Conference on*, 1–5.

Klang, D. M. (2010). The anatomy of the business model: A syntactical review and research agenda. *Summer Conference 2010-Opening Up Innovation*, 1–31.

Krumeich, J. D. (2013). *Interdependencies between Business Model Components – A Literature Analysis*. Retrieved 23.9.2018 from https://aisel.aisnet.org/amcis2013/EnterpriseSystems/GeneralPresentations/7/

Lee, G. (2011). Product portfolio and mobile apps success: Evidence from app store market. *AMCIS*.

Liu, C. Q. (2011). Status and trends of mobile-health applications for iOS devices: A developer's perspective. *Journal of Systems and Software*, *84*(11), 2022–2033.

Liu, C. Z. (2012). An empirical study of the freemium strategy for mobile apps: Evidence from the google play market. *Proceedings of the 33rd International Conference on Information Systems*.

Maitland, C. F. (2002). The European market for mobile data: Evolving value chains and industry structures. *Telecommunications Policy*, *26*(9), 485–504.

Manchiganti, R. (2001). On emerging ecosystems in the mobile phone industry: An evaluation of current and emerging mobile phone. *Master Thesis*, 1–85.

Mancuso, P. (2012). Regulation and efficiency in transition: The case of telecommunications in Italy. *International Journal of Production Economics*, *135*(2), 762–770.

March, S. T. (1995). Design and natural science research on information technology. *Decision Support Systems*, *15*(4), 251–266.

Maxwell, K. (2004). *USA Patent No. US 20040148229A1*. Retrieved 23.9.2018from https://patents.google.com/patent/US20040148229?oq=20040148229A1

Morris, M. S. (2005). The entrepreneur's business model: Toward a unified perspective. *Journal of Business Research*, *58*(6), 726–735.

Müller, R. M. (2011). A comparison of inter-organizational business models of mobile app stores: There is more than open vs. closed. *Journal of Theoretical and Applied Electronic Commerce Research*, *6*(2), 63–76.

Osterwalder, A. Y. (2004). The business model ontology: A proposition in a design science approach. *PhD dissertation*.

Osterwalder, A. Y. (2005). Clarifying business models: Origins, present, and future of the concept. *Communications of the Association for Information Systems*, *16*(1), 1.

Osterwalder, A. Y., and Pigneur, Y. (2010). *Business model generation: A handbook for visionaries, game changers, and challengers*. Hoboken, New Jersey: John Wiley & Sons.

Park, Y. Y. (2012). Toward integration of products and services: Taxonomy and typology. *Journal of Engineering and Technology Management*, *29*(4), 528–545.

Peppard, J. (2006). From value chain to value network: Insights for mobile operators. *European Management Journal*, *24*(2), 128–141.

Perez, S. (2013). *Apple's app store hits 50 billion downloads, 900K apps, $10 billion paid to developers; iTunes now with 575M Accounts*. Retrieved 23.9.2018 from https://techcrunch.com/2013/06/10/apples-app-store-hits-50-billion-downloads-paid-out-10-billion-to-developers/

Pil, F. K. (2006). Evolving from value chain to value grid. *MIT Sloan Management Review, 47*(4), 72.

Pousttchi, K. H. Y. (2011). Value creation in the mobile market: A reference model for the role(s) of the future mobile network operator. *Business & Information Systems Engineering, 3*(5), 299–311.

Pussep, A. M. (2011). The software value chain as an analytical framework for the software industry and its exemplary application for vertical integration measurement. *In AMCIS*.

Raivio, Y. (2011). Mobile networks as a two-sided platform-case open telco. *Journal of Theoretical and Applied Electronic Commerce Research, 6*(2), 77–89.

Rocheska, S. M. (2015). A new methodological approach for designing the software industry value chain. *Ecoforum Journal, 4*(2).

Roma, P. G. (2013). An empirical analysis of revenue drivers in the mobile app market. *In Proceedings of POMS 24th Annual Conference 2013*, 3–6.

Salgatidou, A. (2001). Business models and transactions in mobile electronic commerce: Requirements and properties. *Computer Networks, 37*(2), 221–236.

Shafer, S. M. (2005). The power of business models. *Business Horizons, 48*(3), 199–207.

Son, C. Y. (2013). How to identify the trends of services: GTM-TT service map. *Expert Systems with Applications, 40*(8), 2956–2965.

Sosna, M. R.-R. (2010). Business model innovation through trial-and-error learning: The Naturhouse case. *Long Range Planning, 43*(2), 383–407.

Spohrer, J. (2008). The emergence of service science: Toward systematic service innovations to accelerate co-creation of value. *Production and Operations Management, 17*(3), 238–246.

Sterman, J. D. (2000). *Business dynamics: Systems thinking and modeling for a complex world*. Boston, MA: McGraw-Hill.

Teece, D. J. (2010). Business models, business strategy and innovation. *Long Range Planning, 43*(2), 172–194.

Timmers, P. (1998). Business models for electronic markets. *Electronic Markets, 8*(2), 3–8.

Wang, J. J.-Y.-C. (2015). Value network analysis for complex service systems: A case study on Taiwan's mobile application services. *Service Business, 9*(3), 381–407.

Xia, R. M. (2010). Business models in the mobile ecosystem. *In Mobile Business and 2010 Ninth Global Mobility Roundtable (ICMB-GMR), 2010 Ninth International Conference on*, 1–8.

Yan, Z. C. (2013). Exploring the impact of trust information visualization on mobile application usage. *Personal and Ubiquitous Computing, 17*(6), 1295–1313.

Zhang, X. (2008). Examining the mechanism of the value co-creation with customers. *International Journal of Production Economics, 116*(2), 242–250.

Zott, C. (2007). Business model design and the performance of entrepreneurial firms. *Organization Science, 18*(2), 181–199.

Driver or inhibitor for innovation? Modelling and simulating dynamic aspects and contradictory forces of team diversity

Anja Kreidler[1] and Meike Tilebein[2]

Keywords: *functional diversity; creativity; innovation; new product development teams; simulation; modelling; system dynamics*

Empirical studies show seemingly contradictory forces of team diversity on innovation. Diversity can act either as a driver or as an inhibitor for innovation. There is also evidence of a dynamic change of those diversity effects. However, empirical studies cannot entirely capture the effects of diversity on innovation. Thus, we apply a systemic approach to investigate the contradictory and dynamic effects of diversity by using modelling and simulation. We show an exemplary System Dynamics model of cross-functional diversity in teams and the effects on innovation.

1 Introduction

An ever-changing and highly competitive marketplace and shorter product life cycles force companies to be more innovative and to develop new and improved existing products, technologies, and services continually and fast (Lovelace et al., 2001; Gebert et al., 2006; Chen et al., 2015). Many organizations seek to achieve this with cross-functional new product development teams (NPD teams) since they hope to improve creativity and innovation with such teams (Lovelace et al., 2001; Gebert et al., 2006). Cross-functional teams have a wider variety of perspectives and thought worlds, which can lead to higher innovation (Gebert et al., 2006; Blindenbach-Driessen, 2015). However, cross-functionality can also cause communication barriers, obstruct team work, and lead to diminished innovation (Gebert et al., 2006; Blindenbach-Driessen, 2015). Other diversity attributes such as educational background, interdisciplinarity, different areas and levels of knowledge and capabilities, or tenure can result in similarly inconclusive results and act in opposite ways (Milliken and Martins, 1996; Gebert et al., 2006).

2 Effects of diversity

Numerous studies investigate the relationship between the diversity of team members and innovation (Milliken and Martins, 1996; Kreidler and Tilebein, 2013). The results of those empirical studies lead to inconclusive results, showing evidence of contradictory forces of team diversity (Milliken and Martins, 1996; Blindenbach-Driessen, 2015; Chen et al., 2015): Team diversity can be a driver for innovation through a wider perspective of ideas and a higher ability to react in unpredictable and uncertain situations (Milliken and Martins, 1996; Gebert et al., 2006; Blindenbach-Driessen, 2015). On the other side, team diversity can also be an inhibitor for innovation by hindering communication and promoting conflicts and thus obstructing team work (Milliken and Martins, 1996; Gebert et al., 2006; Blindenbach-Driessen, 2015).

Studies also show that there is a dynamic component to team work: the positive and negative effects of team diversity on innovation change over time (Perry-Smith and Shalley, 2003; Skilton and Dooley, 2010). The creative potential of diverse teams diminishes, while the barriers to team work also decline (Perry-Smith and Shalley, 2003; Skilton and Dooley, 2010). Few studies also indicate that there might be nonlinear effects of some diversity attributes (Chi et al., 2009). Numerous other factors like team size, team and organizational management, trust, or motivation influence the relationship between diversity and innovation (Clercq et al., 2011; Sivasubramaniam et al., 2012). Those effects can also be interconnected and interdependent (Milliken and Martins, 1996). In a previous paper, we have taken a closer look at the direct and indirect effects of team diversity on creativity, innovation, and performance (Kreidler and Tilebein, 2016).

Since most empirical studies in the area of diversity research are cross-sectional, they cannot entirely capture the effects of team diversity on innovation. Therefore, we propose to apply a systemic approach, based on (Davis et al., 2007; Happach and Tilebein, 2015). We use simulation as a complementary research method to investigate the contradictory forces, dynamic change, and nonlinear effects of team diversity on innovation. In previous publications, we have demonstrated that simulation can offer an additional method to gain insight into the complex relationship between diversity and innovation (Kreidler and Tilebein, 2013).

3   Simulation approach

NPD teams can be viewed as complex adaptive systems with individuals as operating agents and performance variables as emergent phenomena (Tilebein, 2006). There are a few agent-based models of heterogeneous teams and the effects on team work, which show the effect of information processing and diversity in a generic agent-based model. Chae et al. (2015) show an agent-based model of team diversity and creativity, and Crowder et al. (2012) provide a general framework for agent-based modelling of teamwork, using a multidisciplinary product team. However, since most empirical studies focus on the behaviour on the team level rather than on individual team members, we conclude that System Dynamics (SD) might be better suited to capture the direct and indirect effects of diversity on innovation (based on Kreidler and Tilebein, 2013). A System Dynamics model can capture the phenomena on a team level, thereby representing the two main contradictory forces of team diversity as a driver or inhibitor for innovation (Kreidler and Tilebein, 2013).

Therefore, we have developed an exemplary System Dynamics model of cross-functional diversity in NPD teams and the contradictory and dynamic effects on innovation.

4 System dynamics model

Figure 101.1 shows the developed System Dynamics model. Higher functional diversity leads to a higher diversity of mental models. On one hand, a higher diversity of mental models increases the development of ideas, making a team more creative. On the other hand, a higher diversity of mental models leads to more conflicts, diminishing synergistic communication and thus obstructing the development and implementation of ideas, decreasing a team's creativity and innovation. The dynamic aspect of team work is added by reduction of diversity of mental models. The longer a team works together successfully, the more team members' mental models assimilate, diminishing effects of diversity as a driver or as an inhibitor for innovation.

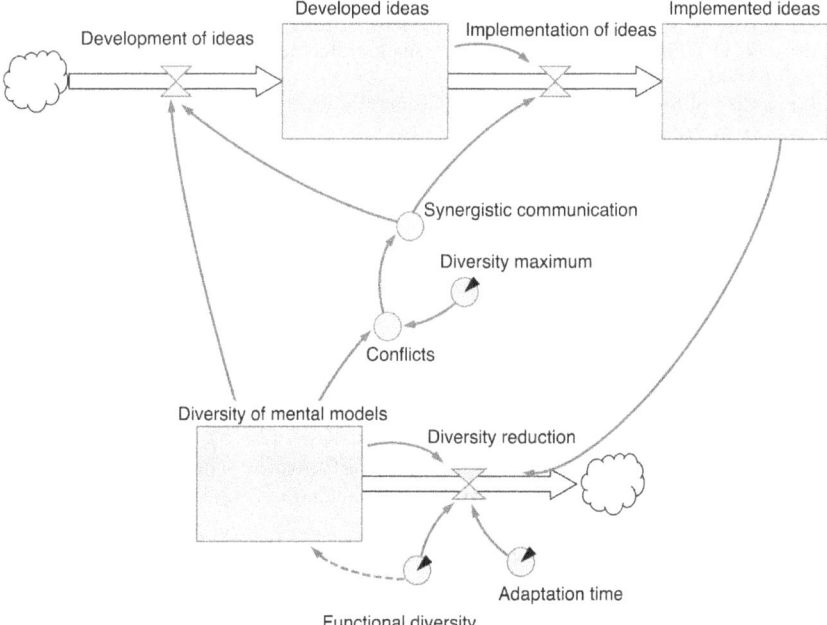

Figure 101.1 SD model of functional diversity and innovation

5   Discussion

By building and using the System Dynamics model, we achieve three goals:

> First, we show the potential of a systemic approach to better understand the effects of diversity in teams. We show the prospects and challenges of modelling and simulation as a relatively new method in the area of diversity research.
>
> Second, we gain further insight into the complex, interconnected, and dynamic effects of team diversity on innovation. The model integrates seemingly contradictory forces of cross-functionality: diversity as a driver or inhibitor for innovation. The model also takes into account the dynamic change of team work and is able to capture nonlinear behaviour. Thus, the model can help to gain a better understanding of the complex effects of diversity in NPD teams.
>
> Third, the model aims to show researchers where further empirical data is needed and helps in developing and testing a new theory. Furthermore, it helps practitioners and managers to gain a deeper insight into underlying team dynamics due to the diversity of team members and ultimately might be a tool to test new strategies for team management before implementing them.

Notes

1 University of Stuttgart, Institute for Diversity Studies in Engineering, Pfaffenwaldring 9, D-70569 Stuttgart, Germany, anja.kreidler@ids.uni-stuttgart.de, +49 711 685 61703.
2 University of Stuttgart, Institute for Diversity Studies in Engineering, Pfaffenwaldring 9, D-70569 Stuttgart.

Bibliography

Blindenbach-Driessen, F. (2015), "The (In)Effectiveness of Cross-Functional Innovation Teams: The Moderating Role of Organizational Context", *IEEE Transactions on Engineering Management*, 62(1), 29–38.

Chae, S.W., Seo, Y.W. and Lee, K.C. (2015), "Task Difficulty and Team Diversity on Team Creativity: Multi-Agent Simulation Approach", *Computers in Human Behavior*, 42, 83–92.

Chen, C.-J., Hsiao, Y.-C., Chu, M.-A. and Hu, K.-K. (2015), "The Relationship between Team Diversity and New Product Performance: The Moderating Role of Organizational Slack", *IEEE Transactions on Engineering Management*, 62(4), 568–577.

Chi, N.-W., Huang, Y.-M. and Lin, S.-C. (2009), "A Double-Edged Sword? Exploring the Curvilinear Relationship between Organizational Tenure Diversity and Team Innovation: The Moderating Role of Team-Oriented HR Practices", *Group & Organization Management*, 34(6), 698–726.

Clercq, D. de, Thongpapanl, N.T. and Dimov, D. (2011), "A Closer Look at Cross-Functional Collaboration and Product Innovativeness: Contingency Effects of

Structural and Relational Context", *Journal of Product Innovation Management*, 28(5), 680–697.

Crowder, R.M., Robinson, M.A., Hughes, H.P.N. and Sim, Y.-W. (2012), "The Development of an Agent-Based Modeling Framework for Simulating Engineering Team Work", *IEEE Transactions on Systems, Man, and Cybernetics, Part A: Systems and Humans*, 42(6), 1425–1439.

Davis, J.P., Eisenhardt, K.M. and Bingham, C.B. (2007), "Developing Theory through Simulation Methods", *Academy of Management Review*, 32(2), 480–499.

Gebert, D., Boerner, S. and Kearney, E. (2006), "Cross-Functionality and Innovation in New Product Development Teams: A Dilemmatic Structure and Its Consequences for the Management of Diversity", *European Journal of Work and Organizational Psychology*, 15(4), 431–458.

Happach, R.M. and Tilebein, M. (2015), "Simulation as Research Method: Modeling Social Interactions in Management Science", in Misselhorn, C. (Ed.), *Collective Agency and Cooperation in Natural and Artificial Systems*, Springer International Publishing, Cham, pp. 239–259.

Kreidler, A. and Tilebein, M. (2013), "Diversity and Innovationess in New Product Development Teams: Adressing Dynamic Aspects with System Dynamics.", in Eberlein, R. and Martínez-Moyano, I.J. (Eds.), *Proceedings of the 31st International Conference of the System Dynamics Society*, Cambridge, MA, USA.

Kreidler, A. and Tilebein, M. (2016), "Modeling the Dynamic Aspects of Team Diversity: A Comparison of System Dynamics and Agent-based Modeling", in Kopainsky, B. and Größler, A. (Eds.), *Proceedings of the 34th International Conference of the System Dynamics Society*, Delft, Netherlands.

Lovelace, K., Shapiro, D.L. and Weingart, L.R. (2001), "Maximizing Cross-Functional New Product Teams' Innovativeness and Constraint Adherence: A Conflict Communications Perspective", *Academy of Management Journal*, 44(4), 779–793.

Milliken, F.J. and Martins, L.L. (1996), "Searching for Common Threads: Understanding the Multiple Effects of Diversity in Organizational Groups", *Academy of Management Review*, 21(2), 402–433.

Perry-Smith, J.E. and Shalley, C.E. (2003), "The Social Side of Creativity: A Static and Dynamic Social Network Perspective", *Academy of Management Review*, 28(1), 89–106.

Sivasubramaniam, N., Liebowitz, S.J. and Lackman, C.L. (2012), "Determinants of New Product Development Team Performance: A Meta-Analytic Review", *Journal of Product Innovation Management*, 29(5), 803–820.

Skilton, P.F. and Dooley, K.J. (2010), "The Effects of Repeat Collaboration on Creative Abrasion", *Academy of Management Review*, 35(1), 118–134.

Tilebein, M. (2006), "A Complex Adaptive Systems Approach to Efficiency and Innovation", *Kybernetes*, 35(7/8), 1087–1099.

Theme VIII

Quantum modelling

François Dubois

On macroscopic intricate states

François Dubois[1]

Keywords: *fractaquantum hypothesis; embryogenesis; acupuncture*

We have proposed in 2002 the fractaquatum hypothesis, motivated by the following remark: Nature seems to be both fractal and quantum. The fractaquantum hypothesis expresses that the quantum approach is relevant for all the Atoms in Nature, whatever their size. In this contribution, an Atom is any natural element whose qualitative properties are modified at least in one subset if we divide it into two parts. The present idea follows in particular the work of Heisenberg (1969), and we refer also to Aerts et al. (2000) for the violation of Bell inequalities at a macroscopic scale.

We can explore the consequences of the fractaquantum hypothesis for simple associations and configurations. Let's bare in mind that the quantum association of two identical particles of spin equal to $1/2$ conducts to a boson of spin equal to zero. Therefore, the anti-symmetric association of two Atoms of matter naturally defines a new relation. Several examples have been proposed in a previous contribution (2006).

A natural question is associated with the fractaquantum hypothesis: does entangled matter, an astonishing quantum phenomenon observed by Aspect and his colleagues (1982) exist at a macroscopic scale? Is it possible to evidence that at macroscopic scale phenomena, a single Atom is present in two distinct loci? Observe here that a quantum computer is a true intricate Atom at a mesoscopic level. Independently, the quantum brain model of Vitiello (1995) shows another example of macroscopic entangled state: the water inside our brain could be macroscopic correlated matter.

With the help of the fractaquantum hypothesis, we propose to construct links between embryology and acupuncture. On the one hand, the embryogenesis. A complex organism such as a human being comes from a single cell that divides many times. At a certain step of embryogenic development, all the embryonic stem cells are a priori interchangeable. After a certain time, they are different, they have been specialized in specific functions in order to promote the development of the entire Atom to the superior scale. On the other hand, empirical knowledge developed in China 3000 years ago with acupuncture. Remember that acupuncture sets up some relations between the internal organs inside the body and some precise locations on the skin, the acupuncture points. Of course, these correlations essentially resist simple explanations through scientific approaches. We have suggested (Dubois, 2014) that relations between acupuncture points and internal organs could be the sign of the existence of macroscopic intricate state.

We can stylize the embryonic process as a binary planar graph. There is a complex dynamic between the two "daughters" cells of the same original cell. This representation provides a hierarchical breakdown of the body's cells. Thus the cells are labeled by a generation number. After several generations, they form a tree, in the sense of graph theory. This tree is interpreted in this contribution as an intricate macroscopic state.

In front of these ideas and strong assumptions, several objections and open questions are formulated. A main drawback of the fractaquantum hypothesis is the contradiction between quantum indiscernibility and macroscopic individuation. Nevertheless, we observe that the common points between two human persons for example are much more important than the different ones. The existence of medicine establishes empirically this fact! Moreover, the explicitation of genomic structure in each human cell shows that two human deoxyribonucleic acid sequences coincide up to 99.99%! There exist also circumstances in quotidian life when two persons can be exchanged. With the example of a crowd, one can consider that a new entity is created, where each human being is reduced to a very primitive component and develops intense internal relations (Freud, 1921). And in hierarchical organization, each Atom is a priori exchangeable and is reduced to specific function.

Understanding cell division is a key point concerning the possibility of intrication. In this case, there is not a single Atom composed by two components. At the contrary, a single cell $|+>$ interacts with its environment and generates a double cell that we can note as $|++>$. Could this absolutely non trivial biological process create a macroscopic entangled state? In this case, one can imagine that during cell division which is the primitive organism, especially during the first cell divisions of the blastocyst, a form of global unity remains persistent. From a mathematical point of view, quantum field theory could be introduced since the total mass of the referring element (one single cell that becomes two cells) is changed. Probably, the co-product of Hopf algebras is a good mathematical tool to describe the process of cell division $|+> \rightarrow |++>$.

Nevertheless, a main difficulty of micro-physics experiments is due to decoherence. When interacting with the environment, mesoscopic quantum systems lose quickly their coherence properties. In consequence, we think that the notion of intrication studied here is not strictly identical to the microscopic one. A generalization of the intrication concept, that we propose to name "weak intrication" in this contribution, has to be considered. What is essential for our purpose, the links between the related components of our body considered as an intricate macroscopic state are not explicit through space framework at a given time, but have to be searched in the time process of creating the entire structure.

Note

1 AFSCET (French Association for Systems Science) and Department of Mathematics, Conservatoire National des Arts et Métiers.

Bibliography

Aerts, D., Aerts, S., Broekaert, J., Gabora, L. (2000). The Violation of Bell Inequalities in the Macroworld. *Foundations of Physics*, 53, 1387–1414.

Aspect, A., Grangier, P., Roger, R. (1982). Experimental Realization of Einstein-Podolsky-Rosen-Bohm Gedankenexperiment: A New Violation of Bell's Inequalities. *Physical Review Letters*, 49(2), 91–94.

Dubois, F. (2002). Hypothèse fractaquantique. Res-Systemica, 2(21), www.res-systemica.org

Dubois, F. (2006). On Fractaquantum Hypothesis. Res-Systemica, 5(55), www.res-systemica.o

Dubois, F. (2014). Acupuncture, embryologie et états macroscopiques intriqués. Res-Systemica, 12(11), www.res-systemica.org

Freud, S. (1921). *Massenpsychologie und Ich-Analyse*. Internationaler Psychoanalytischer Verlag, Wien.

Heisenberg, W. (1969). *Das Teil and das Ganze, Grespräache im Umkreisis des atomphysik*. Piper Verlag, München.

Vitiello, G. (1995). Dissipation and Memory Capacity in the Quantum Brain Model. *International Journal of Modern Physics-B*, 9(8), 973–989.

A quantum-based model for textual semantics and information retrieval

Francesco Galofaro,[1] *Zeno Toffano,*[2]
and Bich-Liên Doan[3]

Keywords: *information retrieval; textual semantics; antonymy; hyponymy; entanglement*

1 Purpose

The present paper proposes a model aimed to detect and to typify semantic relations in each text. According to Semiotics, each text sets up its peculiar semantic relations. For example, according to the definition from Wikipedia, "sand is a naturally occurring granular material composed of finely divided rock and mineral particles". Thus, sand is a hyponym of rock: if "sand", then "rock". However, if we consider a different text, this relation can vary. For example, in Matthew 7:24–27, the wise man builds his home on the rock, while the foolish one builds on the sand. Because of the homology sand/rock = fool/wise – Marsciani (2012: 35–47) – in the Gospel "sand" becomes an antonym of "rock".

Considered as a process, the text progressively determines the type of semantic relation between terms and their strength. Greimas (1984: 105) first noticed an analogy between the process of stabilization of coherent semantic layers (*isotopies*) and the probabilistic models in use in Information Theory. The strength of semantic relations can be weighed: to this purpose, we will adopt a probabilistic point of view. We will retrieve information on semantic relations in the *context* of each word, where the semantic relation between terms should be stronger.

2 State-of-the-art

Our approach is inspired from Quantum Theory (QT), whose mathematical formulation has been applied to Information Retrieval by Van Rijsbergen (2004) to unify vector, logic, and statistical approaches. Language suggests that QT formalism could be adequate to language reality: meaning is noncommutative, and in different cases Leibniz's law seems to fail, as it happens in QT – Galofaro et al. (2016).

Barros et al. (2013) proposed to use the Hyperspace Analogue Language (HAL) method – Lund and Burgess (1996) – to transform a text into a matrix of word-vectors to keep track of the relations of proximity between the words inside the text; to sum all the vectors of the matrix, obtaining a normalized vector document $|\varphi>$; to express $|\varphi>$ through its components in two different bases provided by two query-world vectors u_A, u_B and their respective orthogonal vectors $u_{A\perp}$, $u_{B\perp}$. Thus, a Bell test – Bell (1964) – can be used to measure the degree of entanglement between u_A and u_B; the authors interpret the presence of a relation of entanglement as the presence of a semantic correlation. The present paper proposes a way to typify it, considering the actions of Pauli spin-operators on the document vector $|\varphi>$, and calculating their expectation values, in analogy to QT – see Susskind and Freedman (2014).

3 Approach

Let us say that we want to typify the semantic relation between "green" and "coal" in three different texts *t*, *u*, *w*, which represent three different possible types. We can set the width of the context we are interested in (*window*). To this purpose, we will use two *abstract machines* σ and τ, which are two linear operators: their input vector $|\varphi >$ is the text we are considering. The machine σ modifies the meaning of "green" in each context: the expected outcome of this work can be +1 when – in a certain context – the meaning "green" is modified and –1 when it is not. The machine τ applies the same transformation on the meaning of "clean"; in a similar way, we expect a value of +1 when τ changes the meaning of "clean" and –1 when it does not.

1   First, we consider a text $|u >$, whose topic is "New clean coal technologies". In this text, "green" and "coal" will partially share their meaning. Thus, if we apply the two semiotic machines to this text – $\sigma\tau|u>$ – when σ transforms the first meaning in a certain context, τ will transform the second meaning too. So, we expect that when the result of the transformation σ is +1, the outcome of τ will be +1 too; when the output of σ will be –1, the output of τ will be –1 too. If we multiply the two values (+1 × +1) or (–1 × –1) we will get +1. Thus, +1 will indicate the presence of a hyponymy: if "green" then "coal" or if "coal" then "green".

2   Now we consider a text *t* whose topic is "Pollution caused by coal power". In this report, "coal" is never "green", and vice-versa, since "coal" has an immanent semantic value of "dirty", whereas "green" has a semantic value of "clean". In semantic terms, the two words are *allotopic* – see Rastier (1997). Let us consider $\sigma\tau|u$: when σ transforms the meaning of "coal" in a certain context, τ will not, and vice versa. Thus, we expect that, when the output of σ is +1, the output of τ will be –1 and vice versa. If we multiply the two expected outcomes, we will always get –1. This value means that the correlation between "green" and "coal" are anti-correlated. They are *antonyms*: if "green" then not "coal", if "coal" then not "green".

3   The third text $|w >$ is about "Mining songs", collected by the musicologist Archie Green. Thus, in $\sigma\tau|u$, in some contexts the output of the two machines will be {+1, +1}, while in others it will be {+1, –1}, {–1, +1} {–1, –1}. If we consider many contexts, the average of the outcomes will be 0: this value indicates the absence of semantic correlations.

Since the co-occurrence of two words in a context is a matter of probability, a second goal of our model is to weigh semantic relations. In other terms, $\{+1, 0, -1\}$ are only the maximal values. In most cases, we expect outcomes between them (for example: $+0.98$; -0.1).

4　Example of the method

As we said, we deal with two word-vectors u_A, u_B representing two meanings ("green", "coal"). Vectors can't be identified with the respective "words": they represent their relations with their contexts, and this is a partial representation of their meaning.

When we apply our machines to each vector, the outcome can be a transformation (+) or not (−). Because of this, we deal with a four-state semantic space: ++, + −, − +, − −. From now on, the first sign of each couple will refer to the transformation of u_A, while the second will be referred to u_B. We can represent a document $|\varphi>$ through the respective probabilities ψ_{AB} to find each of the four states after a measure of the abstract machine:

a)　$|\varphi>=\psi_{++}|++>+\psi_{+-}|+->+\psi_{-+}|-+>+\psi_{++}|-->.$

Each ψ_{AB} is a complex number: it measures the amplitude of probability of an outcome.

Now we will construct our *abstract machines* to transform meaning. We will start from Pauli spin-operators. Together with the I operator, Pauli operators can represent every kind of semantic relations: they are a *quaternion*. Amazingly, the algebra of quaternions was considered the general model of linguistic relations according to the founder of contemporary linguistics, Ferdinand de Saussure (2002).

Among the others, the action of Pauli spin operator σ_x is to reverse the values of the input vector (quantum negation). In other terms, our abstract machine will consist of a simple quantum logic gate, applied to meaning.

To obtain a four-state operator, we multiply σ_x by the identity matrix I. Thus we construct two different four-space operators acting respectively on the u_A and on the u_B part of the vector document:

b)　$\sigma_{xA}=\sigma_x \otimes I_B$
c)　$\tau_{xB}=I_A \otimes \sigma_x$

σ_{xA} turns all the u_A-related values of $|\varphi>$ into their opposite value (+ becomes − and vice versa), while τ_{xB} turns all the u_B-related values of $|\varphi>$ into their opposite value.

Now we apply the method described in Barros et al. (2013) to the text t – the report on pollution caused by coal power – and we obtain:

d)　$|t>=0|++>+\dfrac{1}{\sqrt{2}}|+->-\dfrac{1}{\sqrt{2}}|-+>+0|-->$

$=\dfrac{1}{\sqrt{2}}(|+->-|-+>)$

Notice how d) is a case of a). In QT the expectation of an outcome is calculated applying the Born rule:

e) $\quad < t \mid \sigma_{xA} \tau_{xB} \mid t > = < t \mid \sigma_{xA} \tau_{xB} \mid \frac{1}{\sqrt{2}} (\mid +- > - \mid -+ >) +$

$< t \mid \tau_{xB} \mid \frac{1}{\sqrt{2}} (\mid -- > - \mid ++ >) =$

$= < t \mid \frac{1}{\sqrt{2}} (\mid -+ > - \mid +- >) = \frac{1}{2} (< +- \mid - < -+ \mid)(\mid -+ > - \mid +- >) =$

$= \frac{1}{2} (+0 - 1 - 1 + 0) = -1$

The result of –1 indicates that the two meanings u_A, u_B are anti-correlated: in our report on the pollution caused by coal power, "green" and "coal" are antonyms.

5 Discussion

The expectations on the outcome of a measure is related to σ_x, the Pauli operator we applied to the text. Should we use a different operator, for example σ_z, the expectation would be different. In other terms, we have to *transform* meaning to typify the semantic relation and to measure its strength. There's no way to "detect" meaning without transforming it, since meaning is a relational feature. For example, let us consider the feature of "being taller than x". Paul is taller than John, an is shorter than Dick, but he can't be "taller than" in absolute terms.

A second remark: in our example about |t >, we choose on purpose a maximally entangled state to exemplify our method. In real texts, it will be difficult to find this degree of anti-correlation. For example, antonyms do occur in the same contexts, as in the rhetoric figure of the *oxymoron*. However, its frequency is relatively scarce; otherwise it would not be incisive. For example, oxymora occur 3.9 times per 100 lines in Shakespeare's *Richard III* – Keller (2009: 278–279). We expect a remarkably lower value in more prosaic documents, such a report on pollution and coal power.

6 Implications

Deleuze and Guattari (2009) first proposed the notion of *abstract* machine, distinct from its physical implementation, and identified it with a probabilistic Markov chain. At the beginning of the Eighties, Umberto Eco generalized Ross Quillian's notion of *semantic memory* – Quillian (1968) – to interpretative semantics: Eco (1976, 1984). The HAL method converts a document into a semantic memory, which can be retrieved by a *semiotic machine*. Updating classic semiotic models to recent theories, operators will not only "decode", "read" or "interpret" meaning: they will rather *produce* it by reducing the indetermination of documents.

From this point of view, *signification is not the exclusivity of a human subject.* Criticizing the Turing test, the great semiotic and cybernetic scholar Jurij Lotman

(1979) wrote that if we identify "intelligent" and "human" we raise the failings of an actual form of intelligence to the rank of an essential characteristic. On this line, we consider meaning as a feature of social, artificial, and biological systems.

Notes

1 Politecnico di Milano.
2 CentraleSupélec, Gif-sur-Yvette.
3 CentraleSupélec, Gif-sur-Yvette.

Bibliography

Barros, J., Toffano, Z., Meguebli, Y. and Doan, B.L. (2013). *Contextual Query Using Bell Tests*. LNCS 8369, pp. 110–121. Berlin: Springer.

Bell, J. (1964). On the Einstein-Podoklsky-Rosen Paradox. *Physics*, 1(3), pp. 195–200.

Deleuze, G. and Guattari, F., (2009). *Anti-Oedipus: Capitalism and Schizophrenia*. London, UK: Penguin.

Eco, U. (1976). *A Theory of Semiotics*. Bloomington, IN: Indiana University Press.

Eco, U. (1984). *Semiotics and the Philosophy of Language*. Bloomington, IN: Indiana University Press.

Galofaro, F., Doan, B.L. and Toffano, Z. (2016). Linguistics and Quantum Theory: Epistemological Perspectives. 2016 IEEE International Conference on Computational Science and Engineering, IEEE International Conference on Embedded and Ubiquitous Computing, and International Symposium on Distributed Computing and Applications to Business, Engineering and Science, 2016 IEEE. DOI: 10.1109/.116 660 10.1109/CSE-EUC-DCABES.2016.257, pp. 600–607.

Greimas, A.J. (1984). *Structural Semantics: An Attempt at a Method*. Lincoln, NE: University of Nebraska Press.

Keller, S.D. (2009). *The Development of Shakespeare's Rhetoric: A Study of Nine Plays*. Tübingen: Franke Verlag.

Lotman, J. (1979). Culture as Collective Intellect and Problems of Artificial Intelligence. *Russian Poetics in Translation*, 6, pp. 84–96.

Lund, K. and Burgess, C. (1996). Producing High-Dimensional Semantic Spaces from Lexical Co-Occurrence. *Behavior Research Methods, Instruments & Computers*, 28(2), pp. 203–208.

Marsciani, F. (2012). *Ricerche semiotiche II: in fondo al semiotico*. Bologna: Esculapio.

Quillian, M. R. (1968). Semantic Memory. In M. Minsky (Ed.), *Semantic Information Processing* (pp. 227–270). Cambridge, MA: MIT press.

Rastier, F. (1997). *Meaning and Textuality*. Toronto: University of Toronto Press.

Saussure, F. de (2002). *Writings in General Linguistics*. Oxford: Oxford University Press.

Susskind, L. and Freedman, A. (2014). *Quantum Mechanics: The Theoretical Minimum*. London, UK: Penguin.

Van Rijsbergen, C.J. (2004). *The Geometry of Information Retrieval*. Cambridge, UK: Cambridge University Press.

Websites

https://en.wikipedia.org/wiki/Sand

Social laser model – Stimulated Amplification of Social Actions

From color revolutions to Brexit and Trump

Andrei Khrennikov[1]

Keywords: *stimulated amplification of social actions; social energy; Bose-Einstein and Fermi-Dirac statistics; quantum field theory; information field*

1 Outline

This paper is devoted to analysis of assumptions on the information field and human gain medium providing the possibility of creation of Stimulated Amplification of Social Actions (SASA) – a kind of social laser. The model and its analysis are based on the formalism of quantum thermodynamics and field theory (applied outside of physics). SASA is the hot topic in socio-political studies. Evidence of such amplifications is rapidly accumulating, from color revolutions to such democratically structured protest actions as Brexit and the recent election of Donald Trump as the president of the USA. These studies are characterized by diversity of opinions and conclusions. The presented quantum-like model provides the consistent operational model of this complex socio-political phenomenon. This is the conceptual paper aimed at attracting attention of other researchers (both from physics and socio-political science) to the problem of modeling of SASA.

2 State of art

Recently the formalism of quantum mechanics started to be widely explored to describe biological, cognitive, psychological, and socio-political phenomena, see, e.g., the monographs of Khrennikov (2010, 2015). This formalism provides the consistent probabilistic picture of observations performed for systems exhibiting (statistically) quantum-like features: from cells, animals, humans to societies and ecosystems. In particular, in recent papers (Khrennikov, 2015, 2016) there was presented the model of Stimulated Amplification of Social Actions (SASA) describing a kind of a social laser device. In this note we plan to briefly present assumptions on social systems (human gain-medium) which make possible functioning of social lasers.

3   Social lasers in action

We start with color revolutions which are the most sharp exhibitions of SASA. The series of color revolutions started in the territory of the former Soviet Union and Balkans, for example, Yugoslavia's Bulldozer Revolution (2000), Georgia's Rose Revolution (2003), Ukraine's Orange Revolution (2004), Kyrgyzstan's Tulip Revolution (2005), Belarus's Jeans Revolution (2006), Moldova's Grape Revolution (2009). Further examples are presented by color revolutions outside of the former Soviet Union and Balkans, e.g., Lebanon's Cedar Revolution (2005), Kuwait's Blue Revolution (2005), Myanmar's Saffron Revolution (2009). Some researchers trace the root of color revolutions to the 1986 People Power Revolution (also known as the "Yellow Revolution") in the Philippines. Recent years were characterized by the wave of color revolutions at the Middle East: Tunisia's Jasmine Revolution (2011), Egypt's Lotus Revolution (2011), Muslim military uprisings in Iraq, Libya, Syria. We can also mention Iran's Green Revolution (2009) and China's Jasmine Revolution (2011), Ukraine's second color revolution Maidan revolution (2014) following the civil war in Ukraine.

These revolutions were the subject of numerous studies and publications in political and social science; see (Khrennikov 2015, 2016) for references. The color revolutions are definitely the new socio-political phenomenon. However, by recognizing this, different authors treat them in very different ways: from folk uprisings against corruption to a new form of warfare and terrorism. We point out that the diversity of opinions and conclusions is really amazing. For the moment, there is no consistent and commonly acceptable theory of this phenomenon.

In spite of differences in the interpretations, the majority of authors point to common features of these socio-political events: no clear programs and aims; no real leaders (compatible with the leaders of "real revolutions" such as Marx, Lenin, Trotsky and Stalin, Orozco, Villa and Zapata, Ataturk, Mussolini, Hitler, Mao, Castro, Che Guevara, . . .); no ideology (as expressed in Marx's "Capital", Lenin's "Imperialism as the last stage of capitalism", Trotsky's "The Permanent Revolution", or Hitler's "Mein Kampf"); "quick relaxation" – explosion-like characteristics of events.

Similar features have happened in recent events inside the Western democratic system: the Greek bailout referendum (2015), Brexit (2016), and even the election of Donald Trump.

(The latter differs from the previously discussed events, in particular, by the presence of the "leader of the movement". Nevertheless, some commonality can be found: first of all the protest character of this election – protest against the corrupted system.)

As was pointed out, in spite accumulating evidence of SASA (throughout the world, in various social, political, and economic contexts) an adequate theory of this socio-political phenomenon has not yet been created. In this situation, it is natural to explore the methodology and formalism of quantum mechanics. The latter is treated (Khrennikov, 2010) as the operational model describing the outputs of observations. It cannot "explain" behavior of quantum systems (electrons or photons). But it can present the consistent picture of possible observations. This picture is based on representation of states of systems by vectors in complex Hilbert space and observables by Hermitian operators. Thus we do not try to explain the essence of a variety of events related to SASA. We want to elaborate the quantum-like operational formalism which can be used to model (at least qualitatively) behavior of social systems leading to aforementioned social phenomena, from color revolutions to Brexit and Trump's presidency.

4   Features of social systems and information field making possible the functioning of social lasers

Social Laser: basic components: Quantum information field, its excitations – quanta of information; gain medium: humans.

Quantum information field: addivity of social energy: energy transmitted to a person by excitations of the information field is integrated; indistinguishability of information excitations: the content of messages is not analyzed, we simply absorb social energy carried by them; Bose-Einstein statistics of excitations: messages with totally different context can carry the same portion of social energy.

Human gain medium: the discrete spectrum of the social energy; distribution of the social energy in individual's state space follows Fermi-Dirac statistics; an individual can absorb and emit only discrete portions of the social energy (adapted to the differences between the energy levels); individuals belonging to gain medium have the same spectrum; stimulated emission of social energy.

Note

1 International Center for Mathematical Modeling in Physics and Cognitive Sciences, Linnaeus University, Växjö, S-35195, Sweden; email: Andrei.Khrennikov@lnu.se.

Bibliography

Khrennikov, A. (2010), *Ubiquitous Quantum Structure: From Psychology to Finances*, Springer, Berlin-Heidelberg.
Khrennikov, A. (2015), Towards information lasers, *Entropy*, 17, No. 10, 6969–6994.
Khrennikov, A. (2016), Social laser: Action amplification by stimulated emission of social energy, *Philosophical Transactions of the Royal Society*, 374, No 2054, 20150094.

Quantum modelling of the learning curve

Clas-Otto Wene[1]

Keywords: *learning curves; operational closure; entropy production; energy policy*

1 Background and objective

Learning systems (LS) can be described as operationally closed non-trivial machines (Wene, 2007). The learning curve describes the performance of such systems in a competitive environment. Following the machine metaphor, performance is measured as the ratio between output and input.

Operational closure is the key rationale for quantum modelling of learning systems and their performance. In the same way as the particle-wave dualism in quantum mechanical systems, operational closure severely reduces the degrees of freedom of the system and results in eigenstates and eigenbehaviour. The view of LS as a quantum system is consistent with the fractaquantum hypothesis (Dubois, 2002).

Wene (2007) calculated eigenvalues based on the observed functional form of the learning curve and from these eigenvalues the learning rates were deduced. Double closure (von Förster, 2003) manages perturbations and disperses learning rates around eigenvalues (Wene, 2010). The approach reproduces observed distributions of learning rates (Wene, 2010, 2011). The quantum modelling, however, is limited to deriving learning rates by positing a state function and operator to describe the learning system and measurement of eigenvalues. Wene (2013) reproduced the functional form from findings in macroscopic non-equilibrium thermodynamics but could not tie this result to the quantum modelling of eigenvalues and learning rates to provide a comprehensive approach for learning curves.

The objective of this project is to develop a comprehensive explanation of the learning curve, functional form and distribution of learning rates, based on quantum modelling. The approach is inspired by the treatment of entropy proposed in Prigogine (1980). The purpose of this first paper is to set up the quantum model for the LS and explore the formalism for deriving learning rates and the functional form of the learning curve.

2 Approach

The insight that the learning system works as a steady-state non-equilibrium thermodynamic system guides the approach. The task is to find the quantum dynamical counterpart. Following Wene (2007, 2010) the state of the unperturbed learning system in a competitive environment is characterized by

$$| \ LS \ > \ = \ 2^{-1/2}\begin{bmatrix} i \\ 1 \end{bmatrix}$$

The learning system is depicted as a superposition of two equally probable states, which, e.g., in the OADI-SMM model (Kim, 1993) could be representing a reflective state, A+D+SMM, and a physical action state, O+I. In the non-trivial machine, these states correspond to Computation and Production, respectively. The quantum mechanical counterpart would be a spinor. The state |LS> provides the density matrix for the learning system. This matrix together with an entropy operator reflecting that the entropy production is constant in the systems eigen-time provide the starting point for deriving learning rates and the form of the learning curve.

3 Applications

Learning curves are widely used in industry (Jaber, 2010) and since the 1990s also for developing energy policy (IEA, 2000; Junginger et al., 2010). Many efforts have been made to understand the phenomenon (for ref. see e.g., Wene, 2015, 2016) but so far no comprehensive theory has emerged. The lack of a theoretical foundation has raised serious doubts on the reliability of the curves in designing and applying strategies and policies (Nordhaus, 2014). A comprehensive theoretical understanding will settle such issues, which is important to ensure legitimacy and efficiency of government deployment programmes based on learning curve analysis. Learning curves pervade all aspects of competitive industrial activities and thus raise the question if the present global trend towards deregulated markets threatens the long-range effectiveness of the energy system.

Note

1 Lund and Chalmers University of Technology.

Bibliography

Dubois, F. (2002), "Hypothese fractaquantique", *Res-Systemica*, Vol. 2(21). Retrieved from: www.res-systemica.org
IEA (2000). *Experience Curves for Energy Technology Policy*, Paris: International Energy Agency/Organisation for Economic Co-operation and Development.

Jaber, M.Y. (2010), *Learning Curves: Theory, Models, and Applications*, Boca Raton, FL: CRC Press.

Junginger, M., van Sark, W., and Faaij, A. (2010), *Technological Learning in the Energy Sector: Lessons for Policy, Industry and Science*, Cheltenham: EdwardElgar.

Kim, D.H. (1993), "The link between individual and organizational learning", *Sloan Management Review*, Vol. 35, pp. 37–50.

Nordhaus, W.D. (2014), "The perils of the learning model for modeling endogenous technological change", *Energy Journal*, Vol. 35(1), pp. 1–13.

Prigogine, I. (1980), *From Being to Becoming: Time and Complexity in the Physical Sciences*, New York, NY: W.H. Freeman and Company.

von Förster, H. (2003), *Understanding Understanding*, New York, Berlin and Heidelberg: Springer.

Wene, C.-O. (2007), "Technology Learning Systems as non-trivial machines", *Kybernetes*, Vol. 36(3/4), pp. 348–363.

Wene, C.-O. (2010), "Adaptation in Technology Learning Systems", in: *Proceedings of the 11th IEAA European Conference*, Vilnius, Lithuania. Retrieved from: www.wenergy.se/pdf/Wene–IAEE2010.pdf

Wene, C.-O. (2011), "Energy technology learning: Key to transform into a low: Carbon society", *Climate Change: Research and Technology for Adaptation and Mitigation*, Juan Blanco and Houshang Kheradmand (Eds.), ISBN: 978-953-307-621-8, Rijeka: InTech. Retrieved from: www.intechopen.com

Wene, C.-O. (2013), "Learning Curves tracing the optimal path for Technology Learning Systems", in: *Proceedings of the 13th IAEE European Conference*, Düsseldorf, Germany, 18–21 August. Retrieved from: www.iaee.org/en/publications/proceedingssearch.aspx

Wene, C.-O. (2015), "A cybernetic view on learning curves and energy policy", *Kybernetes*, Vol. 44(6/7), pp. 852–865.

Wene, C.-O. (2016), "Future energy system development depends on past learning opportunities", *WIREs Energy Environ*, Vol. 5(1), pp. 16–32. doi: 10.1002/wene.172

Eigenlogic

Interpretable quantum observables with applications to fuzzy behavior of vehicular robots

Zeno Toffano[1] and François Dubois[2]

Keywords: *quantum agents; multivalued logic; fuzzy logic; robots; Braitenberg vehicles*

1 Outline

This work proposes a formulation of propositional logic, named Eigenlogic, using quantum observables as propositions. The eigenvalues of these operators are the truth-values and the associated eigenvectors the interpretations of the propositional system. Fuzzy logic arises naturally when considering vectors outside the eigensystem, the fuzzy membership function is obtained by the Born rule of the logical observable.

This approach is then applied in the context of quantum robots using simple behavioral agents represented by Braitenberg vehicles. Processing with non-classical logic such as multivalued logic, fuzzy logic and the quantum Eigenlogic permit to enlarge the behavior possibilities and the associated decisions of these simple agents.

2 History: Boole: "0" and "1"; von Neumann: quantum projections as propositions

George Boole gave a mathematical symbolism through the two numbers {0,1} representing *resp.* the "false" or "true" character of a proposition (Boole 1847). An *idempotent* symbol x verifies the equation: $x^2 = x$, which admits only two possible values: 0 and 1. This equation was considered by Boole as the "fundamental law of thought", the associated formulation for logic is operational as pointed out in (Halperin 1981) because x acts as a selection operator on classes. As will be emphasized here the algebra of idempotent symbols can also be interpreted as an algebra of commuting projection operators and used for developing propositional logic in a quantum linear-algebraic framework (Toffano 2015).

John von Neumann (Von Neumann 1932) considered projectors as propositions; he also introduced the formalism of the density matrix in quantum mechanics where a pure quantum state $|\psi>$ is be represented by a rank-1 projection operator: $\hat{\rho} = |\psi\rangle\langle\psi|$.

3   Eigenlogic: quantum observable logic

A projection operator in Hilbert space is associated to a logical proposition; the operator being Hermitian it has the properties of a *quantum observable* and is considered here a logical observable. This view is named *Eigenlogic* (Toffano 2015; Dubois and Toffano 2017) and can be summarized:

> eigenvectors in Hilbert space ⇔ interpretations (atomic propositional cases)
> logical observables ⇔ logical connectives
> eigenvalues ⇔ truth values

One can express the logical observable as a development:

$$F = f(0)\Pi_0 + f(1)\Pi_1 = diag\left(f(0), f(1)\right)$$

the terms of the development are the 2-dimensional rank-1 projectors Π_0 and Π_1, the cofactors $f(0)$ and $f(1)$ are the eigenvalues and correspond to the truth-values $\{0,1\}$ of the logical connective. This allows to generate four one-argument logical observables:

$$F_A = 0\Pi_0 + 1\Pi_1 = \Pi, F_{\bar{A}} = 1\Pi_0 + 0\Pi_1 = I_2 - \Pi,$$
$$F_\varnothing = 0\Pi_0 + 0\Pi_1 = \mathbb{O}_2, F_U = 1\Pi_0 + 1\Pi_1 = I_2$$

Then the two-argument logical observables can be developed on the corresponding four rank-1 projectors:

$$F_2 = diag\left[f(0,0), f(0,1), f(1,0), f(1,1)\right]$$

There are 2^{2^n} logical connectives for a n-argument (*arity*) system. For n = 2 this gives the 16 binary logical connectives e.g.: $AND, OR, XOR, \rightarrow, \leftrightarrow, \ldots$

The complete orthonormal basis for a two input quantum states are: |00>, |01>, |10> and |11>. These state vectors are the eigenvectors of the logical observables and correspond to *interpretations* of the logical system. What happens when the quantum state is not one of the eigenvectors of the logical system?

In quantum mechanics one can always express a state vector as a combination on a complete orthonormal basis. In particular on the canonical eigenbasis of the logical observable family:

$$\left|\psi\right\rangle = c_{00}\left|00\right\rangle + c_{01}\left|01\right\rangle + c_{10}\left|10\right\rangle + c_{11}\left|11\right\rangle$$

When only one of the coefficients is not zero, then one has the case of a determined interpretation for the proposition.

Fuzzy logic (Zadeh 1965) deals with truth-values that can take values between 0 and 1, so the truth of a proposition can lie between "completely true" and "completely false". When more than one coefficient in the development of |ψ> is non-zero one can give a "fuzzy" interpretation, and the quantum state $|\psi\rangle$ can be considered as a *quantum superposition of interpretations*.

For a projective observable F measured in the context of a quantum state, $|\psi\rangle$ the mean value (Born rule) gives directly a probability measure by:

$$p_{|\psi\rangle} = <\psi\,|\,F\,|\,\psi> = Tr(\rho \cdot F) \text{ with } \rho = |\,\psi\psi\,| \text{ the } density\ matrix$$

The mean value of the logical projector observable F is thus a fuzzy measure of the truth of a logical proposition in the form of a fuzzy membership function μ.

For one-argument, an arbitrary 2-dimensional quantum state is:

$$|\phi\rangle = \sin\alpha|0\rangle + e^{i\beta}\cos\alpha|1\rangle$$

where $\alpha = \theta/2$ and $\beta = \theta/2$ are real numbers and these angles can be represented on the *Bloch sphere* (see Figure 104.1). The quantum mean value of the logical projector observable $\mathbf{A} = \mathbf{\Pi}$ is then given by:

$$\mu(a) = \langle\phi|\mathbf{\Pi}|\phi\rangle = \cos^2\alpha \qquad \text{representing a probability.}$$

The recently observed revival of interest in applying *multivalued logic* to the description of quantum phenomena is closely connected with fuzzy logic. Multivalued logic is of interest to engineers involved in various aspects of information technology and has a long history in *computer aided design*.

The total number of logical connectives for a system of m values and n arguments is the combinatorial number: m^{m^n}. For a 3-valued 2-argument system: $3^{3^2} = 19683$. Showing that the possibilities of new connectives becomes intractable, but some special ones play an important role.

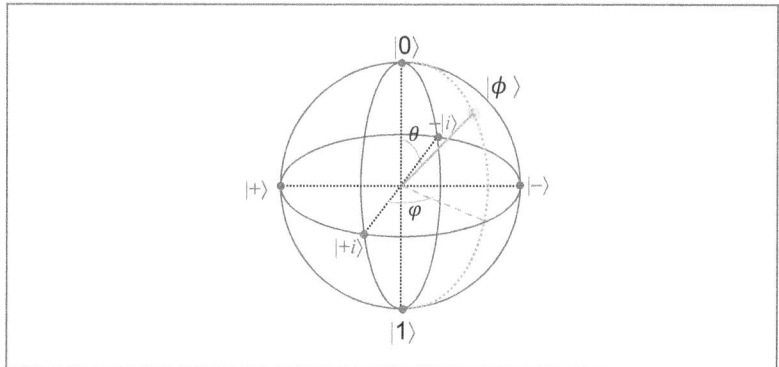

Figure 104.1 Bloch sphere with the general qubit quantum state $|\phi\rangle$ characterized by angles θ and φ

Multivalued logic is naturally associated to quantum angular momentum: the eigenvalues of the z component of the angular momentum observable for $l = 1$ is given by:

$$L_z = \hbar \Lambda = \hbar \, diag \left(+1, 0, -1 \right)$$

with the associated logical truth-values:

"false" $F \equiv +1$, "neutral" $N \equiv 0$, "true" $T \equiv -1$

The corresponding logical observables can be expressed as spectral decompositions over the three rank-1 projectors given by:

$$\Pi_{+1} = \frac{1}{2} \Lambda \left(\Lambda + 1 \right) \quad \Pi_0 = I - \Lambda^2 \quad \Pi_{-1} = \frac{1}{2} \Lambda \left(\Lambda - 1 \right)$$

4   Eigenlogic applied to quantum robot Braitenberg vehicles

Valentino Braitenberg was a Cyberneticist and former director at the Max Planck Institute for Biological Cybernetics in Tübingen. A Braitenberg vehicle (BV) (Braitenberg 1986) is an agent that can autonomously move around based on its light sensor inputs. Depending on the sensor-motor wiring, it appears to achieve certain situations and to avoid others, changing course when situation changes. Several elementary vehicles can be considered:

- BV-2a (**fear**): turns away from the light if one sensor is activated more than the other.
- BV-2b (**aggress**): when the light source is placed near either sensor, the vehicle will go toward it.
- BV-3a (**love**): will go until it finds a light source, then slows to a stop.
- BV-3b (**explore**): goes to the nearby light source, but keeps an eye open to sail to a stronger source.

The vehicles can be designed according the *law of uphill analysis and downhill invention* (Braitenberg 1986), according to which it is far easier to create machines from simple structures that exhibit complex behavior than it is to try to build their structures from behavioral observations.

Practical realization of *BVs* uses generally simple Boolean logic. It is interesting to extend the design to multivalued, fuzzy or probabilistic logic and even quantum logic.

Paul Benioff (Benioff 1998) introduced the theoretical principle of a *quantum robot* as a first step towards a quantum mechanical description of systems that are aware of their environment and make decisions.

Table 104.1 Quantum logical observables for BV actuators

Braitenberg Vehicle\Actuator	ML	MR
BV2a (*fear*)	−Z	−Y
BV2b (*aggress*)	−Y	−Z
BV3a (*love*)	+Z	+Y
BV3b (*explore*)	+Y	+Z

The research team of Marek Perkowski has designed robots, based on *BV*s, using quantum gates and also introducing control, fuzziness and higher than binary valued logic (Raghuvanshi and Perkowski 2010). The potential applications presented here are inspired from these researches.

Considering the binary alphabet {+1, −1} leads to analogies with the vehicle's behavior; it has a natural correspondence with inhibition (negative: −) and excitation (positive :+). In this way the *BV*s sensors *SL* and *SR* (see Figure 104.2) represent the inputs and the actuators *ML* and *MR* are represented by the 2-argument dictators $Z = \mathrm{diag}(1,1,-1,-1)$ and $Z = \mathrm{diag}(1,-1,1,-1)$.

The different possible combinations are given in Table 104.1.

In another configuration the sensors *SL* and *SR* can be represented by tri-valued 2-argument dictators U and V. For this purpose it is interesting to use the three positive values {0,1,2} with the following interpretation:

"no light" ≡ 0 "weak-level light" ≡ 1 "high-level light" ≡ 2

Involved behaviors can thus be described using the *Min* and *Max* connectives. From the formulation given above based on the classical interpolation methods (Dubois and Toffano 2016) it is easy to derive, the expressions for the alphabet {0,1,2}, giving the following logical observables:

$$Min_{3\{0,1,2\}} = U + V + U^2V + V^2U - \frac{1}{2}U^2V^2 - \frac{5}{2}UV$$
$$= diag(0,0,0,0,1,1,0,1,2)$$
$$Max_{3\{0,1,2\}} = \frac{5}{2}UV + \frac{1}{2}U^2V^2 - U^2V - V^2U = diag(0,1,2,1,1,2,2,2,2)$$

One could also combine the multivalued operators above in a fuzzy logic configuration, i.e. when using input states that are not eigenstates. Using fuzzy Eignelogic one can calculate the fuzzy membership functions for complement: $\mu(\bar{a})$, conjunction $\mu(a\wedge b)$ and disjunction $\mu(a\vee b)$. These functions could be implemented in the processor preceding the actuators *ML* and *MR* (see Fuzzy-Demux in Figure 104.2).

Fuzzy membership for logical implication (→) corresponding to motion-decision such as "step backwards", "step forwards", "turn left" and "turn right" can

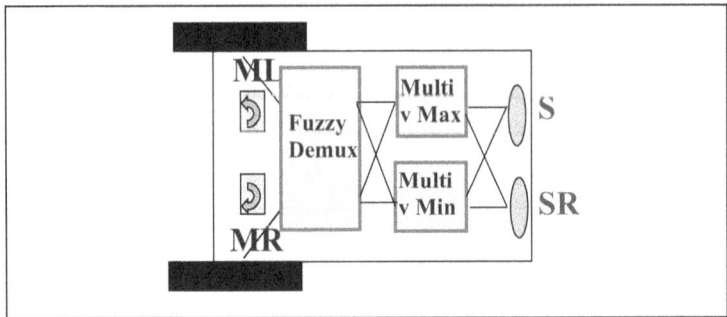

Figure 104.2 Principle of a multivalued fuzzy quantum Braitenberg vehicle

also be evaluated in this model. An idea presented in (Raghuvanshi and Perkowski 2010) consists in mapping emotions onto the Bloch sphere (Figure 104.1); this quantum fuzzy model uses conjunction, disjunction and complement operations. The internal emotional state could be described now by a quantum circuit built from counterparts of fuzzy operators. For instance, all kinds of quantum phase gates could be used combined, associated with recent techniques of *quantum tomography*.

Notes

1 CentraleSupélec, Telecom Dep., Gif-sur-Yvette, Laboratoire des Signaux et Systèmes, L2S – UMR8506-CNRS, Université Paris-Saclay, France.
2 AFSCET (French Association for Systems Science) and Department of Mathematics, Conservatoire National des Arts et Métiers, Paris, France. [WOSC, Rome, January 25–27 2017,subm: March 29, 2017.]

Bibliography

Benioff, P., (1998), "Quantum Robots and Environments", *Physical Review A*, Vol. 58, No. 2, 893–904.

Boole, G., (1847), *The Mathematical Analysis of Logic: Being an Essay to a Calculus of Deductive Reasoning*, Ed. Forgotten Books.

Braitenberg, V. (1986), *Vehicles: Experiments in Synthetic, Psychology*, MIT Press, Cambridge, USA.

Dubois, F. and Toffano, Z. (2016), "Eigenlogic: A Quantum View for Multiple-Valued and Fuzzy Systems", *Quantum Interaction 2016: Lecture Notes in Computer Science*, Vol. 10106, Springer, 239–251, 2017, *arXiv*:1607.03509.

Halperin, T. (1981), "Boole's Algebra Isn't Boolean Algebra: A Description Using Modern Algebra, of What Boole Really Did Create", *Mathematics Magazine*, Vol. 54, No. 4, 172–184.

Raghuvanshi, A. and Perkowski, M. (2010), "Fuzzy Quantum Circuits to Model Emotional Behaviors of Humanoid Robots", *Evolutionary Computation (CEC), IEEE Congr. on Evolutionary Computation*, 18–23 July, 1–8.

Toffano, Z. (2015), "Eigenlogic in the Spirit of George Boole", *ArXiv*:1512.06632.

Von Neumann, J. (1932), *Mathematical Foundations of Quantum Mechanics*. Investigations in Physics, vol. 2, Princeton University Press, Princeton, (transl. 1955).

Zadeh, L.A. (1965), "Fuzzy Sets", *Information and Control*, Vol. 8, No. 3, 338–353.

Theme IX

Reflexivity, second order science and context

Stuart A. Umpleby and Vladimir Lepskiy

The evolution of cybernetics

Classical, non-classical and post-non-classical science[1]

Vladimir Lepskiy[2]

Keywords: *third-order cybernetics; control; classical, non-classical, post-non-classical rationality; reflection; subject-focused approach; self-developing reflexive – active environment*

1 Introduction

In recent decades the Russian philosophy of science has recognized three stages in the development of science (classical, non-classical, and post-non-classical), which were proposed by V. S. Stepin (2005). If we ignore these changes, we risk losing the sight of basic shifts in the sciences of control and in the evolution of cybernetics (Lepskiy 2015). The revision of the general world scientific views was followed by changes in the standard structures of research and in the philosophical foundations of science.

Each of the three science development stages is associated with the dominance of one of three types of scientific rationality – classical, non-classical, and post-non-classical rationality. It is significant to note that the scientific rationalities are not alternative ones. Every subsequent rationality has its own specifics but includes also the previous types of rationality. Post-non-classical scientific rationality integrates all three types of scientific rationality.

2 Configurator of the philosophical and methodological cybernetics analysis

For the analysis of the evolution of cybernetics, the idea of the system configurator offered by V. A. Lefebvre (1967) is used. The idea is that the researcher makes a selection of the most significant points of view on the object of research. The object is projected on several screens. The screens are connected with each other. The researcher can correlate various points of view on an object.

We define structuring positions of the configurator in the context of the traditional points of view of the scientific analysis:

- philosophical level (science philosophy – basic types of scientific rationality);
- methodological level (basic paradigms and objects of a research, methodology of scientific approach);
- theoretical level (the basic providing areas of knowledge);
- methodical level (basic methods, models, technologies).

The configurator for the analysis of cybernetics evolution is presented in Table 105.1 and Table 105.2.

Table 105.1 The generalized results of the philosophical and methodological analysis of the evolution of cybernetics (philosophical, methodological, and theoretical levels)

Philosophical level		Methodological level			Theoretical level
Type of scientific rationality	*Basic philosophical approaches*	*Basic paradigms*	*Basic objects of control. The dominating types of activity*	*Basic scientific approaches*	*Basic areas of knowledge*
Classical	Positivism	"Subject-object"	Complex system Activity in activity	Activity approach Monodisciplinary approach	Cybernetics
Non-classical	Philosophical constructivism	"Subject-subject"	Active systems Communicative activity	Subject-activity approach Interdisciplinary approach	Second-order cybernetics
Post-non-classical	Humanistic interpretation of philosophical constructivism	"Subject-meta-subject" "Self-developing reflexive-active environment"	"Self-developing environments" Reflexive activity	Subject-focused approach Transdisciplinary approach	Third-order cybernetics (post-non-classical cybernetics of self-developing reflexive-active environments)

Source: Authors' elaboration

Table 105.2 The generalized results of the philosophical and methodological analysis of cybernetics evolution (methodical level)

Type of scientific rationality	Methodical level				
	Basic types of control	Basic models	Basic mechanisms and technologies	Basic ideas of knowledge	The dominating ethical regulators
Classical	Classical control	Analytical (mathematical)	Feedback Hierarchical structures	Information	Ethics of domination of target orientation
Non-classical	Reflexive control, manipulations, etc.	Imitating models, business games, etc.	Communication relations, reflexive processes Network structures	Knowledge tied to subjects Personal (hidden) knowledge	Communicative ethics
Post-non-classical	Environmental control	Models of self-developing reflexive-active environment	Control through self-developing environments	Active knowledge Virtual immortality	Ethics of strategic subjects

Source: Authors' elaboration

3   Classical scientific rationality

Classical scientific rationality, focusing attention on the object, seeks to reduce the research to a theoretical explanation and a description of everything that concerns the subject, means, and operations of activity. Such reduction is considered to be a necessary condition of acquiring objective and true knowledge of the world. The traditional idea of control was born in the context of classical science, and it was restricted to a "subject-object" paradigm. First-order cybernetics is "cybernetics of observed systems".

To model control processes various approaches are used: functional, function-structural, axiomatic, informational, operations research, classical game theory, etc. Within this "subject-object" paradigm, the main mechanisms of control are negative and positive feedbacks. The philosophical foundations of the first-order cybernetics are mainly formed within positivism. Reflexive activity is limited to the framework of an activity approach. The dominant ethical representations are determined as target ethics.

4   Non-classical scientific rationality

The non-classical type of scientific rationality takes into consideration interactions between knowledge about object and character of means and operations of activity. Nevertheless, interactions between scientific and social values and the purposes of inquiry remain outside of a scientific reflection, though implicitly they determine the nature of knowledge: what exactly and in what way we single out and grasp something in the world. The results of the scientific research are influenced by comprehending the correlation among the explained characteristics of objects and the features of means and scientific activity operations. The problem "means determine object" is in the center of attention. In such relationships, the researcher becomes the only person in the system of reflexive relations.

This research created the basis for a transition from a paradigm of "subject-object" to a paradigm of "subject-subject". An increase in the role of the subject leads to the need for a revision, leading away from the activity approach domination. In our opinion, adequate to the specifics of a non-classical scientific rationality is the subject-activity approach. While the basis of the classical scientific rationality is activity in action, non-classical rationality along with it includes other forms of activity, in particular, communicative and reflexive activity. Focusing on the activity of the controlled object predetermined the development of second-order cybernetics – cybernetics of observing systems (Foerster 1974).

The philosophical foundations of second-order cybernetics are formed generally within philosophical constructivism (Umpleby 2014). At the same time, the role of reflexive activity has sharply increased. The communicative reflexive activity becomes the leading concern (Müller 2015). An interdisciplinary approach becomes the basic scientific approach. The transition in control from a "subject-object" paradigm to a "subject-subject" paradigm has led to the formation of new types of control: reflexive control, information control, control of active systems. The emphasis on communicative ethics has become an integral part of the research (Lefebvre 2001).

5 Post-non-classical scientific rationality

The post-non-classical type of scientific rationality broadens the field of the reflection on scientific activity. It takes into consideration the correlation of the acquired knowledge about an object not only with the features of activity means and operations but also with valuable and target structures. At the same time, the connection of inner-scientific goals with extra-scientific ones, social values and aims are explicated. Moreover, the problem of their correlation with the comprehension of valuable, and target orientations of the scientific activities subject is also solved.

The paradigm "Subject-Self-Developing Reflexive-Active System (environment)" (Lepskiy 2010) becomes a key paradigm of control and cybernetics. It is important to note that the environment is considered to be the meta-subject. As a result, the paradigm can be presented as "Subject-Meta-Subject". In self-developing reflexive-active environments, new opportunities for convergence of natural and artificial intelligence emerge. This paradigm can be applied to the organization of active knowledge, for reflexive mechanisms of management of complexity, etc. Formation of this paradigm is inseparably linked with the formation of the subject-focused approach (Lepskiy 1998). It is important to note that the radicalism of philosophical constructivism becomes "softer". The influence of communicative processes on restricting the freedom of subjects is taken into account.

In realizing this function, culture is of particular importance. A transdisciplinary approach becomes basic for post-non-classical scientific rationality. The dominant concern is the ethical treatment of the subjects included in any meta-subject (a family, group, organization, country, etc.), the scientist's identification of himself/herself with this meta-subject and regulating interaction while taking into account his/her influence on the meta-subject. Now formation of scientific ensuring control and cybernetics in the context of post-non-classical rationality has begun. In our opinion, an issue of formation of post-non- classical third-order cybernetics is realized. Thus, the main thesis would be "from Observing Systems to Self-Developing Reflexive-Active Environments".

6 Conclusions

The philosophical and methodological analysis of cybernetics evolution proved its connection with the development of scientific rationality (classical, non-classical, post-non-classical). The classical scientific rationality is correlated with the first-order cybernetics. The non- classical scientific rationality is connected with the second-order cybernetics. The cybernetics of self-developing reflexive-active environments (the third-order cybernetics) corresponds to the post-non-classical scientific rationality. The basic characteristics of the third-order cybernetics are described.

Notes

1 This work is partly funded by Russian Science Foundation, project 17-18-01326.
2 Institute of Philosophy of the Russian Academy of Sciences.

Bibliography

Foerster, H. von. (1974) *Cybernetics of Cybernetics*. Urbana, IL: University of Illinois.
Lefebvre, V.A. (1967) *The Conflict Structures*. Moscow, USSR: Vysshaya shkola. (in Russian).
Lefebvre, V.A. (2001) *Algebra of Conscience*. New York: Springer.
Lepskiy, V. (1998) *The Concept of Subject: Oriented Computerization of Control Activity*. Moscow, Russia: Institute of Psychology, RAS. (in Russian).
Lepskiy, V. (2010) *Reflexive and Active Environments of Innovative Development*. Moscow, Russia: Kogito Center Publishing House. (in Russian).
Lepskiy, V. (2015) *Evolution of Concepts about Control (Methodological and Philosophical Analysis)*. Moscow, Russia: Kogito Center Publishing House. (in Russian).
Müller, K.H. (2015) The Multiple Faces of Reflexive Research Designs. *Systemics, Cybernetics and Informatics*, 13 (6), 87–98.
Stepin, V. (2005) *Theoretical Knowledge*. Springer Verlag GMBH.
Umpleby, S.A. (2014) Second Order Science: Logic, Strategies, Methods. *Constructivist Foundations*, 10 (1), 16–23.

Reflexive models of complex activity

Mikhail V. Belov[1] and Dmitry A. Novikov[2]

Keywords: _complex activity; reflexive model; methodology_

The paper presents methodological aspects of the reflexive models of "complex activity" (CA), which is defined as an activity characterized by at least one of the following attributes:

- The variable technology or/and subject or/and role of object in the target context of activity during the period of activity,
- The multiplicity of technologies or/and subjects or/and objects.

The variability and multiplicity of the subject and object forces the consideration of the subject and object in addition to the activity itself, which makes the models reflexive. A unified formal model unit, named the Structural Element of Activity (SEA), and a procedure to form the fractal hierarchies of SEAs that describe CA as a whole are introduced.

The structural components of activity are analyzed and separated into "implementing" and "governing" types. The "implementing" components (Figure 105.1, elements (1)) are used as the fundamental ones of the SEA, while the "governing" components are analyzed in combination with the subject of activity ("organizing superstructure").

CA is a complex system, so system theory is applicable to study such activity. The most influential elements of an external environment (with respect to activity) are considered: the subject who implements the activity (2) and the object of activity (3), which has to be transformed as the result of activity. The subject is exposed to external constraints, translating them into activity; also the subject interprets the stakeholders' needs, performs goal-setting and governs the activity. The subject and object "encapsulate" all interaction between the external environment and activity as a system of interest.

Each element of the Triad (1–2–3) represents a system; therefore, Triad (1–2–3) is a system of systems.

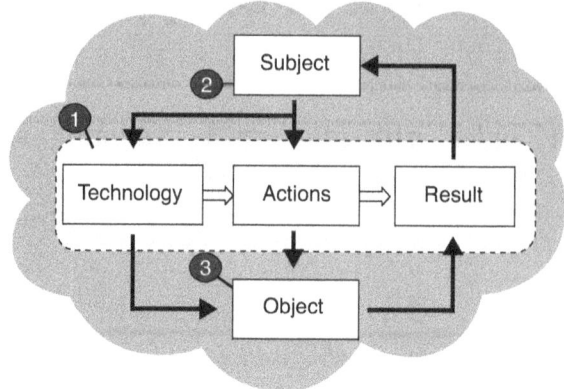

Figure 105.1 Structural Element of Activity (SEA)

SEA is defined as a unit that

A   is established to achieve a given goal or to obtain a given result (to transform the subject of activity);

B   characterizes the goal-focused activity according to a given technology over a given object;

C   has the elements of a certain organizational system as the subject of its actions. An approach is proposed to analyze the temporal and logical structure of activity (i.e., causing the fractality/hierarchies of the SEAs), with the generalization and identification of six basic variants illustrated by Figure 105.2.

The object of lower SEAs in the hierarchy can be any elements of an upper SEA: lower SEAs are established to create/execute/transform the Object/Result, Technology/Actions or Subject (Organizational System) with the parallel (structural) or serial (temporal) decomposition of the elements. Note that other decompositions represent the combinations of these two decompositions. The object of lower SEAs and the selected type of decomposition (parallel or serial) are the foundations of hierarchies/fractality of the SEAs. All variants I–VI correspond to the decomposition of goals (expected results) of SEAs; so goal-setting or the expected result (a required state of the object) is the grounds of the hierarchies/fractality of the SEAs. The implementation of "complex activity" with the course of time is analyzed. As demonstrated, some fragments of CA complete with the meeting the needs, causing no consequences in the sense of activity; other ones generate a new SEA and a hierarchy of SEAs below it. Hence, SEAs (and their hierarchies) disappear as soon as the result is achieved.

As any complex system and system of systems, CA is implemented under uncertainty. We explore and classify different types of uncertainty affecting the technologies, result, and requirements to the activity, the criteria and conditions

	Serial, temporal decomposition	Parallel, structural decomposition
Object/Result	I.	II.
Technologies/Actions	III.	IV.
	V.	VI.

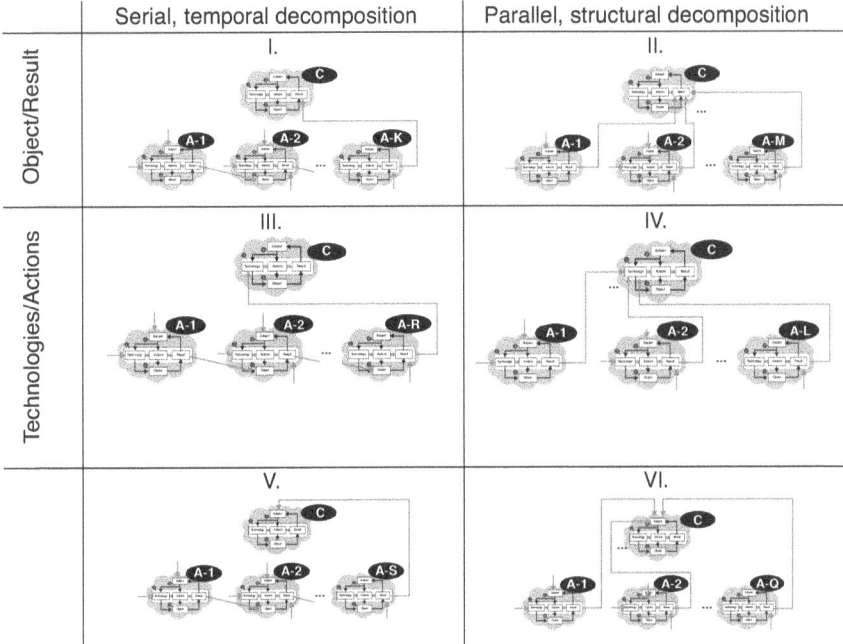

Figure 105.2 Basic hierarchies of SEAs

of the external environment, resources, time, etc., as well as their combination. We employ the well-known notions of measurable uncertainty (describing repeated events when statistical estimates are applicable) and true uncertainty (describing unique/rare events when the statistical approach is not an adequate one).

A classification of the SEAs by several bases is suggested:

- By structurability: the SEAs are divided into "elementary" (that require no detailing) and "complex" (that are detailed as the structures of other SEAs) domains.
- By goal-directedness and value-directedness: the SEAs are divided into "project" SEAs (that describe activity to achieve specific goals in a limited time period) and "process" SEAs (that describe regular activity without specific goals and time limits).
- By the object of activity, SEAs are divided into two classes of SEAs with "tangible/intangible" objects and SEAs with "animate/inanimate" objects:

    a   Material products – goods/systems/objects, accessories, raw materials, energy;
    b   Knowledge and information;
    c   People (individuals, groups, collectives, teams, organizations);

Systems combine classes "a", "b", and "c". "a+b" yields complex technical systems, "b+c" organizational or social systems, and "a+b+c" socio-technical systems.

- By the forms of activity organization:

F1	Elementary operations/works/processes	Elementary cycle of activity
F2	Complex operations/works/processes: several individuals, F2=complex{F1; F2} is a classical hierarchical organization (according to theory of control in organizations)	
F3	Projects and project programs: F3 = complex {F1; F2; F3; F4} (according to project management knowledge area)	Complete cycles of productive activity
F4	Lifecycle: F4 = complex {F1; F2; F3; F4}; (the lifecycles have been existing for a long time; as an active category it has been introduced for the first time in systems engineering knowledge area)	Integral collection of "complete cycles"

- By the discriminant "generate vs non-generate" new "SEAs":

  a "regular SEAs" (associated with "regular activity") that do not create new SEAs;

  b "replicative SEAs" (associated with "replicative activity") that create SEAs and lower hierarchies with a known structure and given fixed technologies;

  c "creative SEAs" (associated with "creative activity") that create SEAs and lower hierarchies with a priori unknown stuff and structure, particularly, with a priori unknown technologies to-be-developed and/or a priori unknown result (in scientific activity or innovative technological activity).

The relations of the superior (upper) and subordinate (lower) SEAs in terms of the type of objects are analyzed and are presented in Table 105.1; the "object => technology => subject" cause-effect relations are studied.

The efficiency analysis of CA using the proposed SEA model is performed.

The applicability of the SEA model is illustrated by examples: we describe exemplar fragments of the activities of a large aerospace corporation, a local industrial enterprise, and a large consulting company.

Table 106.1 The relations of the superior (upper) and subordinate (lower) SEAs

Object =	"Superior SEAs"		
"Subordinate SEAs"	Material product	Knowledge (technology)	Organizational system
Material product	Creation of product components – PBS implementation	Creation of the products in the interests of technology	NO (Indirectly via Knowledge)
Knowledge (technology)	Creation of technological process in the interests of product	Structuring of technology and works or service – WBS implementation	Creation of operation technology for organizational system or technological process in the interests of lifecycle stages of organizational system
Organizational system	NO (Indirectly via knowledge)	Creation of organizational system and employee training for technological process	Transformation of organizational system elements

Notes

1 IBS.
2 Institute of Control Sciences RAS.

Bibliography

Belov, M. How We Engineer Enterprise Systems, INCOSE Italian Chapter Conference on Systems Engineering (CIISE2014) Proceedings, Rome, Italy, November 24–25, 2014. http://ceur-ws.org/Vol-1300/ID13.pdf

Belov, M., IBS Group (2014). Eastern European IT Services: Capability-Based Development for Business Transformation, in *Case Studies in System of Systems, Enterprises, and Complex Systems Engineering*, Gorod, A., B. White, V. Ireland, J. Gandhi, and B. Sauser. (eds.). New York, NY: CRC Press, Taylor & Francis,.

Belov, M., (2015) Systems Engineering and Economical Aspects of Lifecycle Management. *Large-scale Systems Control Journal*, Vol. 56, pp. 6–65. http://ubs.mtas.ru/search/search_results_ubs_new.php?publication_id=19918&IBLOCK_ID=20

John, B., Sauser, B. (2008). *Systems Thinking: Coping with 21st Century Problems*. Boca Raton, FL: CRC Press.

Knight, F.H. (1921). *Risk, Uncertainty, and Profit. Hart, Schaffner, and Marx Prize Essays, no. 31*. Boston and New York: Houghton Mifflin,.

Novikov, A., Novikov D. (2007). *Methodology*. Moscow: Sinteg,. 668 p. (in Russian).

Novikov, D. (2013). *Control Methodology*. New York: Nova Science Publishers,. 76 p.

Novikov, D. (2013). *Theory of Control in Organizations*. New York: Nova Science Publishers,. 341 p.

Novikov, D., Chkhartishvili A. (2014) *Reflexion and Control: Mathematical Models*. London: CRC Press,. 298 p.

Reflexive games of the human mind

People as dolls and dolls as people, people and automats

Victor Borsevici[1]

Keywords: *reflexivity; reflexive systems and control; extended Lefebvre's notation; observing subjects; controlling subjects; animation*

Anthropological and archeological data corroboratively suggest that animism (i.e. human attempt to animate, or to attribute a soul to otherwise inanimate objects and concepts) is the earliest feature of human cultures. And dolls as cult objects play here a special role.

There is not a big difference, as one may wrongly think at first glance, between the earliest finds of coarse stone or clay human and animal figurines, on the one hand, and modern electronic toys, such as Tamagotchi or Pokemon Go electronic characters, which have already generated a mass hysteria, on the other hand.

What is common between these stone, clay and electronic toys? And what is common between the earliest clay figurine of a thoughtful man and the well-known Rodin's Thinker? They all have one thing in common: the capability of the human mind to animate what is inanimate.

And what happens when we see a dancer who reproduces movements of a dancing robot? Or when we see a voter who "automatically" votes for a political leader and who behaves like a "political puppet"?

And what is the meaning of Walt Disney's famous Skeleton Dance and the modern concept of Disneyland in general, including the idea of a "political Disneyland"?

This is a manifestation of something absolutely opposite and unexpected: the human ability to transform themselves into dolls – animate or inanimate objects.

The World of Ideas and the World of People relate to each other from the earliest times via the World of Dolls. A doll is an animated human-like material package of Idea, its Personified Symbol. Once appeared, these Personified Ideas start living their own lives and acquire tremendous power over people. It unleashes a reverse process: people become 'animated automats' programmed by their own creation.

A few centuries ago the great philosopher Baruch Spinoza in his immortal *Ethics, Demonstrated in Geometrical Order*, in a number of wonderful "theorems" showed a human as an "animated automat, driven by the laws of the mind".

And nowadays, another outstanding scientist Vladimir Lefevbre attempted to do the same in his "The Algebra of the Consciousness" based on his revolutionary theory of Reflexivity.

What is in common between these two remarkable personalities, distanced from each other by several centuries, beside the fact that both of them belong to the same great tradition of Moses Maimonid, who became a link between Aristotle and modernity, as well as to the tradition of the first myth about "homunculus" – a clay Golem from Prague, who killed his creator?

The common thread is that one of them used the phenomenon of human Reflexivity implicitly in his studies, helped by the geometric (logical) method, while the other used Reflexivity explicitly, helped by the algebraic method. At the same time, the latter defined Reflexivity as a phenomenon whereas an individual has self and the other's image in "his head", and this other will also have his own and others' images in "his head", etc.

Lefebvre's notation was a remarkable breakthrough in the field of mathematic modeling of Reflexivity. For example, let's have two "observing" subjects (X and Y) and one of many terms in this notation, e.g. Xyx. It means Xy imagined by X, whereas Xy is X as seen by Y.

However, in our opinion, under other conditions, this term can have another interpretation, whereas the term Xyx means X imagined by Yx, whereas Yx is an imaginary Y in X's head. For example, if X are "Europeans" and Y are "Russians", then Xyx is "the imagined Europeans in the heads of the imagined Russians, as thought by the Europeans".

It is noteworthy how concise and precise is Lefebvre's notation for representation of reflexive images (which can be defined by computerized modeling or programs) compared to very sophisticated and fuzzy verbal descriptions.

We offer to extend Lefebvre's notation by complementing the "observing subjects" with the "managing (controlling) subjects" and mark the latter with an apostrophe ('). For example, the term Xy'x will mean the following: subject X managed by subject Y as seen by X. In the latter case it is easy to imagine, for instance, a person realizing that he/she is managed (controlled) by Pokemon Go, or the European Union, realizing that it is managed (controlled) by its overseas partner, and so on.

Thus, our proposal to use the extended notation by Lefebvre enables us to describe and model reflexive systems and processes of conscientization and management (control) with as many subjects as possible, both individual and collective, animated or imaginary. It also enables us to model their cooperation, conflicts and symbiosis within different social, economic, political and cultural interactions.

It seems that behavior of dolls and automats, sometimes, can be more human, than the acts and thoughts of human beings themselves. The twenty-first century will be a century of global ethics and morale. Or it will not exist. This must be the face of true globalization: it must be globalization of justice and humanity. There is no alternative to this.

Note

1 Free University of Moldova (ULIM).

Bibliography

Birshtein, B., Borsevici, V. (2002a). Stratagems of Reflexive Control in Western and Oriental Cultures. *International Interdisciplinary Scientific and Practical Journal "Reflexive Processes and Control"*, vol. 1, nr. 2. Moscow: "Kogito Center" Publishing House.

Birshtein, B., Borsevici, V. (2002b). Theory of Reflexivity by George Soros: Attempt of Critical Analysis. *International Interdisciplinary Scientific and Practical Journal "Reflexive Processes and Control"*, vol. 1, nr. 1. Moscow: "Kogito Center" Publishing House.

Lefebvre, V.A. (1987). The Fundamental Structures of Human Reflexion. *Journal of Social and Biological Structures*, vol. 10, nr. 2, 129–175.

Lefebvre, V.A. (1999). Sketch of Reflexive Game Theory. *Proceedings of Workshop on Multi-Reflexive Models of Agent Behaviour*. Los Alamos, NM: Army Research Laboratory.

Lepsky, V.E. (2010). *Reflexive and Active Environments of Innovative Development*. Moscow: "Kogito Center" Publishing House.

Construction-deconstruction-reconstruction framework

A methodology for interdisciplinary research

Pedro Henrique Juliano Nardelli[1]

Keywords: *interdisciplinary research; self-reference; methodology; systems theory*

This article is an attempt to cope with challenges arising when research activities are carried out within an interdisciplinary group (i.e. a group composed of researchers trained in different disciplines). While many issues could potentially emerge in such a context (e.g. Newell, 2001; Youngblood, 2007), I target only one: the overspecialization process within disciplines that harms the group communication. This may result in isolating subgroups of people with similar backgrounds or misunderstanding the goal to be achieved by the project as a whole. An explicit methodology shared across disciplines may help in overcoming such a problem (although it may not be sufficient). Due to the characteristics of communication processes, I argue that such a methodology needs to be capable of dealing with reflexivity. Therefore, the methodology also becomes an interdisciplinary problem.

I have found in N. Luhmann's systems theory (1995, 2012), E. Morin's complexity thinking (1992, 2006; Malaina, 2015), G. P. Shchedrovitsky's methodological studies (Shchedrovitsky 1966, Shchedrovitsky and Kotel'nikov 1988) and L. Althusser's philosophical position about science (1990) the inspiration – and the first steps – to pursue such a research path. The methodology shall unfold as follows. Theories are constructed within a discipline. Then, they may also be deconstructed in specific ways towards a new reconstruction that fits the totality of the phenomena to be assessed by the interdisciplinary group. I propose that the operations of deconstruction and reconstruction shall map the phenomena into three layers: (i) a physical layer composed by material things and connections, (ii) an informational layer related to symbolic classifications, relations and exchanges, and (iii) a regulatory layer involving decision-making procedures, rules and relations. In the reconstructed form, the interactions occur between elements of the same layer and across different layers. The phenomena are then constituted of (and not reduced to) these layers. This multi-layer approach was described in similar terms by Midgley (1992) and was also employed in computational, agent-based studies that I co-authored with Künhlenz (2016).

As a theory itself, the proposed methodology can also evolve in overspecialized paths. At the same time, because the methodology is also a theory, it can always be assessed in the same way through meta-analysis. The proposed framework then needs to follow its own propositions to indicate the conditions that avoid overspecialized evolutionary paths.

Note

1 University of Oulu.

Bibliography

Althusser, L. (1990). *Philosophy and the spontaneous philosophy of the scientists and other essays*. Verso.

Kühnlenz, F., & Nardelli, P. H. (2016). Dynamics of complex systems built as coupled physical, communication and decision layers. *PloS One*, 11(1), e0145135.

Luhmann, N. (1995). *Social systems*. Stanford University Press.

Luhmann, N. (2012). *Introduction to systems theory*. Cambridge, UK: Polity.

Malaina, A. (2015). Two complexities: The need to link complex thinking and complex adaptive systems science. *Emergence: Complexity and Organization*, 17(1), 1G.

Midgley, G. (1992). Pluralism and the legitimation of systems science. *Systems Practice*, 5(2), 147–172.

Morin, E. (1992). *Method: Towards a study of humankind (volume 1: The nature of nature)*. New York: Peter Lang.

Morin, E. (2007). *Restricted complexity, general complexity. Worldviews, science and us: Philosophy and complexity*. London: World Scientific (pp. 5–29).

Newell, W. H. (2001). A theory of interdisciplinary studies. *Issues in Integrative Studies*, 19(1), 1–25.

Shchedrovitsky, G. P. (1966). Methodological problems of system research. *General Systems*, 11, 27.

Shchedrovitsky, G. P., & Kotel'nikov, S. I. (1988). An organization game as a new form of organizing and a method for developing collective thinking activity. *Soviet Psychology*, 26(4), 57–90.

Youngblood, D. (2007). Multidisciplinarity, interdisciplinarity, and bridging disciplines: A matter of process. *Journal of Research Practice*, 3(2), 18.

Reflexivity as a change in cultural mental models

Carlos Córdoba and César García-Díaz[1]

Keywords: *reflexivity, mental models, culture, agents-based models*

1 Introduction

Physicist and Nobel Prize winner Murray Gell-Mann once said "Imagine how difficult physics would be if electrons could think". This highlights the great challenges social scientists have to face when studying systems composed of human beings, as opposed to natural phenomena. Despite the great advances in understanding these systems during the last 20 years – brought about especially by computer simulations – it remains true that it is still very difficult to deal with this "thinking" capacity in a way amenable to scientific research (Goldspink & Kay, 2007). Human beings have developed important abilities by using this capacity, e.g. making tools to change their environment and communicate with each other through verbal language (Dopfer, 2005). Another interesting aspect of thinking is that of the perceptions humans have about the social phenomena they are part of. These perceptions can have an impact in those phenomena which consequently might alter perceptions through changes in attitudes or behavior. We have decided to term this last ability as *reflexivity*, following one of the most commonly accepted meanings of the word (Beinhocker, 2013; Soros, 2013).

2 Methodology

There are two methodological approaches that could help us to advance in our understanding of this interesting trait. The more traditional – typological – approach (Dopfer, 2005) would start by defining more precisely what reflexivity is. This would certainly imply devising a measure of reflexivity (i.e. what it means to be more or less reflexive), studying how reflexivity is related to other human cognitive abilities (e.g. intelligence and cognition) and how it is influenced by cultural and social factors (e.g. news media or political institutions). Another – more pragmatic – approach would start by asking what mechanisms can trigger reflexivity and in what social contexts a change in our reflexive capacities would be of more significance. The advantage of the latter approach lies in avoiding the troublesome issues involved in pursuing a stable characterization of reflexivity

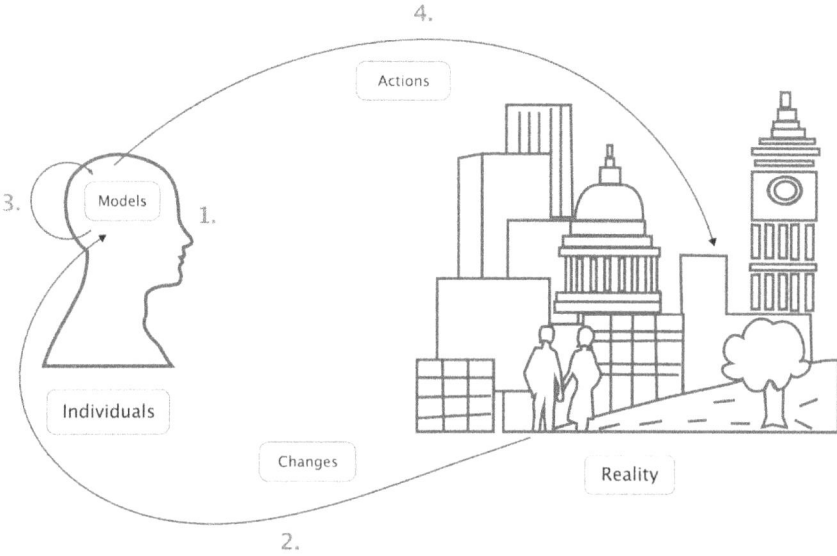

Figure 109.1 Reflexivity described as a process between individuals and reality

(cf. Seibt, 2012) in a similar way as the study of human consciousness has been affected (cf. Siewert, 2002). Instead, reflexivity can be described as a process that takes place between humans and the social phenomena around them. In this way, our understanding of reflexivity will be advanced by learning its relations and consequences, instead of its defining features.

More concretely, we consider a reflexivity process is composed of four stages (as shown in Figure 109.1):

1 Humans create mental models about how different aspects of reality work (Hoff & Stiglitz, 2016). These models are informed by cultural factors such as tradition, news media, scientific theories, religion or life experiences.
2 Changes in reality lead individuals to reflect on the adequacy of their models. At this stage humans receive feedback from reality. This can be positive feedback, which makes them more confident about their models; or negative feedback, which triggers correction or changes in the model.
3 Even if no important changes happen in reality, humans can reflect on new courses of action to improve those aspects that interest them. This is not necessarily given by immediate feedback.
4 After reflecting, people take action on reality based on their updated models. At this stage people feed-forward their models into reality.

At this point it is important to distinguish between reflexivity and related concepts such as single and double-loop learning. Learning consists in acquiring information and skills necessary to solve a domain specific problem. Since single-loop learning is focused in understanding external issues (e.g. how to reduce production costs for the next quarter); it usually does not have an impact on mental models and a corresponding change in behavior. Double-loop learning was defined by Argyris (1991) as a person's ability to reflect critically on her own performance in an organization and modify their work routines if necessary. However, double-loop learning is focused on making people aware of the psychological factors that block self-criticism, so they can appraise for themselves their shortcomings and learn what can be improved in the future. Moreover, double-loop learning appears to reformulate mental schemes that contribute to redesign routines based on prior performance. Again, this type of learning does not affect a person's identity and how they use that identity to behave in the world.

In contrast, reflexivity is a wider concept. A reflexive process – as we understand it – is not necessarily the result of a conscious deliberation (although it might be). Also, *reflexive behavior could be a tendency irrespective of prior performance* (cf. Major et al., 2016). Changes to the social environment alter people's cultural mental models and that is what really affects a change in their behavior. Besides, new mental models acquired during reflexivity can have significant and long-lasting effects on people's lives. For instance, when discussing about the influence of soap operas designed and aired to decrease fertility in Brazil, Hoff and Stiglitz (2016 p 36) argue that

> (t)he soap opera conveys information not so much about how others will respond to the woman, as about what the possibilities are of actions for the woman. . . . The soap opera is changing the women's own sense of identity. And in doing so, it is changing the women and their behavior.

These characteristics make reflexivity a powerful tool to introduce new policies in a community that gets accepted endogenously, instead of seeing to be imposed from above (i.e., "influence" rather than "prediction" and "control" in social systems; cf. Pennock & Rouse, 2016). Besides, since these policies are better internalized than common laws, they have a better chance of survival and acceptance in the community.

3   Conclusions

In order to make our argument stronger, we propose a simple agent-based model (ABM) that renders reflexivity as a change in cultural mental models, based on the work of Demeritt and Hoff (2016) on policy research. Each mental model in our ABM corresponds to a category that defines a set of accepted behaviors, e.g. consider vegetarians and non-vegetarians, whose viewpoints, lifestyles and habits contribute to starkly different assessments in topics such as sustainability and health. The way to make people change from one of these models to another in our ABM is done by introducing one of the following factors in society:

1   Opportunities for people to have or witness new experiences.
2   New role models or alterations in the existing ones.

3　New policies that prohibit a behavior associated with a mental model, but in a way that the policy is culturally related to the behavior object of restrictions.

Note

1 Departament of Industrial Engineering, Universidad de los Andes.

Bibliography

Argyris, C. (1991). Teaching smart people how to learn. *Harvard Business School Publishing, 4*(2), 4-15.

Beinhocker, E. D. (2013). Reflexivity, complexity, and the nature of social science. *Journal of Economic Methodology*, 20(4), 330–342.

Demeritt, A., & Hoff, K. (2016). "Small miracles": Behavioral insights to improve development policy: The world development report 2015. In *Contemporary Issues in Development Economics* (pp. 19–43). UK: Palgrave Macmillan.

Dopfer, K. (2005). Evolutionary economics: A theoretical framework. In *The Evolutionary Foundations of Economics*. Cambridge: Cambridge University Press.

Goldspink, C., & Kay, R. (2007, November). Social emergence: Distinguishing reflexive and non-reflexive modes. In *AAAI Fall Symposium: Emergent Agents and Socialities: Social and Organizational Aspects of Intelligence*, Washington, DC.

Hoff, K., & Stiglitz, J. E. (2016). Striving for balance in economics: Towards a theory of the social determination of behavior. *Journal of Economic Behavior & Organization*, 126, 25–57.

Major, D. L., Maggitti, P. G., Smith, K. G., Grimm, C. M., & Derfus, P. J. (2016). Reflexive and selective competitive behaviors: Inertia, imitation, and interfirm rivalry. *Organization Management Journal*, 13(2), 72–88.

Pennock, M. J., & Rouse, W. B. (2016). The epistemology of enterprises. *Systems Engineering*, 19(1), 24–43.

Seibt, J. (2012). Process philosophy. *Stanford Encyclopedia of Philosophy*. Retreived 23.9.2018 from https://plato.stanford.edu/entries/process-philosophy/

Siewert, C. (2002). Consciousness and intentionality. *Stanford Encyclopedia of Philosophy*. Retreived 23.9.2018 from https://stanford.library.sydney.edu.au/entries/consciousness-intentionality/

Soros, G. (2013). Fallibility, reflexivity, and the nature of social science. *Journal of Economic Methodology*, 20(4), 309–329.

Umpleby, S. (2007). Reflexivity in social systems: The theories of George Soros. *Systems Research and Behavioral Science*, 24(5), 515–522.

Belligerent science versus neutrality

Questioning the role of science in the 21st century

Jose-Rodolfo Hernandez-Carrion[1] *and*
Ignacio Martinez de Lejarza[2]

Keywords: *belligerent science; complexity; emergence; neutrality; economics; knowledge*

There are different types of attitudes. A belligerent person is one with an attitude towards war. Belligerent is a word that comes from the Latin, is formed from the noun bellum; belli means war. Belligerent is an adjective situation that shows aggression or defense of a state that is ready to war; it is a relationship in which a position of struggle is between the different elements. Now we want to focus on science here, to question what kind of scientific knowledge is shown to the public and why. Even if economic affairs could be the explanation for some theories and arguments, we want to explore what is happening in several different areas of our reality today.

The financial collapse around 2007–2008 and the subsequent deep recession on the global economy has consequences for the real economy. Governments are grappling with unprecedented budget deficits and unemployment is over 1 million higher than it was before the recession. This is a crisis for the real economy and for economic policymakers, but it should also be seen as a crisis for the economics profession and for economic theory.

We use and are exposed to nonionizing radiation sources every day. Microwave ovens use microwaves to heat food, toasters use infrared waves to heat and sometimes burn our toast, and we watch television, talk on cell phones and listen to the radio through the use of radio waves. These are all nonionizing forms of radiation. Visible light, radar, laser light and ultraviolet light also fall into this radiation category.

Some forms of nonionizing radiation can damage tissues if we are exposed too much. For instance, too much ultraviolet (UV) light from lying out in the sun is known to cause some skin cancers; even moderate amounts can cause skin burns.

In 1993, Peter Gøtzsche co-founded The Cochrane Collaboration, (the gold standard for the review of medical research data) and The Nordic Cochrane Centre, where he is Managing Director. In 2010 he became professor of Clinical Research Design and Analysis at the University of Copenhagen. He has published more than 70 papers in "the big five" (*BMJ, Lancet, JAMA, Annals of Internal Medicine* and *New England Journal of Medicine*) and his scientific works have been cited over 15000 times. And then he published "Deadly Medicines and

Organised Crime", so the question could be how can you manipulate the masses to believe what you want them to believe or why is Organized Denial so important in today's arena?

Social and biological organizations entail changes involving new types of complex structures, which only is possible if the system remains far from an equilibrium and, moreover, there are 'non-linear' mechanisms acting between the various elements of the system. The history of a system will result from the amplification of certain fluctuations and self-consolidation for certain processes. Self-organizing social systems can be conceived as systems of socio-spatial and temporal, as open systems characterized essentially by internal processes of an economic, social, legal and cultural place in their midst, whose nature is typically unstable in nature, but in turn in constant interaction with the environment. In conclusion, we want to explore how science is evolving today and what kind of knowledge is accessible in the 21st century from a complex and interested approach, and what are the collateral consequences of the present situation in order to plan the future (2017).

Notes

1 Group of Economics and Complexity, University of Valencia (Spain).
2 Group of Economics and Complexity, University of Valencia (Spain).

Bibliography

Abraham, J. (1995). *Science, Politics, and the Pharmaceutical Industry: Controversy and Bias in Drug Regulation*. New York: St. Martin's Press.

Antholis, W., & Talbott, S. (2010). *Fast Forward: Ethics and Politics in the Age of Global Warming*. Washington, DC: Brookings Institution Press.

Gonzalez-Rodriguez, D., & Hernandez-Carrion, J.R. (2014). A Bacterial-Based Algorithm to Simulate Complex Adaptive Systems. *Lecture Notes in Artificial Intelligence (LNAI)*, 8575, 250–259. https://doi.org/10.1007/978-3-319-08864-8_24

Halford, N.G. (2012). Toward Two Decades of Plant Biotechnology: Successes, Failures, and Prospects. *Food and Energy Security*, 1(1), 9–28. http://dx.doi.org/10.1002/fes3.3

Kassirer, J. (2004). *On the Take: How Medicine's Complicity with Big Business Can Endanger Your Health*. Oxford: Oxford University Press.

Kempton, W. (1997). How the Public Views Climate Change. *Environment*, 39, 12–21.

Khan, M.R., & Roberts, J.T. (2013). Adaptation and International Climate Policy. *WIREs Climate Change*, 4(3), 171–189.

LaMattina, J.L. (2008). Drug Truths: Dispelling the Myths about Pharma R&D. http://dx.doi.org/10.1002/9780470434673

Laudisi, F., et al. (2012). Prenatal Exposure to Radiofrequencies: Effects of WiFi Signals on Thymocyte Development and Peripheral T Cell Compartment in an Animal Model. *Bioelectromagnetics*, 33, 652–661.

Lemmens, T., & Freedman, B. (2000). Ethics Review for Sale? Conflict of Interest and Commercial Research Review Boards. *The Milbank Quarterly*, 78(4). Retrieved

October 2016, from www.ncbi.nlm.nih.gov/pmc/articles/PMC2751172/pdf/milq_185.pdf

Markandya, A., & Galarraga, I. (2011). Technologies for Adaptation: An Economic Perspective, 27–42. In L. Christiansen, et al. (eds.). *Technologies for Adaptation: Perspectives and Practical Experiences.* Roskilde, Denmark: UNEP Risø Centre on Energy, Climate and Sustainable Development. Retrieved October 2016, from www.unep.org/pdf/TechnologiesAdaptation_PerspectivesExperiences.pdf

Porter, R.J., & Malone, T.E. (1992). *Biomedical Research: Collaboration and Conflict of Interest.* Baltimore, MD: Johns Hopkins University Press.

Randour, F., et al. (2014). The Cultivation of Genetically Modified Organisms in the European Union: A Necessary Trade-Off? *Journal of Common Market Studies,* 52(6), 1307–1323. http://dx.doi.org/10.1111/jcms.12149

Remling, E., & Persson, A. (2014). Who Is Adaptation for? Vulnerability and Adaptation Benefits in Proposals Approved by the UNFCCC Adaptation Fund. *Climate and Development,* 7(1), 16–34.

Rodríguez-Delgado, R. (1997). *Del Universo al ser humano.* Madrid: McGraw-Hill.

Schwarz, E. (1994). Une modèle générique de l'émergence, de l'évolution et dufonctionnement des systèmes naturels viables, 259–285. In L. Ferrer-Figueras, et al. (eds.). *Tercera Escuela Europea de Sistemas.* Valencia: Ayuntamiento de Valencia.

Schwarz, E. (1995). Where Is the Paradigm? In the People's Mind or in the Social System? *Revista Internacional de Sistemas,* 7(1–3), 5–54.

Schwarz, E. (1997). Toward a Holistic Cybernetics: From Science through Epistemology to Being. *Cybernetics and Human Knowing,* 4(1), 7–49.

Schwarz, E. (2002). Can Real Life Complex Systems Be Interpreted with the Usual Dualist Physicalist Epistemology: Or Is a Holistic Approach Necessary? *Proceedings of the 5th European System Science Congress,* Crete.

SciArt Lab. (2016). 2D Simulator: Hacking Science, Art and Technology. Retrieved October 2016, from www.sciartlab.com/?page_id=517

Wood, F. B. Jr. (1988). The need for systems research on global climate change. *Systems Research,* 5(3), September, 225–240.

The equilibrium at the set of reflexive strategies

Felix Ereshko[1]

Keywords: *control; games, reflexive strategies; hypotheses; decision-making*

1 First basis

The ideas of the author concerning the reflectivity problem are based on the works of I.P. Pavlov, J. von Neumann, Y.B. Germeier, N.N. Moiseev, N.S. Kukushkin, V. Lefebvre, G. Soros, and V.E. Lepsky. In the classical papers by I.P. Pavlov (the 1904 Nobel Prize winner) and his followers, the theoretical postulates of the origin of conditioned and unconditioned reflexes were formulated and reflexive mechanisms were studied experimentally. A formal approach to describing the interconnection of the players viewed from the position of a perfect Observer follows (Y.B. Germeier). The notion of an objective description of a game is used for this purpose.

2 Second inspiration

George Soros's books were a second inspiration for the author. In these books Soros has developed the theory of reflexivity in application to economics as a whole and to stock markets in particular. Soros assumes that "participants' views form a part of the situation, to which they are related". They may influence substantially events and, in turn, are impacted by events. Such a mutual influence of participants Soros names "reflexivity". The present working materials contain the first formalized attempts of the author to study an insufficiently explored problem of reflexivity in the theory of decision-making. In this work descriptions of the reasonings of G. Soros and A.O. Cournot are proposed. The Soros model and a model of sharing a dynamic resource are briefly described. A reflective game model based on hierarchical games theory (Y.B. Germeier) is investigated in more detail. In papers (Ereshko F.I., Y.B. Germeier) reflexive control was presented as a x(y), function – a response of the first player x to the probable control of the second player y.

3 Description of the result

Logical consistency of the task of decision-making demanded the addition of a set of auxiliary conditions concerning the information interaction of the players. A solution to the possible contradictions was found in a way of introducing the notion "first player's right for the first move" and the condition that while choosing the concrete values of his controls the first player already knew the concrete choice of the second player, and the first player utilized his/her advantage in the manner of a dependence strategy of behavior x(y). If the situation differs from the one mentioned above, and the first player has no advantage in obtaining the information, then the use of the strategy x(y) can be considered logically consistent only as a hypothesis of the first player. Our further consideration is based on exactly such a description of the participants in the decision-making process: we admit that every participant acts within the framework of some of his conceptions of the intentions, objectives, and actions of the other participants and obtains information about his decisions a posteriori, i.e. after the act of decision-making and after the action has been accomplished. In this context his conceptions might be far from the real situation. In further speculations we are on the side of an abstract Observer of the events and describe their possible evolution. Unlike I. Newton we, as Observers, let players make up their hypotheses.

So, let the first player assume a hypothesis about the second player's payoff function, g(x,y), calculate the optimal response function for the second player yopt(x)= arg max g(x,y), and then calculate the optimal choice xopt=arg max f(x, yopt(x)).

The second player acts quite similarly, then calculates his optimal choice.

Denote the pair of the optimal player's payoff in this case as Reflex-solution.

If the players' hypotheses prove correct, then the calculation of the players' choices is described similarly. As above, this pair of the optimal player's payoff is called a Real-solution.

Then we discuss the question of how far the Reflex-solution may be from the Real-solution, depending upon how much the players err while hypothesizing about each other's objective function.

Note

1 Head of Department of Information Systems, Dorodnicyn Computing Centre, Russian Academy of Science, Moscow, Russia

Bibliography

Blaug, M. (1985). *Economic Theory in Retrospect*. University of London of Education & University of Buckingham.

Ereshko, F.I. (2001). Modelling of the reflexive strategies in the control systems. Computer Centre of the Russian Academy of the Sciences (pp. 1–48) (Ерешко Ф.И. Моделирование рефлексивных стратегий в управляемых системах. М.: ВЦ РАН, 2001. 48 с. in Russian).

Germeier, Y.B. (1986) Non-antagonistic games (translated by Anatol Rapoport). Theory and decision library, volume 46. (pp. 320–327). Dordrecht, Boston: D. Reidel Pub. Co., Hingham, MA. 1986. xiv+331 pp. Sold and distributed in the U.S.A. and Canada by Kluwer Academic Publishers. Originally published by Nauka,

1976, under the title: Igry s neprotivopolozhnymi interesami, in Russian. Bibliography: p. 320–327.

Lefebvre, V.A. (1991). *The Formula of Man: An Outline of Fundamental Psychology.* University of California, Irvine.

Luce, R.D., Raiffa, H. (1957). *Games and Decisions.* John Wiley and Sons, New York.

Soros, G. (1994). *The Alchemy of Finance: Reading the Mind of the Market.* Jonh Wiley & Sons, Inc.

von Neumann, J., Morgenstern, O. (1953). *Theory of Games and Economic Behavior.* Princeton University Press, Princeton, NJ.

Accelerating technology for self-organizing networked democracy[1]

Alexander Raikov[2]

Keywords: *networked democracy; convergent technology; group decision-making; collective artificial intelligence*

1 Democracy and self-organizing decision-making

For a long time representative democracy has separated the citizens from the necessary conditions of their participation in political life, banking systems and corporations. Now, mass media, information and communication technologies (ICT) and Artificial Intelligence (AI) methods give the capacity to communicate, organize and see the world beyond the spectacle, which has been made by different states, banking and corporate groups (Toret and Callega 2014). The networked communicative movement is developed; the democracy becomes networked. They appear as practices of citizens' control in political, social, technological, economic and communicative spheres. It is important to make sure that the views of all citizens' affected groups are included.

In this condition the "subject-self-developing reflexive-active system (environment)" paradigm (Lepsky 2015) is the new step of development and the post-non-classical type of scientific rationality. In this paradigm modern ICT and AI could be useful for accelerating democracy processes. But ICT and AI do not include a special effective technology to sustain accelerating networked citizens' decision-making in democracy processes, especially if these processes consist of the planning and achieving the strategic goals. The citizens' strategic thinking requires reaching group or crowd consensus. The consensus differs from voting, when some part of the people may have their own opinions that do not match with the opinions of others. Therefore, these processes take a lot of time and may go beyond the deadlines.

For speeding up citizens' decision-making processes this paper proposes to support these processes by organizing the highly qualified networked expert groups, using the collective intelligence technologies created on the base of AI methods (Gubanov et al. 2014), and using the special Convergent methodology of structuring information in decision-making processes (Raikov 2015).

2 Convergent citizens' decision-making

The above-mentioned paradigm in combination with the Convergent methodology and networked expertise becomes a key factor of success in improving the decision-making processes in a networked democracy. It provides the advanced opportunities for ensuring sustained convergence of civil group decision-making, and it could be useful for advancing the theory and practice of direct democracy.

The proposed special convergent methodology is based on the fundamental principles of control thermodynamics, inverse problem solving, holistic discourse approach, cognitive modeling, genetic algorithms, quantum semantic approach and AI (Raikov 2015). The structuring information in a special way during citizens' meetings helps to get groups' insights very fast. For every group of people the process of structuring information may include the following steps: outline the goals, fill out the questionnaire, create cognitive model, verify the model by using Big Data analysis, select some action scenarios and evaluate them with a computer.

The convergent methodology as noticed uses the cognitive modeling method that consists of creating concepts (factors) and their connections. The Cognitive modeling system provides mechanisms to build up relations of the "all to all" kind. Intense matrix representation of the relations is illustrated in the form of directed graphs, where nodes represent factors, and arrows represent corresponding relations. This design process helps in understanding the problem structuring by the decision makers and improves problem understanding by the participants. To verify the cognitive models the Big Data analyses technologies are used (Raikov et al. 2016). The verification is carried out by mapping the factors and connections into the sets of objects such as documents, messages, comments, etc.

For creating a networked expertise system (distributed, e-expertise) (Gubanov et al. 2014) there are several technologies: Expert comments; Electronic brainstorming; Strategic conversation; Virtual collaboration, etc. The practice of using distributed experts' procedures shows that they differ from the group expertise when experts face each other. In network experts there are restrictions for: understanding each other; feeling each other out; reaching the agreement. The discussions in networked experts' procedures tend to divergent dynamics.

Distributed experts and moderators have to possess an additional knowledge in convergent methods for synchronizing collaborations and improving mutual understanding processes. So, the decision-making situation in a social network consists of: dissatisfied people, their requirements, ill-defined goals and problems on the way to the goals. In this situation the task is to solve the problems and find the path to the goals in the shortest amount of time. It is an inverse problem; the participants' thoughts and feelings affect the decision-making process. Consequently, the chaotic crowd behavior implies greater instability of the processes of achievement with the crowd strategic agreement (Figure 112.1).

The convergent methodology has been implemented as software, the e-expertise system (Gubanov et al. 2014; Raikov 2015), and has been tested in the course of decision-making during a Russian e-government activity and for creating

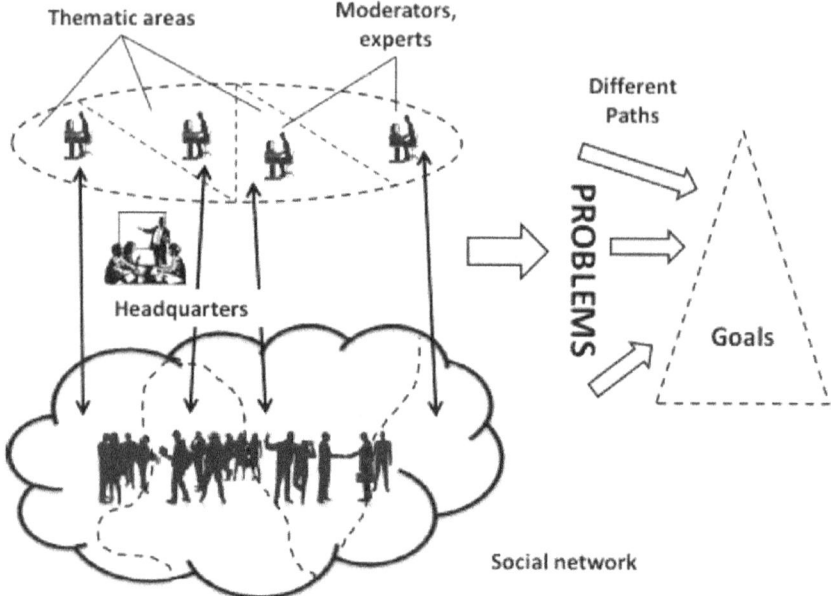

Figure 112.1 The networked citizens' decision-making situation

effective strategies for group actions across regional, national, political and other bodies. It demonstrates how public investment in ICT and AI create public values. The convergent methodology may be useful for supporting targeted forum discussions.

3   Conclusion

For speeding up multi-group self-organizing networked democracy processes it is useful to apply special convergent methodology, including decision-making and the networked expertise support technologies. During decision-making citizens' processes and meetings it helps in structuring information in a special way that makes sure to get a group's insights and strategic consensus in the shortest amount of time.

As was shown in real practice the Convergent methodology in self-organizing networked democracy helps to: increase the level of understanding of decision-making participants; control the behavior and take into account hidden factors in cognitive modeling; analyze visual information; make sure group strategic decision-making is sustainable and purposeful, etc.

The proposed solution helps in creating concepts, strategies and programs in different fields: complex reconstruction and development territories, higher and professional education, public health service, social security, housing and communal services, youth policy, market dynamic assessments, etc. The convergent technology could be applied on the government, municipal and corporate levels, and for non-profit organizations to support strategic group decision-making process.

Notes

1 This work is partly funded by: Russian Science Foundation, grant 17–18–01326
2 Institute of Control Sciences of Russian Academy of Sciences, Leading Researcher.

Bibliography

Gubanov, D., Korgin, N., Novikov, D., and Raikov, A. (2014). E-expertise: Modern collective intelligence, Springer. *Series: Studies in Computational Intelligence*, Volume 558, 18.

Lepsky, V. (2015). *Evolution of concepts about control (methodological and philosophical analysis)*. M: Kogito Center Publishing House. 107 p. (in Russian).

Raikov, A. (2015). Convergent networked decision-making using group insights. *Complex & Intelligent Systems*, December, Volume 1, Issue 1, 57–68.

Raikov, A., Avdeeva, Z., and Ermakov, A. (2016). Big Data Refining on the Base of Cognitive Modeling. *Proceedings of the 1st IFAC Conference on Cyber-Physical&Human-Systems*, Florianopolis, Brazil. 7–9 December. pp. 147–152.

Toret, J., and Callega, A. (2014). Decentralized Citizens Engagement Technologies. FP7 – CAPS. Project no. 610349. D-CENT, 81 p.

Second order science and autoreflexion
Cyber-systemic perspectives[1]

Denis Zhurenkov[2]

Keywords: *social systems; reflexive activity; network-centric systems; post-non-classical scientific rationality*

1 The interdisciplinary approach

Since the second half of the XX century there has been exponential growth of technological advances, along with the differentiation of sciences. Today we can say that fundamental science is ahead of technology, and the resulting backlog is now being implemented in the new technologies. However the lack of mass "request" from the technology to the fundamental science is not conducive to intensive development. An interdisciplinary approach is becoming a major scientific goal, and it must operate for a variety of sciences with useful results and laws.

Crisis management of problems in social systems is impossible without minding all areas of knowledge in the conceptual foundations of management, creating a communicative space for representatives from different management issues and organizing facilitation of their joint activities. These objectives are reflected in the different interpretations of the interdisciplinary approach.

2 The context of scientific rationality

Unlike Wiener's cybernetics, second order cybernetics is conceptual and philosophical. Moreover, today the role of the reflexive activity increases and communicative reflexive activity (V. Lefebvre, S. Umpleby, K. Müller) is becoming a major research problem. Going from the management of a "subject–object" paradigm to a "subject–subject" paradigm led to the formation of new types of management: reflexive control, information management and active control systems.

The emerging "subject-self-developing polysubject environment" is inextricably linked with the development of a subject-oriented approach. This approach is a natural development of the subject-activity approach, with increasing attention to the subjects and their environments, and with a decrease in attention to the activity-related component due to a sharp reduction in the impact of the regulatory component of the actions of the subjects in the conditions of modern reality.

In the context of scientific rationality we observe the evolution of species from the classic management control to "soft" forms of management through social media. Fundamental changes are occurring in management models, particularly striking changes are in the macro social systems – from the dominance of human-dimension models with extensive use of mathematical models.

3 Post-non-classical scientific rationality

The focus of post-non-classical scientific rationality is the ethics of strategic sub-jects' self-developing media focusing particularly on the problem of preserving the integrity of the subjects and their organization. Accordingly, there are three mechanisms to ensure the integrity of the target: classical, non-classical and post-non-classical.

If in the context of classical and non-classical scientific rationality second order cybernetics and cybernetics were providing the control problems then in the context of post-non-classical science the centers of governance issues move into philosophy, synergy, political and economic sciences. Actual problems of the institutionalization of areas of knowledge management to ensure responses per-haps will be replaced by a third order cybernetics (cybernetics of self-developing environments).

For second order cybernetics, the key terms are: recursive, self, reflexion. A promising area of cybernetics development is the development of the theory based on the ideas of autoreflexion.

Third order cybernetics could be extended on the basis of the thesis "of observ-ing systems to self-developing systems." At the same time managing to gradually transform into a wide range of processes to ensure self-development systems of social control, encouragement, support, moderation, organization, "assembly and disassembly" of subjects (V.E. Lepskiy). This approach is one path for the evolution of cybernetics and may become the new mainstream of cybernetics.

Relevant management tasks are: network-centric systems (including military applications), information management systems engineering and cyber security, lifecycle management of complex organizational-technical systems. Promising areas of application are: living systems, social systems, transport, energy.

Notes

1 The research was conducted with financial support of Russian Science Foundation, project 17–18–01326.
2 Central Research Institute of Economy, Informatics and Control Systems of the State Corporation "Rostec," Moscow, Russian Federation

Bibliography

Lepskiy, V.E. (2009) *Sub"ektno-orientirovannyy podkhod k innovatsionnomu razvitiyu [In the Subject-Oriented Approach to Innovative Development]*. M.: Kogito-Tsentr, p. 208.

Lepskiy, V.E. (2014) *Stanovlenie sub"ektno-orientirovannogo podkhoda v kontekste razvitiya predstavleniy o nauchnoy ratsional'nosti [Evolution of Representations about Control in the Contexn of Scientific Rationality]*. M.: Al'fa-M, pp. 392–420.
Moiseev, N.N. (1998) Eshche raz o probleme koevolyutsii [Once Again the Problem of Co-Evolution]. *Voprosy Filosofii*, No. 8, pp. 26–33.
Stepin, V.S. (2003) *Teoreticheskoe znanie [Theoretical Knowledge]*. M.: Progress-Traditsiya, 744 p.

Websites

http://wosc2017rome.asvsa.org/index.php

Change management in social and labor relations during the emergence of a new economy

Tatiana A. Medvedeva[1]

Keywords: *social and labor relations; economic globalization; network economy; dialogue between labor and capital; an extended system approach*

1 A network economy and the possibility of social dialogue between labor and capital

The emergence of new social and labor relations[2] in a time of globalization and a network economy is a difficult and uncertain phenomenon, and therefore many researchers have expressed serious concern about the fate of labor and social and labor relations (Castells 2000; Kaufman, 2004; Chekmarev, 2002; Katukov, Malygin and Smorodinskaya, 2012). Globalization and computerization of the economy, the rapid dissemination of knowledge and the formation of universal interdependence have led to the possibility that while capital-based industries diffuse worldwide, labor is less mobile. Many studies describe the deteriorating situation of workers, the atomization of individuals, individualization of labor and the erosion of social capital (Bobkov et al., 2014; Kolosova, Razumowa and Ludanik 2008; Kochan, 2010). Workers' organizations, established in an era of an industrial economy, are destroyed or weakened. This violates the principle of equality of opportunity for all key players in social and labor relations to be able to represent and protect their interests.

How does the emergence of a new economy alter the social dialogue between labor and capital? Are we seeing the end of a collective consciousness among workers, amid more individualistic behavior? The participants in social and labor relations are now in conditions where they are forced to learn how to operate in network organizations. They now have the opportunity and responsibility to protect their interests in the new economy.

2 The institutional framework and complexity of social and labor relations in a network economy

This article explores how the forming of a network economy influences social and labor relations.

In general, information technologies and a network economy significantly alter and complicate the structure of social and labor relations. Effective partnership and cooperation are becoming essential and increasingly important for actors in social and labor relations. Flexibility and mobility in labor relations are increasing. The intensity and duration of labor communication vary through time. There is a constant openness to the organization of social and labor relations to establish contacts with new partners. Hence, there is a continuing evolution of the structure of relations, which means self-development and self-organization of social and labor relations. Stability in long-term relationships happens only when such relationships go through a test for effectiveness when their advantage over existing alternatives of social and labor relations is unquestioned. Establishment of certain bonds in the field of labor and social–labor relations are increasingly determined by economic expediency. Structures become a part of broader systems interactions and social and labor relations develop when they are more effective than possible alternatives. The hierarchical relationship of subordination is replaced by a variety of horizontal interactions.

The article identifies the organizational foundations and principles of social and labor relations in the emerging new economy.

The first basis for a new form of social and labor relations – organization on geographical principle, which is due to a convenient location of enterprises in relation to each other – leads to the formation of a specific community and more frequent contacts. This means clusters and industrial districts. In this case, the trade unions are the fourth member of the model: the State (represented by local authorities) – Science – Business. Trade unions can act as collaboration institution (American experience).

The second basis for a new form of social and labor relations is the vertical organization of enterprises, which is based on a chain of manufacturers of any product. In this case, the network of social and labor relations can be formed around this manufacturing process. There can be a classic union, uniting workers of this production chain, or some unions of enterprises in this production chain.

Organization of horizontal network enterprises based on similarity of activities leads to broadening the scope of such structures, which may extend beyond the boundaries of nation states and businesses. Here are the most different forms of collaboration.

Combining the workers may also occur on the basis of membership in a particular profession (professional societies, communities of practice) and on the basis of common interests (for example, a female union in Japan).

It seems that for the formation and functioning of the networks of social and labor relations to be effective, the following principles should work: interest by each participant in this collaboration; independence of participants in social and labor relations; possibility of obtaining synergies as a result of communication, expressed in innovative solutions to social and labor problems; etc.

3   The influence of culture on forming networked social and labor relations (Russia as an example)

As pointed out by many researchers, network organization principles are universal. Recognizing the correctness of this conclusion in general, we would like to draw attention to the fact that the forms of the implementation of this general principle are multiple, and they depend on the social and cultural environment, the traditions in the field of social and labor relations, and on the traditions in economic life.

The ways to work with values are (1) Avant-garde: to revolutionalize values, to destroy old traditions and create a new, modern civilization, which would be compatible with the world economy; (2) Archaic: to keep the tradition, originality, even if it is not compatible with the comprehensive modernization; (3) Modern: building on existing informal institutions, to create a comfortable environment of productive existence evolutionarily (Auzan et al., 2011).

The article reviews the influence of cultural values on how networked social and labor relations are formed.

4   Ways to solve problems in the field of social and labor relations

The article considers ways to solve problems in the field of social and labor relations on both organizational and theoretical levels.

The way out of the puzzle, which a globalizing economy creates for employees, is self-organization. What model of social and labor relations countries will adopt depends not only on employers but also on employees and their ability to take advantage of the new organizational forms – the networked enterprise – and, ultimately, on their ability to engage in social learning and self-organization. For the unstable, complex world of a globalizing economy, the national social and labor relations systems must improve their stability and viability due to the growing complexity and instability of the global economy. A relationship between complexity and integrity is needed. Social and cultural factors largely provide this integrity. Ideas, values, beliefs, ideology and goals – in other words, the meaning of social and labor communication – are what binds people in a society.

The theoretical way out is a multidisciplinary approach to the study of social and labor relations based on cybernetics and systems theories. The article promotes a new technique of understanding social and labor relations in an organization based on an extended systems approach (Capra, 2002; Medvedeva, 2012, 2014, 2016; Medvedeva and Umpleby, 2015). An extended systems approach to social and labor relations provides an opportunity to assess the organizational principles of the system of social and labor relations, to build it in such a way that its different parts interact, respond to changes and send feedback signals, affecting the operation of the system.

Notes

1 Siberian State University of Transport, 191, D. Kovalchuk Street, Novosibirsk, 630049, Russia.
2 According to B.E. Kaufman in the US there is a strong tendency to interpret labor relations as only labor-management relations, indicating the deep gap, the conflict that lies between labor and capital (Kaufman, 2004). Scientists thus separated labor relations from the wider range of social relations. Thus the research field "labor relations" and labor policy in the United Kingdom and the United States gradually

acquired a highly specialized, instrumental character separate from other social issues and social policies.

In Russia as well as in Europe, labor relations are conceived holistically, as part of the social system. Great importance is attached to the study of how the social system works. Labor law is an essential part of the social system. In this formulation, it is much more difficult to isolate the labor relations from the general system of social relations. Solutions to problems in employment and labor relations are seen as part of changes of the whole social system.

Bibliography

Auzan, A.A., Afrin, D. (2001). Technology against networks. *Expert*, 38(298). (in Rus.).

Auzan, A.A., Arkhangelsky, A.N., Lungin, P.S., Naishul, V.A. (2011). *Cultural Factors of Modernization*. St. Petersburg: Fund "Strategy 2020". (in Rus.).

Belova, L., Strizhenko, A. (2007). *Information Society: Transformation of Economic Relations in a World Economy: Lomonosov Moscow State University*. Azbuka: Polzunov Altay State Techical University Barnaul. (in Rus.).

Bobkov, V.N., Veredyuk, O.V., Kolosova, R.P., Razumowa, T.O. (2014). *Employment and Social Precarization in Russia: Introduction to Analysis*. Moscow: TEIS.

Capra, F. (2002). *The Hidden Connection: A Science for Sustainable Living*. London: HarperCollins.

Castells, M. (2000). *The Information Age: Economy, Society and Culture*. Moscow: State University – Higher School of Economics. (in Rus.).

Chekmarev, V. (2002). *Economic Space and Its Network Organization*. Kostroma: Nekrasov Kostroma State University. (in Rus.).

Dyatlov, S., Selishcheva, T. (2009). *Regulation of the Economy in the Transition to Innovative Development*. Petersburg: Asterion. (in Rus.).

Espejo, R. (2000). Self-construction of desirable social systems. *Kybernetes*, 29(7–8), 949–963.

Katukov, D., Malygin, V., Smorodinskaya, N. (2012). *Institutional Environment of a Globalized Economy: The Development of Network Interactions: Scientific Presentation*. N. Smorodinskaya (ed.). Moscow: Institute of Economy RAS. (in Rus.).

Kaufman, B.E. (2004). *The Global Evolution of Industrial Relations: Events, Ideas and the IIRA*. Geneva: International Labour Organization.

Kochan, T.A. (2010). Rethinking and Reframing U.S. Policy on Worker Voice and Representation. *Proceedings of the Conference on the 75th Anniversary of the National Labor Relations Act*. Washington, DC, October 28–29.

Kolosova, R., Medvedeva, T. (2015). Social and labor relations in a network economy. *Vestnik MGU: Series "Economika"*, 5, 89–104.

Kolosova, R., Razumowa, T., Ludanik, M. (2008). *Forms of Employment in an Innovative Economy: Textbook*. Moscow: MAKS Press. (in Rus.).

Kolosova, R., Vasiliouk, T., Ludanik, M. (2006). *Distant Employment in Russia*. Moscow: Economic Faculty MSU, TEIS. (in Rus.).

Medvedeva, T.A. (2016). *An Extended Systems Approach to Social and Labor Relations in the Conditions of Globalizing Economy*. Moscow: Economic Faculty MSU, TEIS. (in Rus.).

Medvedeva, T.A. (2015). Interpretation of the system of social and labour relations: SFCP-technique. *Uroven' zhizni naseleniya regionov*, 1(195), 103–111. (in Rus.).

Medvedeva, T.A. (2014, December 30). Developing a new type of social and labor relations system in a time of social and economic change. *Emergence: Complexity and Organization*, 16(4). Retieved 24.9.2018 from https://journal.emergentpublications.com/article/15327000160406/

Medvedeva, T.A. (2012). Developing an innovative style of thinking and innovative behavior. *Systemic Practice and Action Research*, Springer, 25(3), 261–272.

Medvedeva, T.A., Umpleby, S.A. (2015). A multi-disciplinary view of social and labor relations: Changes in management in the U.S. and Russia as examples. *Cybernetics and Systems*, 46(8), 681–697.

Nikolaeva, T. (2001). *Basics of the Information Economy*. St. Petersburg: Peter. (in Rus.).

Senge, P.M. (1990). *The Fifth Discipline: The Art and Practice of the Learning Organization*. New York: Doubleday/Currency.

Wiener, N. (1988). *The Human Use of Human Beings: Cybernetics and Society*. Da Capo Press.

Participant universities

Sapienza University of Rome

Guglielmo Marconi University

Albanian University

Niccolò Cusano University

Second University of Naple

University of Bari Aldo Moro

University of Bergamo

University of Cassino and Southern Lazio

University of Foggia

University of Ibagué

University of Genova

University of Maribor

University of Palerno

Università Telematica Pegaso

University of Salento

University of Salerno

University of Torino

Participant associations

Association Française de Science des Systèmes (AFSCET)

Associazione Italiana per la Ricerca sui Sistemi (AIRS)

American Society for Cybernetics

Association for Research on Viable Systems (ASVSA)

Business Systems Laboratory (BSLab)

Latin American School of Systems Thinking and Design (ELAPDIS)

European Union for Systemics (EUS-UES)

Independent Institute for Organizational Systemic (G.I.R.O.S.)

IASCYS

Italian Association for Sustainability Science (IASS)

System Dynamics Italian Chapter (SYDIC)

UK System Society (UKSS)

World Organization for Systems
and Cybernetics (WOSC)

Associazione Italiana di
Epistemiologia
e Metodologia Sistemiche

Centre CIRRPC of Institute
of Philosophy RAS

Partners

AIDEA – Accademia Italiana di
Economia Aziendale

ASSIOA – Associazione Italiana di
Organizzazione Aziendale

SIMA – Società Italiana di
Management

With the support of

CUEIM – Consorzio Universitario
di Economia Industriale e
Manageriale

SIMAS – Sistemi per l'Innovazione
e Management Sanitario
Università degli Studi di Salerno

Sponsor

Strategic committee

Sergio Barile is Full Professor of Business Management at Sapienza, University of Rome. He is a member of the Board of Directors of the Italian Academy of Business Administration and Management (AIDEA) and of the University Consortium of Industrial and Managerial Economics (CUEIM). He is also a member of the Scientific Committee of the SiMAS at the University of Salerno. His areas of research range from business management to systems, decision and complexity theories.
He funded the Association for Research on Viable Systems (ASVSA) and is one of the main references for the studies on the Viable Systems Approach (VSA). He is member of the editorial board of significant journals dealing with business management and economics science, and author of numerous books and articles in national and international journals, among which are the *European Management Journal, Managing Service Quality, Journal of Service Management, Systems Research and Behavioral Science, Service Science (Informs)*. He was awarded the Evert Gummesson Outstanding Research Award in 2015.

Manlio Del Giudice is Associate Professor of Business Management at the University of Rome "Link Campus", where he serves as Associate Dean of the Faculty of International Business Management. He is also Professor of Management of Biotech Firms at the "Federico II" University of Naples and Affiliate Professor of Business Management and Entrepreneurship at the Paris Business School, in Paris. He holds a PhD in Management from the University of Milano Bicocca and he has taught in
a number of universities worldwide, including Grenoble Graduate School of Business, Twente University and Jonkoping University. Professor Del Giudice has more than 80 publications in mainstream journals and publishers, such as Springer, Palgrave MacMillan, Nature Publishing Group and Elsevier; furthermore, he has active research collaboration programs with more than 30 universities across the globe, including affiliations with celebrated universities and

research centres, from New Zealand to United Arab Emirates. Professor Del Giudice serves on the editorial boards of prestigious peer reviewed academic journals; at the same time he has promoted and managed several academic spin-offs and technology transfer activities. His research interests focus on knowledge management, cross-cultural management, family business management and entrepreneurship. manlio.delgiudice@oasipc.com.

Raul Espejo is President of the World Organisation of Systems and Cybernetics (www.wosc.co) and Director of Syncho Research (www.syncho.com), UK. The main focus of his research is organisational systems, and is author of over 120 academic papers, co-author of three books and co-editor of another three. Professor Espejo has worked with a wide range of organisations worldwide, including national and local governments and small and large enterprises, mainly in aspects of organisational diagnosis and design, transparency and democratic processes in decision making. He has been consultant of organisations like Hoechst AG in Germany, Hydro Aluminium in Norway, 3M in Europe, EdF in France, the Nuclear Inspectorate in Sweden and the National Audit Office, the Ministry of Education and the State Council of Colombia. In 1995 he was appointed Full Professor of Systems and Cybernetics at the University of Lincoln UK. Before that from 1977 to 1994 he was Senior Lecturer at the University of Aston UK. During the mid and late '70s he was a research scholar at the Manchester Business School, UK and at the International Institute for Applied Systems Analysis, Laxenburg, Austria. Professor Espejo has been Visiting Professor at the Universities of Worcester in the UK and Santiago in Chile. From 1971 to 1973 he worked as Operations Director of the CYBERSYN project – the Chilean Government's project for the management of the social economy, under the scientific direction of Professor Stafford Beer.

José Pérez Ríos is Professor of Business Organization at the University of Valladolid (UVA), Spain. His research focuses on the application of management cybernetics and system dynamics to the study of complex systems, and also on developing software tools that can facilitate the application of different systems thinking approaches. He was responsible for the creation of the "Navegador Colón" for the Spanish Foreign Ministry, and the VSMod® software (Viable System Model application). He also served as the Area Director of International Relations at the University of Valladolid (2000–2006). He has worked in multiple research projects and written more than 80 publications. His five books to date include *Design and Diagnosis for Sustainable Organizations. The Viable System Method* (Springer, 2012). Honorary distinctions: "The Kybernetes Research Award" (2006) and the "Honorary HSSS Award as

Distinguished Scientist" (2007). He is also a member of the Board of Directors of the WOSC (World Organisation of Systems and Cybernetics).

Igor Perko is WOSC's Director General and Assistant Professor at the University of Maribor, Faculty of Economics and Business, Slovenia. His PhD thesis was "Intelligent Agents in Management Information Systems". His research interests include business intelligence systems and Big Data, focusing on cooperation support in business ecosystems. He is connecting the use of knowledge management structures, business data sharing, Big Data, cloud technologies and predictive analytics to provide intelligence support of trans-business processes. He is author of professional and scientific papers, actively involved in business ICT development and a member of WOSC.

Francesco Polese is Professor of Business Management, founder and Director of Inderdept. Centre Simas (Innovation Systems and Health Management) at Salerno University. Formerly he was Director of MadiLab and President of CUDH (University Centre of Disabilities) at Cassino University. He has been Visiting Professor at Stockholm University (SWE) and Universidad Torcuato de Tella (AG), and Lecturer at Exteter University (UK) and Mid-sweden University (SWE). He is Co-Chair, with Evert Gummesson and Cristina Mele, of the Naples Forum on Service, an event gathering, every two years since 2009, scholars and researchers from the international service community around the topics Service-Dominant logic, Network and Systems Theories and Service Science (www.naplesforumon service.it). His areas of interests cover systems theories, and specifically the Viable Systems Approach, service, networks and healthcare management. He is the author of several books and numerous contributions in international journals. He is a member of the Editorial Board of *Service Science* and *Journal of Service Theory and Practice* and guest editor of Special Issues of numerous international journals such as *European Management Journal, Journal of Service Management, Journal of Service Theory and Practice, Service Science, International Journal of Quality and Service Sciences.*

Markus Schwaninger is Professor Emeritus of management at the University of St. Gallen, Switzerland. His research and teaching are focused on the management of complex dynamic systems, with a methodological emphasis on System Dynamics and Organizational Cybernetics. Research projects are related to the design, transformation and learning of organisations, organisational intelligence, model-based management, as well as systemic issues of sustainability. Consulting and training mandates

span all sectors – industry, services, health, information, public and NGOs – around the world. He has supervised or co-supervised doctoral projects at leading universities (University of St. Gallen; ETHZ & EPFL-Swiss Polytechnic Institutes, Zurich & Lausanne, Switzerland; University of Bochum, Germany; National University, Singapore; Universidad de los Andes, Bogotá, Colombia; etc.). Markus is the author of more than 200 publications, in six languages, including *Intelligent Organizations* (Springer). He has lectured widely, on four continents, and is involved in international, transdisciplinary research projects. Markus founded the System Dynamics Group at the University of St. Gallen. He is on the board of directors of the World Organisation of Systems and Cybernetics, a managing editor of the System Dynamics Review and a member of various advisory boards.

Marie Noelle Sarget, Institute of Political Science of Paris (International Relations), doctorate of Economy, PhD (Political Sociology). She was Professor for 12 years at the University Paris III (Institute of South American Studies), and then researcher in the EHESS in Paris and Paris X University. Main publications and papers are about economic development processes, political system of Chile, political effects of globalisation and systems thinking. Books: *Système Politique et Parti Socialiste au Chili: un essai d'analyse systémique*, Paris, L'Harmattan, 1994; *Histoire du Chili de la conquête à nos jours*, Paris, l'Harmattan, 1996; Director of the collection: "Practice of Systems Thinking" in the Harmattan Editions in Paris presently, and now for 12 years, mainly an abstract painter and photographer.

Marialuisa Saviano, PhD, is Full Professor of Business Management at the University of Salerno, Italy, where she teaches courses of pharmacy management, business management, service marketing and healthcare marketing. She is Vice Director of the Pharma_nomics Interdepartmental Research Centre and a Member of the Board of Directors of the S.I.Mas. She is also: President of the ASVSA, Association for Research on Viable Systems; Vice President of the IASS, Italian Association for Sustainability Science; Faculty Member of PhD Courses in Marketing & Communication at the University of Salerno. She has participated in several researches and studies contributing to the development of the Viable Systems Approach (VSA). Her main research interests include the Viable Systems Approach (VSA), Service & Retail Marketing, Healthcare and Pharmacy Management, Sustainability and Cultural Heritage Management. She has published several books and articles in national and international journals, among which include the *European Management Journal, Managing Service Quality, Journal of Service Management, Sinergie, Italian Journal of Management,*

Service Science Informs, Journal of Business Market Management, Australasian Marketing Journal, World Wide Web Journal. She received two Best Paper Awards (at the 2011 Naples Forum on Service Conference and at the 2012 XXIV Sinergie Annual Conference). She was also a finalist at the 2012/2013 Emerald/EMRBI Business Research Award for Emerging Researchers. She can be contacted at msaviano@unisa.it.

About the editors

Sergio Barile is Full Professor of Business Management at Sapienza, University of Rome. He is a member of the Board of Directors of the Italian Academy of Business Administration and Management (AIDEA) and of the University Consortium of Industrial and Managerial Economics (CUEIM). He is also a member of the Scientific Committee of the SiMAS at the University of Salerno. His areas of research range from business management to systems, decision and complexity theories.
He funded the Association for Research on Viable Systems (ASVSA) and is one of the main reference for the studies on the Viable Systems Approach (VSA). He is member of the editorial board of significant journals dealing with business management and economics science and author of numerous books and articles in national and international journals, among which are the *European Management Journal*, *Managing Service Quality*, *Journal of Service Management*, *Systems Research and Behavioral Science*, *Service Science (Informs)*. He was awarded the Evert Gummesson Outstanding Research Award in 2015.

Francesco Caputo, PhD, is Research Fellow at Department of Pharmacy – Pharma_nomics Interdepartmental Research Centre, University of Salerno, Italy. He is a member of the Scientific Board of Reald Summer School (University of Reald Vlore – Albania) and secretariat of ASVSA, Association for Research on Viable Systems. He has been Professor of Gestione della conoscenza di impresa at the Department of Informatics, University of Bari "Aldo Moro" (Italy) and Professor
of Service and Systems Thinking and of Knowledge Management at the Department of Computer Systems and Communications of Masaryk University (Czech Republic). He is member of the Editorial Boards of several international journals and he serves as reviewers for several Italian and international journals. He was finalist at the 2012/2013 Emerald/EMRBI Business Research Award and he has won the best presentation award at the

2016 B.S.Lab. Symposium, the best paper award at the 19th Toulon-Verona (ICQSS) Conference, the 2017 Outstanding Reviewer for the journal, and the best commended paper at the 2017 WOSC Congress. His main research interests include but they are not limited to complexity, knowledge management, systems thinking and healthcare management. He can be contacted at the following email: fcaputo@unisa.it.

Raul Espejo is President of the World Organisation of Systems and Cybernetics (www.wosc.co) and Director of Syncho Research (www.syncho.com), UK. The main focus of his research is organisational systems, and is author of over 120 academic papers, co-author of three books and co-editor of another three. Professor Espejo has worked with a wide range of organisations worldwide, including national and local governments and small and large enterprises, mainly in aspects of organisational diagnosis and design, transparency and democratic processes in decision making. He has been consultant of organisations like Hoechst AG in Germany, Hydro Aluminium in Norway, 3M in Europe, EdF in France, the Nuclear Inspectorate in Sweden and the National Audit Office, the Ministry of Education and the State Council of Colombia. In 1995 he was appointed Full Professor of Systems and Cybernetics at the University of Lincoln UK. Before that from 1977 to 1994 he was Senior Lecturer at the University of Aston UK. During the mid and late '70s he was a research scholar at the Manchester Business School, UK and at the International Institute for Applied Systems Analysis, Laxenburg, Austria. Prof Espejo has been Visiting Professor at the Universities of Worcester in the UK and Santiago in Chile. From 1971 to 1973 he worked as Operations Director of the CYBERSYN project – the Chilean Government's project for the management of the social economy, under the scientific direction of Professor Stafford Beer.

Igor Perko is Director General and assistant professor at the University of Maribor, Faculty of Economics and Business, Slovenia; he is the WOSC Director General and a co-editor for *Kybernetes*, an JCR indexed scientific journal. His professional backgrounds include development of information systems in the financial industry, focusing on business intelligence systems, predictive analytics and risk management systems. In his academic work he upgrades these with the concepts of management information systems, systems thinking and cybernetics. His PhD thesis was "Intelligent Agents in Management Information Systems". In his work with the students, he introduces management information systems, business intelligence tools and concepts based on: OLAP, predictive analytics, balanced scorecards and cyber-systems thinking. Currently he is running a Jean Monnet funded international summer school: "Big Data EU Business

Implications". In his research work he is connecting the use of systems thinking, cybernetics, Big Data, predictive analytics, cloud technologies, intelligent agents, knowledge management structures and business data sharing to provide ICT services for the business processes support across organisations. As the WOSC director general, he is trying to connect researchers and practitioners in systems thinking and cybernetics. He is an author of multiple professional and scientific papers, among them: "Behaviour-Based Short-Term Invoice Probability of Default Evaluation", recently published in the *European Journal of Operational Research* (2017).

Marialuisa Saviano, PhD, is Full Professor of Business Management at the University of Salerno, Italy, where she teaches courses of pharmacy management, business management, service marketing and healthcare marketing. She is Vice Director of the Pharma_nomics Interdepartmental Research Centre and a Member of the Board of Directors of the S.I.Mas. She is also: President of the ASVSA, Association for Research on Viable Systems; Vice President of the IASS, Italian Association for Sustain-
ability Science; Faculty Member of PhD Courses in Marketing & Communication at the University of Salerno. She has participated in several researches and studies contributing to the development of the Viable Systems Approach (VSA). Her main research interests include the Viable Systems Approach (VSA), Service & Retail Marketing, Healthcare and Pharmacy Management, Sustainability and Cultural Heritage Management. She has published several books and articles in national and international journals, among which the *European Management Journal, Managing Service Quality, Journal of Service Management, Sinergie, Italian Journal of Management, Service Science Informs, Journal of Business Market Management, Australasian Marketing Journal, World Wide Web Journal*. She received two Best Paper Awards (at the 2011 Naples Forum on Service Conference and at the 2012 XXIV Sinergie Annual Conference). She was also a finalist at the 2012/2013 Emerald/EMRBI Business Research Award for Emerging Researchers. She can be contacted at msaviano@unisa.it.

About the keynotes

Elias G. Carayannis is Full Professor of science, technology, innovation and entrepreneurship, as well as Co-Founder and Co-Director of the Global and Entrepreneurial Finance Research Institute (GEFRI) and Director of Research on science, technology, innovation and entrepreneurship, European Union Research Center, (EURC) at the School of Business of the George Washington University in Washington, D.C. Carayannis's teaching and research activities focus on the areas  of strategic government-university-industry R&D partnerships, technology road-mapping, technology transfer and commercialisation, international science and technology policy, technological entrepreneurship and regional economic development. He is fluent in English, French, German, Greek and has a working knowledge of Spanish.

Ray Ison has been Professor of Systems at The Open University (OU), UK since 1994, where he is part of a group responsible for a successful MSc in Systems Thinking in Practice (see www.open.ac.uk/choose/ou/systemsthinking) which currently has just under 1500 alumni actively engaged in a LinkedIn community. From 2008–2015 Ray was also Professor at the Monash Sustainability Institute, Monash University, Australia where he developed and led the Systemic Governance Research  Program, an interdisciplinary, systems-based research program focusing on water governance, climate change adaptation and social learning. At the Open University he has through various commissioned projects and initiatives, usually entailing collaborative research, demonstrated how social learning, including systemic inquiry, can be employed as an alternative governance mechanism for managing in complex situations such as water governance, program and project governance, climate change adaptation, food security research, social learning and the purposeful creation of communities of practice. He is the author of the book (2010): *Systems Practice: How to Act in a Climate-change*

World (Springer & OU). Most of his major research publications can be seen or accessed here: http://oro.open.ac.uk/view/person/rli2.html. He was in 2015–2016 President of the ISSS (International Society for Systems Sciences).

Alexandre Perez Casares is Co-founder and Director of The Altius Society at Oxford, as well as venture capitalist with a broader interest on issues of international political economy. He has been a private equity investment executive at Kohlberg Kravis Roberts in London, an investment-banking associate at Goldman Sachs New York, and strategy consultant at McKinsey Europe and Brazil. From 2009 to 2012, Alexandre attended the Stanford Graduate School of Business, where he earned his MBA, and the Harvard Kennedy School of Government, where he graduated with a Master in Public Administration. He also holds an MSc and BSc in Industrial Engineering from the University of Valladolid, graduating summa cum laude and earning the National Prize by the Spanish Minister of Education, awarded to the top three engineers in Spain. He has been awarded the Rafael del Pino Foundation, Carolina Foundation, Caja Madrid and Fulbright scholarships. Additionally, he has published policy and opinion pieces in publications such as the *Financial Times* and the *New York Times*.

Alfonso Reyes is a Physicist and a Systems Engineer from Los An des University in Colombia. He has a MSc in Computer Science from the University of Maryland (USA) and a PhD in Management Cybernetics from the University of Humberside in England. He did postdoctoral studies in Organizational Learning with Professor Raúl Espejo at the University of Lincoln in England. He has worked for the last 25 years in addressing organisational problems in the public sector, specially in the administration of justice. He is a former adviser of the Ministry of Justice and the Attorney General in the use of ICT to improve the efficiency of criminal cases in Colombia and was the chairman of a governmental Committee to propose normative recommendations to improve the administration of justice as part of the new Colombian Constitution in 1991. He has been an international consultant for the Interamerican Development Bank and the Agency for International Development (USA) in applying management cybernetics to the public sector. He was a Lecturer in the Department of Industrial Engineering at Los Andes University (Colombia) and has published several papers about individual and organisational learning and self-organiation from the point of view of second-order cybernetics. In March 2009 he became president of the Universidad de Ibagué in Colombia. Currently he is the Dean of the Engineering School at Los Andes University.

Kazuhiko Takeuchi is Senior Visiting Professor of UNU-IAS, Director and Professor of IR3S of University of Tokyo. He has served, among others, as a member of the Science Council of Japan and Editor-in-Chief of the journal *Sustainability Science* (Springer). Educated and trained as a geographer and landscape ecologist at the University of Tokyo, he engages in research and education on creating eco-friendly environments for a harmonious coexistence of people and nature.
He leads the Satoyama Initiative as well as climate and ecosystem change research in Asia and Africa.

Stuart A. Umpleby is one of the originators of "second order science," an effort to expand the conception of science in the direction of a more holistic point of view. His work has been described as a contribution to the unification of science. Stuart is Professor Emeritus in the School of Business at George Washington University, Washington, DC. He is a past president of the American Society for Cybernetics and Associate Editor of the journal Cybernetics and Systems. He is now serving as
president of the Executive Committee of the International Academy for Systems and Cybernetic Sciences, an honor society created by the International Federation for Systems Research.

Contributor index

Subject index